ANALOGUE AND NUMERICAL MODELLING OF SEDIMENTARY SYSTEMS: FROM UNDERSTANDING TO PREDICTION

Other publications of the International Association of Sedimentologists

SPECIAL PUBLICATIONS

39 Glacial Sedimentary Processes and Products
Edited by M.J. Hambrey, P. Christoffersen, N.F. Glasser and B. Hubbard
2007, 416 pages, 181 illustrations

38 Sedimentary Processes, Environments and Basins
A Tribute to Peter Friend
Edited by G. Nichols, E. Williams and C. Paola
2007, 648 pages, 329 illustrations

37 Continental Margin Sedimentation
From Sediment Transport to Sequence Stratigraphy
Edited by C.A. Nittrouer, J.A. Austin, M.E. Field, J.H. Kravitz, J.P.M. Syvitski and P.L. Wiberg
2007, 549 pages, 178 illustrations

36 Braided Rivers
Process, Deposits, Ecology and Management
Edited by G.H. Sambrook Smith, J.L. Best, C.S. Bristow and G.E. Petts
2006, 390 pages, 197 illustrations

35 Fluvial Sedimentology VII
Edited by M.D. Blum, S.B. Marriott and S.F. Leclair
2005, 589 pages, 319 illustrations

34 Clay Mineral Cements in Sandstones
Edited by R.H. Worden and S. Morad
2003, 512 pages, 246 illustrations

33 Precambrian Sedimentary Environments
A Modern Approach to Ancient Depositional Systems
Edited by W. Altermann and P.L. Corcoran
2002, 464 pages, 194 illustrations

32 Flood and Megaflood Processes and Deposits
Recent and Ancient Examples
Edited by I.P. Martini, V.R. Baker and G. Garzón
2002, 320 pages, 281 illustrations

31 Particulate Gravity Currents
Edited by W.D. McCaffrey, B.C. Kneller and J. Peakall
2001, 320 pages, 222 illustrations

30 Volcaniclastic Sedimentation in Lacustrine Settings
Edited by J.D.L. White and N.R. Riggs
2001, 312 pages, 155 illustrations

29 Quartz Cementation in Sandstones
Edited by R.H. Worden and S. Morad
2000, 352 pages, 231 illustrations

28 Fluvial Sedimentology VI
Edited by N.D. Smith and J. Rogers
1999, 328 pages, 280 illustrations

27 Palaeoweathering, Palaeosurfaces and Related Continental Deposits
Edited by M. Thiry and R. Simon Coinçon
1999, 408 pages, 238 illustrations

26 Carbonate Cementation in Sandstones
Edited by S. Morad
1998, 576 pages, 297 illustrations

25 Reefs and Carbonate Platforms in the Pacific and Indian Oceans
Edited by G.F. Camoin and P.J. Davies
1998, 336 pages, 170 illustrations

24 Tidal Signatures in Modern and Ancient Sediments
Edited by B.W. Flemming and A. Bartholomä
1995, 368 pages, 259 illustrations

23 Carbonate Mud-mounds
Their Origin and Evolution
Edited by C.L.V. Monty, D.W.J. Bosence, P.H. Bridges and B.R. Pratt
1995, 543 pages, 330 illustrations

16 Aeolian Sediments
Ancient and Modern
Edited by K. Pye and N. Lancaster
1993, 175 pages, 116 illustrations

3 The Seaward Margin of Belize Barrier and Atoll Reefs
Edited by N.P. James and R.N. Ginsburg
1980, 203 pages, 110 illustrations

1 Pelagic Sediments on Land and Under the Sea
Edited by K.J. Hsu and H.C. Jenkyns
1975, 448 pages, 200 illustrations

REPRINT SERIES

4 Sandstone Diagenesis: Recent and Ancient
Edited by S.D. Burley and R.H. Worden
2003, 648 pages, 223 illustrations

3 Deep-water Turbidite Systems
Edited by D.A.V. Stow
1992, 479 pages, 278 illustrations

2 Calcretes
Edited by V.P. Wright and M.E. Tucker
1991, 360 pages, 190 illustrations

Special Publication Number 40 of the International Association of Sedimentologists

Analogue and Numerical Modelling of Sedimentary Systems: from Understanding to Prediction

EDITED BY

Poppe de Boer
Utrecht University

George Postma
Utrecht University

Kees van der Zwan
Shell Research, Rijswijk, Netherlands

Peter Burgess
Shell Research, Rijswijk, Netherlands

Peter Kukla
Aachen University

SERIES EDITOR

Isabel Montanez
University of California, Davis

Blackwell Publishing was acquired by John Wiley & Sons in February 2007. Blackwell's publishing program has been merged with Wiley's global Scientific, Technical and Medical business to form Wiley-Blackwell.

Registered office
John Wiley & Sons Ltd, The Atrium, Southern Gate, Chichester, West Sussex, PO19 8SQ, UK

Editorial offices
9600 Garsington Road, Oxford, OX4 2DQ, UK
The Atrium, Southern Gate, Chichester, West Sussex, PO19 8SQ, UK
111 River Street, Hoboken, NJ 07030-5774, USA

For details of our global editorial offices, for customer services and for information about how to apply for permission to reuse the copyright material in this book please see our website at www.wiley.com/wiley-blackwell

Library of Congress Cataloguing-in-Publication Data

Analogue and numerical modelling of sedimentary systems : from understanding to prediction / edited by Poppe de Boer *et al.*
p. cm. – (Special publication number 40 of the International Association of Sedimentologists)
Includes bibliographical references and index.
ISBN 978-1-4051-8930-9 (hardcover : alk. paper)
1. Sedimentary basins–Italy–Dolomite Alps–Congresses. 2. Rocks, Carbonate–Italy–Dolomite Alps–Congress. 3. Sedimentary structures–Italy–Dolomite Alps–Mathematical models–Congresses. 4. Sequence stratigraphy–Congresses. 5. Geology, Stratigraphic–Mesozoic–Congresses. 6. Geology–Italy–Dolomite Alps–Congresses. I. Boer, Poppe Lubberts de, 1949-

QE615.5.I8A53 2008
552′.5–dc22

2008021705

A catalogue record for this book is available from the British Library.

Set in 10/12 pt Melior by Newgen Imaging Systems (P) Ltd, Chennai, India
Printed and bound in Malaysia by Vivar Printing Sdn Bhd

1 2008

Contents

Preface

The conference *Analogue and numerical forward modelling of sedimentary systems; from understanding to prediction* was held in Utrecht on 9–11 October 2003 to make a contribution towards developing and applying models for basin fill. Understanding basin-fill evolution and the origin of stratal architectures has traditionally been based on studies of outcrops, well and seismic data, studies of and inferences on qualitative geological processes, and to a lesser extent by quantitative observations of modern and ancient sedimentary environments. Insight gained on the basis of these studies can be tested and also extended through the application of numerical and analogue forward models. For example, numerical forward models are used to make quantitative predictions of reservoir and seal distributions away from points constrained by well data. Similar models are also used to predict the response of modern coastal systems to a rise of sea level, taking into account external forcing, as well as the internal dynamics of sedimentary systems with their inherent non-linear and chaotic effects.

The present-day approach to stratigraphic forward modelling broadly follows two lines: (1) the deterministic process-based approach, ideally with resolution of the fundamental equations of fluid and sediment motion at all scales, and (2) the stochastic approach. The *process-based* approach leads to improved understanding of the dynamics (physics) of the system, increasing our predictive power of how systems evolve under various forcing conditions unless the system is highly non-linear and hence difficult or perhaps even impossible to predict. The *stochastic approach* is more direct and relatively simple. The behaviour of systems is established through empirical studies of modern systems in relation to controls such as subsidence, drainage-basin area, geology of the area, climate (change), as well as other variables. Process-based models, more than stochastic ones, are directly limited by the diversity of temporal and spatial scales and the very incomplete knowledge of how processes operate and interact at these various scales.

Both modelling approaches have disadvantages. Fluid motion, fluid–sediment interaction and sediment transport and deposition are fairly well understood and described in a qualitative sense, but quantitative predictions based on empirical formulae may differ greatly from specific real-world cases. Analogue, physical models are hampered by a lack of universally applicable scaling strategies for modelling at a sedimentary systems scale. The set of dimensionless numbers developed by engineers that define fluid flow and related sediment transport do not allow downscaling of spatial dimensions to much more than 1:50, so that many stratal features cannot be directly studied by conventional, engineering-type flume models. For a stochastic approach, trends in systems development in relation to external controls are used, but since such trends are defined on the basis of specific, often local empirical relationships, they hardly contribute to understanding the physics of a system and are of limited value to predict outside the data region. Based on the aforementioned issues, another more general problem emerges, which concerns calibration and validation of numerical models.

To come up with sensible forward numerical models of sedimentary systems there is a need for prototypes or 'standards' that modellers can use to calibrate the time-averaged sediment transport rates in their models. There are hardly any well researched prototypes from the real world available that link stratigraphic architecture with independent input values that can be used to calibrate numerical models. Hence modelling results are often very poorly constrained. Input values for the specific stratigraphic architecture of real-world prototypes are generally unknown and at best available for the last hundred years or so, too short in duration to be of value in stratigraphic studies. Moreover, they cover a period where morphology has had a significant anthropogenic influence. An alternative is to set up prototypes on the basis of laboratory studies and to relate the laboratory studies to real-world situations. Analogue and numerical forward modelling is a daunting task for the twenty-first century sedimentologist!

We thank the authors for their cooperation and willingness to follow the comments and suggestions of the reviewers, who – with their comments and recommendations – made invaluable contributions to this volume. They are: *José-Javier Alvaro, Yuval Bartov, John Bridge, Quintijn Clevis, Didier Granjeon, Caroline Hern, Matthias Hinderer, Albert Kettner, Mike Leeder, Andrew D. Miall, Juan Pablo Milana, Bruce Nelson, Ulf Nordlund, Phillip Playford, Guy Plint, Sadat Kolonic, John Reijmer, Marco Stefani, Esther Stouthamer, Orsi Sztanó, Daniel Tetzlaff, Frans van Buchem, Peter van der Beek, Michael Wagreich, Dave Waltham, George Warrlich, Lynn Watney and Johannus Wendebourg*, as well as some reviewers who wished to remain anonymous. Many thanks to them all!

September 2006

POPPE DE BOER AND GEORGE POSTMA
Utrecht University
KEES VAN DER ZWAN AND PETER BURGESS
Shell Research, Rijswijk
PETER KUKLA
Aachen University

Spec. Publ. Int. Assoc. Sedimentol. (2008) **40**, 1–36

Numerical simulation of the syn- to post-depositional history of a prograding carbonate platform: the Rosengarten, Middle Triassic, Dolomites, Italy

AXEL EMMERICH*†, ROBERT TSCHERNY‡§, THILO BECHSTÄDT*, CARSTEN BÜKER‡¶, ULLRICH A. GLASMACHER*, RALF LITTKE‡ and RAINER ZÜHLKE*

**Geologisch-Paläontologisches Institut, Universität Heidelberg, Im Neuenheimer Feld 234, 69120 Heidelberg, Germany (E-mail: axel.emmerich@sap.com)*

†RWE Dea AG, Überseering 40, 22297 Hamburg, Germany

‡Lehrstuhl für Geologie, Geochemie und Lagerstätten des Erdöls und der Kohle, RWTH Aachen, Lochnerstraße 4–20, 52056 Aachen, Germany

§IES GmbH, Ritterstraße 23, 52072 Aachen, Germany

¶Shell International E&P Inc., 200 North Dairy Ashford Street, Houston, TX 77079, USA

ABSTRACT

A combination of thermal history, numerical basin-reverse and sequence-stratigraphic forward modelling is applied to the Mesozoic outcrop analogue of the Rosengarten carbonate platform area in the Dolomites of northern Italy. This integrated multidisciplinary approach of numerical simulation quantifies the thermal, subsidence, geometrical and subsequent facies evolution of the area. Calibration data during modelling were vitrinite reflectance (VR) and apatite fission-track (FT) analyses as well as detailed outcrop studies. Vitrinite reflectance values in strata underlying the carbonate platform vary between 0.5 and 0.8% VR_r; apatites from these formations reveal cooling ages of around 165.6 Ma and track lengths of approximately 9.8 μm. This low thermal maturity combined with the FT data in apatites indicates a relatively cool (<110°C), protracted (between 250 and 30 Ma) and shallow burial (thickness of eroded strata overlying present-day topography is <1100 m), as well as a fast exhumation from the Middle Miocene onward. Maximum temperatures are reached during the Middle/Late Triassic, when the basal heat flow was elevated owing to regional volcanic and hydrothermal activity. Local anomalies in vitrinite reflectance of up to 1.1% VR_r in the immediate surroundings of the Predazzo/Monzoni volcanic centre show that its thermal influence decreased rapidly with increasing distance. The geometrical evolution of the Middle Triassic (Anisian/Ladinian) Rosengarten platform is twofold: the first stage reveals aggradation, the second progradation of the platform margin. Basin-reverse modelling results indicate that these two intervals originate from a temporal change in tectonic subsidence. Spatial variations in flexural and tectonic subsidence along the 6 km transect are insignificant due to the rigidity of the basement (up to 2500 m of Late Permian ignimbrites). During the first stage of platform evolution, high pulse-like total subsidence rates of up to 820 m Myr^{-1} led to aggradation, whereas the subsequent drop to 100 m Myr^{-1} initiated platform progradation. The short-spanned subsidence peak was linked to block movements in a strike-slip tectonic setting (Cima Bocche Anticline-Stava Line approximately 10 km southeast of the study area). Stratigraphic forward modelling quantifies the sediment volumes involved in the geometrical evolution of the platform. In order to replicate platform architecture, constant carbonate accumulation rates between 900 and 1000 m Myr^{-1} – increasing from periplatform environments to the slope – have to be assumed throughout the existence (approximately 5.8 Myr) of the Rosengarten. As the carbonate factory successfully keeps up with the modelled accommodation rates, it must have completely recovered from the Permian–Triassic biotic crisis during the onset of platform growth in latest Anisian times despite the low biotic diversity of the platform succession seen elsewhere in the Dolomites. Our forward modelling confirms that the main carbonate factory was situated on the slope at water depths from shallow subtidal to 300 m ('slope-shedding') and that it therefore switched on during all possible stages of accommodation change.

Keywords Basin analysis, numerical simulation, subsidence, thermal maturity, thermochronology, carbonate platform, aggradation, progradation, sedimentation rates, Triassic, Southern Alps, Dolomites.

INTRODUCTION

The Dolomites of northern Italy (Fig. 1) have long been a study area for carbonate platforms and their reef communities. Since Mojsisovics termed the word 'Überguss-Schichtung' back in 1879 (i.e., clinostratification), many authors have worked on platform-to-basin transitions of carbonate build-ups in the Dolomites (Hummel, 1928, 1932; Pia, 1937; Leonardi, 1962, 1967; Bosellini, 1984, 1988). In particular, the Rosengarten has served as a reference model for progradational geometries (Bosellini & Stefani, 1991; Bosellini *et al.*, 1996). However, assessing the evolution of carbonate platforms and their clinoforms has been mainly of a qualitative nature. Quantitative approaches of subsidence and carbonate accumulation of Middle Triassic platforms in the Dolomites have so far been scarce (Schlager, 1981; Doglioni & Goldhammer, 1988; Schlager *et al.*, 1991; Maurer, 1999, 2000; Keim & Schlager, 2001) and lack unbiased numerical modelling techniques. The age of these platforms is usually constrained by coeval basinal sediments containing abundant biostratigraphic information (Buchenstein Fm; Brack & Rieber, 1993, 1994). Recently, age-diagnostic airborne tuff layers in basinal and lagoonal strata were used to synchronize bio-, cyclo- and chronostratigraphy (basinal Buchenstein Fm at Seceda/Geisler Group, western Dolomites: Mundil *et al.*, 1996; lagoonal Schlern Dolomite Fm 1 at Latemar, western Dolomites: Mundil *et al.*, 2003; for locations see Fig. 2) providing a high-resolution database for numerical simulation. The particular feature of the Rosengarten platform is that some of these dated tuff layers can be physically correlated to coeval slope deposits (Maurer, 1999, 2000).

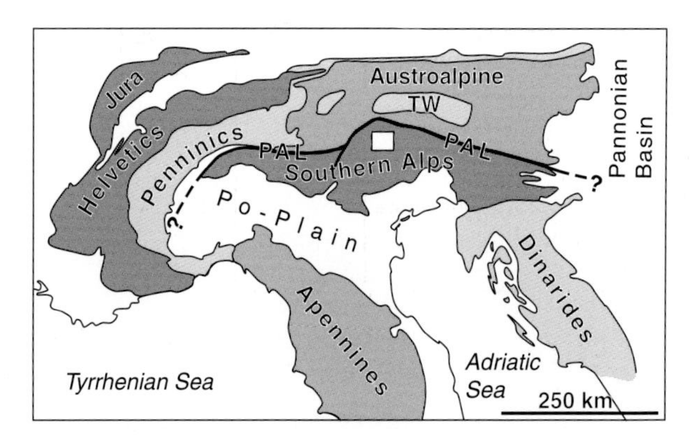

Fig. 1. Schematic tectonic map of the Alps. The white rectangle marks the location of the study area. Abbreviations: TW, Tauern window; PAL, Periadriatic lineament.

The aim of this paper is the quantification of the development of the Rosengarten platform and the assessment of its primary controlling factors during platform growth. This is realized by an integrated approach of basin-reverse and stratigraphic-forward modelling combined with thermal basin modelling. Datasets for all modelling procedures are derived from existing studies and from new detailed analyses on allo-/sequence stratigraphy and facies architecture, thermal maturity and apatite fission tracks in strata underlying the platform body.

BASIN AND CARBONATE-PLATFORM DEVELOPMENT

The southwestern Dolomites (for location within the Alps, Fig. 1) are located on the Adriatic Plate between former Laurussia and Gondwana (Dercourt *et al.*, 2000). Throughout the Triassic, this area represents the eastern margin of a highly differentiated passive continental margin with mixed siliciclastic–carbonate sedimentation (Blendinger, 1985; Doglioni, 1987). First, carbonate ramps (early Anisian/Aegean) and small reef mounds (early in the late Anisian/Pelsonian) developed in the Dolomites (Fois & Gaetani, 1984; Senowbari-Daryan *et al.*, 1993) after the carbonate factory had eventually recovered from the severe faunal crisis at the close of the Permian. From the late Anisian into the late Ladinian, a considerable submarine relief with local subaerial highs prevailed in the western Dolomites. Middle Anisian transpressive–transtensive tectonics dismembered the continental shelf and created strong regional differences in facies (the so-called 'Facies Heteropie' *sensu* Bechstädt & Brandner, 1970; see also Zühlke, 2000). Deep marine, stagnant basins with fine-grained chert- and organic-matter-rich sediments (Anisian: Moena Fm and Anisian/Ladinian: Buchenstein Fm; Figs 2 and 3) existed alongside shallow marine subtidal carbonate ramps and platforms (Anisian: Contrin Fm and Anisian/Ladinian: Schlern Dolomite Fm 1; Figs 2 and 3). Structural highs of the dismembered carbonate ramp (Contrin Fm) represent the nuclei of the Schlern Dolomite Fm 1 platforms in the Late Anisian (Masetti & Neri, 1980; Gaetani *et al.*, 1981; Bosellini, 1989). Evolution of the Ladinian carbonate platforms such as the Rosengarten/Schlern, Monte Agnello and

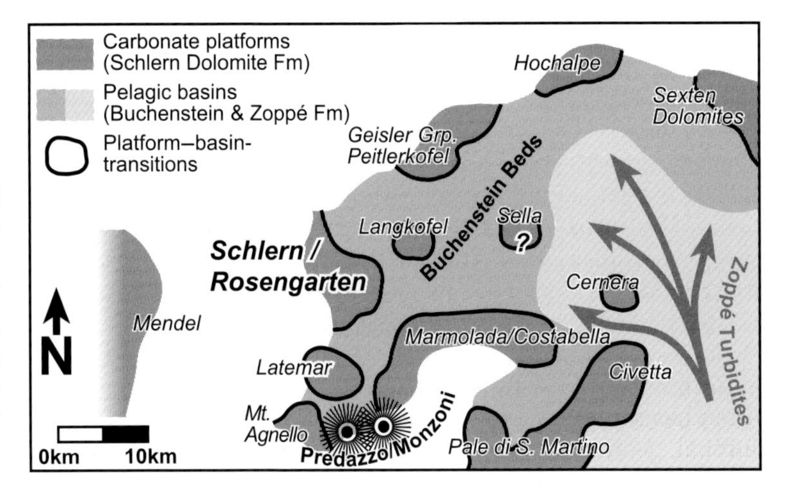

Fig. 2. Schematic palaeogeographical map of the western Dolomites during the Middle Triassic (late Anisian to early Ladinian) highlighting the distribution of platforms and basins. Legend of lithostratigraphic units in the upper-left corner, influx of Zoppé turbidite sands marked by large arrows. The volcanic centre of Predazzo–Monzoni in the immediate surroundings of the Rosengarten–Catinaccio platform is sketched with radial lines.

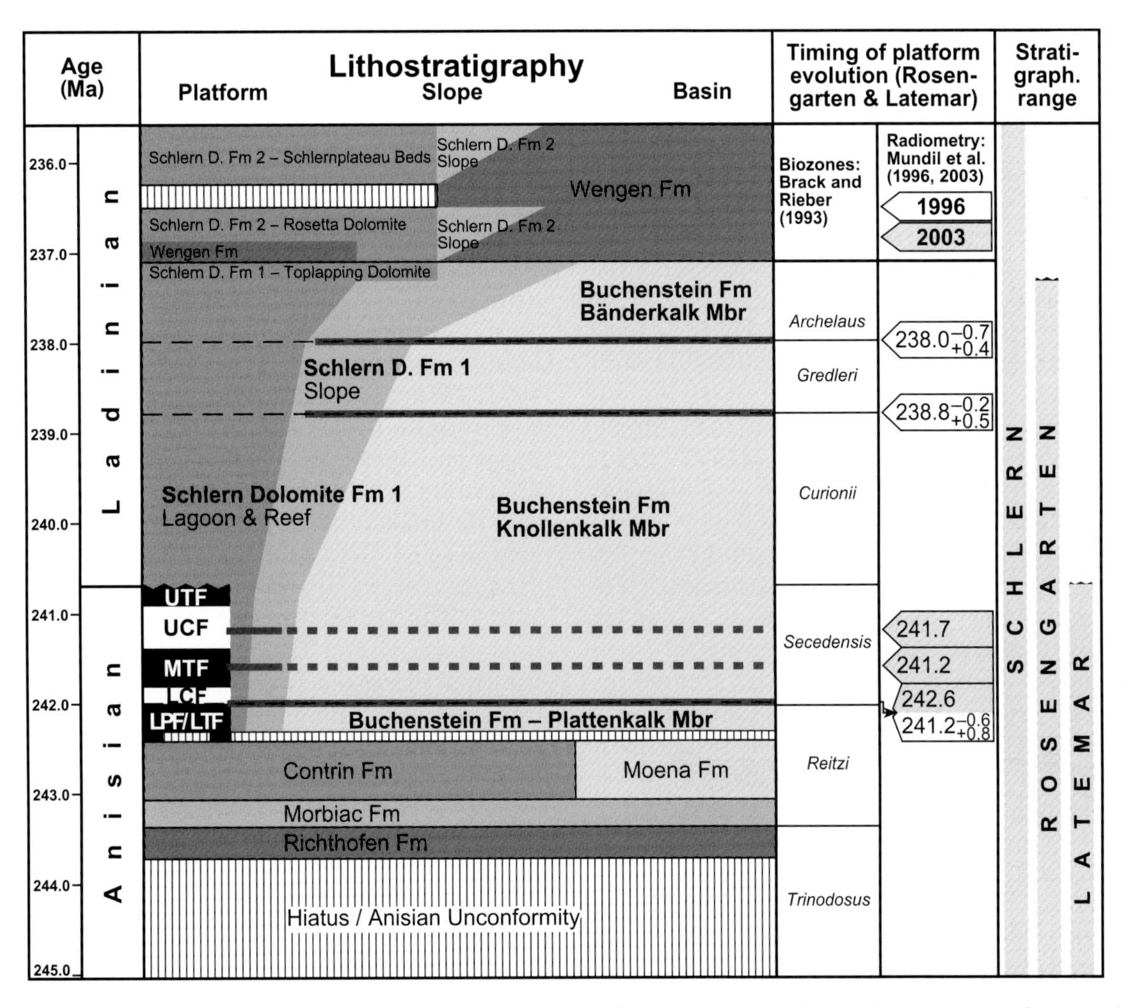

Fig. 3. Detailed Anisian–Ladinian lithostratigraphic succession of the study area (Schlern–Rosengarten and Latemar). The stratigraphic range of the carbonate platforms is illustrated on the right. Chronostratigraphy according to Lehrmann *et al.* (2002) and Mundil *et al.* (1996, 2003). Biozones and position of Anisian–Ladinian boundary according to Brack & Rieber (1993, 1994). Correlation of age-diagnostic tuff layers from basin to slope according to Maurer (1999, 2000). Hollow arrows in the radiometry column indicate ages from tuff layers in the basinal Buchenstein Fm at Seceda (Geisler Group, western Dolomites; see Fig. 2; Mundil *et al.*, 1996) whereas light grey indicates ages from tuff layers in the lagoonal Schlern Fm at Latemar (western Dolomites; see Fig. 2; Mundil *et al.*, 2003).

possibly also the Latemar was terminated by the extrusion of the Longobardian Wengen Group volcanics (Mojsisovics, 1879; Viel, 1979a, 1979b; De Zanche *et al.*, 1995; Fig. 3). The volcanic centre at Predazzo/Monzoni was nourished by a source linked to a deep-reaching fracture zone (Cima Bocche Anticline/Stava Line; Fig. 4; Blendinger, 1985). Tectonics also played a crucial role in platform development in the southwestern Dolomites as regional subsidence and accommodation development were controlled by downward movements along faults and upward movements through magmatic updoming (Doglioni, 1983, 1984, 1987).

Owing to its excellent, laterally continuous seismic and sub-seismic scale outcrop (Fig. 5a and b), the Rosengarten part of the Rosengarten/Schlern platform is ideally suited for a study on the geometric development of a carbonate platform and its accumulation rates. The platform top of the Rosengarten passes laterally into a platform slope interfingering with basinal sediments. The maximum north to south progradation of the Rosengarten slope was approximately 6 km. Lagoon, reef and slope facies of the build-up are all part of the Schlern Dolomite Formation 1, whereas coeval basinal sediments belong to the Buchenstein Formation. Bio- and chronostratigraphic data (Brack & Rieber, 1993, 1994; Mundil *et al.*, 1996, 2003; Maurer, 1999, 2000; Fig. 4) indicate the onset of platform growth in the upper *Reitzi*-biozone (Anisian; Middle Triassic stages after Brack & Rieber, 1993, 1994). According to Maurer (1999, 2000), the slope of the Rosengarten records five ammonite biozones (Fig. 5b). During the first two – *Reitzi* and *Secedensis* – biozones of platform existence, aggradation occurred. This first stage of platform evolution was followed by a second stage of progradational clinoforms. The preserved record of carbonate sedimentation lasted at least until the basal *Archelaus*-zone (middle/late Ladinian; Maurer, 1999, 2000; Fig. 5b). The maximum thickness of the Rosengarten – and therefore also its growth mode – can be inferred only by projecting stratigraphical information from the Schlern platform (Bosellini & Stefani, 1991; Fig. 6). At Schlern, 850 m of cyclically arranged platform carbonates are partially covered by Wengen Group volcanics (Fig. 4) preserving the maximum thickness of the Schlern Dolomite Fm 1 (Fig. 3). Using all available biostratigraphic data, Maurer (1999, 2000) estimated compacted carbonate accumulation rates for the first aggradational phase of 200 m Myr^{-1} increasing during the second progradational phase.

As the oldest rocks in the study area that have been preserved belong to the uppermost Ladinian, the geological evolution from Late Triassic times onward can be derived only by studying younger successions in other parts of the Dolomites (e.g. Sella platform, Fig. 2) and the Southern Alps (e.g. Trento platform). Late Triassic volcaniclastics and carbonates (Wengen Group; Mastandrea *et al.*, 1997) filled the basins. An extensive carbonate platform developed with the onset of a period of tectonic quiescence. The so-called Trento platform comprises the entire central segment of the Southern Alps on the Adriatic Plate (Dolomia Principale Fm and Calcari Grigi Fm; Leonardi, 1967; Bosellini & Broglio Loriga, 1971; Bosellini & Hardie, 1985; Trevisani, 1991; Boomer *et al.*, 2001). From Middle Jurassic times onward, the Trento platform started to subside and eventually drowned. A phase of deep marine sedimentation began (Ammonitico Rosso Fm; Winterer & Bosellini, 1981; Martire, 1996; Winterer, 1998) and lasted until Late Cretaceous times (Marne del Puez Fm; Claps *et al.*, 1991; Antruilles Fm; Stock, 1996). Water depths decreased when the tectonic regime switched from extensional to compressional and the collision of the Adriatic plate with Europe began with Late Cretaceous subduction of oceanic crust (Hsü, 1971; Smith, 1971; Trümpy, 1982; Laubscher & Bernoulli, 1982; 'eoalpine' *sensu* Doglioni & Bosellini, 1987; Hsü, 1989). Towards the east, the Southern Alps were strongly involved in the Dinaric orogeny during the Late Eocene, but the compression front is thought to have extended even into the Dolomites (Doglioni & Bosellini, 1987). This interval of convergence was shortly interrupted by an Oligocene extensional phase recorded along the Periadriatic Line by plutonic intrusions (e.g. Adamello pluton) and dykes as well as by effusive basalts at the southwest termination of the Trento Plateau (Zattin *et al.*, 2006). Upper Oligocene shallow-marine conglomerates (Monte Parei Fm) record uplifted source areas during ongoing or renewed continent collision in the eastern Dolomites (Cros, 1966; Mair *et al.*, 1996). Although steady state or episodic exhumation of parts of the Alps remains debated (Bernet *et al.*, 2001; Carrapa *et al.*, 2003; Kuhlemann *et al.*, 2006), there seems to be increasing consensus on the timing of exhumation. At least three stages of exhumation are observed: rapid exhumation before approximately 35 Ma, slower exhumation until ~15 Ma and very rapid exhumation to present-day positions from then onward

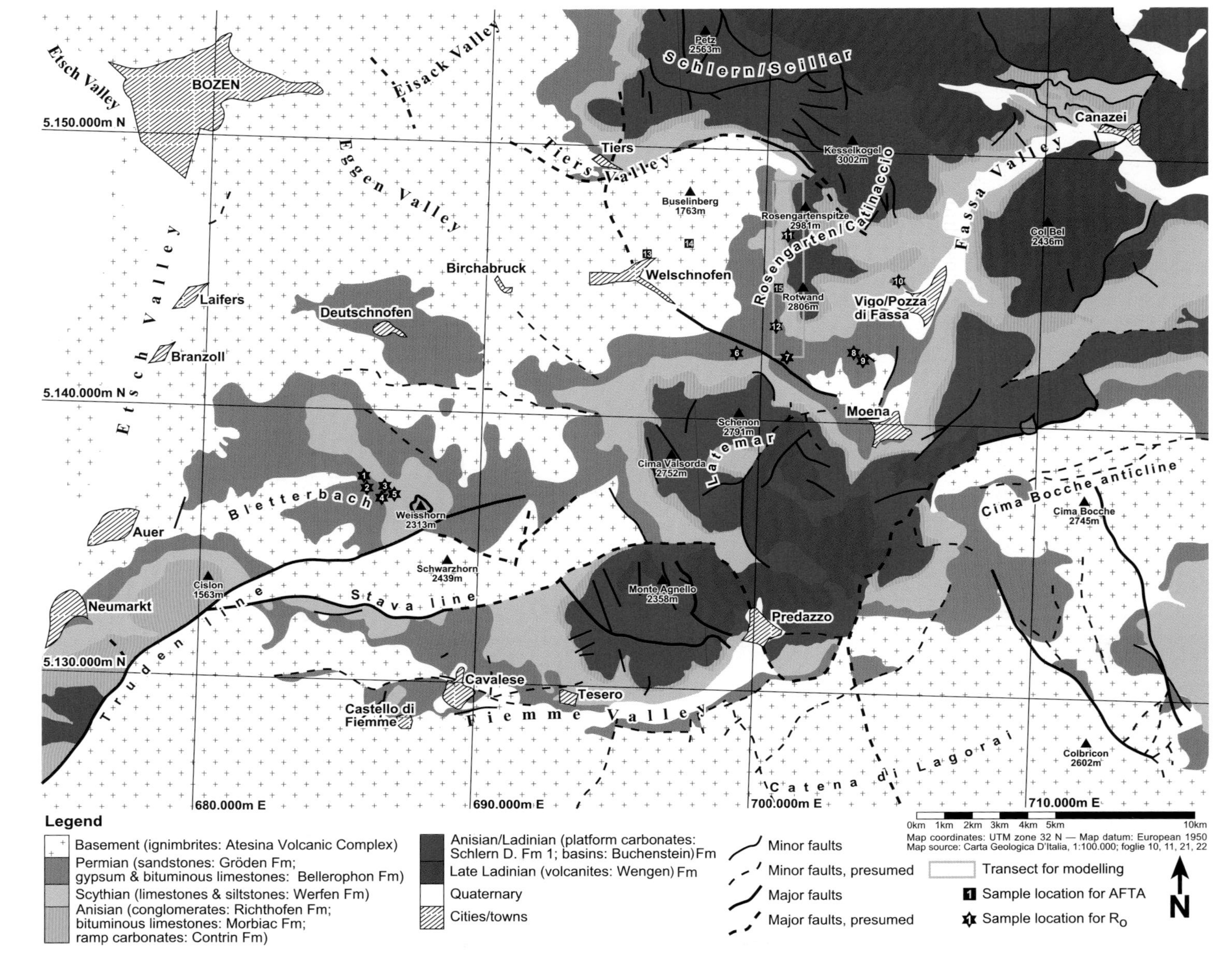

Fig. 4. Simplified geological map of the study area and its surroundings. Note the presence of the Cima Bocche Anticline/Stava Line (major faults) and the late Ladinian Wengen Group volcanics in the eastern and northeastern part of the area. Legend of the geological map in the lower part. Sample locations for vitrinite-reflectance analyses are marked by stars, those for apatite fission-track analyses by squares.

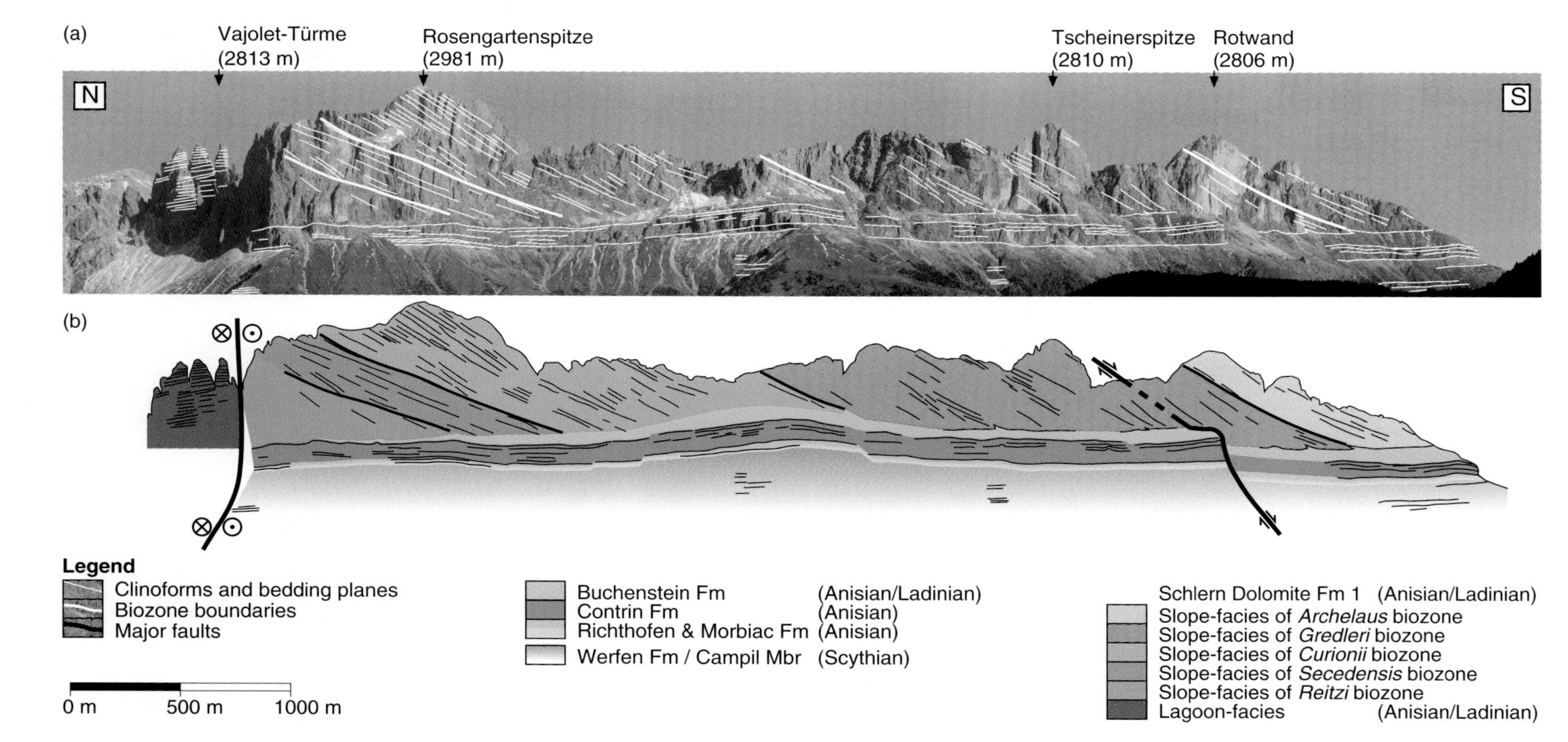

Fig. 5. Sedimentological interpretation of the Rosengarten platform. (a) The Rosengarten platform viewed from the west with an interpretation of stratal lines and clinoforms. The location of the transect is marked in Fig. 4. Legend underneath Fig. 5b. (b) Formations underneath the Rosengarten platform and correlation of carbonate slope deposits with biozones of the basinal Buchenstein Fm (according to Maurer, 1999, 2000). The transect is tectonically undisturbed, major faults are present at the platform interior only ('Vajolet-Türme'/'Torri del Vajolet').

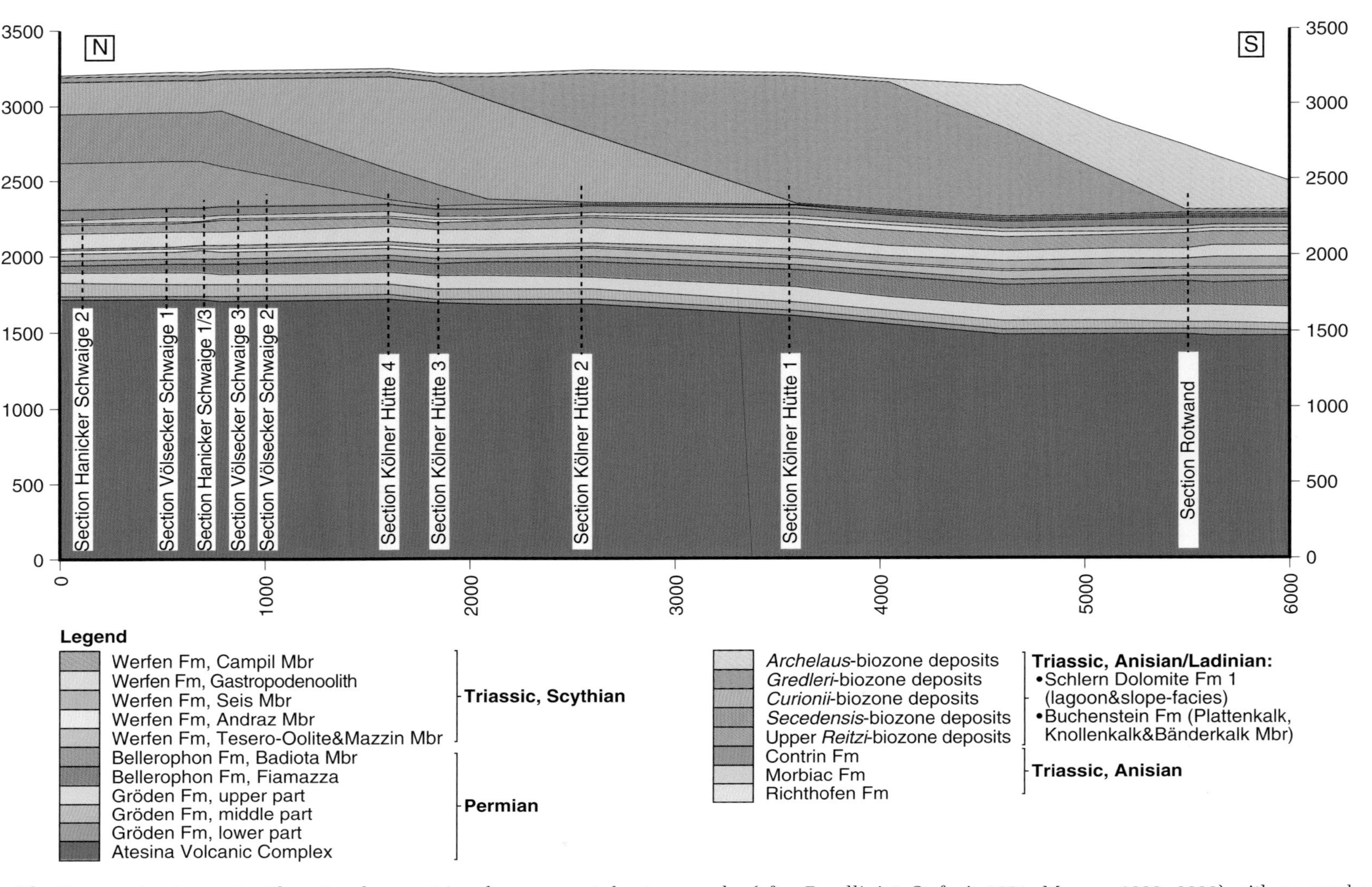

Fig. 6. The Rosengarten transect with restored geometries above present-day topography (after Bosellini & Stefani, 1991; Maurer, 1999, 2000) with an overlay of formations, and intervals of platform growth (colour code in the lower part). The sedimentological sections logged through the underlying strata of the Schlern Dolomite Fm I during the course of this study are indicated. Vertically exaggerated; units of the x-axis are metres along the transect beginning at the platform interior; units of the y-axis are metres above mean present-day sea level.

(Carrapa *et al.*, 2003; Zattin *et al.*, 2003; Bertotti *et al.*, 2006).

METHODS AND DATABASE

Sedimentological analyses

Detailed sedimentological analyses (logging, facies mapping, lateral tracing of physical surfaces, thin sections) have been carried out on the underlying strata of the Rosengarten platform (Fig. 6). Ten sections/sedimentological logs cover the entire basin fill from basement (Atesina Volcanic Complex, AVC) to the basal Schlern Dolomite Fm 1. Additional data on the upper Anisian succession and the Buchenstein Fm were taken from the literature (Bosellini & Stefani, 1991; Maurer, 1999; Zühlke, 2000). These analyses are necessary to obtain datasets on thicknesses, lithologies and palaeowater depths. The latter is based on the integration of all sedimentological evidence from the outcrops (channels, ripples, exposure surfaces, bioturbation, etc.) combined with microfacies analyses of the 283 thin-sections.

Stratigraphy and timescale

Within the past years, there has been some controversy about the duration of Schlern Dolomite Fm 1, informally known as the 'Latemar controversy' (Brack *et al.*, 1997; Hardie & Hinnov, 1997). Several studies (Goldhammer & Harris, 1989; Hinnov & Goldhammer, 1991; Preto *et al.*, 2001) proposed a specific type of orbital forcing for the cyclically arranged lagoonal interior of the Schlern Dolomite Fm 1. Subsequent studies (Brack & Rieber, 1993; Brack *et al.*, 1996; Mundil *et al.*, 1996) questioned this Milankovitch model because of its incompatibility with bio- and chronostratigraphic data. Radiometric age dating on detrital zircons in airborne tuff layers intercalated within the lagoonal sediments of the Latemar helped to solve the controversy (Mundil *et al.*, 2003). Subsequent time-series analyses and numerical simulation based upon *in situ* bio- and chronostratigraphic data indicate a much higher frequency of the cycles and thus a much shorter duration of the entire cyclic succession (Zühlke *et al.*, 2003; Zühlke, 2004; Emmerich *et al.*, 2005a). This study follows the stratigraphic concept laid down in Emmerich *et al.* (2005a).

Upper Triassic formations were evaluated and thicknesses projected from the western and central Dolomites (Schlern, Sella and Gardenaccia platform), Jurassic strata from the Trento platform, and Cretaceous and Tertiary formations from the eastern Dolomites (Table 1). Lithology information was combined with published data on chronostratigraphy and palaeobathymetry. Initial Jurassic palaeowater depths were calculated with a subsidence curve for the Trento platform proposed by Winterer & Bosellini (1981) and Winterer (1998). If necessary, thickness of eroded stratigraphic units and palaeowater depth were adapted within a range of values provided by studies on the regional geology (Table 1) in order to fit the simulated time–temperature evolution.

The timescale applied during modelling (Table 1) was determined by subdividing known chronostratigraphic intervals by biostratigraphical information (for details on this method see Bosence *et al.*, 1994; Aurell *et al.*, 1998). This means that existing studies on the temporal evolution of the Upper Permian (basement: Barth & Mohr, 1994; Barth *et al.*, 1993, 1994; sedimentary cover: Massari & Neri, 1997), Lower Triassic (Broglio Loriga *et al.*, 1983) and Middle Triassic (Brack & Rieber, 1993; Brack *et al.*, 1996; Maurer, 2000; Zühlke, 2000; Zühlke *et al.*, 2003) in the study area were recalibrated to latest radiometric age measurements (Permian: Yugan *et al.*, 1997; Permian–Triassic boundary: Mundil *et al.*, 2001; Lower Triassic: Lehrmann *et al.*, 2002; Middle Triassic: Mundil *et al.*, 1996, 2003). The relative duration of the bio-, sequence- and lithostratigraphic timesteps of the existing studies was left unchanged. Younger chronostratigraphic ages were derived from the *Geological Time Scale 2004* (Gradstein *et al.*, 2004).

Vitrinite reflectance and apatite fission-tracks

Vitrinite reflectance (VR) is the most frequently used parameter in assessing regional thermal maturity and is furthermore widely used as a calibration parameter for numerical simulation of thermal basin development. Vitrinite reflectance data on seven sample locations were previously published (Buggisch, 1978; Schulz & Fuchs, 1991; Bielefeld, 1998; Table 2). One location was included where T_{max} data were available that complemented the published dataset on thermal maturity (Zattin *et al.*, 2006; Table 2). In addition to this dataset, 11 samples from Upper Permian sandstones (Gröden Fm) to Middle Triassic basinal shaly limestones (Buchenstein Fm) were analysed in this study. Vitrinite reflectance was determined

Table 1. List of formations/timesteps (column 2) applied during modelling with ages at tops of formations/timesteps (column 3), thicknesses (column 4), indication of erosion (column 5) and lithologies (column 6). The last column (column 7) shows the main references for these formations with respect to the information shown in this table. The chronostratigraphic framework is derived from the following literature: Permian (Yugan *et al.*, 1997), Triassic (Lehrmann *et al.*, 2002; Mundil *et al.*, 1996, 2001, 2003), Jurassic, Cretaceous and Cenozoic/Tertiary (Gradstein *et al.*, 2004). Strata overlying present-day topography were projected from the western and central Dolomites (Middle and Upper Triassic; see also Fig. 4), the Trento platform (Jurassic) and eastern Dolomites (Cretaceous and Tertiary).

1	2	3 Age (Ma)		4	5	6	7
Era	Formation/timestep	From	To	Thickness (m)	Erosion	Lithology	Reference
TERTIARY	Monte Parei Fm	28.0	24.0	90	Yes	Conglomerate with interbedded sst	8,9
	Hiatus	80.0	28.0	–	Yes	–	8,9
	Antruilles Fm	98.9	80.0	70	Yes	lst with interbedded sst and clay	10
	Marne del Puez Fm	127.0	98.9	60	Yes	Marl with interbedded fine-grained sst	11
	Hiatus	147.5	127.0	–	Yes	–	12,13
JURASSIC	Ammonitico Rosso Fm, upper part (RAS)	154.0	147.5	10	Yes	lst with interbedded silt	12,13,14
	Hiatus	155.5	154.0	–	Yes	–	12,13,14
	Ammonitico Rosso Fm, middle part (RAM)	157.0	155.5	5	Yes	lst with interbedded silt	12,13,14
	Hiatus	160.0	157.0	–	Yes	–	12,13,14
	Ammonitico Rosso Fm, lower part (RAI)	167.0	160.0	15	Yes	lst with interbedded silt	12,13,14
	Hiatus	169.0	167.0	–	Yes	–	12,13,14
	Calcare a Filamenti Fm	172.0	169.0	20	Yes	lst	15
	Hiatus	174.0	172.0	–	Yes	–	15
	San Vigilio Oolite Mbr	178.0	174.0	70	Yes	lst	15
	Calcari Grigi Fm, Grigno Mbr	183.6	178.0	90	Yes	lst	15
	Calcari Grigi Fm, Rotzo Mbr	190.0	183.6	85	Yes	lst	15,16
	Calcari Grigi Fm, Middle Mbr	194.5	190.0	35	Yes	lst	16,17
	Calcari Grigi Fm, Lower Mbr	200.0	194.5	40	Yes	lst	16,17
TRIASSIC	Dolomia Principale Fm, upper part	207.9	200.0	40	Yes	Dolomite	18,19
	Dolomia Principale Fm, middle part	215.8	207.9	175	Yes	Dolomite	18,19
	Dolomia Principale Fm, lower part	223.8	215.8	100	Yes	Dolomite	18,19
	Raibl Fm	229.9	223.8	40	Yes	Sandy lst	18,20,21,22
	Hiatus	232.0	229.9	–	Yes	–	18,20,21,22
	Wengen Fm, Schlern Dolomite Fm 2, Cassian Fm	237.1	232.0	100–730	Partially	Volcanics with lst boulders, dolomite, shale	21,22,23,24
	Archelaus biozone deposits (Schlern Dolomite Fm 1 & Buchenstein Fm)	238.0	237.1	20–400	Partially	dolomite	22,25,26,27,28
	Gredleri biozone deposits (Schlern Dolomite Fm 1 & Buchenstein Fm)	238.8	238.0	18–800	Partially	Dolomite	22,25,26,27,28
	Curionii biozone deposits (Schlern Dolomite Fm 1 & Buchenstein Fm)	241.0	238.8	12–640	Partially	Dolomite	22,25,26,27,28

(*Continued.*)

Table 1. Continued.

1	2	3 Age (Ma)		4	5	6	7
Era	Formation/timestep	From	To	Thickness (m)	Erosion	Lithology	Reference
TRIASSIC (*Continued*)	*Secedensis* biozone deposits (Schlern Dolomite Fm 1 & Buchenstein Fm)	242.3	241.0	10–300	Partially	Dolomite	22,25,26,27,28
	Reitzi biozone deposits (Schlern Dolomite Fm 1 & Buchenstein Fm)	242.9	242.3	10–300	No	Dolomite	22,25,26,27,28
	Contrin Fm	243.5	242.9	40–55	No	Dolomite	22,29
	Morbiac Fm	243.8	243.5	25	No	Marl with interbedded lst	22,29
	Richthofen Fm	244.2	243.8	7–25	No	Conglomerate with interbedded coarse-grained sst	22,29
	Hiatus	247.5	244.2	–	No	–	22,29
	Werfen Fm, Cencenighe Mbr	248.1	247.5	80	Yes	lst with interbedded marl	30
	Werfen Fm, Val Badia Mbr	248.7	248.1	60–80	Partially	Marly lst	30
	Werfen Fm, Campil Mbr	250.0	248.7	45–90	Partially	Fine-grained sandstone with interbedded silt	22,30
	Werfen Fm, Gastropodenoolith Mbr	250.6	250.0	60–100	No	Intercalations of lst, marl and silt	22,30
	Werfen Fm, Seis Mbr	251.5	250.6	12–55	No	Intercalations of lst and marl	22,30
	Werfen Fm, Andraz Mbr	252.0	251.5	12–20	No	Sandy dolomite	22,30
	Werfen Fm, Tesero Oolite & Mazzin Mbr	253.0	252.0	40–50	No	Intercalations of lst, marl and silt	22,30
PERMIAN	Bellerophon Fm, Badiota Mbr	253.8	253.0	25–35	No	Intercalations of lst and marl	22,31,32,33,34
	Bellerophon Fm, Fiamazza Mbr	255.3	253.8	50–150	No	Gypsum with interbedded dolomitic marl	22,31,32,33,34
	Gröden Fm, upper part	256.6	255.3	65–105	No	Silt-rich medium- to fine-grained sst	22,32,33,34
	Gröden Fm, middle part	258.0	256.6	45–75	No	Silt-rich coarse-grained sst	22,32,33,34
	Gröden Fm, lower part	260.0	258.0	25–35	No	Litharenitic sst	22,32,33,34
	Hiatus	267.0	260.0	–	No	–	22,32,33,34
	Atesina Volcanic Complex	276.0	267.0	>2000	Partially	Rhyolitic ignimbrites, sometimes interbedded sediments	35,36,37,38

Key to column 7 (reference):
8: Cros 1966; 9: Mair *et al.* 1996; 10: Stock 1996; 11: Claps *et al.* 1991; 12: Winterer & Bosellini 1981; 13: Winterer 1998; 14: Martire 1996; 15: Trevisani 1991; 16: Bosellini & Broglio Loriga 1971; 17: Boomer *et al.* 2001; 18: Leonardi 1967; 19: Bosellini & Hardie 1985; 20: Doglioni & Goldhammer 1988; 21: Brandner 1991; 22: this study; 23: Schlager *et al.* 1991; 24: Mastandrea *et al.* 1997; 25: Brack & Rieber 1993; 26: Brack & Rieber 1994; 27: Maurer 1999; 28: Maurer 2000; 29: Zühlke 2000; 30: Broglio Loriga *et al.* 1983; 31: Buggisch & Noé 1986; 32: Massari *et al.* 1988; 33: Massari *et al.* 1994; 34: Massari & Neri 1997; 35: D'Amico & Del Moro 1988; 36: Barth *et al.* 1993; 37: Barth *et al.* 1994; 38: Barth & Mohr 1994.

Fm, Formation; Mbr, Member; sst, sandstone; lst, limestone; RAS: rosso ammonitico superiore; RAM: rosso ammonitico mediore; RAI: rosso ammonitico inferiore.

Table 2. List of vitrinite reflectance, thermal maturity and apatite fission-track data. The numbers in the first column correspond to the locations in Fig. 4. The source of the data and their sample numbers are shown in columns 2 and 3. UTM coordinates (column 4) and elevations above present-day sea level (column 5) correspond to the system and map datum used in Fig. 4. The formations from which the samples were taken are listed in column 6. The lithology of the samples is shown in column 7. Columns 8–10 list data on thermal maturity. Palaeotemperature values in column 9 are calculated from the measured thermal maturity data (% VR_r values in column 8) after the equation of Barker & Pawlewicz (1986). The last columns (11, central age; 12, mean track length) refer to the apatite fission-track analysis carried out in Emmerich *et al.* (2005).

1	2	3	4		5	6	7	8	9	10	11	12
Loc.	Reference	Sample number	UTM coordinates x	UTM coordinates y	Elevation (m)	Formation member	Lithology*	VR_r (%)	Calculated T (°C)	T_{max} (°C)	Central age (Ma)	Mean track length (µm)
1	Buggisch 1978	Bb40	685800	5137100	1520	Gröden, middle part	Sandstone	0.73	110			
2	Bielefeld 1998	15e, 15eII	685800	5137100	1520	Gröden, middle part	Sandstone	0.60 ± 0.09	94 ± 11			
3	Bielefeld 1998	Butter2, But	686750	5136900	1650	Gröden, upper part	Sandstone	$0.68^{\dagger}$	104			
4	Buggisch 1978	Bb, Bb10	686750	5136900	1650	Gröden, upper part	Sandstone	$0.62^{\dagger}$	97			
5	Schulz & Fuchs 1991	–	686750	5136900	1650	Gröden, upper part	Sandstone	0.88	125			
6	Bielefeld 1998	20a, 20aII	698900	5142550	1710	Gröden, upper part	Sandstone	0.76 ± 0.10	113 ± 10			
7	Bielefeld 1998	N6a, N6b, N6bII	701000	5142350	1620	Gröden, upper part	Sandstone	0.67 ± 0.14	103 ± 15			
8	This study	groed 1	703493	5142526	1600	Gröden, upper part	Sandstone	$0.9 \pm 0.07^{\ddagger}$	127 ± 6			
9	This study	groed 2	703493	5142526	1600	Gröden, upper part	Sandstone	$1.1 \pm 0.14^{\ddagger}$	143 ± 10			
10	This study	belle	705014	5145331	1615	Bellerophon, Badiota	Limestone	0.52 ± 0.02	83 ± 3			
11	This study	070	700900	5148006	2330	Morbiac	Limestone	$0.77 \pm 0.02^{\S}$	114 ± 2	462		
12	Zattin *et al.* 2006	12	700500	5144460	1520	Moena	Limestone			429		
13	Emmerich *et al.* 2005b	AE14	695807	5145764	1275	Atesina Volcanic Complex	Ignimbrite, rhyolitic				$84.2 \pm 12.7^{\P}$	10.6 ± 1.5
14	Emmerich *et al.* 2005b	AE18	697212	5146488	1450	Atesina Volcanic Complex	Ignimbrite, rhyolitic				165.6 ± 7.3	9.8 ± 1.6
15	Emmerich *et al.* 2005b	AE10	701149	5144079	2160	Werfen, Gastropoden-Oolith	Sandstone				$74.7 \pm 8.1^{\P}$	10.8 ± 1.9

* Coarse to medium-grained sandstones, except sample AE10, which is fine-medium grained. The total organic carbon of limestone is <1%.
† Organic matter not *in situ* or altered.
‡ Proximity to a volcanic dyke, altered organic matter.
§ Number of measurements too low, no statistical significance.
¶ Number of grains too low, no statistical significance.

by microscopic analysis of percentages of light reflected from polished organic particles and calibrated against isotropic standards. The results are given as mean random reflectance (% VR_r, for details on methodology and measurement see Stach *et al.*, 1982; Taylor *et al.*, 1998). Only four of our samples – mostly Permian sandstones of the Gröden Fm – contained measurable 'vitrinite-like' organic matter (locations are shown in Fig. 4 and data are documented in Table 2). Several VR_r values had to be excluded as calibration parameters in thermal history modelling because of altered organic matter and/or statistically insignificant measurements (see also Table 2). As the time component is lacking in vitrinites, apatite fission-track (FT) thermochronology is commonly used to determine the magnitude and timing of cooling, exhumation and rock uplift from shallow crustal levels (Fitzgerald *et al.*, 1995; Tippett & Kamp, 1995). An extensive study on the surroundings of the Rosengarten has already been carried out by Emmerich *et al.* (2005a). Time–temperature paths from this study were used to calibrate the thermal history of the basin (Table 2).

Workflow of the integrated numerical modelling approach

The challenge during numerical simulation of present-day subaerially exposed areas is the determination of the amount and lithology of eroded overburden. Usually this is realized by extrapolating known successions to the study area but this dataset has to be verified iteratively during thermal modelling calibrated with data on thermal maturity and thermochronology. Hence, numerical simulation in this project started with modelling the burial and thermal history of the Rosengarten transect and the reconstruction of its overburden thickness with PetroMod™ (IES GmbH, Aachen, Germany). Essential input parameters were lithology, sediment–water interface temperatures (SWI, Fig. 8a), palaeowater depths (PWD, Fig. 8b) and heat-flow development (Fig. 8c). Results on the eroded rock column were obtained by fitting calculated and measured vitrinite reflectance data (Fig. 9) as well as by forward modelling t–T paths, fission-track distributions and cooling ages in apatites (Fig. 10). The eroded thickness derived from numerical simulation was entered with all other necessary input parameters (Fig. 7) into the inverse modelling routine of PHIL™ (Petrodynamics Inc., Houston, USA) in order to determine tectonic-, flexural- and compaction-induced subsidence rates for forward simulation of sedimentation. The stratigraphic forward modelling module of PHIL™ was used to calculate best-fit stratal patterns and sedimentation rates. At the end of this workflow, simulated minimum–maximum models were ultimately checked against the real-world/outcrop data. Calibration of the simulated sedimentation rates and stratal patterns was hereby performed by visual comparison with observed geometries along the Rosengarten transect and by measurement of stratal thicknesses.

Thermal modelling

Thermal modelling (in addition to petroleum systems or basin modelling) is now routinely used in petroleum exploration (Welte *et al.*, 1997; Makhous & Galushkin, 2004). Numerical simulation software usually incorporates geological, petrophysical, geophysical and geochemical data into integrated four-dimensional (4D) frameworks that allow testing of burial and basin-fill concepts. The geohistory of basins is subsequently recreated by these 4D simulations in three spatial dimensions through geological time. The accuracy of numerical simulations is limited by restricted knowledge of or uncertainty about the values of the input parameters used to constrain the basin history (Peters *et al.*, 2006).

The detailed analysis of the regional thermal maturity (coalification) pattern was combined with fission-track analyses in apatites in order to narrow down these uncertainties of thermal boundary conditions (heat-flow history). The timing and magnitude of intervals characterized by subaerial exposure and erosion were determined (i.e., burial history or geohistory of the basin). Input parameters for the numerical models were thicknesses of stratigraphic units, lithologies (Table 1), petrophysical parameters (e.g. thermal conductivities; see Büker (1996) and Hertle & Littke (2000) for a more detailed description of the calculation of physical properties of stratified sediment bodies), temperature at the sediment–water interface (Fig. 8a), palaeowater depths (Fig. 8b) and heat flow at the base of the succession (Fig. 8c). Subsidence history and temperature field through time were subsequently calculated. Burial and heat-flow histories were continuously calibrated using measured vitrinite reflectance values (Fig. 9) and fission-track data derived from apatites (Fig. 10b). The model was

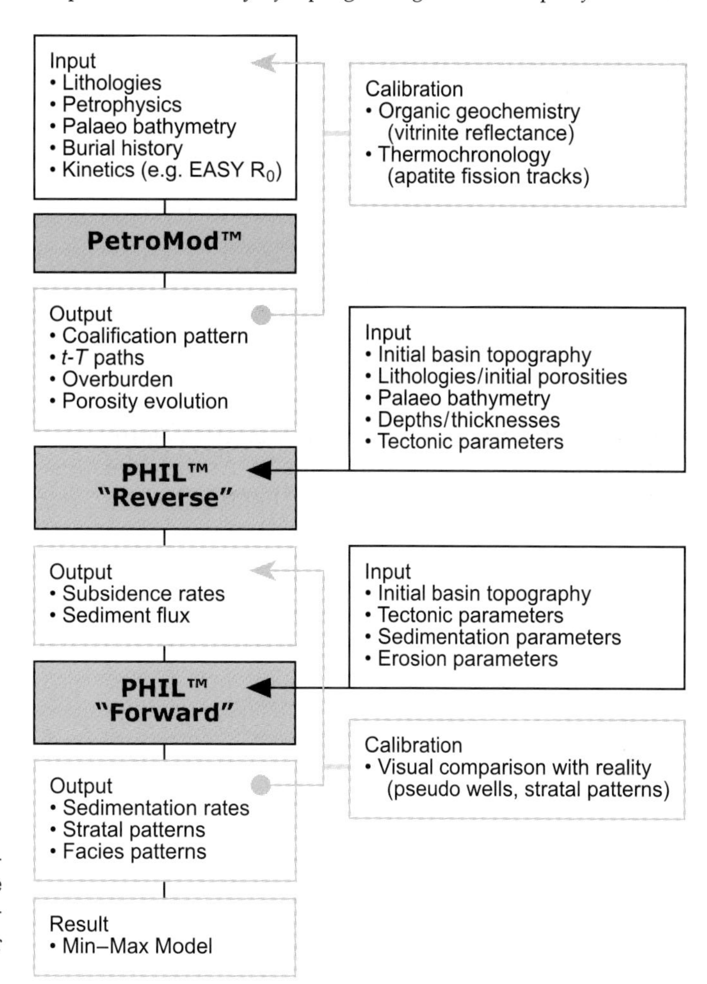

Fig. 7. Sketch illustrating the workflow during modelling; workflow starts at the top and ends at the bottom. Input data are arrows towards a simulator (PetroMod™ or PHIL™); output data used for calibration are lines away from simulators.

then fine-tuned by modifying all input parameters until a satisfactory fit between measured and calculated calibration data was achieved. Vitrinite reflectance was calculated using the EASY% R_0-algorithm of Sweeney & Burnham (1990; Sachsenhofer & Littke, 1993; Littke *et al.*, 1994; Leischner, 1994; Sachsenhofer *et al.*, 2002; Fig. 9). Fission-track data in apatites (age and track length distribution) were forward modelled with the algorithm of the AFTSolve® routine (Ketcham *et al.*, 2000) now incorporated in the latest version of the basin simulator PetroMod™ (version 9.x; www.ies.de).

Reverse-basin modelling

Inverse modelling or reverse-basin modelling in this study followed the sequence-stratigraphy concept, which considers the creation/destruction of accommodation space and its infill as the two main controls on sedimentary systems and basins. Reverse-basin modelling was carried out to determine (1) all components of total subsidence (termed 'subsidence' in the following text) which are flexural, tectonic and compaction-induced subsidence (Fig. 11) and (2) decompacted sediment flux rates in time. These two datasets are essential input parameters for the subsequent forward modelling of the sedimentary system where sedimentation rates are fine-tuned.

Input parameters for the inverse modelling are lithology (initial porosity; after Bowman & Vail, 1999), geometries (measured and projected bed thicknesses), palaeowater depth (derived from facies analysis) and crustal parameters (e.g. effective elastic thickness T_e, plate-end-boundary distance and density of the mantle). Sea-level oscillations were initially applied during the inverse modelling (second-order sea-level curve after Hardenbol *et al.* (1998) recalibrated to the timescale used in this study) but this frequency proved to be too low especially for the studied interval of platform growth at Rosengarten (5.8 Myr; Fig. 3 and Table 1) and was subsequently

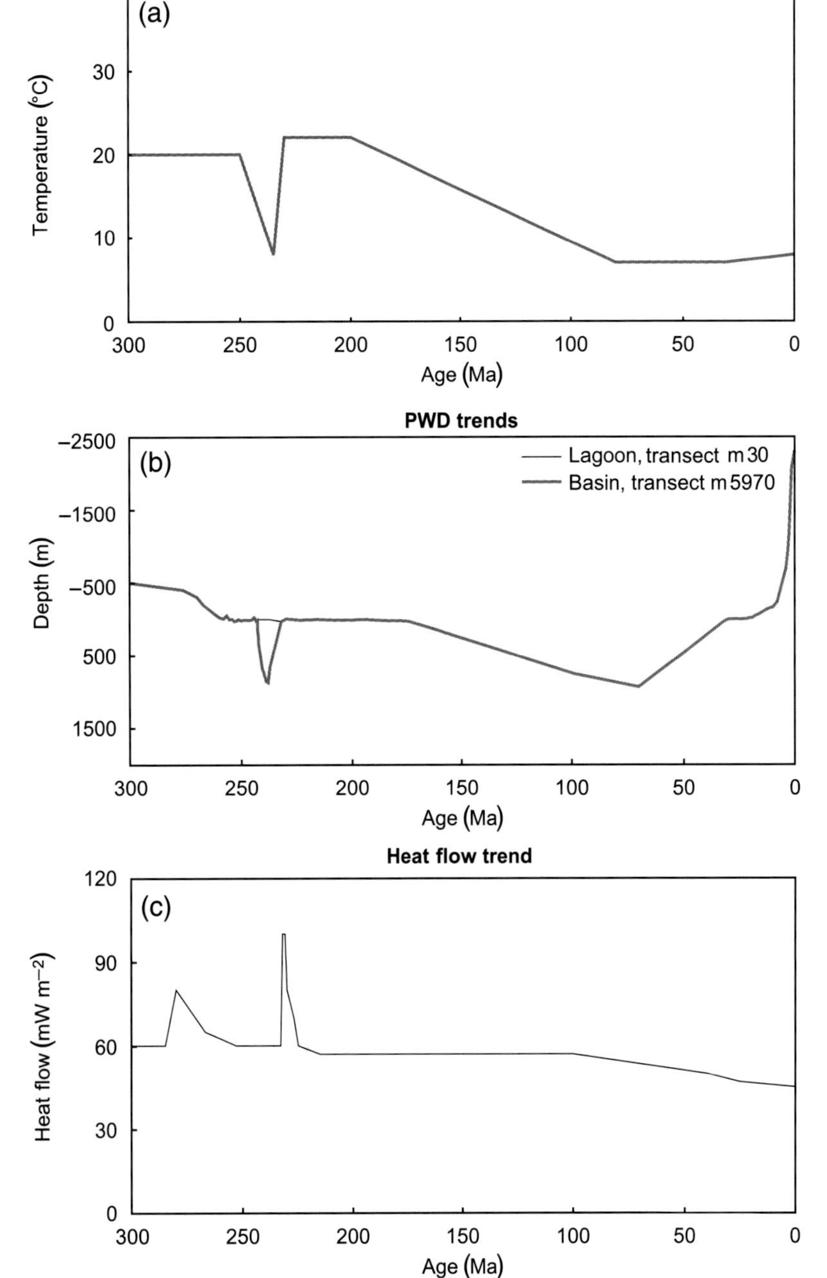

Fig. 8. Input data for thermal modelling. (a) Temperatures at the sediment–water interface (i.e., sea bottom) in time. (b) Palaeowater depth (PWD) of sea bottom in time as applied during modelling with PetroMod™ and Phil™. The two curves show PWD trends for locations of a proximal lagoon (i.e. transect metre 30) and a distal slope setting (i.e. transect metre 5970). The curve measured at a proximal location runs exactly parallel to the curve measured at a distal location apart from the time when the basin deepens (i.e., the lagoonal curve remains at constant water depths). (c) Heat-flow history as applied during modelling with PetroMod™.

neglected during further modelling runs. Third-order sea-level oscillations have not been considered because their timing and amplitude in the latest Anisian/earliest Ladinian is controversially discussed in the literature (Rüffer & Zühlke, 1995; Gianolla & Jacquin, 1998). Hence, eventual accommodation changes related to sea-level oscillations are incorporated within the calculated subsidence rates. Owing to the shortness of the transect (approximately 6 km) and the rigidity of the underlying rheological basement (2000–2500 m thick Permian AVC), crustal parameters and flexural subsidence play a minor role. Nevertheless, recent data on the effective elastic thickness (T_e of approximately 20 km) of the lithosphere of the Venetian basin (Barbieri *et al.*, 2004) were applied (Table 3).

Subsidence component, accommodation and sediment-flux histories were calculated for each of the timesteps identified in the Permian to

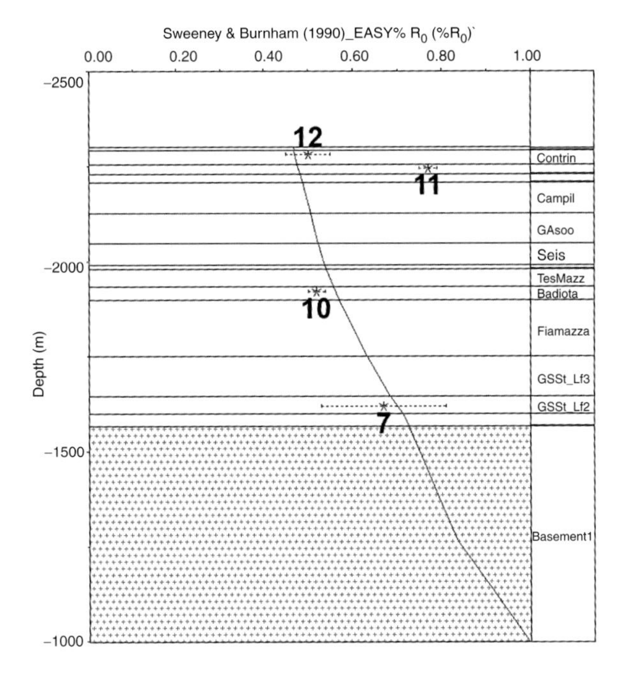

Fig. 9. Calibration of the subsidence and thermal history at Rosengarten with measured VR_r values. x-axis, vitrinite reflectance in EASY%_VR_0; y-axis, depth in metres (negative values indicate an elevation above present-day sea level). Formations are indicated at the right-hand side. The asterisks reflect measured values, the numbers correspond to sample locations (see Fig. 4 and Table 2). The curve illustrates the calculated maturity values of the present-day situation at Rosengarten. The maturity vs. depth curve was extracted from the southern part of the transect, i.e., the part where the density of calibration parameters is at a maximum.

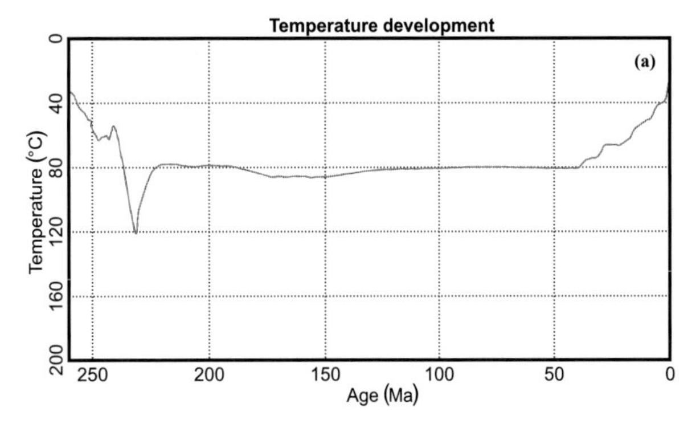

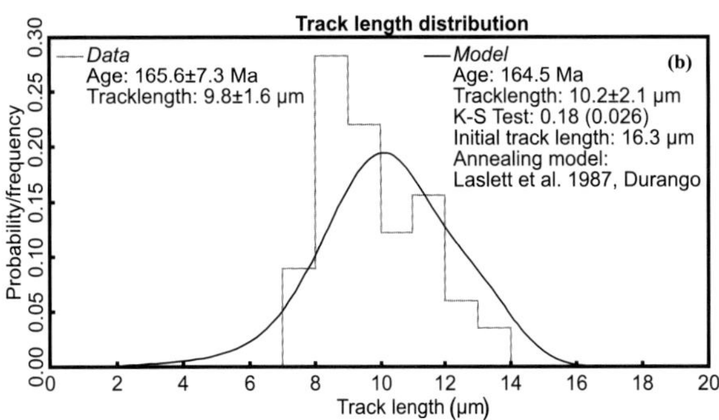

Fig. 10. Calibration of the subsidence and thermal history at Rosengarten with measured apatite-fission-track data. (a) Calculated time–temperature path at Rosengarten as derived from PetroMod™. x-axis, age in million years; y-axis, temperature in degrees Celsius. (b) Comparison of calculated with measured apatite fission tracks. Data values in grey, calculated values in black.

Late Triassic basin fill (Table 1). Basin reverse modelling runs in the opposite direction of sedimentation: the process starts at time t_2 and runs backwards in time to t_0 when all sediment layers have subsequently been removed ('backstripped'). Each timestep (t_0, t_1, t_2) is characterized by a distinct vector of tectonic subsidence (TS_0 to TS_2), flexural and compaction-induced subsidence as well as a change in palaeowater depth (PWD_0 to PWD_2). During each step of removal, the hypothetical depth of the basin floor is calculated without being loaded and the current depositional surface is adjusted to predefined palaeobathymetry (Fig. 11). Rates are calculated for each time layer after the removal of flexural loading effects, changes in palaeobathymetry, changes in sea level and compaction. The flexural backstripping procedure applied in the reverse-basin modelling of this study is based on the equations introduced by Turcotte & Schubert (1982, 2002) and Dickinson *et al.* (1987). The backstripping procedure with the applied software – PHIL™ (Marco Polo Software Inc., Houston, USA) – is also described by Bowman & Vail (1999).

Sequence-stratigraphic forward modelling

Stratigraphic forward modelling in this project was mainly used to quantify carbonate accumulation rates. The simulation of facies patterns and the identification of processes operating on the platform slope were a minor goal, as well as the determination of siliciclastic sedimentation rates in the Wuchiapingian to Anisian formations. Therefore this chapter – as well as the entire study – focuses on the methodology of simulating carbonate sediments. Input parameters for sequence-stratigraphic forward modelling were tectonic subsidence, sediment parameters (Fig. 12 and Table 4) and depth-dependant carbonate-accumulation rates ('carbonate-production curve'; Fig. 13). As published by Bowman & Vail (1999), PHIL™ 'represents the carbonate production

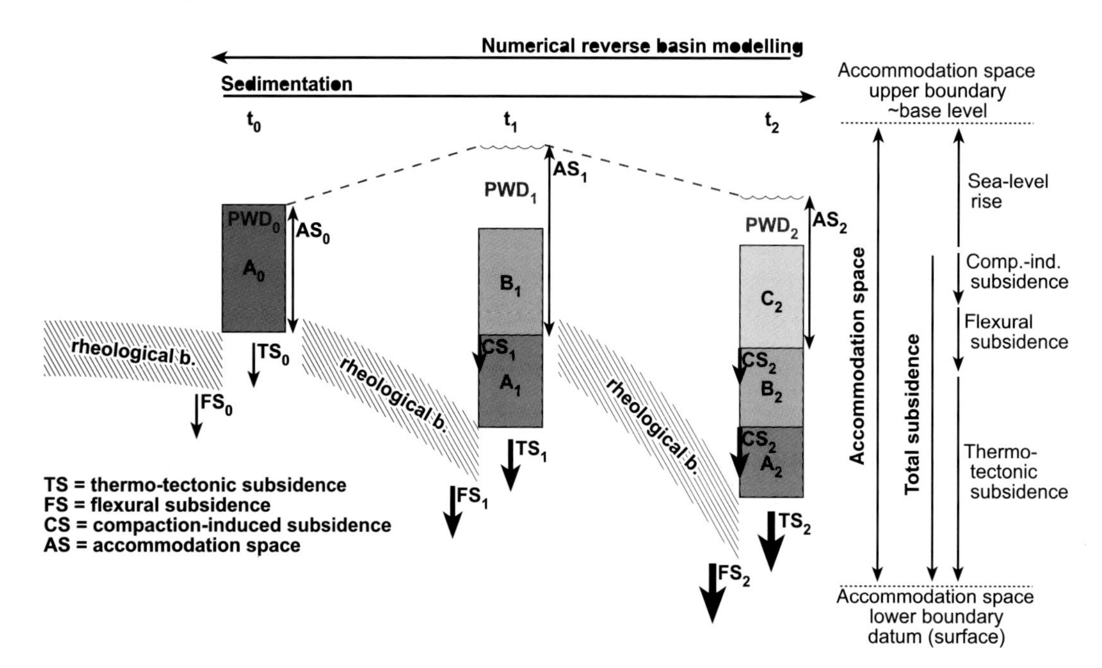

Fig. 11. Sketch illustrating the basin reverse modelling process ('backstripping'). *x*-axis, time; *y*-axis, burial depth; stratigraphic units are marked by letters A to C and different greyscales; arrows below the stratigraphic columns illustrate vectors of tectonic and flexural subsidence (thick arrow, high subsidence; thin arrow, low subsidence). The principle of development of accommodation space is illustrated on the right. Numerical values used for the flexural modelling process are shown in Table 3.

system with four water-depth-dependant functions' (Fig. 13) 'including (1) coarse-grained traction-load production composed of a laterally unrestricted factory and a laterally restricted shelf-margin factory, and (2) fine-grained suspension-load production composed of an unrestricted fine-grained factory and a pelagic factory'. However, as redeposition is included in the 'production' *sensu* Bowman & Vail (1999), the modelled rates reflect accumulation rates, not production rates. We therefore prefer to use the term accumulation rate *sensu* Bosscher & Schlager (1993). The carbonate accumulation functions used in PHIL™ (Bowman & Vail, 1999: 'productivity functions') are normal distribution curves with a specified width and a maximum accumulation at a specified bathymetry (Fig. 13). Unrestricted traction and fine-grained accumulation will occur on any surface below sea level. Shelf-margin accumulation is centred about an optimal location for accumulation with respect to the open basin and exponentially reduced as a function of distance from that location by a specified factor:

$$A_{\text{depth}} = M \cdot t \cdot R \cdot \mathrm{e}^{(-(B-D_{\text{ma}})^2/W^2)}$$

where A_{depth} is the accumulation for a cell during the time increment (*sensu* Bosscher & Schlager, 1993), M is the maximum accumulation rate (in m Myr^{-1}), t is time (in Myr), R is a siliciclastic reduction factor, B is the bathymetry (in m), D_{ma} is the maximum accumulation bathymetry and W the width of the productivity function (in m; after Bowman & Vail, 1999).

Sediment parameters and sedimentation/accumulation rates were adjusted in an iterative process within a geologically reasonable range and in accordance with measured data (Table 4) until a match with present-day geometries was achieved. During this iterative course of forward modelling, a sensitivity analysis was carried out in order to determine the influence of different carbonate-specific sedimentation parameters. Bowman & Vail (1999) have described in further detail the stratigraphic forward modelling process with PHIL™.

RESULTS

Thermal modelling

The combination of vitrinite reflectance recording the maximum temperatures during basin evolution and fission-track data in apatites recording the temporal development of heating and cooling enabled a highly fine-tuned thermal model of the

Table 3. Input parameters for flexural inverse modelling.

Sediment type	Sedimentary parameters	Ranges in Bowman & Vail (1999)	This study
Siliciclastics	Fluvial plain gradient	0.001–0.00001	0.001
	Coastal plain gradient	0.0–0.00001	0.00001
	Shoreface gradient	0.01–0.001	0.008
	Depositional front gradient	0.1–0.01	0.08
	Coastal plain width	0–200 km	20 km
	Rollover width	1–5 km	3 km
	Depth of offlap break	10–20 m	10 m
	Barrier island height	0–10 m	0.1 m
	Barrier island width	0–15 km	2 km
	Depth of fairweather wave base	5–20 m	10 m
	Prodelta suspension distance	5–100 km	15 km
	Suspension mixing depth limit	1–200 m	5 m
	Traction fraction	0–100%	100%
	Coarse sand fraction	0–100%	10%
Carbonates	Sabkha gradient	0.0–0.00001	0.000001
	Tidal flat gradient	0.0001–0.0006	0.001
	Tidal range	0–15 m	1 m
	Lagoon gradient		0.01
	Back-reef gradient	0.1–0.001	0.02
	Depth of reef crest	1–20 m	10 m
	Width of margin production	0.1–5.0 km	0.2 km
	Rollover width	0.5–3.0 km	1.55 km
	Slope gradient	0.01–1.1	0.7 (35°)
	Suspension distance of lagoonal fine-grained carbonates	1–50 km	0.1 km
	Maximum suspension depth of fine-grained carbonates	3–40 m	3 m
	Siliciclastic damping limit	1–150 m Myr^{-1}	1 m Myr^{-1}
	Production time increment	0.001–0.025 Myr	0.001 Myr
Gravity flow	Minimum bathymetric relief	100–400 m	8 m
	Slope-fan threshold depth	0–30 m	30 m
	Basin-floor fan gradient	0.001–0.0001	0.001
	Slope-fan gradient	0.03–0.001	0.001
	Relative water-level trigger factor	0 to –30 m Myr^{-1}	–10 m
	Turbidite volume factor	0–1	1

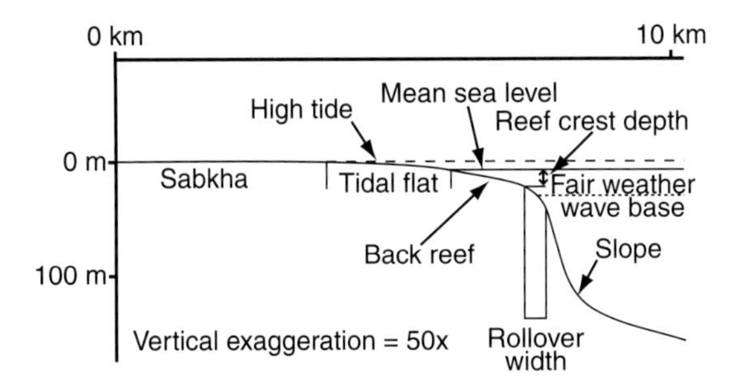

Fig. 12. Sketch illustrating the depositional environments and parameters used during the forward-modelling procedure of PHIL™ (modified after Bowman & Vail, 1999). Numerical values of the input parameters in Table 4.

Table 4. Input parameters for sequence-stratigraphic forward modelling.

Flexural parameters	This study
Flexural wavelength (km)	55.0
Effective elastic thickness (km)	20.2
Mantle density (kg m^{-3})	3340.0
Water density (kg m^{-3})	1030.0
Left taper limit (km)	50.0
Right taper limit (km)	500.0

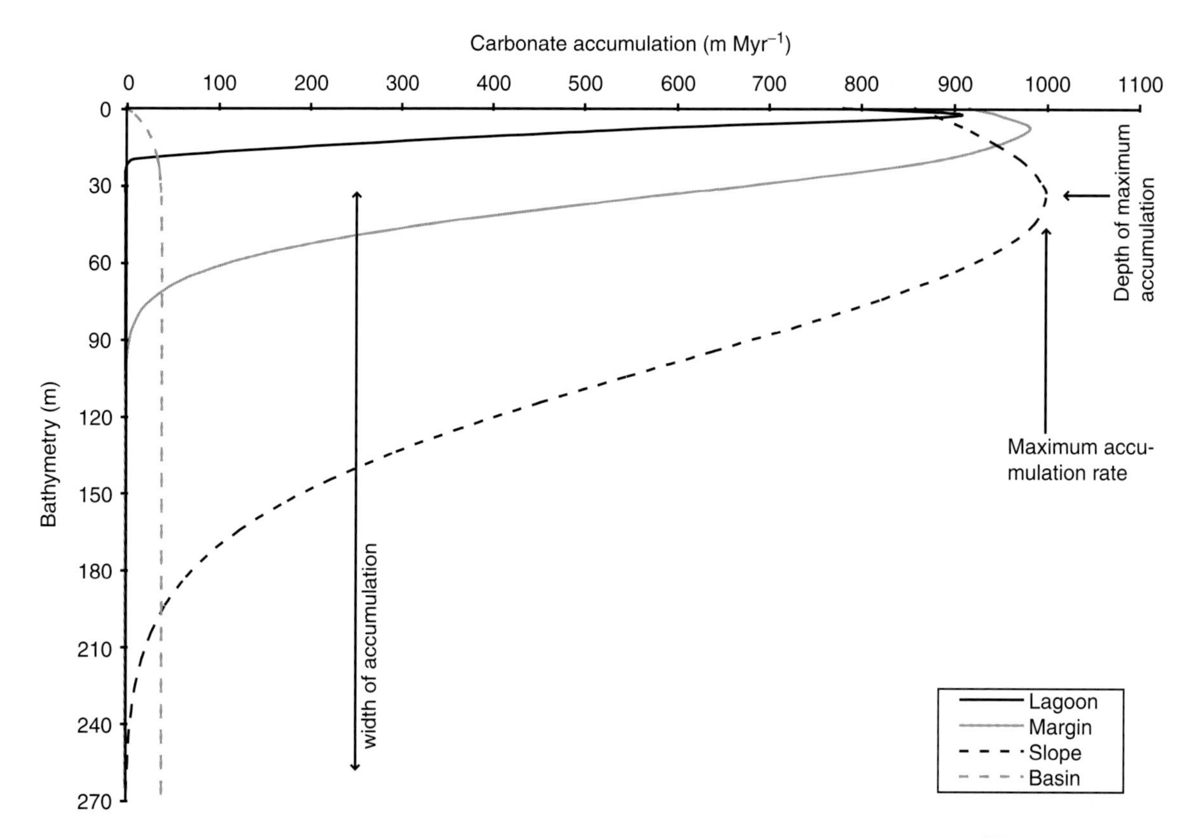

Fig. 13. Carbonate-production rates as a function of water depth during simulation with PHIL^{TM}. *x*-axis, carbonate-production rates in m Myr^{-1}; *y*-axis, water depth in metres. Legend to functions to the bottom right. The most important variables (depth, maximum and width of production) used in the carbonate-productivity functions of PHIL^{TM} (Bowman & Vail, 1999) are indicated for the slope function (dashed black line). In order to reproduce present-day geometries, the production of the carbonate factory had to be set to values between 900 and 1000 m Myr^{-1}. Production increases from periplatform environments (900 m Myr^{-1}) across the margin (950 m Myr^{-1}) to the slope (1000 m Myr^{-1}). Pelagic production (i.e., Buchenstein Fm) was limited to 50 m Myr^{-1}.

southwest Dolomites. Modelling was stopped at a deviation of less than 5% with respect to the calibration data. In other words, the difference between iterative changes of input parameters such as: (1) the magnitude and timing of heating or cooling; (2) changes in sediment–water-interface temperatures; and (3) overall stratigraphic thickness was less than 0.3%.

Time–temperature history

A constant heat flow of around 60 mW m^{-2} is a typical value for continental crust (Allen & Allen, 1990) and hence it also was the starting point of this numerical simulation. However, after numerous runs of iteratively calibrated geohistory simulation it turned out that two intervals of elevated basal heat flow (Fig. 8c) are required in this part of the southwest Dolomites in order to match measured apatite fission-track data (Fig. 10). The first interval between 285 and 253 Ma corresponds to the generation, extrusion and cooling of the Permian Atesina Volcanic Complex (AVC); the second interval between 233 and 225 Ma corresponds to a time when strong hydrothermal activities are recorded in the Val di Fassa area. This modelling result is in accordance with suggestions by Greber *et al.* (1997), who proposed that heat flow reached its peak (90 mW m^{-2}) at 236 Ma when an intrusion of pegmatites and an appearance of tuffs and localized lava flows are observed across the central Southalpine domain. This is also backed up by Bertotti *et al.* (1997, 1999) but with different estimates for time and extent of the elevated basal heat flow. All aforementioned authors concluded that the heat flow decreased to values of approximately 70 mW m^{-2} by 220 Ma. In contrast, Zattin *et al.* (2006) suggested that heat flow remained at high values until the end of Jurassic rifting. We believe that the observations of Barth *et al.* (1993, 1994) – a hydrothermal event in the Val di Fassa area resetting isotopic clocks at around 230–225 Ma – much better reflects the thermal evolution in our study area. The time interval between ~100 Ma until today is characterized by a slowly decreasing basal heat flow down to values of 45 mW m^{-2}, as for example measured in wells in the Belluno basin (Sedico-1; Zattin *et al.*, 2006).

Thermal modelling indicates that maximum thermal maturity is reached during the Late Triassic (at the beginning of stage 4 in Fig. 14a and in Fig. 15b) shortly after the temperature anomaly during the extrusion of the Wengen Group volcanics (around 230 Ma; Fig. 14b) rather than during maximum burial (Fig. 15c). Formation temperatures return to background levels ~220 Ma and even seem to display a slightly negative anomaly (Fig. 14b). Thermal maturity remains unchanged up to the present day as the strata remain in the same temperature regime (Figs 14a and 15c and d); in other words, formation temperatures are constant despite increasing burial to deep-sea settings at the end of the Cretaceous (stage 4, Fig. 14b). Formation temperatures do not decrease during continent–continent collision (stage 5, Fig. 14b) and drop only as the Dolomites are exhumed to a subtle foreland-basin setting (stage 6, Fig. 14b). Fast exhumation to present-day settings decreases formation temperatures rapidly (stage 7, Fig. 14b).

Basin geohistory

Sedimentation on the exposed basement starts at around 260 Ma (Table 1 and Fig. 14), leading to a continuous subsidence to depths at the top basement of about 800 m (stage 1 in Fig. 14). Uplift in the southwest Dolomites triggers subaerial erosion during stage 2 and the subsequent development of the Anisian unconformity (Fig. 14). The third stage (Fig. 14) is characterized by rapid subsidence until Middle/Late Triassic times; this episode corresponds to the accumulation of the Schlern Dolomite Fm 1 carbonate platform. The top of the youngest strata preserved at Rosengarten is indicated by the dashed line in Fig. 14. The thick sedimentary package deposited directly afterwards represents the airborne and submarine volcaniclastics of the Wengen Fm.

In order to keep platform tops and basin floors horizontal and to maintain a common gradient of slope deposits throughout all growth stages (Fig. 16), water-depth maps of more than 800 m in the deepest settings had to be applied. These palaeobathymetric values are in accordance with assumptions from Bosellini & Stefani (1991) and Brack & Rieber (1993). Compaction of strata underneath the Rosengarten platform moves basinward together with its prograding carbonate slope, reducing porosity by its overburden (Fig. 16a–f; cf. Hunt & Fitchen, 1999; Permian Delaware and Midlands basins, USA). Highest porosities are preserved where the overlying sediments are at a minimum until the extrusion of the Wengen Group fills the basins (Fig. 16g). This interval is followed by a long period of steady subsidence (formation of the Trento platform, stage 4 in Fig. 14). Maximum thermal maturity of the transect is reached during Late Triassic times when basal heat flow is at a maximum (Figs 14a and 15a,b). The subsequent burial and subsidence of the Trento platform to deep-marine environments does not increase the thermal maturity significantly as heat flow diminishes during this period (Figs 8c and 15b,c). Maximum burial of the succession is reached during Late Cretaceous time (80 Ma; Fig. 14) when the polarity of the tectonic regime changes from extension to compression (Hsü, 1989; Dercourt *et al.*, 2000). Continuous uplift until about 32 Ma (stage 5 in Fig. 14, corresponding temperatures in Fig. 10a) followed by subaerial exposure and erosion of the youngest strata of the succession allows for a cooling of the basal strata below 80°C as required by the FT data (stage 6 in Fig. 14). Deposition of late Oligocene conglomerates (Monte Parei Fm, stage 6 in Fig. 14) leads to short-lived subsidence until major uplift occurs from 12 Ma onward (stage 7 in Fig. 14).

Another important result of thermal modelling is the quantification of eroded thicknesses above the present-day stratigraphy. The integrated modelling approach constrains the overburden to less than 1100 m (see bed thicknesses in Table 1). Both thermal maturity and apatite fission-track data clearly rule out a thick Mesozoic and Cenozoic sedimentary cover, which is also supported by recent studies on the burial history of the Trento platform (Zattin *et al.*, 2006). This furthermore confirms that there is no extensive Miocene cover during a flysch or foreland basin stadium as postulated by studies from other basins in the Alps (Venetian basin; Massari *et al.*, 1986; Eastern Alps: Winkler, 1988). The VR_r values from Permian sandstones in our study area are too low (Table 2) and the forward-modelled fission-track ages and t–T development (Fig. 10) indicate temperatures of around 80°C from 220 to 40 Ma and <80°C in the Permian succession since 40 Ma. The latter feature points to

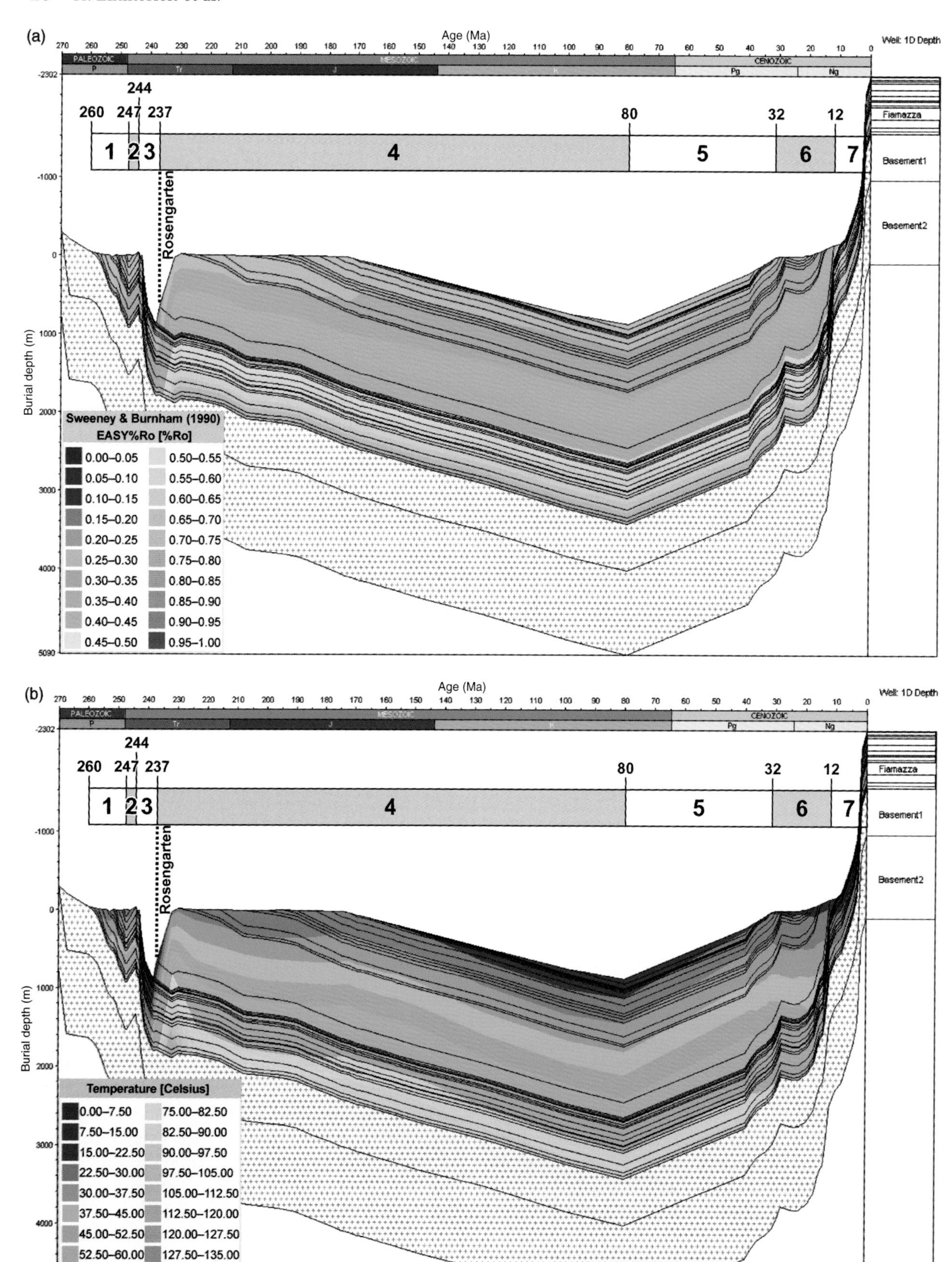
(a)
Age (Ma)
Well: 1D Depth
PALEOZOIC
MESOZOIC
CENOZOIC
P
Tr
J
K
Pg
Ng
244
260
247
237
80
32
12
1
2
3
4
5
6
7
Rosengarten
Fiamazza
Basement1
Basement2
Burial depth (m)
Sweeney & Burnham (1990)
EASY%Ro [%Ro]
0.00–0.05
0.05–0.10
0.10–0.15
0.15–0.20
0.20–0.25
0.25–0.30
0.30–0.35
0.35–0.40
0.40–0.45
0.45–0.50
0.50–0.55
0.55–0.60
0.60–0.65
0.65–0.70
0.70–0.75
0.75–0.80
0.80–0.85
0.85–0.90
0.90–0.95
0.95–1.00
(b)
Age (Ma)
Well: 1D Depth
Rosengarten
Fiamazza
Basement1
Basement2
Burial depth (m)
Temperature [Celsius]
0.00–7.50
7.50–15.00
15.00–22.50
22.50–30.00
30.00–37.50
37.50–45.00
45.00–52.50
52.50–60.00
60.00–67.50
67.50–75.00
75.00–82.50
82.50–90.00
90.00–97.50
97.50–105.00
105.00–112.50
112.50–120.00
120.00–127.50
127.50–135.00
135.00–142.50
142.50–150.00

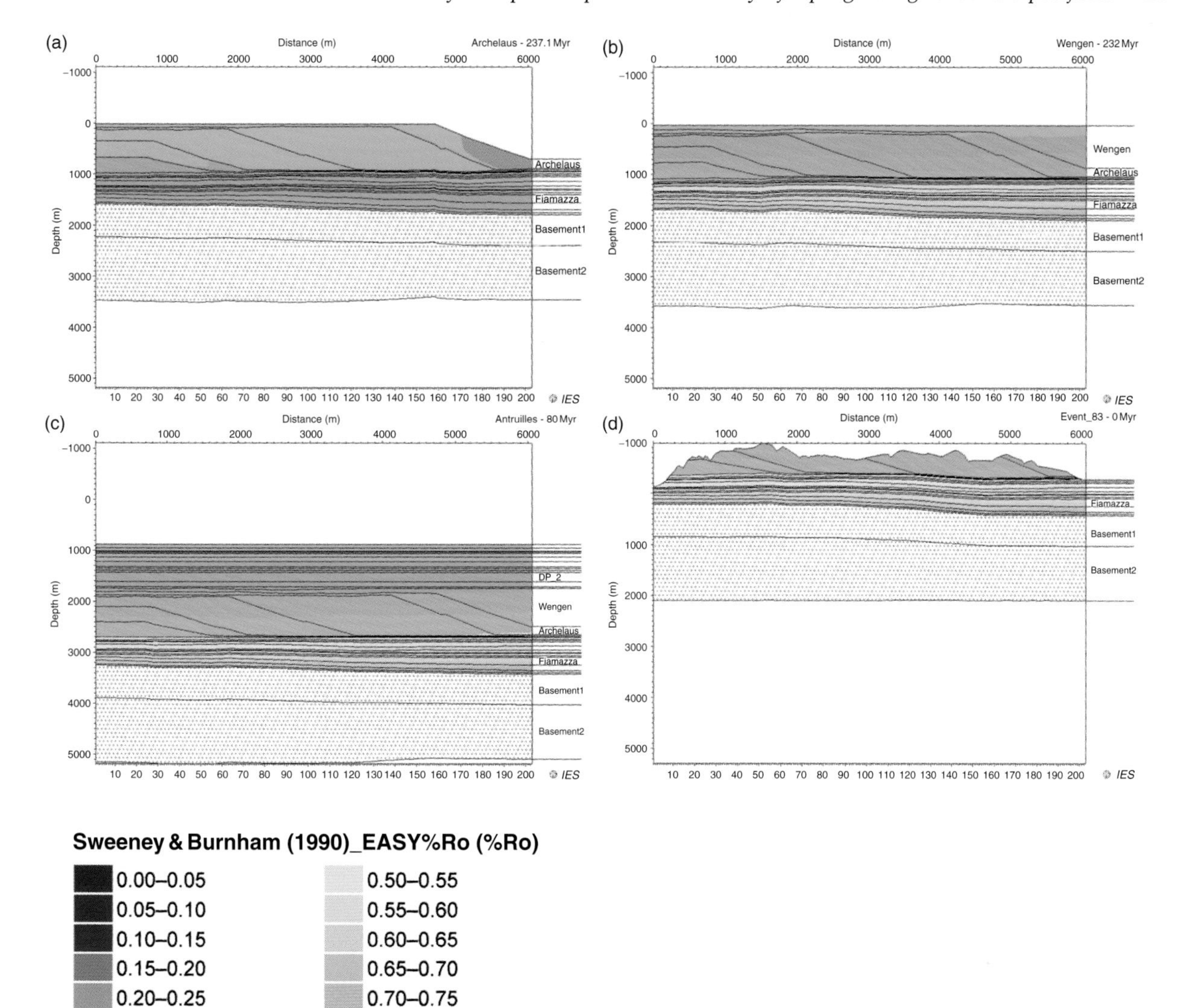

Fig. 15. Basin evolution at Rosengarten (PetroMod™). *x*-axis, distance along the transect in metres; *y*-axis, depth/elevation in metres; formations are indicated at the right-hand side; name and age of timestep in the upper right corner. Vertically exaggerated. Overlay: calculated maturity values (%VR_r), legend to the lower left. Increasing temperature during burial and rising basal heat flow create higher thermal maturity preserved until today. Fig. (d) shows the present-day situation where maximum thermal maturity is recorded by organic matter in coarse- to medium-grained sandstones of the Permian Gröden Fm (see Table 1). (a) Timestep after *Archelaus* biozone (Middle Triassic), i.e., after the last stage of platform progradation. (b) Timestep after deposition of the Wengen Fm (Middle Triassic), i.e., after maximum heating. (c) Timestep after deposition of the Antruilles Fm (Upper Cretaceous), i.e., after maximum burial. (d) Present-day situation.

Fig. 14. (Opposite page) Burial history of a pseudo well through the Rosengarten transect as calculated by PetroMod™; same location as the calibration plot in Fig. 9. *x*-axis, age in Ma (abbreviations: P, Permian; Tr, Triassic; K, Cretaceous; Pg, Paleogene; Ng, Neogene); *y*-axis, burial depth in metres; black lines are boundaries of formations and their subdivisions. Bar at the top (numbers 1–7) indicates the main stages of basin evolution with their corresponding ages. The top of the preserved stratigraphy at Rosengarten is indicated by a dashed line. (a) Overlay: calculated vitrinite reflectance (%VR_r) indicating the thermal history (legend/colour code to the lower left) of the basin. (b) Overlay: calculated temperature (°C) indicating the thermal history (legend/colour code to the lower left) of the section at Rosengarten.

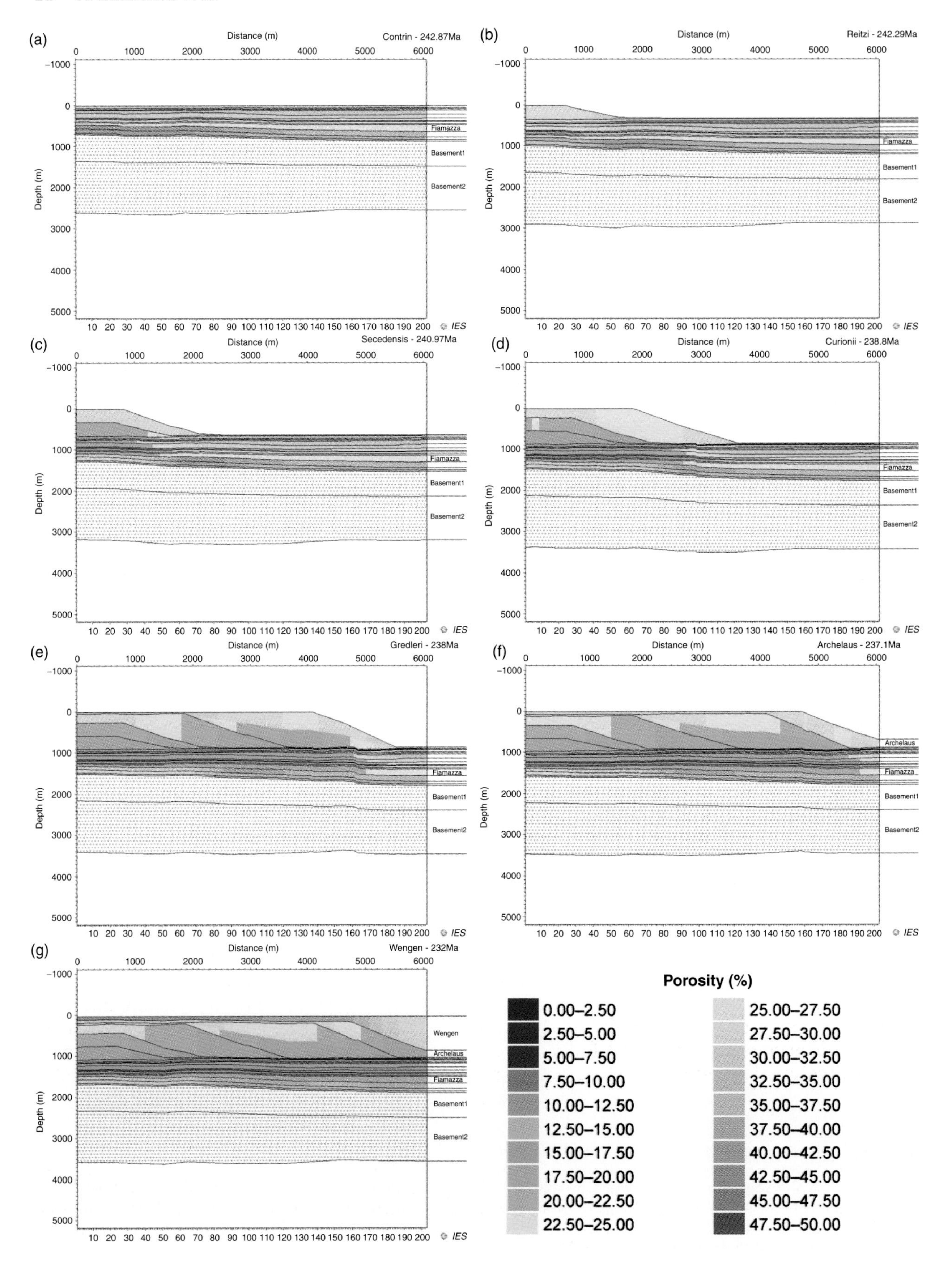

(a)
Distance (m)
Contrin - 242.87Ma
Depth (m)
Fiamazza
Basement1
Basement2
IES
(b)
Reitzi - 242.29Ma
(c)
Secedensis - 240.97Ma
(d)
Curionii - 238.8Ma
(e)
Gredleri - 238Ma
(f)
Archelaus - 237.1Ma
Archelaus
(g)
Wengen - 232Ma
Wengen
Porosity (%)
0.00–2.50
2.50–5.00
5.00–7.50
7.50–10.00
10.00–12.50
12.50–15.00
15.00–17.50
17.50–20.00
20.00–22.50
22.50–25.00
25.00–27.50
27.50–30.00
30.00–32.50
32.50–35.00
35.00–37.50
37.50–40.00
40.00–42.50
42.50–45.00
45.00–47.50
47.50–50.00

a rapid uplift and exhumation of the succession as observed in other areas of the Alps and their surroundings – whether steady state or episodic (Bernet *et al.*, 2001; Carrapa *et al.*, 2003; Zattin *et al.*, 2003; Bertotti *et al.*, 2006).

Basin reverse modelling

The basin reverse modelling routine of PHIL™ (Fig. 11) calculates subsidence rates for every grid point along the transect. Two sets of subsidence rates are presented in Fig. 17, one set is derived from a proximal location within the platform interior (transect metre 30) and another one from a distal location in the basin (transect metre 5970). Owing to the rigidity of the underlying Permian AVC (>2000 m thickness) and the shortness of the transect (<10 km length), there are no distinct differences in flexural subsidence between a proximal and a distal setting. Differences in subsidence in the beginning of basin evolution – during late Wuchiapingian (uppermost part of the Gröden Fm and Fiamazza Member of the Bellerophon Fm) and late Scythian (Seis Mbr of the Werfen Fm) times – are attributed to lateral changes in bed thickness owing to relative movements along small faults. The differences in subsidence between a distal and proximal setting on the transect during late Anisian/early Ladinian times (Schlern Dolomite Fm 1, *Reitzi* and *Secedensis* biozone) originate mainly from differences in compaction-induced subsidence as larger amounts of carbonate sediment are accommodated in a proximal, lagoonal realm. As this paper focuses on the Anisian–Ladinian platform evolution, a detailed description of the pre-Anisian subsidence history is omitted for clarity.

Anisian to Late Triassic subsidence rates

Subsidence commenced with the deposition of the Richthofen Fm (conglomerates and evaporites) and increased during Morbiac Fm sedimentation (bituminous limestones and marls). The shallow marine carbonate ramp of the Contrin Fm represents an interval of decreasing subsidence. On top of the Contrin Fm, a basinwide correlatable unconformity developed due to a subsidence peak with rates of up to 820 m Myr^{-1}. These values decrease significantly to 200–300 m Myr^{-1} during the *Secedensis* biozone, before subsidence drops down to values around 130 m Myr^{-1} (*Curionii* biozone) or less (50–60 m Myr^{-1}; *Gredleri* and *Archelaus* biozone) and eventually Wengen volcanics terminated platform growth. These high subsidence rates during the upper *Reitzi* and entire *Secedensis* biozone are responsible for the aggradational behaviour of the Rosengarten platform. As subsidence drops to 100 m Myr^{-1}, progradational patterns develop. As mentioned earlier and discussed by several authors (Blendinger, 1985; Doglioni, 1987), the study area was located near an active strike-slip fault system during most of the Mesozoic. Hence it seems likely that movements along the Cima Bocche Line/Stava Anticline triggered these temporal changes in subsidence (Emmerich *et al.*, 2005a).

Stratigraphic forward modelling

In order to adequately simulate the Rosengarten platform with a larger platform interior to the north and a larger basin to the south, the transect had to be extended by 2 km on each side. Essential results from previous modelling steps – such as tectonic subsidence rates and sediment flux (Fig. 7) – were incorporated in the input for the stratigraphic forward simulator of PHIL™. Sediment parameters and sedimentation rates were adjusted as explained earlier and discussed in Bowman & Vail (1999). Subaerial erosion was neglected during the forward simulation but carbonate redistribution to slope and basin was accounted for. Hence, the rates presented in this study correspond to best-fit accumulation rates (in the sense of calibration with the present-day outcrop). Real (i.e., fossil) accumulation rates would be higher because subaerial erosion and bioerosion have to be compensated for by carbonate production and sedimentation. However, the Anisian and Ladinian sea-level oscillations – indicated by the record in the neighbouring Latemar platform (Zühlke,

Fig. 16. (Opposite page) Modelled porosity evolution during progradation at Rosengarten (PetroMod™). *x*-axis, distance along the transect; *y*-axis, depth/elevation; overlay, porosity (%); legends bottom right. The progradation of the carbonate platform reduces porosity in the underlying strata. For further explanation refer to the text. (a) After deposition of the Contrin Fm (242.87 Ma). (b) After the *Reitzi* biozone (242.29 Ma). (c) After the *Secedensis* biozone (240.97 Ma). (d) After the *Curionii* biozone (238.80 Ma). (e) After the *Gredleri* biozone (238.00 Ma). (f) After the *Archelaus* biozone (237.10 Ma). (g) After deposition of the Wengen Fm (232.00 Ma).

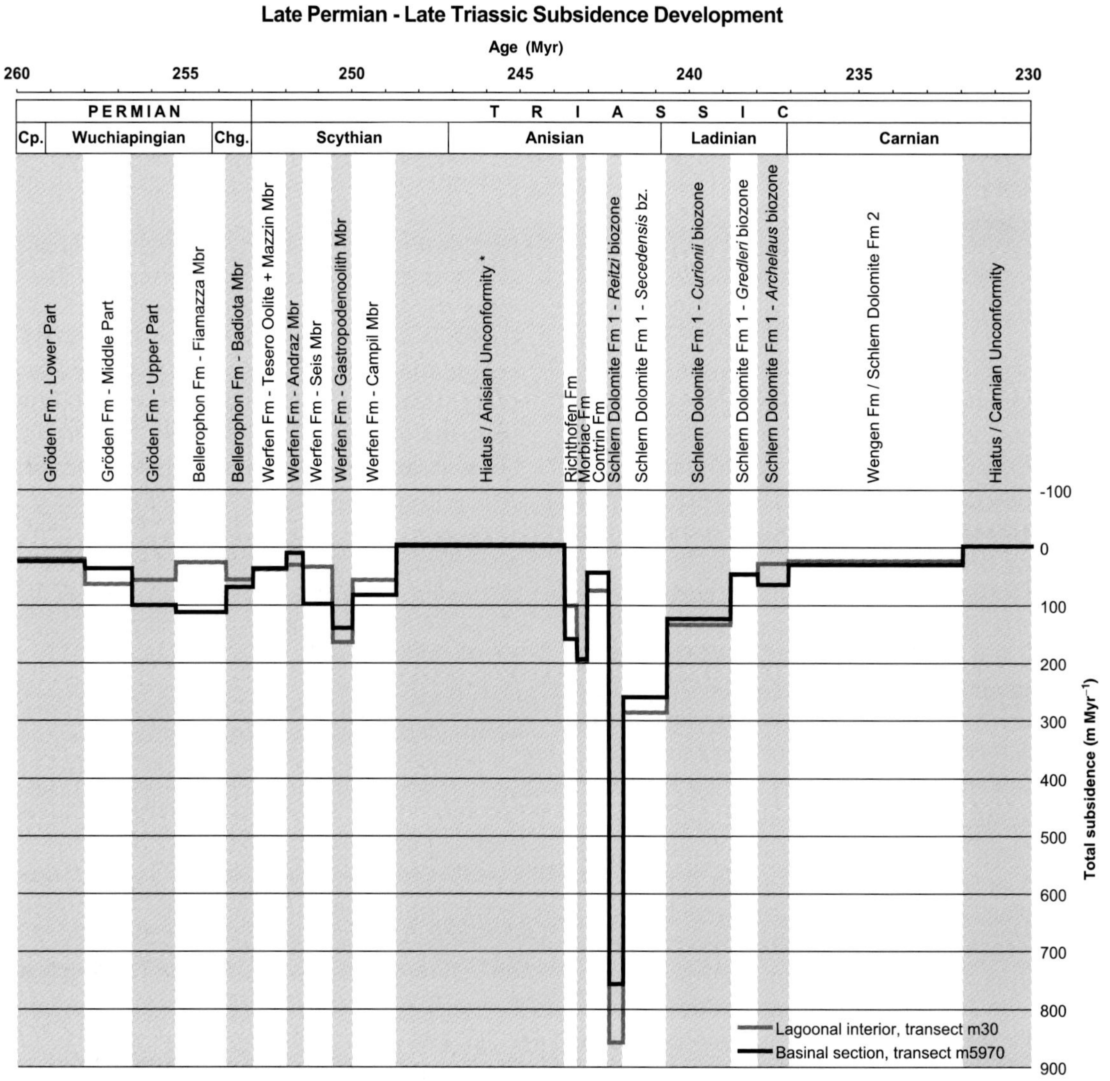

Fig. 17. Calculated total-subsidence development at the study area until the extrusion of Wengen Group volcanics (basin-reverse modelling module of PHIL™). *x*-axis, age in Ma; *y*-axis, subsidence rates in m Myr^{-1}. Subsidence was calculated at a proximal (transect metre 30; grey line) and a distal point (transect metre 5970; black line). Differences in total subsidence at proximal and distal settings are mainly attributed to differences in compaction-induced subsidence (thicker succession in distal parts). Timescale after Lehrmann *et al.* (2002), Mundil *et al.* (1996, 2001, 2003) and Yugan *et al.* (1997). Permian stages after Yugan *et al.* (1997), Triassic stages after Brack & Rieber (1993, 1994). *The backstripping process of PHIL™ is not capable of calculating subsidence values of nowadays eroded formations. Therefore, the timespan of the Anisian Unconformity pictured in this figure comprises also the interval of deposition of the Val Badia and Cencenighe Members of the Werfen Fm (Table 1).

2004) – are of extraordinary high frequency and low amplitude. Therefore, the exposure time, as shown at the Latemar by dolomitic caps on top of shallowing-upward cycles (Goldhammer & Harris, 1989), of the platform top is short, and the influence of subaerial erosion can be neglected. This also implies that the carbonate factory of the platform margin and slope remains switched on during all stages of accommodation change, as seen in studies on the composition of calciturbidites derived from Middle Triassic platforms in the Dolomites (Reijmer *et al.*, 1991; Reijmer, 1998; Maurer & Schlager, 2003; Maurer *et al.*, 2003). The influence of bioerosion is, however, difficult to quantify because it depends on many factors which are impossible to derive from the geological record. As mentioned earlier, accommodation changes are included in total subsidence rates and thus also constitute a small proportion of the forward modelled carbonate accumulation rates.

During the course of forward modelling, the entire carbonate system was set very sensitive to siliciclastic poisoning (i.e., its lower limit; Table 4) and accumulation time increments of the output files were 1 kyr (smaller time increments were limited by computing power) such that the temporal platform development could be simulated in great detail (approximately 5800 timesteps for the entire Rosengarten history).

Sensitivity analysis

A sensitivity analysis was carried out in order to quantify the influence of all input parameters (Table 4) of the PHIL™ simulator. The main results of this analysis are summarized in Fig. 18a–d. More than 1200 different runs with different numerical values of the input parameters were calculated in order to achieve a best-fit model. Among the 13 carbonate-specific ones, two parameters emerged as the most important: (1) suspension width of fine-grained carbonates and (2) rollover width (influencing the distribution of boundstone facies on the slope; for parameter definition see Fig. 12; for numerical values see Table 4). The challenge of the forward model was to match slope inclination, growth mode and distribution of boundstones on the slope. The correct pattern of aggradation followed by progradation, the extent of boundstones on the slope and the observed geometry of steep, straight to exponentially curved clinoforms was successfully modelled. This required the unique combination of rollover width set to 1.55 km and a very short suspension distance of lagoonal fine-grained carbonates of 0.1 km.

The reason why the suspension distance of fine-grained material from the platform top was set to such low values (Table 4) is the incorrect deposition of carbonate on the upper slope with higher values of the suspension distance (cf. bulge of the upper slope in Fig. 18a). This peculiar behaviour of the modelled slope system is related to the 'carbonate productivity function' of PHIL™ (Fig. 13) which already includes redeposition (Bowman & Vail, 1999). If the suspension distance was larger than the extent of the platform slope and all material was exported to basinal settings, progradation could not be modelled (Fig. 18b).

As discussed in the literature (Reijmer *et al.*, 1991; Blendinger, 1994; Reijmer, 1998; Kenter *et al.*, 2002, 2005; della Porta *et al.*, 2004; Seeling *et al.*, 2005), the main carbonate factory is located on the upper slope and within water depths between 0 and 300 m. This specific, slope–productivity-driven growth characteristic is especially true for prograding carbonate platforms (Bosellini, 1989; Boni *et al.*, 1994; Seeling *et al.*, 2005). Similarly, Kenter *et al.* (2005) and della Porta *et al.* (2004) attribute the progradation of Carboniferous carbonate platforms in the Cantabrian Mountains of Spain mainly to an increased productivity of the cement-rich boundstone belt. The only possibility to simulate restricted *in situ* carbonate accumulation on the upper slope with PHIL™ is to model the shelf-margin factory, i.e., the lithology 'boundstone' (Bowman & Vail, 1999, p. 129). All other carbonate sediments on the slope are modelled by PHIL™ with a significant amount of allochthonous material such as lithoclastic rud- to wackestones (cf. Bowman & Vail, 1999; refer also to Section on 'Sequence stratigraphic forward modelling').

Best-fit model

The best-fit model is presented in Fig. 19a and b, where Fig. 19a shows the modelled transect after the last timestep of platform growth (*Archelaus* biozone, 237.10 Ma) with an overlay of lithofacies, and Fig. 19b shows the same model with an overlay of palaeobathymetry. The formations above the Permian basement and underneath the carbonate platform such as Gröden Fm (siliciclastics above the basement), Bellerophon Fm (evaporites and carbonates underneath the thin light blue line marking grainstones of the Tesero-Oolite Member at the P/T boundary), Werfen Fm (carbonates and clastics underneath the thin red line marking coarse-grained siliciclastics of the Richthofen Fm) and all other Anisian strata (fine-grained carbonates above the Richthofen Fm) are easy to discern in Fig. 19a. Formation of the Rosengarten platform post-date these Anisian strata. Platform nucleation on top of the Contrin Fm carbonate ramp occurs in response to various input parameters/boundary conditions: (1) a narrow margin combined with a restricted rollover width (Table 4); (2) higher accumulation rates on the upper slope than on the platform top (Fig. 13); and (3) subtle differences in basal topography owing to fault movement and/or lateral differences in subsidence (Fig. 19a). Minor back and forth adjustments of the margin during the first stage of platform development are attributed to difficulties of the carbonate system in establishing slope and keeping up with the rapid accommodation increase during the *Reitzi* biozone (sediment input in time is nearly outpaced

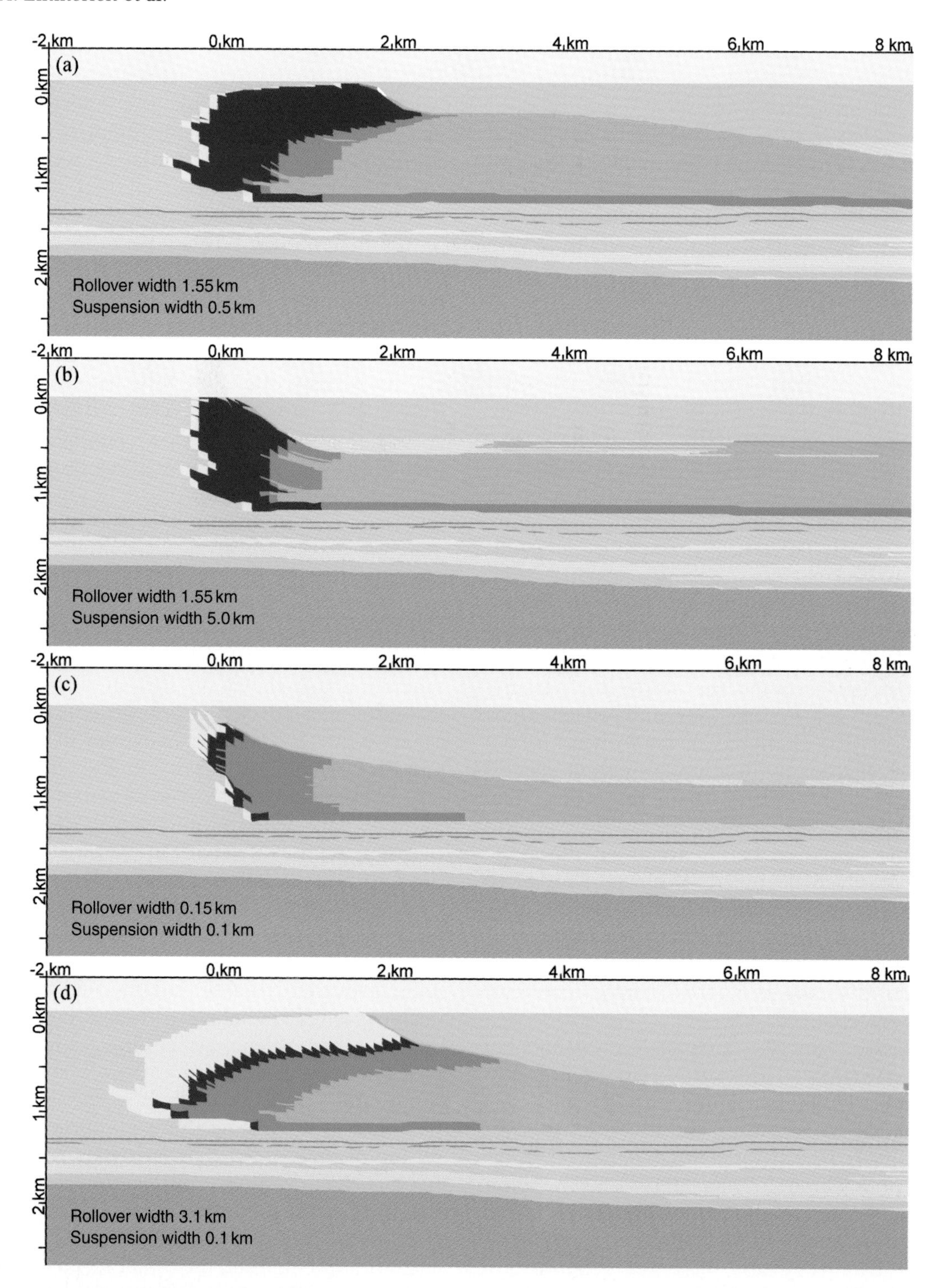

Fig. 18. Sensitivity analysis of the forward model (pictured platform development after the *Archelaus* biozone, 237.10 Ma). (a) Rollover width 1.55 km (wide boundstone belt) and suspension width 0.5 km (bulge-like deposition of fine-grained material in an intermediate position and steep proximal clinoforms) resulting in weak, late-stage progradation. (b) Rollover width 1.55 km (wide boundstone belt) and suspension width 5.0 km (deposition of fine-grained material in a distal position and steep proximal clinoforms) resulting in aggradation, late-stage progradation. (c) Rollover width 0.15 km (narrow boundstone belt) and suspension width 0.1 km (deposition of fine-grained material within the bounds of the slope and exponentially curved clinoforms) resulting in weak retrogradation. (d) Rollover width 3.1 km (wide back-reef belt and narrow boundstone belt) and suspension width 0.1 km (deposition of fine-grained material within the bounds of the slope and exponentially curved to nearly straight clinoforms) resulting in progradation. For legend see Fig. 19.

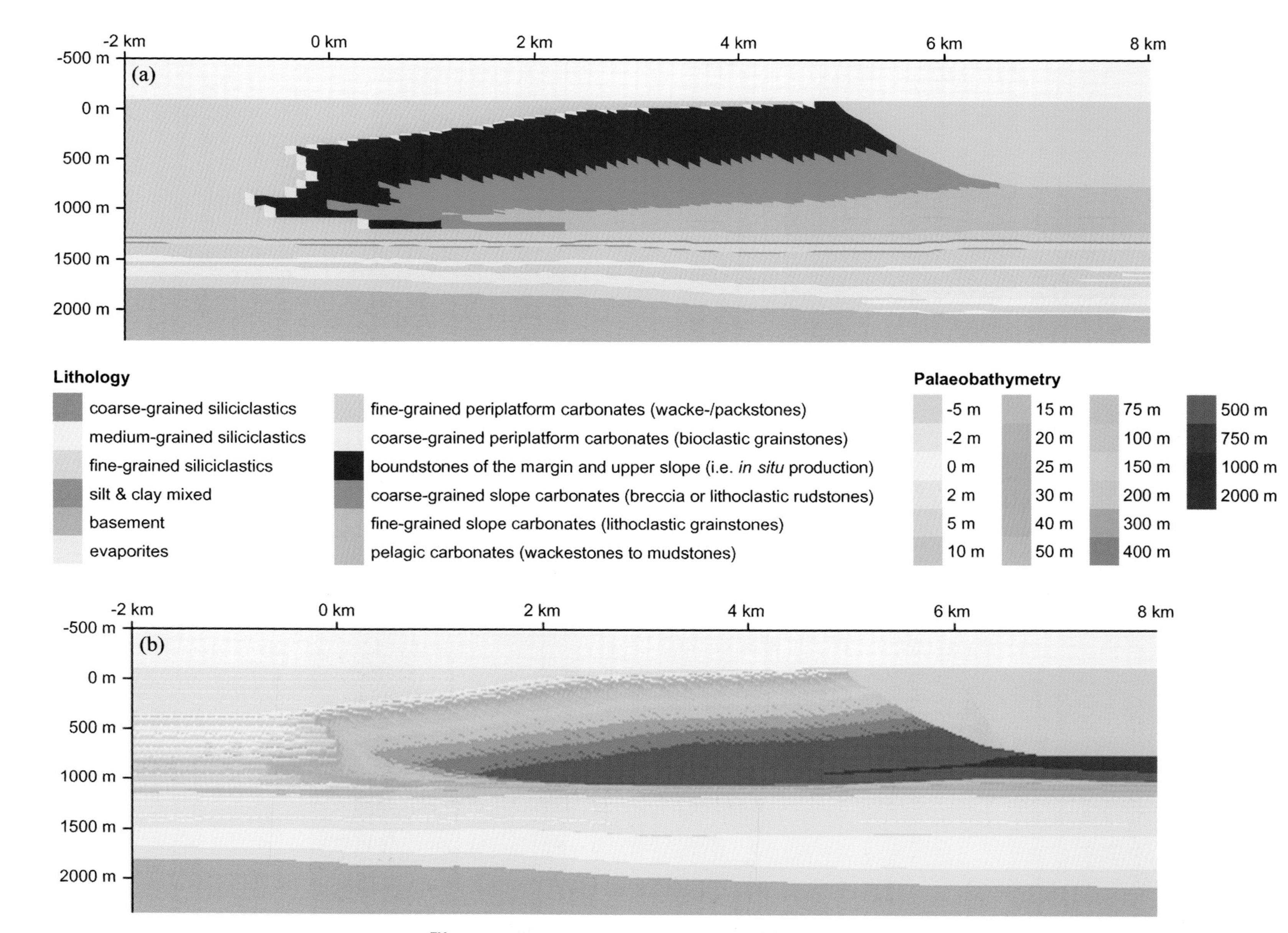

Fig. 19. Rosengarten transect as simulated with PHIL™ after the last timestep of platform progradation (*Archelaus* biozone, 237.10 Ma). *x*-axis, distance along transect in kilometres; *y*-axis, elevation/depth in metres (negative values, elevation above sea level of the respective timestep; positive values, depth below sea level); no vertical exaggeration. (a) Overlay: lithologies. Owing to the similarity in grain size, fine-grained toe-of-slope deposits (Schlern Fm) have the same colour code as the fine-grained calciturbiditic deposits of the basinal Buchenstein Fm. (b) Overlay: palaeobathymetry. This overlay illustrates the Anisian transgressive surface, where deep marine sediments of the Buchenstein Fm (dark blue) unconformably overlie the shallow-marine carbonate ramp of the Contrin Fm (light green).

by accommodation change, i.e., $A' \rightarrow S'$; *sensu* Seeling *et al.*, 2005). Nevertheless, the aggradation trend is clearly identifiable. The palaeowater depth overlay facilitates the recognition of this first phase and especially highlights the transgressive surface in basinal settings developed on top of the Contrin Fm (Fig. 19b).

The resulting trajectories of the margin are illustrated in Fig. 20a and b. The first stage, characterized by a steep, nearly vertical trajectory (1 in Fig. 20; the present-day situation is shown in Fig. 20c for comparison), is followed by an interval of progradation (2 in Fig. 20). The second stage – i.e., the *Curionii* biozone – is typified by a shallow to intermediate trajectory of the margin until fast progradation with very shallow trajectories occurs during the last stages (3 in Fig. 20). A comparison with the growth mode of the Rosengarten platform according to Maurer (1999, 2000) reveals almost the same arrangement of trajectories

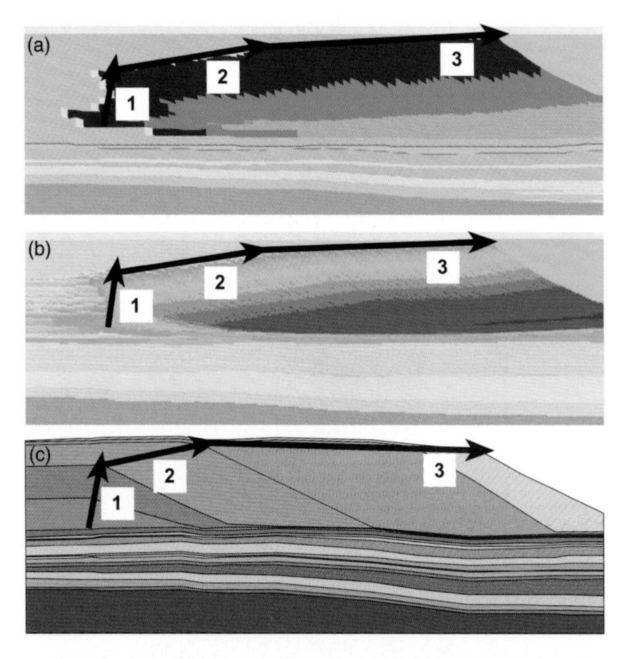

Fig. 20. Simplified comparison of the simulated transect with reality. *x*-axis, distance along transect; *y*-axis, elevation/depth; no vertical exaggeration. (a) and (b) Details from Fig. 19a (lithology) and 19b (palaeobathymetry). The three different growth stages are marked with arrows and numbers (1: aggradation during *Reitzi* and *Secedensis* biozone; 2: progradation during *Curionii* biozone; 3: rapid progradation during *Gredleri* and *Archelaus* biozone). (c) Detail from reconstructed transect after Bosellini & Stefani (1991) and Maurer (1999, 2000; Fig. 6). The three different growth stages are marked in the same way as in (a) and (b). The last stage of the platform development appears to be dipping to the right (south) owing to the post-sedimentary, fault-related dip of the underlying strata.

(Fig. 20c). In the model, the initial phase (stage 1) is a bit shorter whereas the growth stage of the *Curionii* biozone (stage 2) seems a bit longer. This indicates that the duration of the *Curionii* biozone may in reality have been shorter than modelled in this study (i.e., <2.2 Myr as derived from the chronostratigraphic data of Mundil *et al.*, 1996, 2003).

Decompacted carbonate accumulation rates without subaerial erosion and bioerosion rates but including redeposition in some facies ('production rates'; Fig. 13) increase from lagoon ('periplatform production') to slope ('platform slope production'). The growth mode of the Rosengarten platform is matched with constant accumulation rates throughout all biozones of platform growth: 900 m Myr^{-1} was applied for periplatform accumulation, 980 m Myr^{-1} for platform margin accumulation, 1000 m Myr^{-1} for platform slope accumulation and 50 m Myr^{-1} for pelagic accumulation. The width of the carbonate functions applied in the forward model corresponds to the following facts: (1) the main carbonate factory is found on the slope (Blendinger, 1994); and (2) boundstone carbonate production on the slope takes place until ~300 m water depth (della Porta *et al.*, 2004).

DISCUSSION

Burial and thermal history

Low vitrinite reflectance values together with short apatite fission tracks indicate a long, shallow burial of the Rosengarten succession until the Oligocene, followed by fast uplift/cooling (Figs 10 and 14). Values of vitrinite reflectance between 0.5% VR_r and 0.8% VR_r indicate maximum temperatures during burial of the order of 110°C. As illustrated by the burial history plot in Fig. 10b, maximum temperatures were reached by Late Triassic time when a high basal heat flow of 110 mW m^{-2} was reached for a short while around 230 Ma. This was necessary in order to reproduce (1) the observed thermal maturity; and (2) the measured ages of the apatites. This elevated basal heat flow during the Late Triassic is associated with violent magmatic events in the vicinity of the Predazzo–Monzoni area. Many studies confirm this observation by either reset isotopic clocks from the Fassa Valley (Barth *et al.*, 1993, 1994) or thermal modelling of sedimentary successions in the Southern Alps (Greber *et al.*, 1997). However, our integrated 4D basin simulation contradicts the thermal history postulated by Zattin *et al.* (2006) for the Dolomites. The raw data presented in this study as well as the modelling results show no indications for an elevated Jurassic heat-flow due to rifting as reported by Zattin *et al.* (2006).

The Late Jurassic to Cretaceous subsidence trend calculated by Winterer (1998) for the Trento platform had to be lowered by ~15%. Otherwise, the Rosengarten would have been buried too deeply and the forward modelled time–temperature history would not have matched the measured fission-track data. Uplift occurred when the polarity of plate movement was reversed (approximately 80 Ma), fast uplift prevailed from 32 Ma onward and the succession eventually moved above the 80°C isotherm. This rapid exhumation–cooling rate is confirmed by recent fission-track data of the Southern and Western Alps (Carrapa *et al.*, 2003), as well as by provenance analyses (Dunkl *et al.*, 2001; Zattin *et al.*, 2003) and apatite (U–Th)/He thermochronology (Bertotti *et al.*, 2006).

Porosity evolution

The porosity evolution of the strata underlying the carbonate platform reflects the growth mode of the Rosengarten. During the last, progradational stage of platform evolution, maximum porosities in the Permian Gröden Fm are preserved in distal, basinal areas (Fig. 15e and f). The decrease in porosity underneath the slope sediments mimics their progradation – i.e., an area of low porosity is moving basinwards propagated by compacting slope deposits. This has implications for (1) early diagenetic fluid-flow dynamics and (2) shallow siliciclastic hydrocarbon reservoirs below prograding (carbonate) slopes. Concerning (1), the migration of fluids during the earliest stages of diagenesis might thus be rather basinward than directed towards the platform owing to the pressure differences exerted by the different timing of sedimentary loading. With respect to model 2, better reservoir characteristics are subsequently preserved underneath basinal strata, not underneath carbonate platforms. However, basin infill and diagenesis during deeper burial will most probably level lateral differences in the underlying strata (Fig. 15g).

Carbonate accumulation rates

Previous studies on reef communities and carbonate platforms in the Dolomites (Fois & Gaetani, 1984; Senowbari-Daryan *et al.*, 1993) have stressed the Anisian recovery of reef builders from the end-Permian faunal crisis when 62% of marine invertebrate families (McKinney, 1985) and up to 96% of species (Raup, 1979) were extinguished. New investigations on trace-fossil abundance in the Lower Triassic Werfen Fm (Fig. 3) demonstrate a gradual reappearance of taxa throughout the Lower Triassic of the Dolomites (Werfen Fm) and a complete recovery in the Uppermost Scythian (represented by the top of the Werfen Fm; Twitchett, 1999). In addition, palaeontological studies from other Triassic carbonate platforms around the world also indicate a fast recovery of the calcimicrobial carbonate factory (for complete discussion and references see Flügel, 2002).

These different observations are confirmed by the modelled sedimentation rates: accumulation of up to 1000 m Myr^{-1} of carbonate sediment is necessary in order to: (1) keep-up the platform during the rising A'/S' conditions at the beginning; and (2) to simulate the fast progradation during the decreasing A'/S' conditions in the second stage of platform development (*sensu* Seeling *et al.*, 2005). The best fit with present-day geometries was achieved using a constant rate of carbonate accumulation from 900 to 1000 m Myr^{-1}, increasing from platform top to upper slope. Hence, the main *in situ* carbonate factory is located on the margin and upper slope ranging from shallow subtidal to 200–300 m water depth. This is in accordance with studies on carbonate platforms from the Upper Carboniferous of the Asturian and Cantabrian mountains in Spain (Kenter *et al.*, 2002, 2005; della Porta *et al.*, 2004) and confirmed at platforms where $A' > 0$ and $A'/S' \rightarrow 0$ (i.e., 'pathologically prograding' Bosellini, 1989; Seeling *et al.*, 2005). In the Dolomites, the studies of Keim & Schlager (2001) and Maurer *et al.* (2003) highlight the importance of the automicrite factory on the slope for the growth of Middle Triassic carbonate platforms. The boundstone facies with laterally restricted and hence *in situ* accumulation, modelled in this project, is the driving mechanism behind (1) initial platform growth and (2) late-stage progradation.

Previous studies have calculated/estimated compacted accumulation rates of 200–500 m Myr^{-1} (Dürrenstein: Schlager, 1981; Schlager *et al.*, 1991), 600 m Myr^{-1} (Sella: Keim & Schlager, 2001) and 800 m Myr^{-1} (the Anisian part of the Latemar succession: Egenhoff *et al.*, 1999) for Middle to Late Triassic carbonate platforms in the Dolomites. The rates simulated during this study are higher because they represent decompacted accumulation (up to 1000 m Myr^{-1} of accumulation in order to keep up with up to 820 m Myr^{-1} of subsidence). However, these values are within the range of the slope-shedding model of Kenter *et al.* (2005), who postulated ~1000 m Myr^{-1} of *in situ* boundstone growth for prograding platforms in the Carboniferous (Asturias, Spain and North Caspian Basin, Kazakhstan). This adds further evidence to observations of a rapid recovery of the carbonate factory some 10 Myr after the faunal crisis at the end of the Permian (following the timescale in Table 1). Hence, the results of our study question the hypothesis of mounded geometries owing to low carbonate productivity at the coeval but drowned Monte Cernera carbonate platform (Blendinger *et al.*, 2004). It is much more likely that the drowning is related to upwelling of cold water masses into the western Tethys as proposed by Preto *et al.* (2005) than to low productivity of a not fully recovered carbonate factory.

The difference of 80–180 m Myr^{-1} between subsidence rates as determined by the basin-reverse modelling and the minimum carbonate accumulation rates as required in the stratigraphic forward modelling results for two reasons: (1) timesteps in basin-reverse modelling are much longer than in forward modelling and they thus reflect longer intermediate values (10^6–10^5 vs. 10^3–10^2 years); (2) carbonate sediments are redistributed out of the transect (debris flows, calciturbidites) and subject to bioerosion and dissolution. These effects have subsequently to be added on top of the subsidence rates.

Controls on platform evolution at Rosengarten

The slope of the Rosengarten is characterized by two stages of platform evolution as recognized by Bosellini (1984), Bosellini & Stefani (1991) and Maurer (1999, 2000). The initial aggradational phase is followed by a progradational period towards the end of platform development. This study shows that a significant increase in carbonate accumulation rates during the development of the Rosengarten platform as postulated by Maurer (1999, 2000) is not necessary for the explanation of this behaviour and can most likely be ruled out as a possible explanation for the fast progradation.

(1) The onset of platform formation occurs in the late Anisian. By this time, the carbonate factory in the Dolomites is fully recovered from the faunal crisis at the Permian–Triassic boundary as evidenced by early Anisian reef communities (Fois & Gaetani, 1984; Senowbari-Daryan *et al.*, 1993).
(2) The existence of the Rosengarten platform covers less than five biozones (i.e., 5.77 Myr) – an interval during which changes in intrinsic features (*sensu* Schlager, 2000; i.e., changes of the biotic and abiotic carbonate factory) are not observed on coeval platforms in the vicinity (Emmerich *et al.*, 2005b, and references therein).
(3) According to Schlager (1999), the accumulation rate of carbonate platforms is much more likely to decrease in the million-year range due to changing environmental factors – a fact which would contradict increasing accumulation rates.
(4) The palaeogeographical configuration of the southwest termination of the Tethys is more or less stable during the Anisian and Ladinian, there are no indications for climate changes (Dercourt *et al.*, 2000).

Therefore, temporal changes in tectonic subsidence have to be assumed as the driving force for this two-phased growth. A short-lived, pulse-like peak of up to 820 m Myr^{-1} subsidence during the *Reitzi* biozone decreased to 200–300 m Myr^{-1} during the *Secedensis* biozone, resulting in an aggrading platform. As soon as the *Curionii* biozone was reached, subsidence dropped to 100 m Myr^{-1} resulting in strong progradation of the platform lasting until the termination of platform development via the extrusion of Wengen Group volcanics. Changes in wind or wave direction as the driving mechanisms behind platform progradation are considered less likely for two reasons.

(1) This extrinsic factor must have first hindered (during the aggradational phase) and then promoted progradation of the slope. This means that environmental factors must have changed by 180° during the growth of the Rosengarten platform. This radical change is not supported by any other observations from the Dolomites (Egenhoff *et al.*, 1999).
(2) The main carbonate factory is located on the upper slope and is especially active during progradational phases (Blendinger, 1994; Reijmer, 1998; Della Porta *et al.*, 2004; Seeling *et al.*, 2005). This behaviour was coined 'slope-shedding' by Kenter *et al.* (2005).

The subsidence peak of the late *Reitzi* to early *Curionii* biozone can be observed throughout the western Dolomites (Bechstädt & Brandner, 1970; Bechstädt *et al.*, 1978; Rüffer & Zühlke, 1995) and is most probably connected with strike-slip tectonics at the transpressive–transtensive passive continental margin (Doglioni, 1983, 1984; Blendinger, 1985). As the southeast side of the Rosengarten is located very close to the Stava Line/Cima Bocche Anticline (approximately 10 km), this two-phased growth is most probably caused by tectonic movements and stillstands along this line. Sudden, in the order of 1 Myr or less, movements cause peaks in tectonic subsidence (intervals with aggradational clinoforms) whereas during times of tectonic inactivity and/or updoming of the Predazzo magmatic chambers subsidence stopped (intervals with progradational clinoforms).

CONCLUSIONS

The chosen methods of an integrated basin simulation are a prerequisite for modelling subaerially exposed sedimentary systems. The quantification of sedimentary hiatuses, erosion and burial by thermal subsidence modelling corrects for compaction and stratigraphic bias before sequence-stratigraphic modelling is realized. Integrated basin simulation must include the reconstruction of burial history. Calibration of thermal modelling with vitrinite reflectance and FT measurements revealed that Neogene flysch or molasse-type sediments above the present-day topography, as inferred for other Alpine basins (Massari *et al.*, 1986; Winkler, 1988), did not affect temperature and burial history. In other words, these foreland-basin sediments were either very thin or not present at all. The thinness of eroded stratigraphy above the Rosengarten transect contrasts with the regional coalification pattern of the eastern Dolomites (Zattin *et al.*, 2006). The higher thermal maturity in that area requires either a significantly higher heat flow or greater thicknesses of the now eroded Cretaceous to Cenozoic overburden. Whereas the latest cooling phase of the eastern Dolomites seems to be similar to that of the Rosengarten area, higher thermal maturity in the eastern and central Dolomites further highlights the importance of the Trento platform for the thermal evolution of the western Dolomites.

Owing to numerous and high-quality chrono-, bio- and cyclostratigraphic data, the Middle Triassic Rosengarten carbonate platform is an ideal area for assessing carbonate accumulation rates after the Permian–Triassic crisis and the response of platforms to temporal changes in subsidence. An integrated approach of thermal and stratigraphic modelling reveals that the Rosengarten platform keeps up successfully with subsidence rates of up to 820 m Myr^{-1}. Both stages of platform growth – first aggradation and later progradation – originate in temporal variations in total subsidence. In the case of the Rosengarten, tectonic subsidence and its variations have been discussed as the major extrinsic (*sensu* Schlager, 2000) factors for platform and slope evolution. Other parameters such as sea-level oscillations and palaeowind or -wave directions were significantly less important and/or heavily overprinted by variations in subsidence. High-frequency sea-level oscillations as recorded by the accommodation history of the lagoonal interior at Torri del Vajolet (Fig. 5a) have not been preserved or recorded in the development of the platform slope (Harris, 1994; Reijmer, 1998; Emmerich *et al.*, 2005b). The best-fit carbonate accumulation rates of 900–1000 m Myr^{-1} reach the values of subrecent carbonate platforms (Enos, 1991). Additionally, sequence-stratigraphic forward modelling allowed further constraints on the palaeowater depths of the Buchenstein Fm. Best-fit simulation indicates water depths of up to 800 m at the sediment–water interface of the distal succession. The key considerations of this study are: (1) the *locus* of the carbonate factory on the upper slope; (2) its high productivity after the Permian–Triassic faunal crisis; and (3) the independence of production from sea-level and ultimately accommodation change.

ACKNOWLEDGEMENTS

Axel Emmerich thanks the ‘Studienstiftung des deutschen Volkes’ (German National Academic Foundation, Bonn) and the International Postgraduate Programme (IPP, University of Heidelberg) for financial support during this study. This article benefited from thorough reviews by John Reijmer (Marseille, France), Marco Stefani (Ferrara, Italy) and Georg Warrlich (Muscat, Oman).

REFERENCES

Allen, P.A. and **Allen, J.R.** (1990) *Basin Analysis, Principles and Applications.* Blackwell Science, Oxford, 451 pp.

Aurell, M., Badenas, B., Bosence, D.W.J. and **Waltham, D.A.** (1998) Carbonate production and offshore transport on a Late Jurassic carbonate ramp (Kimmeridgian, Iberian Basin, NE Spain); evidence from outcrops and computer modelling. In: *Carbonate Ramps* (Eds V.P. Wright and T.P. Burchette), *Geol. Soc. Spec. Publ.*, **149**, 137–161.

Barbieri, C., **Bertotti, G.**, **Di Giulio, A.**, **Fantoni, R.** and **Zoetemeijer, R.** (2004) Flexural response of the Venetian foreland to the Southalpine tectonics along the TRANSALP profile. *Terra Nova*, **16**, 273–280.

Barker, C.E. and **Pawlewicz, M.J.** (1986) The correlation of vitrinite with maximum temperature in humic kerogen. In: *Paleogeothermics* (Eds G. Buntebarth and L. Stegena), pp. 79–93. Springer, New York.

Barth, S. and **Mohr, B.A.R.** (1994) Palynostratigraphically determined age of the Tregiovo sedimentary complex in relation to radiometric emplacement ages of the Atesina volcanic complex (Permian, Southern Alps, N Italy). *Neues Jb. Geol. Paläontol. Abh.*, **192**, 273–292.

Barth, S., **Oberli, F.**, **Meier, M.**, **Blattner, P.**, **Bargossi, G.M.** and **Di Battistini, G.** (1993) The evolution of a calc-alkaline basic to silicic magma system: Geochemical and Rb–Sr, Sm–Nd and $^{18}O/^{16}O$ isotopic evidence from the Late Hercynian Atesina-Cima d'Asta volcano-plutonic complex, Northern Italy. *Geochim. Cosmochim. Acta*, **57**, 4285–4300.

Barth, S., **Oberli, F.** and **Meier, M.** (1994) Th–Pb versus U–Pb isotope systematics in allanite from co-genetic rhyolite and granodiorite: Implications for geochronology. *Earth Planet. Sci. Lett.*, **124**, 149–159.

Bechstädt, T. and **Brandner, R.** (1970) Das Anis zwischen St. Vigil und dem Höhlensteintal (Pragser und Olanger Dolomiten, Südtirol). *Festband Geol. Inst., 300-Jahr-Feier Univ. Innsbruck*, Innsbruck, 9–103.

Bechstädt, T., **Brandner, R.**, **Mostler, H.** and **Schmidt, K.** (1978) Aborted rifting in the Triassic of the Eastern and Southern Alps. *N. Jb. Geol. Paläont. Abh.*, **156**, 157–178.

Bernet, M., **Zattin, M.**, **Garver, J.I.**, **Brandon, M.T.** and **Vance, J.A.** (2001) Steady-state exhumation of the European Alps. *Geology*, **29**, 35–38.

Bertotti, G., **ter Voorde, M.**, **Cloething, S.** and **Picotti, V.** (1997) Thermomechanical evolution of the South Alpine rifted margin (North Italy): Constraints on the strength of passive continental margins. *Earth Planet. Sci. Lett.*, **146**, 181–193.

Bertotti, G., **Seward, D.**, **Wijbrans, J.**, **ter Voorde, M.** and **Hurford, A.J.** (1999) Crustal thermal regime prior to, during, and after rifting: A geochronological and modelling study of the Mesozoic South Alpine rifted margin. *Tectonics*, **18/2**, 185–200.

Bertotti, G., **Mosca, P.**, **Juez, J.**, **Polino, R.** and **Dunai, T.** (2006) Oligocene to Present kilometres scale subsidence and exhumation of the Ligurian Alps and the Tertiary Piedmont Basin (NW Italy) revealed by apatite (U–Th)/He thermochronology: Correlation with regional tectonics. *Terra Nova*, **18**, 18–25.

Bielefeld, D. (1998) *Reifebestimmungen an Kohlen des Grödner Sandsteins in Südtirol (Norditalien).* Master's thesis, Geol. Inst. Univ. Cologne, Germany, 87 pp.

Blendinger, W. (1985) Middle Triassic strike-slip tectonics and igneous activity of the Dolomites (Southern Alps). *Tectonophysics*, **113**, 105–121.

Blendinger, W. (1994) The carbonate factory of Middle Triassic buildups in the Dolomites, Italy: A quantitative analysis. *Sedimentology*, **41**, 1147–1159.

Blendinger, W., **Brack, P.**, **Norborg, A.K.** and **Wulff-Pedersen, E.** (2004) Three-dimensional modelling of an isolated carbonate buildup (Triassic, Dolomites, Italy). *Sedimentology*, **51**, 297–314.

Boni, M., **Iannace, A.**, **Torre, M.** and **Zamparelli V.** (1994) The Ladinian-Carnian reef facies of Monte Caramolo (Calabria Southern Italy). *Facies*, **30**, 101–118.

Boomer, I., **Whatley, R.**, **Bassi, D.**, **Fugagnoli, A.** and **Loriga, C.** (2001) An Early Jurassic oligohaline ostracod assemblage within the marine carbonate platform sequence of the Venetian Prealps, NE Italy. *Palaeogeogr. Palaeoclimatol. Palaeoecol.*, **166**, 331–344.

Bosellini, A. (1984) Progradation geometries of carbonate platforms: Examples from the Triassic of the Dolomites, Northern Italy. *Sedimentology*, **31**, 1–24.

Bosellini, A. (1988) Outcrop models for seismic stratigraphy: Examples from the Triassic of the Dolomites. In: *Atlas of Seismic Stratigraphy*, Vol. 2 (Ed. A.W. Bally), *AAPG Stud. Geol.*, **27**, 194–203.

Bosellini, A. (1989) Dynamics of Tethyan carbonate platforms. In: *Controls on Carbonate Platform and Basin Development* (Eds P.D. Crevello, J.J. Wilson, J.F. Sarg and J.F. Read), *SEPM Spec. Publ.*, **44**, 3–13.

Bosellini, A. and **Broglio Loriga, C.** (1971) I "Calcari Grigi" di Rotzo (Giurassico Inferiore, Altopiano d'Asiago) e loro inquadramento nella paleogeografia e nella evoluzione tettono-sedimentaria delle Prealpi Venete). *Ann. Univ. Ferrara, N.S., Sez. IX Sci. Geol. Paleontol.*, **V/1**, 1–61.

Bosellini, A. and **Hardie, L.A.** (1985) Facies e cicli della Dolomia Principale delle Alpi Venete. *Mem. Soc. Geol. Ital.*, **30**, 245–266.

Bosellini, A. and **Stefani, M.** (1991) *The Rosengarten: A Platform-to-Basin Carbonate Section (Middle Triassic, Dolomites, Italy).* Dolomieu Conference on Carbonate Platforms and Dolomitization, Guidebook Excursion **C**, 24 pp.

Bosellini, A., **Neri, C.** and **Stefani, M.** (1996) *Geometrie deposizionali e stratigrafia fisica a grande scala di piattaforme carbonatiche triassiche.* 78a riunione estiva, Soc. Geol. Ital., San Cassiano Guidebook, 36 pp.

Bosence, D.W.J., **Pomar, L.**, **Waltham, D.A.** and **Lankester, T.H.G.** (1994) Computer modelling a Miocene carbonate platform, Mallorca, Spain. *AAPG Bull.*, **78**, 247–266.

Bosscher, H. and **Schlager, W.** (1993) Accumulation rates of carbonate platforms. *J. Geol.*, **101/3**, 345–355.

Bowman, S.A. and **Vail, P.R.** (1999) Interpreting the stratigraphy of the Baltimore Canyon section, offshore New Jersey with Phil, a stratigraphic simulator. In: *Numerical Experiments in Stratigraphy: Recent Advances in Stratigraphic and Sedimentologic Computer Simulations* (Eds J.W. Harbaugh, W.L. Watney, E.C. Rankey, R. Slingerland, R.H. Goldstein and E.K. Franseen), *SEPM Spec. Publ.*, **62**, 117–138.

Brack, P. and **Rieber, H.** (1993) Towards a better definition of the Anisian/Ladinian boundary: New biostratigraphical data and correlations of boundary sections from the Southern Alps. *Eclogae Geol. Helv.*, **86**, 415–527.

Brack, P. and **Rieber, H.** (1994) The Anisian/Ladinian boundary: Retrospective and new constraints. *Albertiana*, **13**, 25–36.

Brack, P., Mundil, R., Oberli, F., Meier, M. and **Rieber, H.** (1996) Biostratigraphic and radiometric age data question the Milankovitch characteristics of the Latemar cycles (Southern Alps, Italy). *Geology*, **24,** 371–375.

Brack, P., Mundil, R., Oberli, F., Meier, M. and **Rieber, H.** (1997) Biostratigraphic and radiometric age data question the Milankovitch characteristics of the Latemar cycles (Southern Alps, Italy). Reply. *Geology*, **25,** 471–472.

Brandner, R. (1991) Geological setting and stratigraphy of the Schlern–Rosengarten buildup and Seiser Alm basin. In: *The Northern Margin of the Schlern/Sciliar Rosengarten/Catinaccio Platform* (Eds R. Brandner, E. Flügel, R. Koch and L.A. Yose), Dolomieu Conference on Carbonate Platforms and Dolomitization, Guidebook Excursion A, pp. 4–16.

Broglio Loriga, C., Masetti, D. and **Neri, C.** (1983) La formazione di Werfen (Scitico) delle Dolomiti occidentali: Sedimentologia e biostratigrafia. *Riv. Ital. Paleontol. Stratigr.*, **88,** 501–598.

Buggisch, W. (1978) Die Grödener Schichten (Perm, Südalpen). Sedimentologische und geochemische Untersuchungen zur Unterscheidung mariner und kontinentaler Sedimente. *Geol. Rundsch.*, **67,** 149–180.

Buggisch, W. and **Noé, S.** (1986) Upper Permian and Permian-Triassic Boundary of the Carnia (Bellerophon Formation, Tesero Horizon, Northern Italy). *Mem. Soc. Geol. Ital.*, **34,** 91–106.

Büker, C. (1996) *Absenkungs-, Erosions und Wärmeflussgeschichte des Ruhr-Beckens und des nordöstlichen Rechtsrheinischen Schiefergebirges.* Ber. Forschz. Jülich, **Jül-3319,** 212 pp.

Carrapa, B., Wijbrans, J. and **Bertotti, G.** (2003) Episodic exhumation in the Western Alps. *Geology*, **31,** 601–604.

Claps, M., Masetti, D., Pedrielli, F. and **Lucchi Garavello, A.** (1991) Analisi spettrale e cicli di Milankovitch in successioni cretaciche del sudalpino orientale. *Riv. Ital. Pal. Stratigr.*, **97,** 153–174.

Cros, P. (1966) Age Oligocène Supérieur d'un poudingue (du Monte Parei) dans les Dolomites centrales italiennes. *CR Somm. Soc. Géol. Fr.*, **7,** 250–252.

D'Amico, C. and **Del Moro, A.** (1988) Permian and Triassic Rb-Sr dating in the Permian rhyodacitic ignimbrites of Trentino (Southern Alps). *Rend. Soc. Ital. Mineral. Petrol.*, **43,** 171–180.

De Zanche, V., Gianolla, P., Manfrin, S., Mietto, P. and **Roghi, G.** (1995) A Middle Triassic back-stepping carbonate platform in the Dolomites (Italy): Sequence stratigraphy and biochronostratigraphy. *Mem. Sci. Geol.*, **47,** 135–155.

Della Porta, G., Kenter, J.A.M. and **Bahamonde, J.R.** (2004) Depositional facies and stratal geometry of an Upper Carboniferous prograding and aggrading high-relief carbonate platform (Cantabrian Mountains, N Spain). *Sedimentology*, **51,** 267–295.

Dercourt, J., Gaetani, M., Vrielynck, B., Barrier, E., Biju-Duval, B., Brunet, M.F., Cadet, J.P., Crasquin, S. and **Sandulescu, M.** (2000) *Atlas Peri-Tethys, Palaeogeographical Maps.* CCGM/CGMW, Paris: 24 maps and explanatory notes: I–XX, 269 pp.

Dickinson, W.R., Armin, R.A., Beckvar, T.C., Goodlin, S.U., Janecke, R.A., Mark, R.D., Norris, G., Radel, G. and **Wortman, A.A.** (1987) Geohistory analysis of rates of sediment accumulation and subsidence for selected California Basins. In: *Cenozoic Basin Development of Coastal California* (Eds V.R. Ingersoll and W.G. Ernst), pp. 1–23. Rubey Volume VI, Prentice-Hall, Englewood Cliffs, NJ.

Doglioni, C. (1983) Duomo Medio-Triassico nelle Dolomiti. *Rend. Soc. Geol. Ital.*, **6,** 13–16.

Doglioni, C. (1984) Triassic diapiric structures in the Central Dolomites. *Eclogae Geol. Helv.*, **77,** 261–285.

Doglioni, C. (1987) Tectonics of the Dolomites. *J. Struct. Geol.*, **9,** 181–193.

Doglioni, C. and **Bosellini, A.** (1987) Eoalpine and mesoalpine tectonics in the Southern Alps. *Geol. Rundsch.*, **76,** 735–754.

Doglioni, C. and **Goldhammer, R.K.** (1988) Compaction-induced subsidence in the margin of a carbonate platform. *Basin Res.*, **1,** 237–246.

Dunkl, I., di Giulio, A. and **Kuhlemann, J.** (2001) Combination of single-grain fission-track chronology and morphological analysis of detrital zircon crystals in provenance studies sources of the Macigno Formation (Apennines, Italy). *J. Sed. Res.*, **71,** 516–525.

Egenhoff, S.O., Peterhänsel, A., Bechstädt, T., Zühlke, R. and **Grötsch, J.** (1999) Facies architecture of an isolated carbonate platform: Tracing the cycles of the Latemar (Middle Triassic, Northern Italy). *Sedimentology*, **46,** 893–912.

Emmerich, A., Glasmacher, U.A., Bauer, F., Bechstädt, T. and **Zühlke, R.** (2005a) Meso-/Cenozoic basin and carbonate platform development in the SW-Dolomites unraveled by basin modelling and apatite FT analysis: Rosengarten and Latemar (Northern Italy). *Sed. Geol.*, **175,** 415–438.

Emmerich, A., Zamparelli, V., Bechstädt, T. and **Zühlke, R.** (2005b) The reefal margin and slope of a Middle Triassic carbonate platform: The Latemar (Dolomites, Italy). *Facies*, **50,** 573–614.

Enos, P. (1991) Sedimentary parameters for computer modelling. In: *Sedimentary Modelling: Computer Simulations and Methods for Improved Parameter Definition* (Eds E.K. Franseen, W.L. Watney, C.G.St.C. Kendall and W. Ross), *Kansas Geol. Surv. Bull.*, **233,** 63–99.

Fitzgerald, P.G., Sorkhabi, R.B., Redfield T.F. and **Stump, E.** (1995) Uplift and denudation of the central Alaska Range: A case study in the use of apatite fission track thermochronology to determine absolute uplift parameters. *J. Geophys. Res.*, **100,** 20175–20191.

Flügel, E. (2002) Triassic reef patterns. In: *Phanerozoic Reef Patterns* (Eds W. Kiessling, E. Flügel and J. Golonka), *SEPM Spec. Publ.*, **72,** 391–463.

Fois, E. and **Gaetani, M.** (1984) The recovery of reef-building communities and the role of Cnidarians in carbonate sequences of the Middle Triassic (Anisian) in the Italian Dolomites. *Palaeontogr. Am.*, **54,** 191–200.

Gaetani, M., Fois, E., Jadoul, F. and **Nicora, A.** (1981) Nature and evolution of the Middle Triassic carbonate buildups in the Dolomites (Italy). *Mar. Geol.*, **44,** 25–57.

Gianolla, P. and **Jacquin, T.** (1998) Triassic sequence stratigraphic framework of western European basins. In: *Mesozoic and Cenozoic Sequence Stratigraphy of European Basins* (Eds P.-C. de Graciansky, J. Hardenbol, T. Jacquin and P.R. Vail), *SEPM Spec. Publ.*, **60,** 643–650.

Goldhammer, R.K. and **Harris, M.T.** (1989) Eustatic controls on the stratigraphy and geometry of the Latemar buildup (Middle Triassic), the Dolomites of Northern Italy. In: *Controls on Carbonate Platform and Basin Development* (Eds P.D. Crevello, J.J. Wilson, J.F. Sarg and J.F. Read), *SEPM Spec. Publ.*, **44**, 323–338.

Gradstein, F.M., Ogg, J.G., Smith, A.G. ***et al.*** (Eds) (2004) *A Geologic Time Scale 2004.* Cambridge University Press, Cambridge, 589 pp.

Greber, E., Leu., W., Bernoulli, D., Schuhmacher, M. and **Wyss, R.** (1997) Hydrocarbon provinces in the Swiss Southern Alps – a gas geochemistry and basin modelling study. *Mar. Petrol. Geol.*, **14**, 3–25.

Hardenbol, J., Thierry, J., Farley, M. B., Jacquin, T., de Graciansky, P.-C. and **Vail, P. R.** (1998) Mesozoic and Cenozoic sequence chronostratigraphic framework of European basins. In: *Mesozoic and Cenozoic sequence stratigraphy of European Basins* (Eds P.-C. de Graciansky, J. Hardenbol, T. Jacquin and P.R. Vail), *SEPM Spec. Publ.*, **60**, 3–15.

Hardie, L.A. and **Hinnov, L.** (1997) Biostratigraphic and radiometric age data question the Milankovitch characteristics of the Latemar cycles (Southern Alps, Italy). Comment. *Geology*, **25**, 470–471.

Harris, M.T. (1994) The foreslope and toe-of-slope facies of the Middle Triassic Latemar buildup (Dolomites, Northern Italy). *J. Sed. Res.*, **B64**, 132–145.

Hertle, M. and **Littke, R.** (2000) Coalification pattern and thermal modelling of the Permo-Carboniferous Saar basin (SW-Germany). *Int. J. Coal Geol.*, **42**, 273–296.

Hinnov, L. and **Goldhammer, R.K.** (1991) Spectral analysis of the Middle Triassic Latemar limestone. *J. Sediment. Petrol.*, **61**, 1173–1193.

Hsü, K.J. (1971) Origin of the Alps and the western Mediterranean. *Nature*, **233**, 44–48.

Hsü, K.J. (1989) Time and place in Alpine orogenesis – the Fermor Lecture. In: *Alpine Tectonics* (Eds M.P. Coward, D. Dietrich and R.G. Park), *Geol. Soc. London Spec. Publ.*, **45**, 421–443.

Hummel, K. (1928) Das Problem des Fazieswechsels in der Mitteltrias der Südtiroler Dolomiten. *Geol. Rundsch.*, **19**, 223–228.

Hummel, K. (1932) Zur Stratigraphie und Faziesentwicklung der südalpinen Mitteltrias. *N. Jb. Mineral. Geol. Paläontol. Suppl.*, **68 B**, 403–462.

Hunt, D. and **Fitchen, W.M.** (1999) Compaction and the dynamics of carbonate-platform development: Insights from the Permian Delaware and Midlands Basins, Southeast New Mexico and West Texas, U.S.A. In: *Advances in Carbonate Sequence Stratigraphy: Application to Reservoirs, Outcrops and Models* (Eds P.M. Harris, A.H. Saller and J.A. Simo), *SEPM Spec. Publ.*, **63**, 75–106.

Keim, L. and **Schlager, W.** (2001) Quantitative compositional analysis of a Triassic carbonate platform (Southern Alps, Italy). *Sediment. Geol.*, **139**, 261–283.

Kenter, J.A.M., Harris, P.M.M. and **Porta, G.D.** (2005) Steep microbial boundstone-dominated platform margins – examples and implications. *Sediment. Geol.*, **178**, 5–30.

Kenter, J.A.M., van Hoeflaken, F., Bahamonde, J.R., Bracco Gartner, Guido L., Keim, L. and **Besems, R.E.** (2002) Anatomy and lithofacies of an intact and seismic-scale Carboniferous carbonate platform (Asturias, NW Spain): Analogues of hydrocarbon reservoirs in the Pricaspian Basin (Kazakhstan). In: *Paleozoic Carbonates of the Commonwealth of Independent States (CIS): Subsurface Reservoirs and Outcrop Analogues* (Eds W.G. Zempolich and H.E. Cook), *SEPM Spec. Publ.*, **74**, 181–203.

Ketcham, R.A., Donelick, R.A. and **Donelick, M.B.** (2000) AFTSolve: A program for multikinetic modelling of apatite fission-track data. *Geol. Mater. Res.*, **2(1)**, (electronic: 18 pages, 2 tables, 12 figures). Mineral. Soc. Am., Washington, DC.

Kuhlemann, J., Dunkl, I., Brügel, A., Spiegel, C. and **Frisch, W.** (2006) From source terrains of the Eastern Alps to the Molasse Basin: Detrital record of non-steady-state exhumation. *Tectonophysics*, **413**, 301–316.

Laslett, G.M., Green, P.F., Duddy, I.R. and **Gleadow, A.J.W.** (1987) Thermal annealing of fission tracks in apatite, 2. A quantitative analysis. *Chem. Geol.*, **65**, 1–13.

Laubscher, H. and **Bernoulli, D.** (1982) History and deformation of the Alps. In: *Mountain Building Processes* (Ed. K.J. Hsü), pp. 169–180.Academic Press, London.

Lehrmann, D., Enos, P., Montgomery, P., Payne, J., Orchard, M., Bowring, S., Ramezani, J., Martin, M., Jiayong, W., HongMei, W., YouYi, Y., Jiafei, X. and **Rongxi, L.** (2002) Integrated biostratigraphy, magnetostratigraphy and geochronology of the Olenikian-Anisian boundary in marine strata of Guandao section, Nanpanjiang Basin, South China: Implications for timing of biotic recovery from the end-Permian extinction. In: *I.U.G.S. Subcommission on Triassic Stratigraphy* (Ed. O. Piros), STS/IGCP 467 Field Meeting, Veszprém, Hungary, pp. 7–8.

Leischner, K. (1994) *Kalibration simulierter Temperaturgeschichten von Sedimentgesteinen.* Ber. Forschz. Jülich, **Jül-2909**, 309 pp.

Leonardi, P. (1962) Il gruppo dello Sciliar e le scogliere coralligene dolomitiche. *Ann. Univ. Ferrara, N.S., sez. IX*, **3**, 1–83.

Leonardi, P. (1967) *Le Dolomiti: Geologia dei Monti tra Isarco e Piave.* Manfrini, Rovereto, **1**, 1019 pp.

Littke, R., Büker, C., Lückge, A., Sachsenhofer, R.F. and **Welte, D.H.** (1994) A new evaluation of palaeo-heatflows and eroded thicknesses for the Carboniferous Ruhr basin, western Germany. *Int. J. Coal Geol.*, **26**, 155–183.

Mair, V., Stingl, V., Krois, P. and **Keim, L.** (1996) Die Bedeutung andesitischer und dazitischer Gerölle im Unterinntal-Tertiär (Tirol, Österreich) und im Tertiär des Mte. Parei (Dolomiten, Italien). *N. Jb. Geol. Paläontol. Abh.*, **199**, 369–394.

Makhous, M. and **Galushkin, Y.I.** (2004) *Basin Analysis and Modeling of the Burial, Thermal and Maturation Histories in Sedimentary Basins.* Éditions Technip, Paris, 400 pp.

Martire, L. (1996) Stratigraphy, facies and synsedimentary tectonics in the Jurassic Rosso Ammonitico Veronese (Altopiano di Asiago, NE Italy). *Facies*, **35**, 209–236.

Masetti, M. and **Neri, C.** (1980) L'Anisico della Val di Fassa (Dolomiti occidentali): Sedimentologia e paleogeografia. *Ann. Univ. Ferrara, N.S., sez. IX*, **7**, 1–19.

Massari, F. and **Neri, C.** (1997) The infill of a supradetachment(?) basin: The continental to shallow-marine Upper Permian succession in the Dolomites and Carnia (Italy). *Sediment. Geol.*, **110**, 181–221.

Massari, F., Grandesso, P., Stefani, C. and **Jobstraibizer, P.G.** (1986) A small polyhistory foreland basin evolving in a context of oblique convergence: The Venetian basin (Chattian to Recent, Southern Alps, Italy). In: *Foreland Basins* (Eds P.A. Allen and P. Homewood), *Int. Assoc. Sedimentol. Spec. Publ.*, **8**, 141–168.

Massari, F., Conti, M.A., Fontana, D., Helmold, K., Mariotti, N., Neri, C., Nicosia, U., Ori, G.G., Pasini, M. and **Pittau, P.** (1988) The Val Gardena sandstone and Bellerophon formation in the Bletterbach gorge (Alto Adige, Italy): Biostratigraphy and sedimentology. *Mem. Sci. Geol.*, **40**, 229–273.

Massari, F., Neri, C., Pittau, P., Fontana, D. and **Stefani, C.** (1994) Sedimentology, palynostratigraphy and sequence stratigraphy of a continental to shallow-marine rift-related succession: Upper Permian of the Eastern Southern Alps (Italy). *Mem. Sci. Geol.*, **46**, 119–243.

Mastandrea, A., Neri, C. and **Russo, F.** (1997) Conodont biostratigraphy of the S. Cassiano Formation surrounding the Sella Massif (Dolomites, Italy): Implications for sequence stratigraphic models of the Triassic of the Southern Alps. *Riv. Ital. Paleontol. Stratigr.*, **103**, 39–52.

Maurer, F. (1999) Wachstumsanalyse einer mitteltriadischen Karbonatplattform in den westlichen Dolomiten (Südalpen). *Eclogae Geol. Helv.*, **92**, 361–378.

Maurer, F. (2000) Growth mode of middle Triassic carbonate platforms in the Western Dolomites (Southern Alps, Italy). *Sediment. Geol.*, **134**, 275–286.

Maurer, F. and **Schlager, W.** (2003) Lateral variations in sediment composition and bedding in Middle Triassic interplatform basin (Buchenstein Formation, southern Alps, Italy). *Sedimentology*, **50**, 1–22.

Maurer, F., Reijmer, J.J.G. and **Schlager, W.** (2003) Quantification of input and compositional variations of calciturbidites in a Middle Triassic basinal succession (Seceda, Dolomites, Southern Alps). *Int. J. Earth Sci. (Geol. Rundsch.)*, **92**, 593–609.

McKinney, M.L. (1985) Mass extinction patterns of marine invertebrate groups and some implications for a causal phenomenon. *Paleobiology*, **11**, 227–233.

Mojsisovics, E. (1879) *Die Dolomitriffe von Südtirol und Venetien: Beiträge zur Bildungsgeschichte der Alpen.* Holder, Wien, 522 pp.

Mundil, R., Brack, P., Meier, M., Rieber, H. and **Oberli, F.** (1996) High resolution U/Pb dating of Middle Triassic volcaniclastics: Time-scale calibration and verification of tuning parameters for carbonate sedimentation. *Earth Planet. Sci. Lett.*, **141**, 137–151.

Mundil, R., Metcalfe, I., Ludwig, K.R., Renne, P.R., Oberli, F. and **Nicoll, R.S.** (2001) Timing of the Permian–Triassic biotic crisis: Implications from new zircon U/Pb age data (and their limitations). *Earth Planet. Sci. Lett.*, **187**, 131–145.

Mundil, R., Zühlke, R., Bechstädt, T., Brack, P., Egenhoff, P., Meier, M., Oberli, F., Peterhänsel, A. and **Rieber, H.** (2003) Cyclicities in Triassic platform carbonates: Synchronizing radio-isotopic and orbital clocks. *Terra Nova*, **15**, 81–87.

Peters, K.E., Walters, C.C. and **Mankiewicz, P.J.** (2006) Evaluation of kinetic uncertainty in numerical models of petroleum generation. *AAPG Bull.*, **90**, 387–403.

Pia, J. (1937) *Stratigraphie und Tektonik der Pragser Dolomiten in Südtirol.* Pia, Wien, 248 pp.

Preto, N., Hinnov, L.A., Hardie, L.A. and **De Zanche, V.** (2001) Middle Triassic orbital signature recorded in the shallow-marine Latemar carbonate buildup (Dolomites, Italy). *Geology*, **29**, 1123–1126.

Preto, N., Spötl, C., Mietto, P., Gianaolla, P., Riva, A. and **Manfrin, S.** (2005) Aragonite dissolution, sedimentation rates and carbon isotopes in deep-water hemipelagites (Livinallongo Formation, Middle Triassic, northern Italy). *Sed. Geol.*, **181**, 173–194.

Raup, D.M. (1979) Size of the Permo-Triassic bottleneck and its evolutionary implications. *Science*, **206**, 217–218.

Reijmer, J.J.G. (1998) Compositional variations during phases of progradation and retrogradation of a Triassic carbonate platform (Picco di Vallandro/Dürrenstein, Dolomites, Italy). *Geol. Rundsch.*, **87**, 436–448.

Reijmer, J.J.G., ten Kate, W.G.H.Z., Sprenger, A. and **Schlager, W.** (1991) Calciturbidite composition related to the exposure and flooding of a carbonate platform (Triassic, Eastern Alps). *Sedimentology*, **38**, 1059–1074.

Rüffer, T. and **Zühlke, R.** (1995) Sequence stratigraphy and sea-level changes in the Early to Middle Triassic of the Alps: A global comparison. In: *Sequence Stratigraphy and Depositional Response to Eustatic, Tectonic and Climatic Forcing* (Ed. B.U. Haq), pp. 161–207. Kluwer, Amsterdam.

Sachsenhofer, R.F. and **Littke, R.** (1993) Vergleich und Bewertung verschiedener Methoden zur Berechnung der Vitrinitreflexion am Beispiel von Bohrungen im Steirischen Tertiärbecken. *Zbl. Geol. Paläont.*, **1992/1**, 597–610.

Sachsenhofer, R.F., Privalov, V.A., Zhykalyak, M.V., Büker, C., Panova, E.A., Rainer, T., Shymanovskyy, V.A. and **Stephenson, R.** (2002) The Donets basin (Ukraine/Russia): Coalification and thermal history. *Int. J. Coal Geol.*, **49**, 33–55.

Schlager, W. (1981) The paradox of drowned reefs and carbonate platforms. *Geol. Soc. Am. Bull.*, **92**, 197–211.

Schlager, W. (1999) Scaling of sedimentation rates and drowning of reefs and carbonate platforms. *Geology*, **27**, 183–186.

Schlager, W. (2000) Sedimentation rates and growth potential of tropical, cool-water and mud-mound carbonate systems. In: *Carbonate Platform Systems: Components and Interactions* (Eds E. Insalaco, P.W. Skelton and T.J. Palmer), *Geol. Soc. Spec. Publ.*, **178**, 217–227.

Schlager, W., Biddle, K.T. and **Stafleu, J.** (1991) *Picco di Vallandro (Dürrenstein) a platform-basin transition in outcrop and seismic model.* Dolomieu Conference on Carbonate Platforms and Dolomitization, Guidebook Excursion **D**, 22 pp.

Schulz, O. and **Fuchs, H.W.** (1991) Kohle in Tirol: Eine historische kohlenpetrologische und lagerstättenkundliche Betrachtung. *Arch. Lagerst.forsch. Geol. B.-A.*, **13**, 123–213.

Seeling, M., Emmerich, A., Bechstädt, T. and **Zühlke, R.** (2005) Accommodation/sedimentation development and massive early marine cementation: Latemar vs. Concarena (Middle/Upper Triassic, Southern Alps). *Sed. Geol.*, **175**, 439–457.

Senowbari-Daryan, B., **Zühlke, R.**, **Bechstädt, T.** and **Flügel, E.** (1993) Anisian (Middle Triassic) buildups of the Northern Dolomites (Italy): The recovery of reef communities after the Permian/Triassic crisis. *Facies*, **28**, 181–256.

Smith, A.G. (1971) Alpine deformation and the oceanic areas of the Tethys, Mediterranean and Atlantic. *Geol. Soc. Am. Bull.*, **82**, 2039–2070.

Stach, E., **Mackowsky, M.**, **Teichmüller, M.**, **Taylor, G.H.**, **Chandra, D.** and **Teichmüller, R.** (1982) *Stach's Textbook of Coal Petrology*. Borntraeger, Berlin, 535 pp.

Stock, H.W. (1996) Planktonische Foraminiferen aus der Oberkreide der nordöstlichen Dolomiten (Italien). *Revue Paléobiol.*, **15**, 155–182.

Sweeney, J.J. and **Burnham, A.K.** (1990) Evaluation of a simple model of vitrinite reflectance based on chemical kinetics. *AAPG Bull.*, **74**, 1559–1570.

Taylor, G.H., **Teichmüller, M.**, **Davis, A.**, **Diessel, C.F.K.**, **Littke, R.** and **Robert, P.** (1998) *Organic Petrology*. Borntraeger, Berlin, Stuttgart, 704 pp.

Tippett, J.M. and **Kamp, P.J.J.** (1995) Quantitative relationships between uplift and relief parameters for the Southern Alps, New Zealand, as determined by fission track analysis. *Earth Surf. Proc. Land.*, **20**, 153–175.

Turcotte, D.L. and **Schubert, G.** (1982) *Geodynamics – Applications of Continuum Physics to Geological Problems*. Wiley, New York, 450 pp.

Turcotte, D.L. and **Schubert, G.** (2002) *Geodynamics*. University Press, Cambridge, 456 pp.

Trevisani, E. (1991) Il Toarciano-Aaleniano nei settori centro-orientali della piattaforma di Trento (Prealpi Venete). *Riv. Ital. Paleontol. Stratigraf.*, **97**, 99–124.

Trümpy, R. (1982) Alpine palaeogeography: A reappraisal. In: *Mountain Building Processes* (Ed. K.J. Hsü), pp. 149–156. Academic Press, London.

Twitchett, R.J. (1999) Palaeoenvironments and faunal recovery after the end-Permian mass extinction. *Palaeogeogr. Palaeoclimatol. Palaeoecol.*, **154**, 27–37.

Viel, G. (1979a) Litostratigrafia Ladinica: Una revisione. Ricostruzione paleogeografica e paleostrutturale dell'area Dolomitico-Cadorina (Alpi Meridionali). I. Parte. *Riv. Ital. Paleontol. Stratigraf.*, **85**, 88–125.

Viel, G. (1979b) Litostratigrafia Ladinica: Una revisione. Ricostruzione paleogeografica e paleostrutturale dell'area Dolomitico-Cadorina (Alpi Meridionali). II. Parte. *Riv. Ital. Paleontol. Stratigraf.*, **85**, 297–352.

Welte, D.H., **Horsfield, B.** and **Baker, D.R.** (1997) *Petroleum and Basin Evolution*. Springer, Berlin, 535 pp.

Winkler, W. (1988) Mid- to Early Cretaceous flysch and melange formations in the western part of the Eastern Alps, palaeotectonic implications. *Jb. Geol. Bundesanst.*, **131**, 341–389.

Winterer, E.L. (1998) Palaeobathymetry of mediterranean Tethyan Jurassic pelagic sediments. *Mem. Soc. Geol. Ital.*, **53**, 97–131.

Winterer, E.L. and **Bosellini, A.** (1981) Subsidence and sedimentation on Jurassic passive continental margin, Southern Alps, Italy. *AAPG Bull.*, **65**, 394–421.

Yugan, J., **Wardlaw, B.R.**, **Glenister, B.F.** and **Kotlyar, G.V.** (1997) Permian chronostratigraphic subdivisions. *Episodes*, **20**, 10–15.

Zattin, M., **Stefani, C.** and **Martin, S.** (2003) Detrital fission-track analysis and petrography as keys of Alpine exhumation: The examples of the Veneto foreland (Souther Alps, Italy). *J. Sediment. Res.*, **7**, 1051–1061.

Zattin, M., **Cuman, A.**, **Fantoni, R.**, **Martin, S.**, **Scotti, P.** and **Stefani, C.** (2006) From Middle Jurassic heating to Neogene cooling: The thermochronological evolution of the southern Alps. *Tectonophysics*, **414**, 191–202.

Zühlke, R. (2000) Fazies, hochauflösende Sequenzstratigraphie und Beckenentwicklung im Anis (Mittlere Trias) der Dolomiten (Südalpin, Italien). *Gaea heidelberg.*, **6**, CD-ROM.

Zühlke, R. (2004) Integrated cyclostratigraphy of a model Mesozoic carbonate platform – the Latemar (Middle Triassic, Italy). In: *Cyclostratigraphy: Approaches and Case Histories* (Eds B. D'Argenio, A. Fisher, I. Premoli Silva, H. Weissert and V. Ferreri), *SEPM Spec. Publ.*, **81**, 183–211.

Zühlke, R., **Bechstädt, T.** and **Mundil, R.** (2003) Sub-Milankovitch and Milankovitch forcing on a model Mesozoic carbonate platform – the Latemar (Middle Triassic, Italy). *Terra Nova*, **15**, 69–80.

Spec. Publ. Int. Assoc. Sedimentol. (2008) **40**, 37–64

Fine-scale forward modelling of a Devonian patch reef, Canning Basin, Western Australia

CLAUDE-ALAIN HASLER*[1], ERWIN W. ADAMS†[2], RACHEL A. WOOD‡[3] and TONY DICKSON*

**Department of Earth Sciences, University of Cambridge, Downing Street, Cambridge CB2 3EQ, UK*

†Department of Earth, Atmospheric, and Planetary Sciences, Massachusetts Institute of Technology, 77 Massachusetts Avenue, Cambridge, MA 02139, USA

‡Schlumberger Cambridge Research, High Cross, Madingley Road, Cambridge CB2 0EL, UK

ABSTRACT

Although factors that control the distribution of carbonate sediments are relatively well understood, their interaction in producing known geometries remains poorly known. A fine-scale forward model was developed to analyse the controls on decimetre-scale heterogeneity and carbonate facies distribution within larger-scale sediment bodies. The model was tested on a Frasnian (Late Devonian) patch reef, part of the so-called Glenister Knolls reef complex in the Canning Basin, Western Australia. The evolution of Glenister Knolls was digitally reconstructed using digital surveying technologies that resulted in a high-resolution dataset containing geological information from centimetre-scale to outcrop dimensions. The three-dimensional digital outcrop model of Glenister Knolls was used to evaluate predictions from the fine-scale forward modelling results.

The fine-scale forward model is based on a stochastic cellular automaton that simulates the spatial distribution of carbonate facies through time. The modelled carbonate facies is based on two main factors, stromatoporoid growth and sediment transport, and it successfully predicts the distribution of the basic elements – reef fabric and sediment – that compose the patch reef. The fine-scale forward model also successfully reproduces all major depositional relationships observed in the field such as backstepping, downlapping and interfingering contacts. The external structure of the reef is very sensitive to subsidence and production rates, since these directly influence accommodation space so constraining the vertical growth of the patch reef. However, the internal structure predicted by this model does not seem to be affected by one specific parameter, and displays the same sensitivity to other parameters, such as current intensity and wave amplitude.

Keywords Forward modelling, cellular automaton, digital field mapping, patch reef, Devonian, Canning Basin, Western Australia.

INTRODUCTION

Over the last few decades, forward modelling has been used to explain the factors that influence carbonate depositional systems and their interactions by simulating carbonate production and sedimentary processes (Aigner *et al.*, 1989; Bosence & Waltham, 1990; Bosscher & Schlager, 1992; Eberli *et al.*, 1994; Paola, 2000; Boylan *et al.*, 2002). In particular, cyclicity and stacking patterns of carbonates have received considerable attention (Drummond & Dugan, 1999; Tipper, 2000; Burgess & Wright, 2003). However, knowledge of the factors that control decimetre-scale heterogeneity of carbonate facies within larger-scale sediment bodies is still rudimentary. In order to better understand and predict the internal facies architecture, it is important to use tools that work at the scale of the heterogeneities. The

[1]Present address: Department of Geology and Paleontology, University of Geneva, Rue des Maraîchers 13, CH-1205 Geneva, Switzerland (Claude-Alain.Hasler@terre.unige.ch).

[2]Present address: Carbonate Development Team, Shell International Exploration and Production B.V., Kessler Park 1, 2288 GS Rijswijk, The Netherlands.

[3]Present address: School of Geo-Sciences, Grant Institute, University of Edinburgh, Kings Buildings, West Mains Road, Edinburgh EH9 3JW, UK.

heterogeneities within reef bodies, for example, are related primarily to the nature of the reef-building organisms. As it was doubtful that existing forward models could efficiently be adapted to work at such a fine scale, new software was developed and is presented here.

Ecological systems are generally considered to be very complex because they are characterized by a large number of diverse components as well as by non-linear interactions, scale multiplicity, and spatial heterogeneity. For this reason, simple rule-based models such as cellular automata should be used to approach the complexity of such systems (Wolfram, 2002). Over the last 20 years, cellular automata have been used in a great number of disciplines, for instance, physics (Chopard & Droz, 1998), natural sciences (Gaylord & Nishidate, 1996), urbanism (White & Engelen, 1993) and, relevant to our case, platform and reef development (Tipper, 1997; Blanchon & Blackeway, 2003) and species competition (Silvertown *et al.*, 1992).

The fine-scale forward model presented here is based on a 2D stochastic cellular automaton in which we combined both sedimentary and biological processes to predict the development of specific reef-building organisms on a high-resolution decimetre scale. Instead of only predicting palaeoenvironments, such as reef-margin or backreef environments, the result of our forward model is a decimetre-scale 3D distribution of facies, which may be hosted in several depositional environments. Fine-scale forward modelling such as proposed here, as well as the sensitivity tests made on the major parameters used in the model, may aid in the understanding of the development of heterogeneities within carbonate bodies.

Much is known about the sedimentology, stratigraphic development, and facies evolution of Devonian reef complexes of the Canning Basin, Western Australia (Playford, 1980, 2002; Kennard *et al.*, 1992; Wood, 2000a; George & Chow, 2002; George *et al.*, 2002; Stephens & Sumner, 2003). The existing geological framework and the exhumed topography of the well exposed outcrops allow precise 3D quantification of carbonate platform and internal facies geometries. Digital surveying technologies, which have proved to be useful in reconstructing the architecture of carbonate platforms (Adams *et al.*, 2004, 2005; Verwer *et al.*, 2004; Bellian *et al.*, 2005; McCaffrey *et al.*, 2005) were employed in the field. The resulting quantitative, high-resolution digital datasets were then used to extract spatial information on which our digital reconstructions of Devonian reef complexes were based. The high-resolution outcrop data constrained inferences of environmental parameters and served as input for fine-scale forward models, allowing decimetre-scale prediction of facies distributions.

The result of this study is a fine-scale forward model and 3D outcrop model of a Devonian patch reef of the Canning Basin, Western Australia. We first present the lithofacies, 3D mapping, and digital reconstruction of the patch reef in the field. This is followed by a description of the fine-scale forward modelling technique. Then, the results of sensitivity tests performed on the major parameters are given, and finally we discuss and compare, both qualitatively and quantitatively, the predicted model with the outcrop data.

GEOLOGICAL SETTING

Devonian reef complexes crop out over a length of 350 km and a width of up to 50 km along the northeastern edge of the intracratonic sedimentary Ordovician to Cretaceous Canning Basin (Fig. 1; Playford, 1980; Forman & Wales, 1981; Copp, 2000). The reef complexes fringed and developed on the fault-controlled Precambrian landmass of the Kimberley block in an extensional regime (Playford, 1980). Reef growth started at the flanks of the rapidly subsiding Fitzroy Trough (Begg, 1987; Drummond *et al.*, 1988) in the Givetian and ceased in the Famennian, leaving a rock record of approximately 15 Myr (Playford, 1980; Wood,

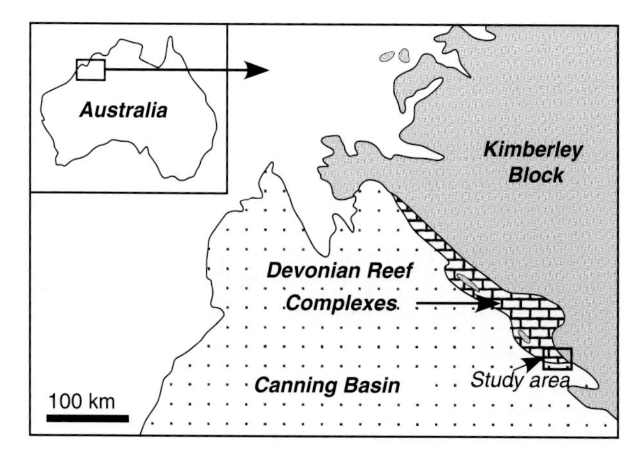

Fig. 1. Simplified bedrock geological map showing the Canning Basin and the Devonian reef complexes that developed along the Precambrian landmass of the Kimberley Block. Box indicates the southernmost tip where Devonian rocks are exposed and coincides with the study area.

2000b). The carbonate platforms of the Canning Basin are mainly reef-rimmed shelves with both accretionary and bypass margins (Read, 1982; Playford *et al.*, 1989). However, isolated platforms and atolls are also present. It is the wide variety of platform types and minor tectonic disturbance in combination with well exposed and well preserved outcrops that make the Devonian reef complexes ideal for studying carbonate depositional environments.

The southernmost Givetian to Frasnian outcrops of the Devonian reef complexes are exposed in the Bugle Gap area where limestone ranges representing exhumed platforms are separated by valleys that coincide with basinal deposits (Fig. 2). As a result, the present-day topography mirrors the depositional profile of the Devonian sea floor (Fig. 3). The carbonate platforms (Pillara Limestone) are surrounded by well developed slopes (Sadler Limestone) that are separated by intra-shelf basins (Gogo Formation; Playford, 1980). The Pillara and Sadler Limestones are well exposed; the Gogo Formation is mostly covered. Based on the present-day geomorphological setting of limestone ranges, the Canning Basin had a connection to the open ocean in the south (Playford, 1980; Playford & Hocking, 1998). Backstepping relationships, which are interpreted to have developed during times of increased creation of accommodation space, can be observed and

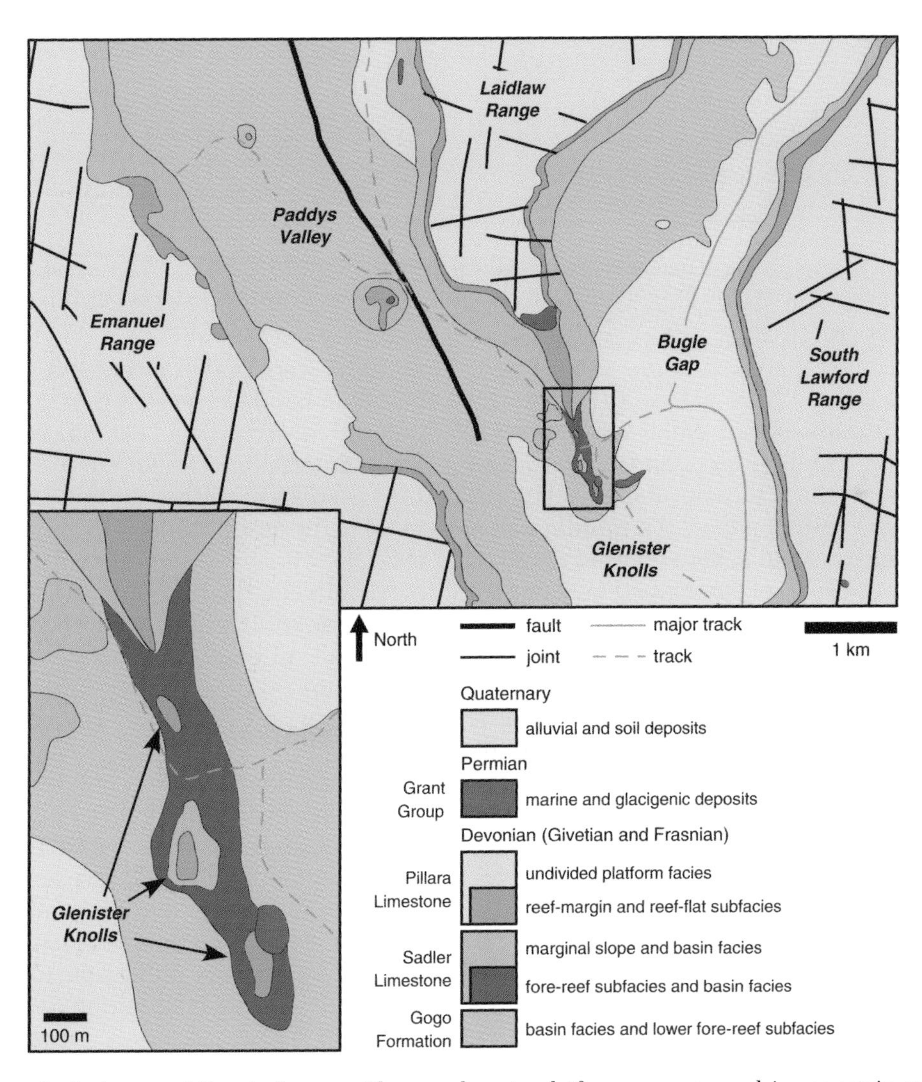

Fig. 2. Detailed geological map of the study area. Three carbonate platforms are exposed in mountain ranges and are separated by valleys representing intra-shelf basins. A connection to the open ocean was probably present in the south (see text for discussion). The inset (see box for location) illustrates the setting of the Glenister Knolls. Palaeogeographically the Glenister Knolls are an extension of the southern tip of the Laidlaw Range carbonate platform and formed a set of isolated patch reefs. Modified after Playford & Hocking (1998).

Fig. 3. Oblique aerial photograph of the Glenister Knolls. The arrow indicates the major knoll, which represents the middle and best-developed patch reef. The field of view covers roughly the same area as the inset of Fig. 2.

studied in this area (Playford *et al.*, 1989; Ward, 1999).

A group of three isolated patch reefs, the so-called Glenister Knolls (Figs 2 and 3), form the southern tip of the Laidlaw Range Platform (Playford, 1980). In this paper, the middle patch reef of the Glenister Knolls, here referred to as the major knoll, is the focus of our study (Figs 2 and 3). The major knoll is recognized as a 20–30 m high, 200 m wide, and 300 m long well exposed outcrop displaying Frasnian carbonate rocks of both the Pillara and the Sadler Limestones (Fig. 3). It has a central depression resulting from glaciations and karstification during the Mid-Carboniferous to Early Permian (Fig. 3; Playford, 2001, 2002). The edges of the Devonian patch reef are dissected by several steep normal faults with a few metres of displacement.

METHODS

Traditional field mapping

Outcrop observations defined stratigraphic contacts, depositional boundaries, and sedimentary bodies for mapping. Fifty hand samples were collected from the study area, from which 28 thin sections were made. Lithofacies descriptions are based on hand samples, thin-section analyses and outcrop descriptions. To obtain more detailed facies information, other comparable outcrops in the proximity of the study area were incorporated in the descriptions (Fig. 2). Two stratigraphic sections were measured digitally (see next section for details) on the northern and southern flanks of the major knoll, allowing verification of the internal details of platform lithology and stratigraphy.

Digital field mapping

The depositional contacts of interest were mapped by physically walking on the contacts while collecting spatial data. The stratigraphic evolution was captured by digitally measuring stratigraphic sections including the dip and strike of each bed to compute section thicknesses. For 3D digital geological data collection we used real-time kinematic Global Positioning System (RTK GPS) receivers (Trimble 4700 GPS receiver) with a relative positional accuracy of approximately 2–5 cm horizontally and 5–10 cm vertically (cf. Adams *et al.*, 2005 and references therein). In world map coordinates, the absolute positional accuracy of the dataset is approximately 20 cm and based on the use of static GPS measurements collected at a base station over a period of 8–10 h. Each mappable geological feature, identified by outcrop observation, was assigned a specific attribute, for example character of stratigraphic contact, lithofacies, lithology, fossil species, and so on. These attributes with their associated 3D spatial locations were stored in a data collector (Trimble TSC1 data collector), transferred from the data collectors to a standard PC laptop, and visualized in a Windows-based software package (Petrel; a software package for subsurface reservoir modelling and a trademark of Schlumberger Limited) together with a digital elevation model (DEM) of the study area. The DEM was constructed from data points (~4000) that were systematically collected every 2 m across the outcrop. The DEM has a horizontal grid spacing of 1 m and a vertical resolution similar to the mapping accuracy, that is, 5–10 cm.

Fine-scale forward modelling

The fine-scale forward modelling presented in this paper is based on a 2D stochastic cellular automaton, which simulates the distribution of carbonate facies on the sea floor through time. Each cell composing the 2D surface can adopt any one of the possible discrete states (facies). The predicted facies are defined as either reef fabric (resulting from the development of reef-building organisms) or unconsolidated sediments (resulting from erosion or *in situ* production of skeletal components). Each cell interacts with the eight neighbouring cells following specific rules in order to simulate the expansion of reef fabric (stromatoporoid species growth) while taking into account processes such as competition between species and sediment transport. Because the system is based on a discrete time dynamic, the state of each cell is updated every timestep (iteration) according to the rules given, which integrate four constantly changing inferred environmental parameters (water depth, shear stress on the sea floor, turbidity, and light incident on the sea floor). Inferences were drawn from geological observations on all scales and in the vicinity, but not directly on the major knoll.

The vertical evolution of the reef is constrained by eustatic and subsidence evolution, as well as the ability of the reef-building organisms to grow and of sediment to be produced. The outcome of the simulation is a 3D distribution of both reef fabric and unconsolidated sediments resulting from the vertical evolution of the 2D stochastic cellular automaton. The reconstructed 3D digital outcrop model of the major knoll was used for the validation of and comparison with the fine-scale forward model.

LITHOFACIES

The Glenister Knolls outcrop was recognized by Playford (1980) as isolated patch reefs. Based on lithofacies associations as well as stratal and depositional relationships, we subdivided this patch reef into five major depositional environments, including: backreef, reef margin, bioherm, slope, and basin environments. The reef-margin, bioherm, and slope environments are well exposed. The backreef environment is partially covered; the basin environment is exposed locally along the edges of the outcropping patch reefs. Each depositional environment is subdivided into several lithofacies. Additionally, stromatolites developed at several locations on the patch reefs.

Backreef environment

Two major lithofacies types were recognized in the backreef environment and based on the type of stromatoporoid: dendroid-rich stromatoporoid boundstone (Fig. 4a) and bulbous-rich stromatoporoid boundstone (Fig. 4b). Both lithofacies bear metre-scale beds and are mud-rich. Corals and brachiopods are present. The dendroid stromatoporoids are mainly *Stachyodes costulata,* with rare *Amphipora rudis* (Cockbain, 1984). *Actinostroma windjanicum, Actinostroma papillosum* and *Hermatostroma* (*schlueteri?*) are the dominant bulbous stromatoporoid types, with average heights and diameters of less then 50 cm. Laminar/tabular stromatoporoids (*Hermastostroma ambiguum?*) are usually associated with the bulbous lithofacies.

The dendroid stromatoporoid-rich beds (*Stachyodes* spp. and especially *Amphipora rudis*) are considered to represent platform-interior environments in the Canning Basin (Playford, 1980), and therefore are also held to be the shallowest environment of the major knoll patch reef. Evidence of subaerial exposure has not been observed. Other lithofacies assigned to backreef environments found elsewhere in the Canning Basin such as fenestral, oolite and oncolite lithofacies (Playford & Lowry, 1966) are not observed.

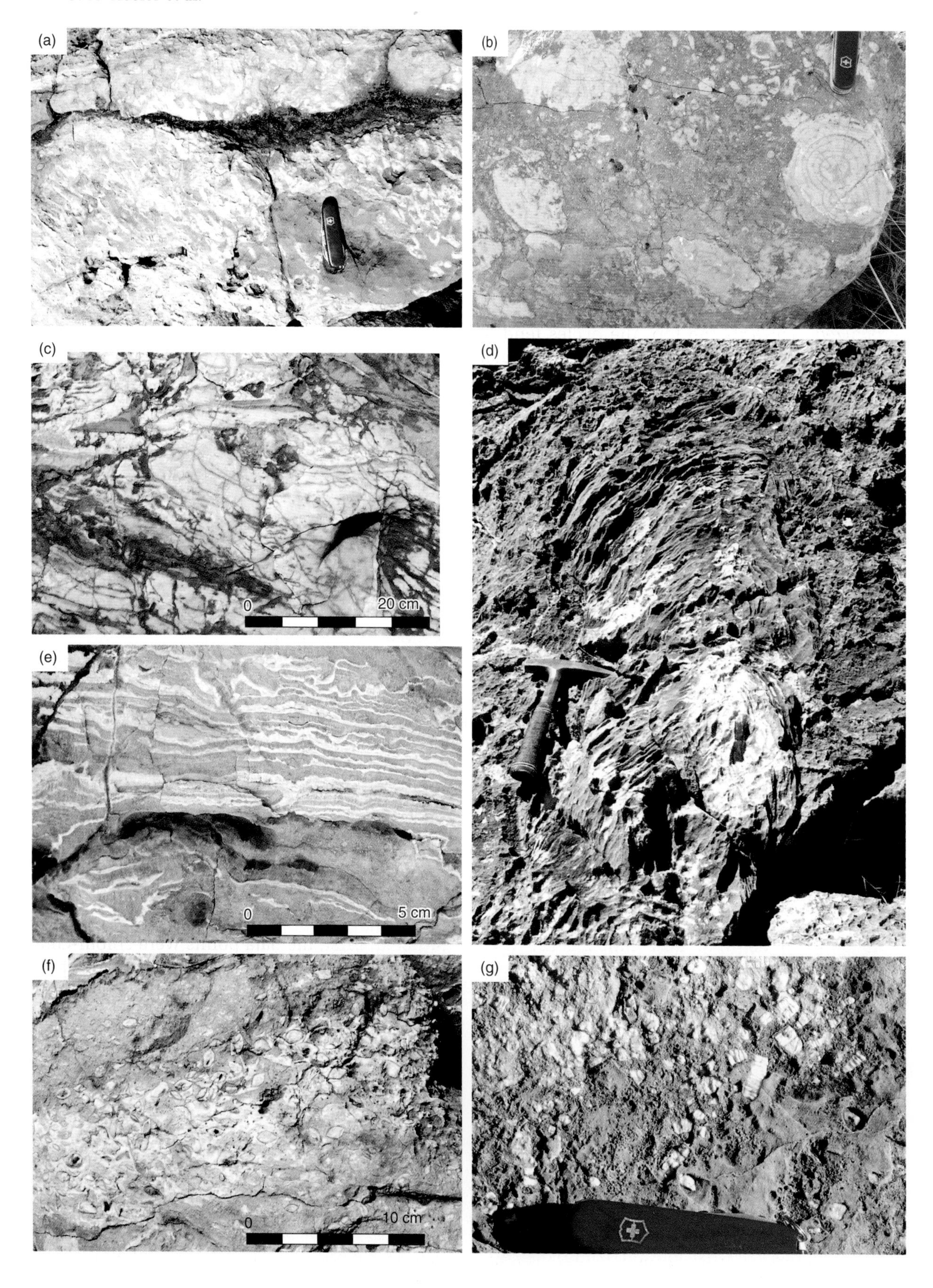
(a)
(b)
(c)
0
20 cm
(d)
(e)
0
5 cm
(f)
0
10 cm
(g)

Reef-margin environment

The reef-margin environment is characterized by stromatoporoid-rich boundstone (Fig. 4c) dominated by tabular morphologies (*Clathrocoilona*, *Actinostroma* and *Hermatostroma*). Bulbous, domal (Fig. 4d) and laminar stromatoporoids (Fig. 4e) as well as tabular corals (*Thamnopora*) are often present. Brachiopods are usually found in this facies association, but are mostly limited to localized pockets.

Several cycles of approximately 3 m thick are recognized. Towards the top, the cycles tend to decrease in brachiopod and increase in tabular stromatoporoid abundance. The cycles are interpreted as shallowing-upward cycles with probably higher current activity toward the top.

Bioherm environment

The bioherm environment consists of relatively massive and poorly bedded, metre-scale, crinoid-rich mounds flanked by decimetre-scale crinoidal rudstone beds. The youngest beds flanking the mounds are rich in nautiloids and goniatites. Brachiopods, lithistid sponges, receptaculitids, and large gastropods are abundant (Wood, 1999). Large cavities, up to a metre in diameter, preserved by early fibrous cements, are filled with dolomitized geopetal sediment. Stromatoporoids are rare within the bioherm complex.

Slope environment

The slope lithofacies is dominated by skeletal rudstone (Fig. 4f). Crinoids, sponges, and brachiopods (mainly atrypids; Grey, 1978) dominate the skeletal components. Slope angles consistently range between 35° and 40°. Slope beds, with a metre-scale thickness, have a planar geometry interpreted to represent the angle of repose (Adams *et al.*, 2002). The upper slope contains both reworked and *in situ* laminar stromatoporoids. The toe of slope contains crinoid-rich packstone–grainstone (Fig. 4g).

Basin environment

Basinal lithofacies include organic-rich shale and thin mudstone–packstone beds. The basin environment is identified as the Gogo Formation and is well known for its rich pelagic faunal content including exceptionally well preserved fish (Long, 1988, 1995), goniatites (Glenister, 1958; Becker *et al.*, 1993), conodonts (Glenister & Klapper, 1966; Seddon, 1970; Druce, 1976), crustaceans, radiolarians, and tentaculitids (Playford, 1980).

Stromatolites

Branching columnar stromatolites are observed at Glenister Knolls (Wood, 1999). The first stage of stromatolite growth is characterized by a few centimetre-thick flat layers of bacterial growth parallel to the substratum. These are generally overlain by centimetre-scale bifurcating columns. They developed on and in direct contact with the reef-margin environment. Stromatolites are described throughout the Canning Basin (Playford *et al.*, 1976; George, 1999), where stromatolite growth is interpreted to occur in areas of low sedimentation, during both periods of lowstand and highstand (George, 1999).

DIGITAL RECONSTRUCTION

The relationship between mapped depositional contacts and outcrop topography was visualized by combining the digital elevation model with digitally collected outcrop data. Figure 5 illustrates the step-by-step procedure by which a 3D geological model of the major knoll was constructed. First, the data points were colour-coded according to the type of feature or contact that was mapped (Fig. 5b). Polygons were created by linking data points that mark the boundaries between depositional environments. The polygons therefore represent the intersection of surfaces defining depositional environments with the

Fig. 4. (Opposite page) Representative photographs of the major lithofacies observed in the field. (a) Dendroid-rich stromatoporoid boundstone (*Stachyodes* sp.) of the backreef environment. The knife is 9 cm long. (b) Bulbous-rich stromatoporoid boundstone of the backreef environment. The knife is 9 cm long. (c) Tabular stromatoporoids of the reef-margin environment. Fabric contains cavities filled by geopetal sediment and cement. (d) Large domal stromatoporoid of the reef-margin environment. Hammer is approximately 30 cm long. (e) Reworked and *in situ* laminar stromatoporoids in an upper slope setting. (f) Brachiopod rudstone of the slope environment. (g) Crinoid-rich packstone of the toe-of-slope environment. The knife is 9 cm long.

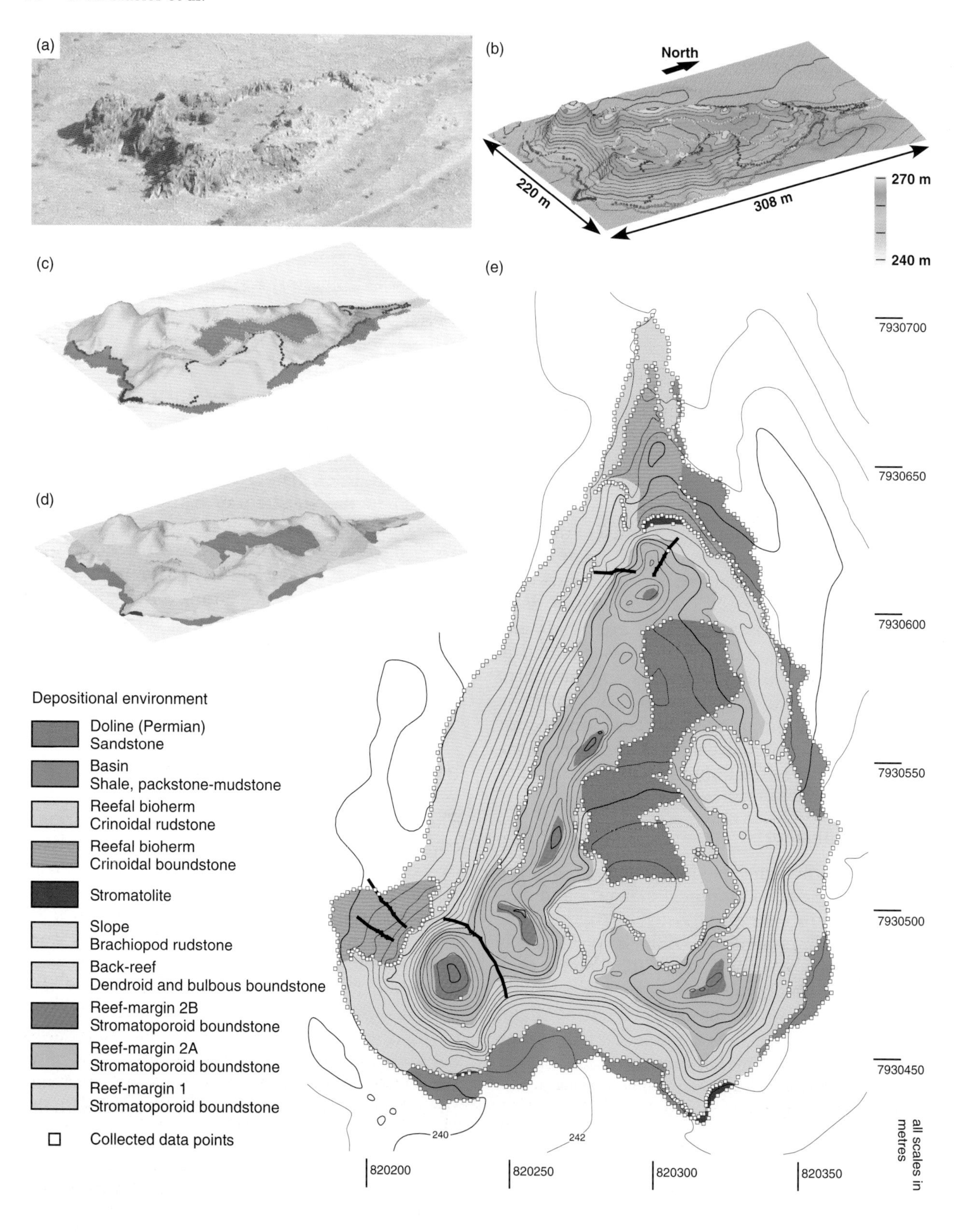
(a)
(b)
North
220 m
308 m
270 m
240 m
(c)
(d)
(e)
7930700
7930650
7930600
7930550
7930500
7930450
240
242
820200
820250
820300
820350
all scales in metres
Depositional environment
Doline (Permian)
Sandstone
Basin
Shale, packstone-mudstone
Reefal bioherm
Crinoidal rudstone
Reefal bioherm
Crinoidal boundstone
Stromatolite
Slope
Brachiopod rudstone
Back-reef
Dendroid and bulbous boundstone
Reef-margin 2B
Stromatoporoid boundstone
Reef-margin 2A
Stromatoporoid boundstone
Reef-margin 1
Stromatoporoid boundstone
Collected data points

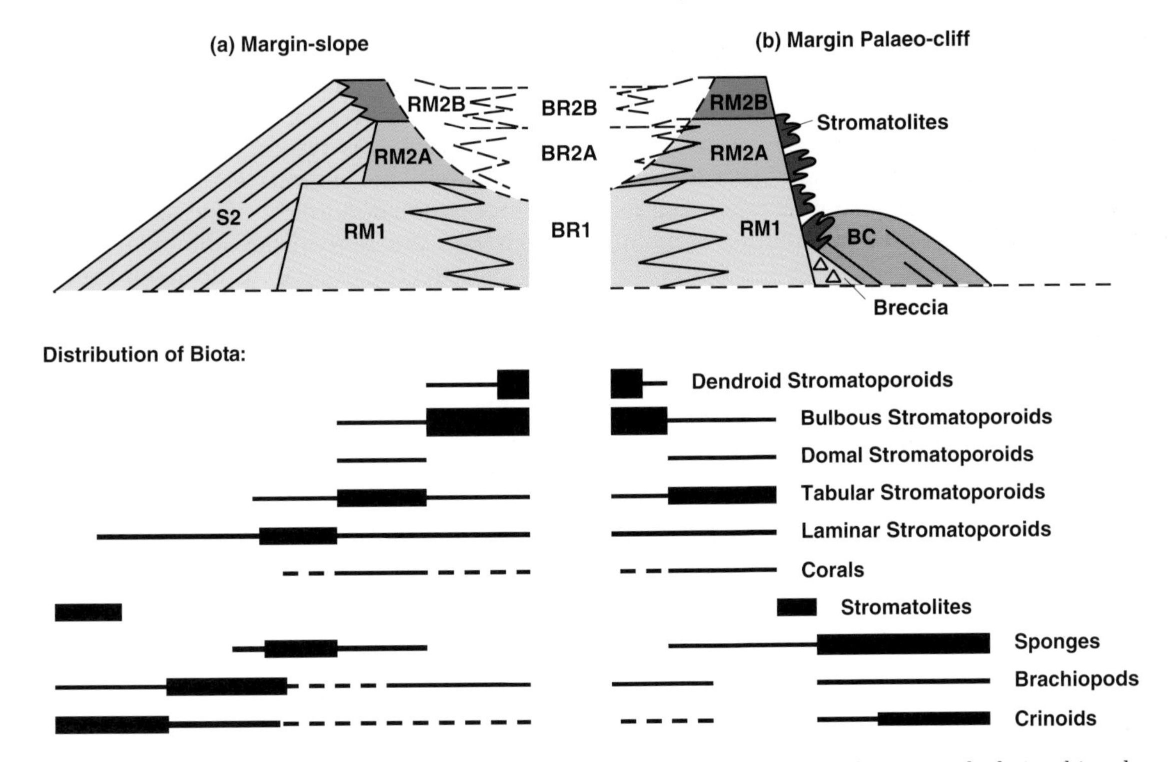

Fig. 6. Schematic depositional model for Glenister Knolls illustrating the evolution and depositional relationships observed in the field. For comparison of depositional relationships and asymmetry of the Glenister Knolls major patch reef see Fig. 5e. The central part of the structure is symmetrical with backreef environment (BR) surrounded by reef margin (RM). The reef-margin-to-basin transition exhibits several configurations depending on the absence or presence of slope deposits (S) or a bioherm complex (BC). Stromatolites developed at locations where slope sediments were not shedded. A relative abundance and distribution is shown for the major biotic components (data from Playford, 1980). Numbers 1, 2, 2A, and 2B at the environment notations BR, RM, and S relate to platform stages.

present-day topography. Consequently, a subset of surfaces was created and colour-coded according to depositional environment (Fig. 5c).

In order to model the vertical stratigraphy of the reef-margin environment (grey surface in Fig. 5c and d), data points collected along stratigraphic contacts (e.g. contact between different stages of reef development) were interpolated by fitting surfaces with a minimum curvature algorithm (Isaaks & Srivastava, 1990). A digital 3D geological model was created by combining these surfaces and it represents with volume the different depositional environments (Fig. 5e).

DEPOSITIONAL MODEL

Based on stacking patterns and depositional relationships, the major knoll was subdivided into three platform stages. Overall, the patch reef shrank in aerial extent through time and shows a clear backstepping geometry (Fig. 6; Playford,

Fig. 5. (Opposite page) Three-dimensional geological reconstruction of the Glenister Knolls major patch reef. (a) Oblique aerial outcrop photograph (see also Fig. 3) of the major knoll for comparison with digital models. (b) Digital elevation model (DEM) of the outcrop with superimposed digitally mapped data points mapped along depositional contacts. Data points are colour-coded according to kind of mapped contact (c) Polygons (not shown in the figure) were created by linking data points that mark the boundaries between depositional environments and represent the intersection of surfaces defining depositional environments with the present-day topography. The surfaces are colour-coded according to type of depositional environment (see legend). (d) The undefined reef-margin environment (grey surface) was stratigraphically subdivided by surfaces (e.g. the transparent yellow), which were constructed by fitting between mapped stratigraphic data points. (e) Map view of final 3D geological model of the major knoll patch reef representing the intersection of surfaces defining depositional environments with the present-day topography. Each data point contains a spatial coordinate and information on the kind of contact that was mapped, for example a Quaternary-to-basin contact. Vertical and horizontal ticks represent UTM x and y location in metres and are spaced every 50 m; contour interval is 2 m.

1980; George *et al.*, 2002). Unfortunately, it is impossible to relate the three platform stages to a precise timescale since no precise markers have been observed. The reef-stage subdivisions are listed in evolutionary order. Figure 6 illustrates schematically the depositional relationships observed in the field and the distribution of biota.

Platform stage 1

Platform stage 1 is recognized by a well developed reef margin as well as backreef environment. The reef-margin environment consists of massive and chaotic beds several metres thick dominated by tabular and bulbous stromatoporoids (mainly *Actinostroma*) with common corals. The maximum thickness of the massive reef margin is 10.5 m with three cycles of roughly similar thickness. The geometry and synoptic relief of the reef margin with respect to the backreef could not be established; the backreef environment has been eroded at most places. No interfingering has been observed between reef-margin lithofacies and slope deposits (Fig. 6). Slope deposits are only observed with a sharp, onlapping contact with the reef-margin lithofacies of Platform stage 1. These relationships are typical of many platform systems in the Canning Basin (Playford *et al.*, 1989).

The contact between Platforms 1 and 2 is sharp. Evidence of subaerial exposure at this contact was not found. A backstepping relationship is observed and illustrated by (1) Platform stage 2 slope deposits that downlap on reef-margin lithofacies of stage 1 (Fig. 7a), and (2) reef-margin lithofacies of stage 2 that sits directly on top of stage 1 backreef lithofacies in the centre of the major knoll (Fig. 7b). The backstepping that occurs between Platform stages 1 and 2 probably records a rapid increase in accommodation space.

Platform stage 2

Platform stage 2 is subdivided into two units based on the vertical development of reef-margin lithofacies measured along stratigraphic sections. The backreef environment has been mostly removed by erosion and consequently the stratal relationship between backreef and reef margin could not be established (Fig. 6). Overall, the second stage consists of metre-scale beds, dominated by tabular and domal stromatoporoids (*Actinostroma*, *Clathrocoilona*) with some corals. Brachiopods (mainly atrypids) are abundant.

Unit 2A is composed of massive-bedded reef-margin lithofacies containing three cycles with a total thickness of 13 m. The presence of domal stromatoporoids is typical of Unit 2A. The reef margin Unit 2B contains abundant tabular and laminar stromatoporoids. The bedding is well developed. Brachiopods are abundant in Unit 2B and form pockets as well as decimetre-scale continuous layers. The upward increase in brachiopod abundance (from Stage 2A to 2B) probably indicates either a temporary increase in brachiopod production or less effective sorting and transport of shells. Overall, the increase in abundance of brachiopods, decrease in abundance of stromatoporoids as well as the overall decrease in the lateral extent of the atoll between stages 1 and 2, probably indicate a deepening trend.

Slope deposits consisting of brachiopod rudstones interfinger with reef margin 2B (Fig. 7b). The slope deposits are well developed, forming planar inclined beds at angles of 35°. At locations where slope sediments do not occur vertical cliffs are observed that were locally covered by stromatolite growth (Fig. 6). The stromatolite columns developed perpendicular to the cliff surface and record a depositional break during carbonate sedimentation (George, 1999).

Platform stage 3

Crinoid bioherms are typical of Platform stage 3 (Fig. 6) and contain crinoids, receptaculitids, and gastropods (Wood, 1999). The bioherms developed on the slope or grew directly on stromatolites of Platform stage 2. All biota (i.e., crinoids, nautiloids, goniatites) indicate deeper water than the previous platform stages. The break in shallow-water carbonate production underlined by the lack of *in situ* stromatoporoids and deposition of deeper water facies during this stage is probably the result of a major regional drowning event (Playford, 2002). The bioherm complex is well exposed in the northern part of the knoll.

FINE-SCALE FORWARD MODELLING

The overall dimensions and limits of the two environments that compose the main edifice of the patch reef were digitally reconstructed (Fig. 5e). The fine-scale forward modelling is therefore focused on the distribution of the backreef and reef-margin environments of Platform stages 1 and

Fig. 7. Outcrop photographs illustrating stratal relationships. (a) Slope deposits associated with Platform stage 2 downlap on stage 1 reef-margin lithofacies. Note the steep (restored slope angle of 35°) and planar geometry of the slope deposits. (b) Platform stage 2A reef-margin lithofacies are found stratigraphically directly on top of backreef lithofacies. Slope deposits are found adjacent to stage 2 reef-margin deposits; the contact is massive but not sharp. An interfingering relationship is interpreted between the massive reef margin of stage 2 and the well developed slope deposits.

2. The facies relationships between the main reef-margin units and the slope are predicted by the model but are not evaluated quantitatively because they cannot be confirmed by outcrop data. The exposed slope is probably coeval with Platform stage 2 (see Fig. 7) but this cannot be confirmed for all slope deposits. Stromatolite development and the youngest depositional Platform stage 3 do not contribute to the main platform development (see Fig. 5e), and, as data on their distribution are fragmentary, have been disregarded in the modelling.

This section starts with an explanation of the initial conditions for the modelling process, followed by a description of the simulation parameters. Finally, we describe in detail the computer simulation processes.

Initial conditions

Dimension of the model area

Each cell represents an area of 0.25 m^2. The model is based on a rectangular area of 300 by 400 cells. The total area has a size of 30 000 m^2 (see Fig. 5e; UTM system; zone 51 South; datum WGS 1984; geoid model EGM96 global). The reef fabrics are allowed to grow vertically by 10-cm steps. Therefore, the final voxel size of the 3D forward model is 50 by 50 by 10 cm.

Initial topography

The initial topography used in the model represents a dome-like surface whose dimensions are related to observations made in the field. The morphology of the base of Platform stage 1 cannot be observed and reconstructed. Consequently, the initial top of the dome-like surface is modelled as being flat. The original water depth of the flat-topped surface is set at 4 m and corresponds to the approximate depth estimated for the development of backreef stromatoporoid species according to Playford (1980). The flanks of the surface are dipping with an angle similar to the angle of repose observed in the field, that is 35° (Fig. 7a). The lateral extent of the reef is broader in the southern part and narrows towards the north. The substratum bears no inheritance from previous deposits and corresponds to non-specified hard rock. Consequently, the initial value for all cells of the stochastic automaton is set at *null value* (see below).

Initiation of platform growth

Twenty-five iterations are conducted before the start of the true computer simulation (t0) in order to populate the initial flat surface and create synoptic relief between the reef margin and back-reef environment. During these first 25 iterations, eustatic sea-level fluctuations are not used but a constant subsidence rate is applied (0.077 m per iteration) that corresponds to the subsidence rate observed in the field for Platform stage 1 (see below for details on how the subsidence rate was inferred). First, reef fabrics randomly appear on the complete surface. If a reef fabric appears in an area where the reef-building organism corresponding to this fabric is not able to survive (production rate = 0; see below for details on production rate function), the fabric will disappear and leave the cell empty until another reef fabric occupies the cell. Then, the fabrics grow (laterally or vertically) according to their production rate functions.

Simulation parameters

Protagonists

Two types of protagonist were defined: (a) reef fabric, and (b) unconsolidated sediment. The protagonists represent the different lithofacies described in previous sections. Two groups of protagonists are represented in the following way.

(a) *Reef fabric* is represented by a cell in which the major constituents are reef-building organisms. Three reef fabrics are used: (1) laminar and tabular stromatoporoids, (2) bulbous stromatoporoids, and (3) dendroid stromatoporoids. All these organisms have the ability to grow both laterally and vertically.
(b) *Unconsolidated sediment* corresponds to a cell in which the major constituents are not constructed and may be moved by sediment transport. Two sediment types have been used in the simulation: (1) unspecified carbonate sediment with a grain size of 1 mm, and (2) coarse brachiopod-rich sediment with a grain size of 1 cm. The fine sediment is an erosional product of the reef. In contrast, the brachiopod-rich sediment can be produced *in situ* in the same manner as a reef fabric. The physical properties of these two types of sediment are summarized in Table 1.

In addition, a *null value* is introduced that corresponds neither to a sediment (i.e., no particles can

Table 1. Range of parameters and values together with source and assumptions used in the modelling of Glenister Knolls major patch reef.

Parameter	Parameter value	Source
Grid size	300 × 400 cells	
Cell size	50 × 50 cm	
Three-dimensional voxel size	50 × 50 × 10 cm	
Optimal production rate	14.2 cm per 10^6 years	Bosscher & Schlager (1993) after Playford (1980)
Extinction coefficient (k)	0.04	Bosscher & Schlager (1992)
Main current direction	North	Based on Bugle Gap morphology and orientation
Main current intensity	0.5 m s^{-1}	
Main waves amplitude	1	Based on the Paddys Valley morphology, using Eqs 3-39 and 3-40 of the Shore Protection Manual (Coastal Engineering Research Center, 1984)
Main waves frequency	4	
Main direction of waves propagation	North-east	
Fine carbonate sediment		Carbonate porous grains (20%) filled with water
Size	0.1 × 0.1 cm	
Density	2.32 g cm^{-3}	
Coarse carbonate sediment (Brachipods)		Carbonate porous grains (50%) filled with water
Size	1 × 1 cm	
Density	1.83 g cm^{-3}	
Average angle of repose	35°	Based on outcrop measurements
Eustatic cycle duration (possible fourth order)	45 iterations	
Subsidence rates		
Reef stage 1	0.077 m per iteration	Based on outcrop measurements
Reef stage 1	0.096 m per iteration	Based on outcrop measurements

be removed), nor to a reef fabric with living reef-building organisms (i.e., no *in situ* growth). This value represents neutral hard rock; it is used at the beginning of the simulation or when a specific reef-building organism does not survive.

Time scale

The cyclicity observed in the Pillara Formation of the Canning Basin is still equivocal (Hocking & Playford, 2003). Metre-scale cycles are the product of repeated, short term relative sea-level fluctuations, with shallowing-upward phases separated by rapid deepenings. Greenhouse conditions (Read, 1995) can be inferred from the relatively small variations in bathymetry throughout the cyclic succession. However, a tectonic control over cyclicity is also possible (Read, 1973; Hardie *et al.*, 1991; Read *et al.*, 1991).

In the present model we arbitrarily favour the hypothesis of high-order Milankovich cyclicity. This choice is not justified by geological observations made in the field but by the wish to combine both subsidence and eustatic cyclicity in the modelling. The period of the cycle we used in the simulation is 22.5 kyr, subdivided in 45 iterations (although this is arbitrary it is compatible with efficient computation), corresponding to 0.5 kyr. The total simulation reproduces seven cycles corresponding to 315 iterations for approximately 150 kyr.

Hydrodynamic parameters

During the simulation an average current with a constant value is used. The wave and current action in the Paddy's Valley is interpreted as coming from the north (Playford, 1980). However, because of the palaeogeographical setting of the Bugle Gap area (see geological setting; Fig. 2) including a platform in the north and a connection to the open ocean toward the south, the current is modelled as originating in the south. No direct observations were made in order to reconstruct current velocities. The intensity of the current was chosen to be 0.5 m s^{-1} and is based on the calibre of the sediments involved and transported in the simulation. The direction and intensity of the current are the starting values applied to the southern boundary of the model. The evolution of the current is subsequently calculated by the program,

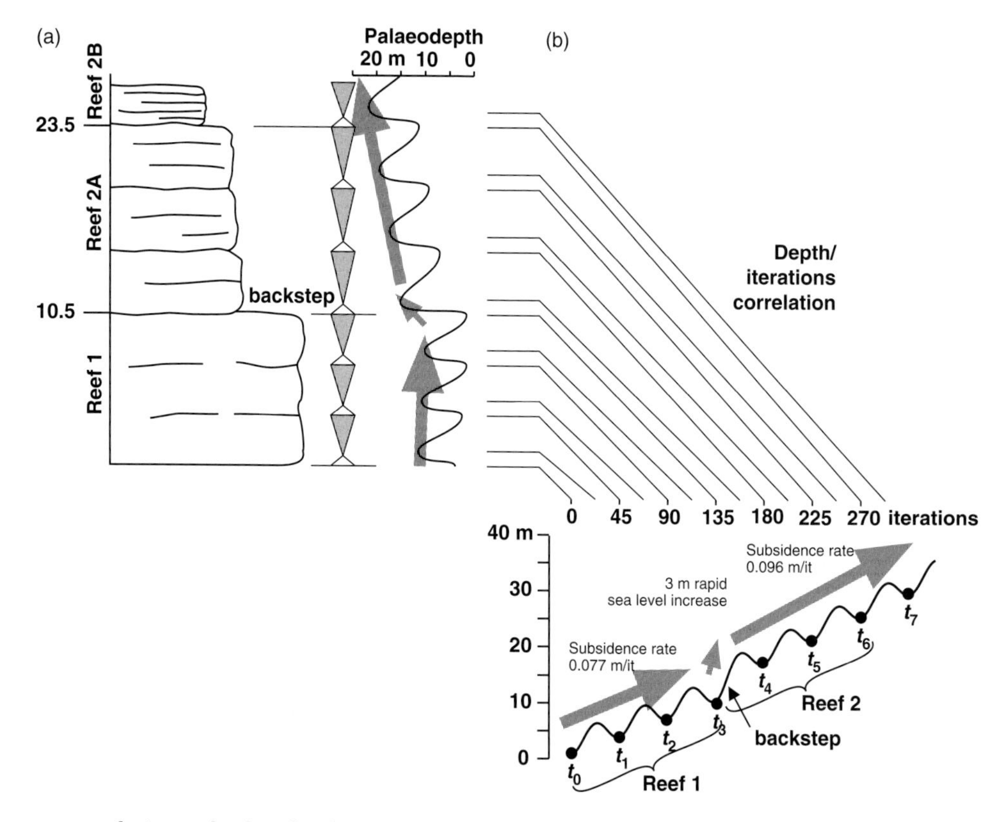

Fig. 8. The accommodation and palaeodepth evolution and relative sea-level curve used in the fine-scale forward modelling. (a) Accommodation and palaeodepth evolution of the reef-margin environment. The evolution is based on four measured stratigraphic sections in the Bugle Gap area. (b) The relative sea-level curve inferred from these stratigraphic sections. The curve is interpreted as a combination of both subsidence and eustatism. The eustatic curve is responsible for the asymmetry of the cycles. Subsidence changes instantly after Platform stage 1, creating a backstep. Platform stage 2 has a greater subsidence rate than that of stage 1.

which takes into account the effect of changing topography.

Based on the dimensions of Paddy's Valley, which represents the main area with water deep enough to generate waves, the average frequency and amplitude of waves affecting the area have been set at 1 m and 4 s. The following parameters have been used to determine the wave amplitude and frequency: wind speed of $15\,m\,s^{-1}$ from the south-west, average depth of 50 m for Paddy's Valley, fetch of 5 km (using the *sedx* package available on the US Geological Survey web site, based on the Shore Protection Manual, Equations 3–39 and 3–40; Coastal Engineering Research Center, 1984). The angle between the wave propagation direction and the currents is 45°.

Evolution of accommodation space

The evolution of accommodation space and depth of the reef-margin environment in the Bugle Gap area are illustrated in Fig. 8. The succession is based on four stratigraphic sections measured near Glenister Knolls but not directly from the studied area in order to avoid circular reasoning since we used the digital outcrop model later to compare and validate the forward model. Constructed boundstone is one of the major facies involved in the modelling for which compaction effects are likely to be small. Therefore, thickness measurements conducted in the field are compared directly with thicknesses obtained from fine-scale forward modelling without correction for compaction.

Platform stage 1 is characterized by three asymmetric shallowing-upward cycles with an aggrading stacking pattern. The boundary between Platform stages 1 and 2 is underlined by a regional backstep of the reef margin. Platform stage 2 is similar to stage 1 and records three shallowing-upward cycles with a transgressive trend. The relative sea-level curve reconstructed from this succession is interpreted as being the result of combined eustasy and subsidence, where

Table 2. Parameter estimates for the catastrophic events simulating periodic collapse of the patch reef margin.

Event number	Iteration (when the collapse occurs)	Direction of collapse	Maximum width	Maximum thickness
1	45	N80°W	10	4
2	90	N45°E	10	4
3	135	N80°W	10	4
4	180	N45°E	10	4
5	270	South	10	4

both produced the stacking trend and major backstep.

The eustatic cycles are simulated by a sinusoidal curve with two metres amplitude and 45 iterations per eustatic cycle (Fig. 8). During stage 1, the subsidence rate is constant (0.077 m per iteration) and related to the thickness of Platform stage 1 formed during three cycles. At the end of the third cycle a rapid relative sea-level rise of 3 m is applied, in order to induce the backstepping of the margin observed between stages 1 and 2 (Fig. 8). Stage 2A, much in the same way as stage 1, is composed of three cycles, but with a higher subsidence rate (0.096 m per iteration). This subsidence rate is also applied to Platform stage 2B. The increase in subsidence rate simulates the gradual drowning of the platform.

Catastrophic events

Five events representing margin collapses have been applied during the simulation. The events are related to lowstand periods. The characteristics of these erosion events are set out in Table 2. Due to the lack of information available in the field on the volume of margin rocks that were removed, each event is estimated to have removed equal amounts of material. Therefore, the morphology of the rockslide scars is similar for each collapse event. However, the position and the direction differ. Three major directions for syndepositional fissures and faults have been observed in the field; the directions of collapse (listed in Table 2) are perpendicular to these observed locations. The location of collapse events is determined at the reef margin that is tangential to the synsedimentary fissures and faults.

Computer simulation

The forward modelling is based on a 2D stochastic cellular automaton, for which each cell composing the grid contains five values. One value is given by the type of protagonist (related to the nature of the substratum, which may be one of the reef fabrics or one of the sediment types). Four values are given by the ecological parameters water depth, shear stress on the sea floor, turbidity of the water column, and light incident on the sea floor. The ecological parameters are used to determine, for each cell, the production rate of the three reef fabrics and for one of the types of unconsolidated sediment (brachiopod-rich sediment). Four processes, which drove the evolution of the reef, were simulated and occurred in all of the 315 iterations (seven cycles of 45 iterations) in the modelling. These four processes are: (1) the appearance of reef fabric; (2) the growth of reef fabric; (3) erosion and production of sediment; and (4) sediment transport. The outcome of the simulation is a distribution of reef fabric and unconsolidated sediments resulting from the facies succession modelled using the stochastic cellular automaton.

Ecological parameters

The simulation is based on the assumption that four ecological parameters control reef growth. The first parameter, which is updated every iteration, is water depth. To speed up the calculation, the last three parameters – shear stress on the sea floor, turbidity, and light incident on the sea floor – were updated only once every five iterations.

The *water depth* is derived from the original topography, the given relative sea-level curve (eustatics and subsidence, Fig. 8) as well as the thickness of the carbonate produced or reworked by currents and margin collapse.

The second parameter is *shear stress on the sea floor* and is given by current and wave conditions interacting with the changing topography. The shear stress is calculated for each cell using the iterative method and formula proposed by

Madsen (1994), where the shear stress is a function of the current intensity and direction, and wave characteristics (amplitude, frequency and direction of propagation).

The third parameter is *turbidity* related to sediment transport by suspension in the water column in each cell. In this case, the sediment types are each composed of carbonate grains of a unique size and density. For the fine unspecified carbonate sediment we used a sphere with 1 mm radius and 20% of the porosity is filled with sea water. The brachiopod-rich sediment has carbonate spheres with 1 cm radius and 50% of the porosity is filled by sea water. The physical value of the grains is used to determine the roughness of the sea floor (Madsen, 1994) and the minimum shear stress needed to place the sediment grains into suspension (Le Roux, 1997).

The fourth parameter is the *light incident on the sea floor* (I_z), which is directly dependent on depth (Beer–Lambert's law) and turbidity (Bosscher & Schlager, 1992):

$$I_z = I_0 \times \mathrm{e}^{-k(T) \times z} \quad (1)$$

where I_0 is the light intensity at the surface (fixed at 100% for the simulation); $k(T)$ is the extinction coefficient k ($k = 0.04$, Bosscher and Schlager, 1992) weighted by the turbidity factor (T) and z the water depth.

Production rate function

For each reef fabric representing the type of stromatoporoid as well as the brachiopod facies, the production rate function (P_f) is given by the following equation:

$$P_\mathrm{f} = (M_\mathrm{pf}/\mathrm{Voxel}_\mathrm{height}) \times D_\mathrm{f} \times S_\mathrm{f} \times T_\mathrm{f} \times L_\mathrm{f} \quad (2)$$

where M_pf is the maximal production rate in centimetres per iteration for the facies f, and D_f, S_f, T_f and L_f are weighting functions related to the ecological parameters depth, shear stress on the sea floor, turbidity and luminosity. The weighting functions used in the simulation presented hereafter are plotted in Fig. 9. The maximal production rates (M_pf) for the stromatoporoid species have been established using the space between *latilaminae* (Meyer, 1981) available in the literature (Cockbain, 1984) and are also based on our observations. Using *latilaminae* as growth banding may not be absolutely accurate but they provide reasonable growth rates for reef-building organisms (Kershaw, 1998).

Reef-fabric appearance

To simulate the reproduction of a species of a particular reef fabric, the model allocates a specific number to randomly located cells in which the reef fabric could appear. The program assigned a fabric to a cell if both of the following rules were true: (1) the production rate for the calculated fabric using the environmental parameters of the cell is greater than zero; (2) where a cell is already occupied by another fabric, the production rate of the new fabric exceeds that of the pre-existing one in order to take into account competition between species.

Reef-fabric growth

Each reef fabric can grow either vertically or horizontally. The growth rate of the fabric is a direct function of the production rate, because the growth rate in effect corresponds to the probability of fabric growth. In the same way as for reef-fabric appearance, a particular reef fabric is able to grow over another one but only if the production rate of the new fabric (both are constantly updated) exceeds that of the pre-existing one in order to take into account competition between species. If a specific vertical succession of fabrics is observed in the field, it is also possible to restrict the lateral extent or appearance of a fabric overlying another one by following Markov-chain properties (Harbaugh & Bonham-Carter, 1970). Markov's chain has not been applied here and therefore, the fabrics are allowed to extend freely and depend only on the value of the ecological parameters and not on the nature of the underlying fabric.

The direction of growth may be randomly chosen or forced in specific directions (e.g. vertical growth can be favoured over lateral extent, or the growth can be related to the main direction of the current). In our case, we did not use any preferred direction of lateral extent; all species may grow towards all of the eight neighbouring cells with the same probability. Laminar and tabular stromatoporoids grow faster laterally than vertically (Meyer, 1981). In our model, lateral growth is three times faster than vertical growth. In the case of the bulbous species, due to their hemispherical shape, this fabric is allowed to grow in all directions (laterally and vertically) with the same probability. We assumed that the dendroid fabric could grow faster vertically than laterally (2 versus 1). The brachiopod facies cannot grow

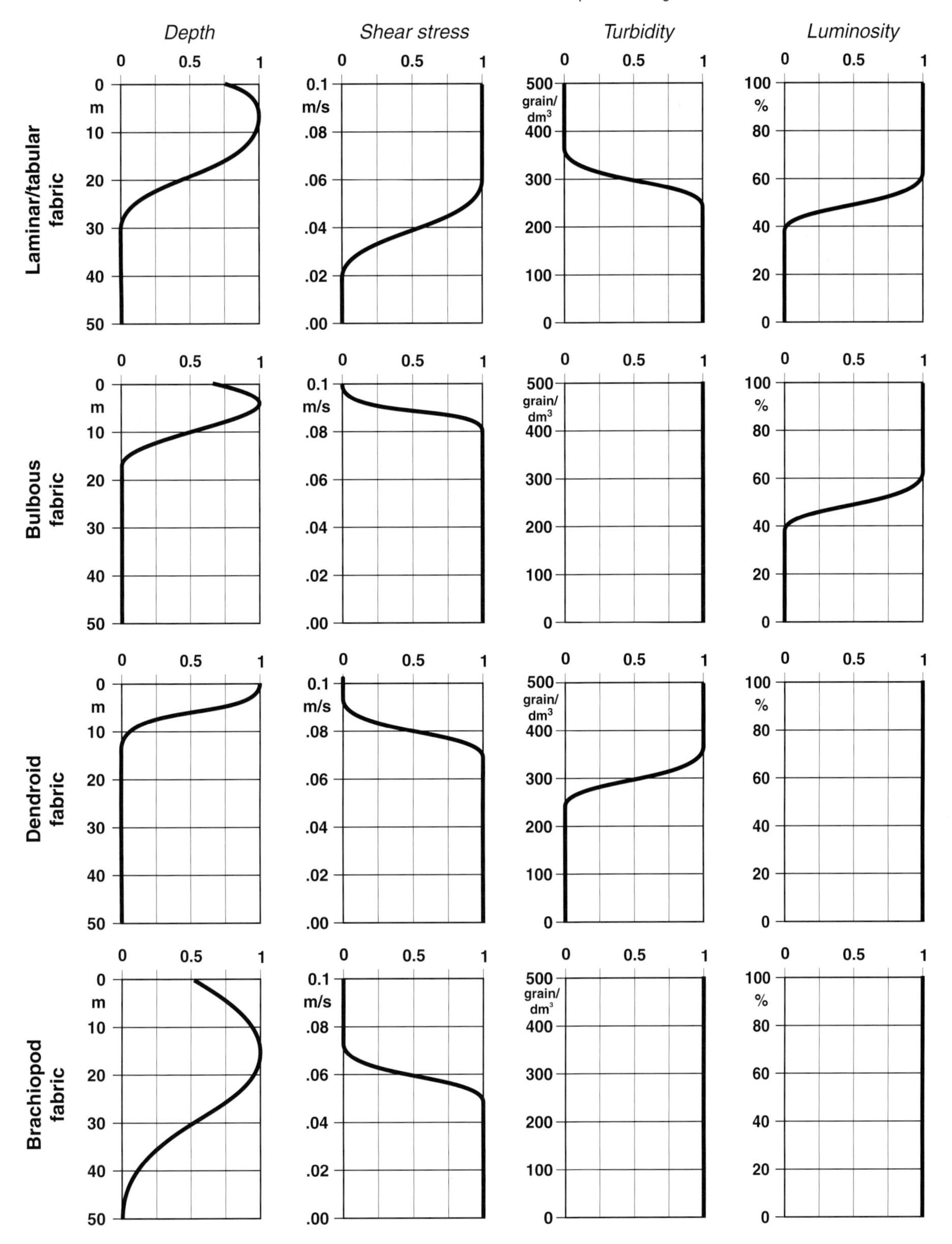

Fig. 9. Weighting functions D_f, S_f, L_f, and T_f for three types of reef fabric and one type of unconsolidated sediment (brachiopod-rich sediment) used in the simulation to calculate the production rate. D_f, S_f, L_f, and T_f are related to the ecological parameters depth, shear stress, luminosity on the sea floor, and turbidity.

vertically since it is not a reef-building organism and any relief created would be reworked by currents.

Erosion and sediment production

The four ecological parameters, which determine the production rate, vary over time in each cell because the topography of the sea floor constantly changes throughout the simulation. The water depth is updated every iteration and the three other parameters (shear stress on the sea floor, turbidity, and light incident on the sea floor) are recalculated every five iterations taking into account the newly developed topography. If the production rate of a particular fabric in a cell is unfavourably affected by one such variation (i.e., when the production rate falls to zero), the organisms corresponding to the fabric do not survive. The fabric turns into non-specified hard rock (null value), and the program changes the value of the corresponding cells accordingly. The hard rock then may be eroded, which in turn produces clastic sediment.

The program also allows the simulation of sediment production from a living reef fabric, albeit with a lower probability. In these instances, the probability of sediment production is calculated as being inversely proportional to the production rate of the fabric: a low fabric production rate is indicative of precarious development conditions and therefore of a higher probability of fabric alteration and sediment production. Diagenetic features such as early cementation, which may prevent erosion and reworking of carbonate sediment, are not taken into account in the simulation.

Sediment transport

The size and density of grains together with the shear stress on the sea floor were used to determine whether the sediment is able to be entrained in suspension and transported according to the direction and intensity of the current. Sediment transport also occurs if the dip of the slope at which the sediments rest becomes higher than the angle of repose. We used an angle of repose of 35° for both types of unconsolidated sediment because both types are non-cohesive and have a sand-sized grain size (Kenter, 1990; Adams *et al.*, 2002).

MODELLING RESULTS

After 315 iterations (seven cycles of 45 iterations, i.e., 157.5 kyr), the simulation reproduced seven accommodation cycles and the most significant observations made in the field. The first part of the simulation corresponds to Platform stage 1 (Top Platform stage 1, t_3 = 67.5 kyr in Fig. 10b). It shows neither direct contact nor any interfingering between the slope deposits and the margin. This is the result of the relatively shallow average water depth of the reef during the first 135 iterations (67.5 kyr), facilitating the development of stromatoporoids at the expense of brachiopods, that is, the brachiopods are transported and form the slope deposits, whereas the stromatoporoids build the margin. The three cycles form an aggradational platform unit. The margin is represented as being massive and well developed and is strongly dominated by laminar/tabular fabric. In the inner part of the platform, stage 1 shows an alternation of dendroid and bulbous-rich beds. In the model, dendroid fabric develops during times of low accommodation (due to a higher production rate at low depth), whereas bulbous fabric develops during highstands.

For Platform stage 2 (Fig. 10c and d), the simulation shows a strong increase in the production of brachiopods towards the top of the reef (t_7 = 157.5 kyr). This event causes the development of

Fig. 10. (Opposite page) Sections and 3D palaeo-topographies of the Glenister Knolls major patch reef, produced by the forward model, at the end of each major Platform stage. (a) Initial conditions of the modelling. Twenty-five iterations are conducted before the start of the true computer simulation (t_0) in order to populate the initial flat surface and to create synoptic relief between the reef-margin and backreef environment. (b) Platform stage 1 after 135 iterations (t_3 = 67.5 kyr). Three cycles form an aggradational platform unit. The margin is massive and dominated by laminar/tabular fabric. In the inner part of the platform, an alternation of dendroid- and bulbous-rich beds developed. (c) Platform stage 2 after 270 iterations (t_6 = 135 kyr) shows a strong increase in the abundance of brachiopods towards the top. The simulation shows the backstepping of the reef margin related to the rapid relative sea-level rise at t_3 and the collapse of the margin. (d) End of the simulation after 315 iterations (t_7 = 157.5 kyr). The production of brachiopods is still high. Brachiopod-rich slope deposits downlap onto reef-margin deposits of Platform stage 1. (e) The relative sea-level curve used during the simulation (see Fig. 8 for details).

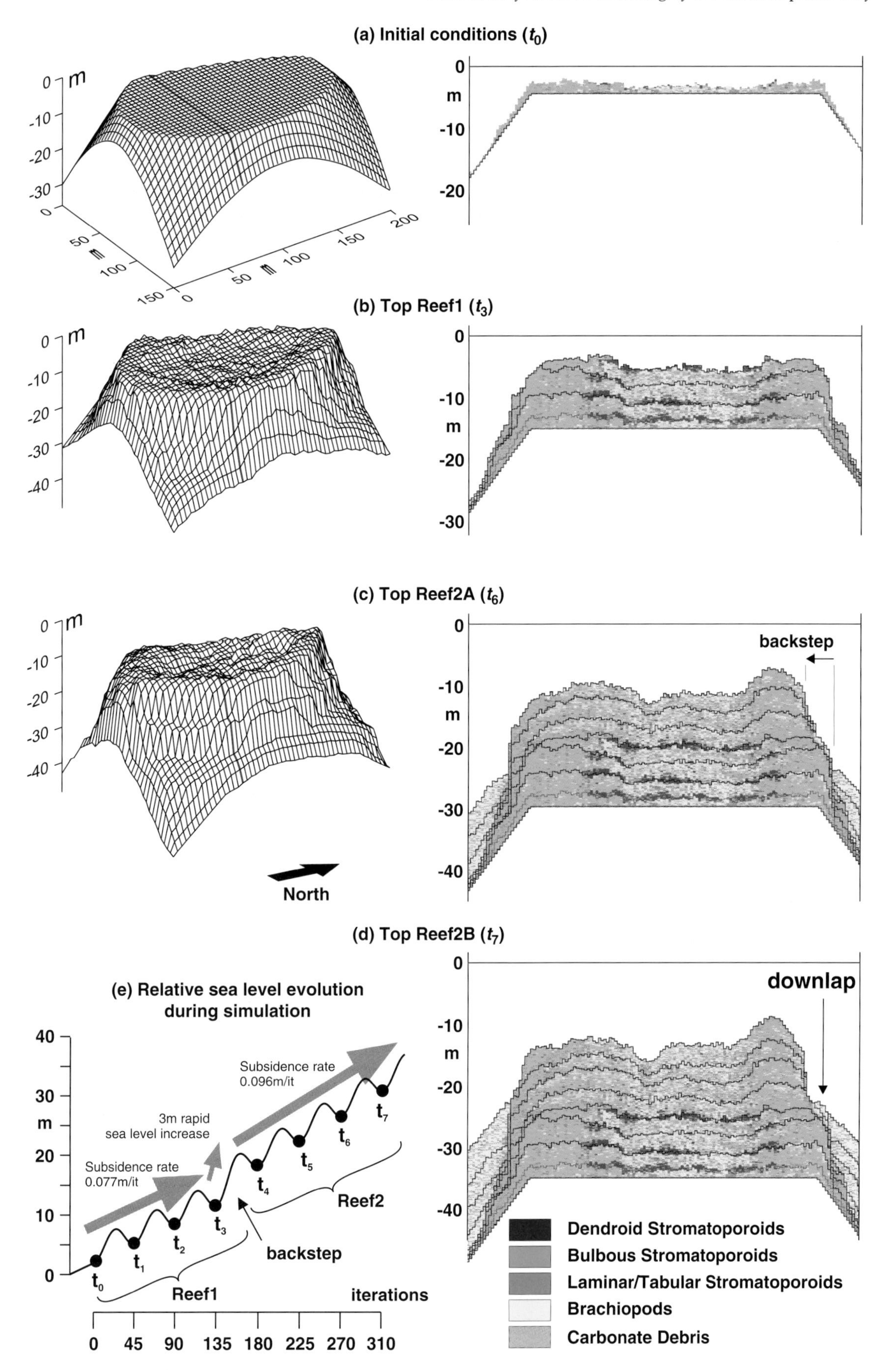
(a) Initial conditions (t_0)
(b) Top Reef1 (t_3)
(c) Top Reef2A (t_6)
(d) Top Reef2B (t_7)
backstep
downlap
North
(e) Relative sea level evolution during simulation
Subsidence rate 0.077m/it
3m rapid sea level increase
Subsidence rate 0.096m/it
backstep
Reef1
Reef2
iterations
0 45 90 135 180 225 270 310
Dendroid Stromatoporoids
Bulbous Stromatoporoids
Laminar/Tabular Stromatoporoids
Brachiopods
Carbonate Debris

thicker slope deposits as well as a more significant presence of the brachiopod fabric in the reef margin. Furthermore, the simulation shows the backstep of the reef margin related to the rapid relative sea-level rise in t_3 (67.5 kyr) and the collapse of the margin. The simulation also produced the downlap of slope deposits associated with Platform stage 2 on stage 1. The dendroid-rich beds disappear as a result of the weaker light on the sea floor due to the increased mean water depth.

MODEL COMPARISONS

To compare quantitatively the model predictions with outcrop observations, the 3D fine-scale forward model has been tilted and eroded in order to match the same topographic profile as obtained from the major knoll. Figure 11 shows the distribution of fabric predicted by the forward modelling along the present-day erosion surface in plan view. This map can be directly compared with the field data (Fig. 5e). Three comparisons were made. The first test compares the thickness of the cycles, the second the 3D prediction of the Platform stage boundaries, and the last one the prediction of depositional environments.

Cycle thickness

The main thickness values obtained with the 3D model are very close to the thicknesses observed in the field (see normal distribution functions in Fig. 12). On the major knoll, the thickness of the cycles can be observed only in the reef-margin environment. A belt corresponding to the reef margin has been extracted from the model and the thickness of each cycle has been measured vertically along every cell (526 values). Figure 12 shows the modelled thickness values and the normal function computed for the modelled distribution of thicknesses, compared with the thickness measured on the outcrop.

In some cases the predicted thickness values show a bimodal distribution (middle bed of stage 2A for instance). This distribution is due to the inheritance of the asymmetric margin, which is better developed in the south-eastern part of the patch reef (Fig. 10). This asymmetry probably developed due to the major current related to the Bugle Gap orientation and to the connection with the Fitzroy Trough and the basin in the South. Accommodation space is still the same in both the southern and northern part of the patch reef but under drowning conditions, due to fast subsidence, favourable conditions to produce carbonate are restricted to the windward part of the patch. The less productive leeward part is not able to keep up with the apparent rising sea level. The same observations have been made in other 3D forward stratigraphic modelling programs (Warrlich *et al.*, 2002).

Platform stage boundaries

Three stages of platform development have been observed and mapped in the Glenister Knolls major patch reef. The platform-stage boundary test compares two maps showing the spatial extent of platform stages. In contrast with the previous test where values without specific position were compared, here we test the accuracy of the model to predict in 3D the extent of platform stages. Figure 13 presents the result of the image analysis between the predicted extent of platform stages (Fig. 13a) and the real extent of these stages measured in the field (Fig. 13b). In each cell, the prediction is accurate if the stages in both images are identical (green cells in Fig. 13c) and inaccurate if the prediction is different from the platform stage observed in the field (red cells in Fig. 13c). In this case, the accuracy reaches 88.5% (6689/7558 pixels).

Depositional environments

The depositional-environment test is based on the comparison between two images. This comparison cannot be made directly since the forward modelling does not predict depositional environments but rather the distribution of reef fabric (Figs 11 and 14a). Therefore, it is important to transform the facies distribution map predicted by the forward model into a depositional environment map. This is easily accomplished since we are only interested in the core of the reef, composed only of reef-margin and backreef environments. The following rule has been used to determine the extent of the backreef environment: a cell belongs to the backreef if the majority (more than four out of nine) of the cell plus its eight neighbouring cells shows one of the backreef-characterizing fabrics (dendroid and bulbous fabrics). If this is not the case, the cell belongs to the reef margin (Fig. 14b). This map can be compared directly to the one based on the outcrop mapping (Fig. 14c). In this case, the accuracy reaches 89.9% (7656/8513 pixels).

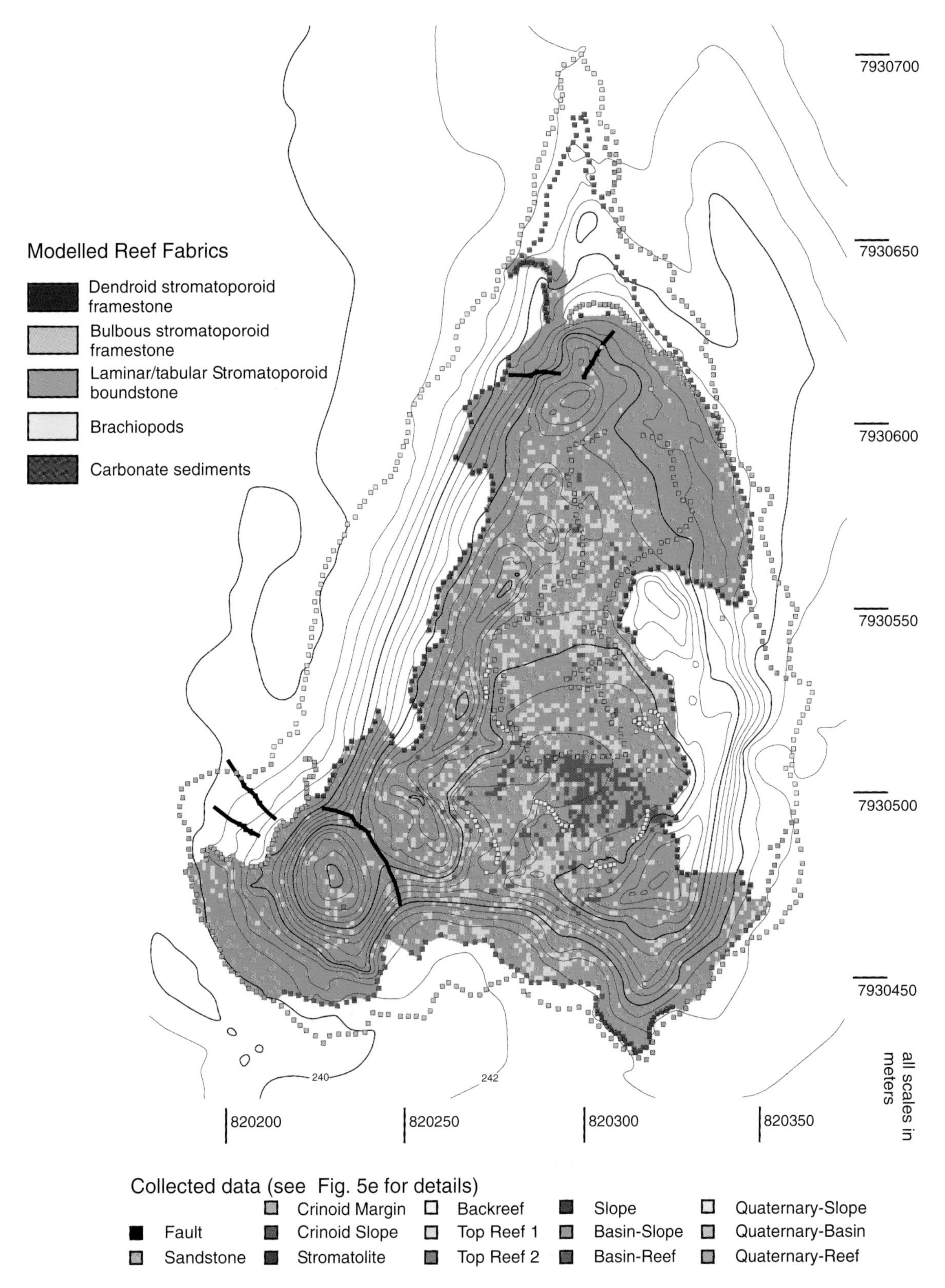

Fig. 11. Map view of 3D fine-scale forward model of the major knoll. The model was tilted and eroded to match the topographic profile of one of the outcrop analogues. The distribution of the reef fabric is shown only in the inner part, which corresponds to the reef-margin and backreef environments. The slope environment is not plotted here.

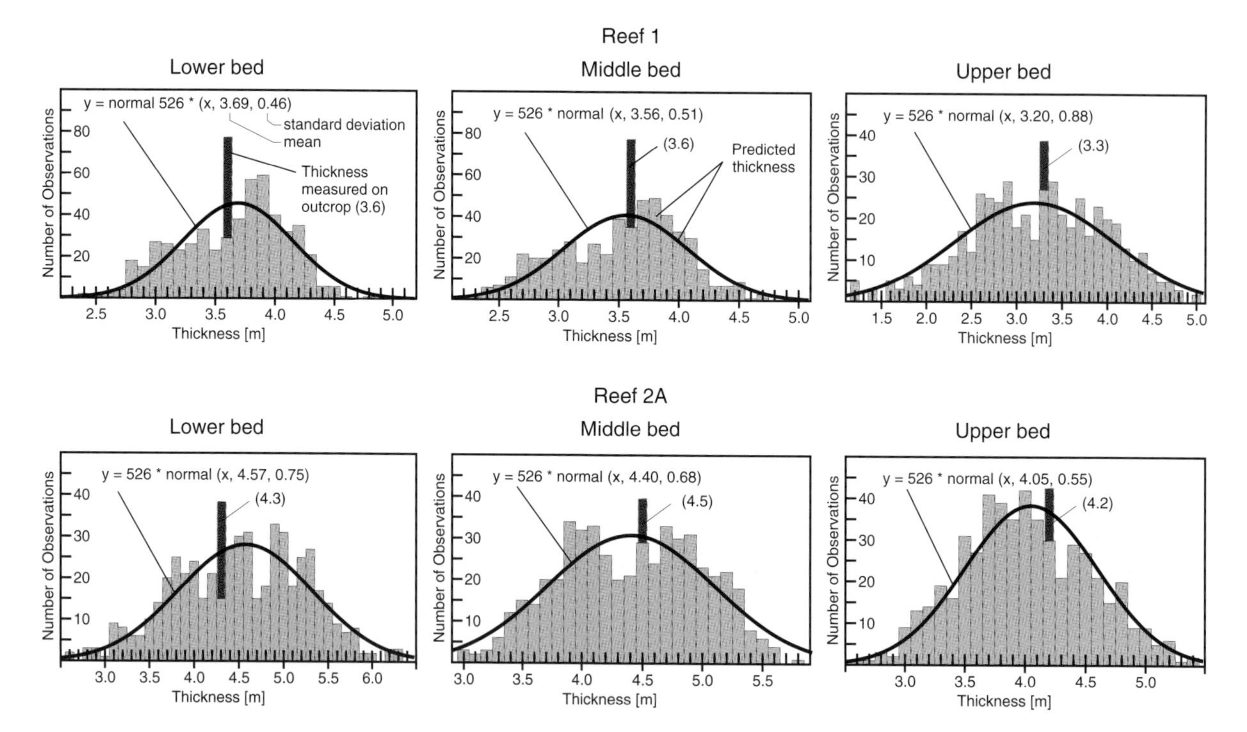

Fig. 12. Bed thickness of the reef-margin environment predicted by the forward modelling compared with field observations for Platform Stages 1 and 2A. Platform stage 2B is not shown because of its incompleteness. For each bed, corresponding to one cycle, the predicted thickness values (histograms) and the normal function related to the distribution of these values (black curves) are plotted and compared with the thickness measured at the outcrop (single dark vertical lines). The equation of all normal functions is also given and includes the mean and standard deviation of the predicted thickness values.

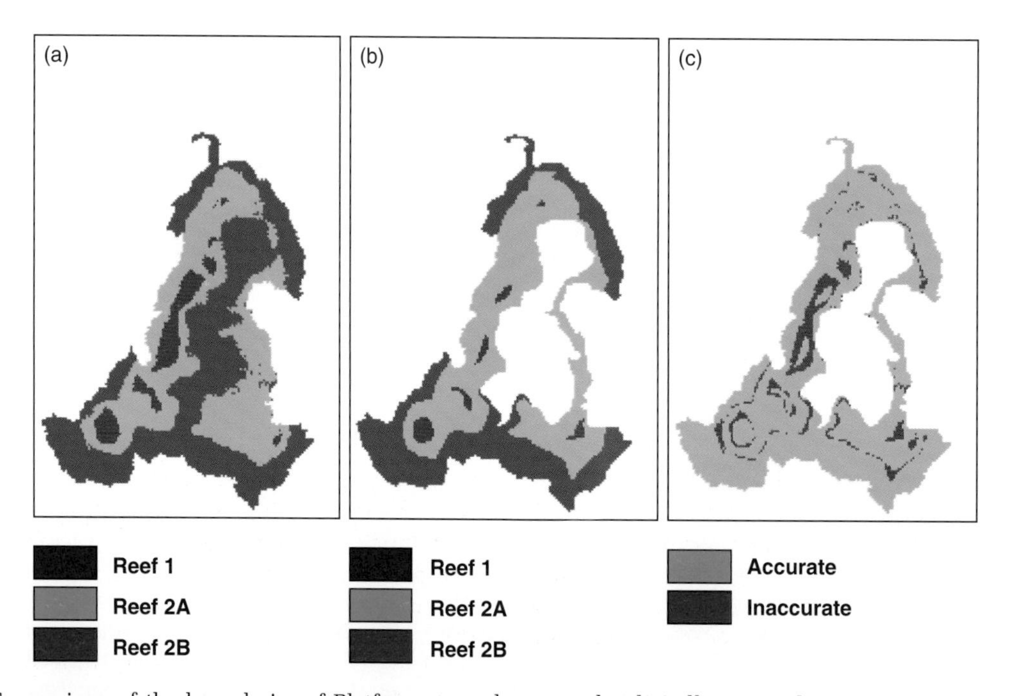

Fig. 13. Comparison of the boundaries of Platform stages between the digitally mapped outcrop model and fine-scale forward model. (a) Platform stage map obtained from the forward model (restored to the surface corresponding with the present-day topography of Glenister Knolls). (b) Platform stage map based on digitally mapped outcrop model. (c) Platform stage comparison map evaluating the accuracy of fit between the two models. The prediction is accurate if the stages in both images are identical (green cells) and inaccurate if the stages differ (red cells). In this case, the accuracy of fit reaches 88.5% (6689/7558 pixels).

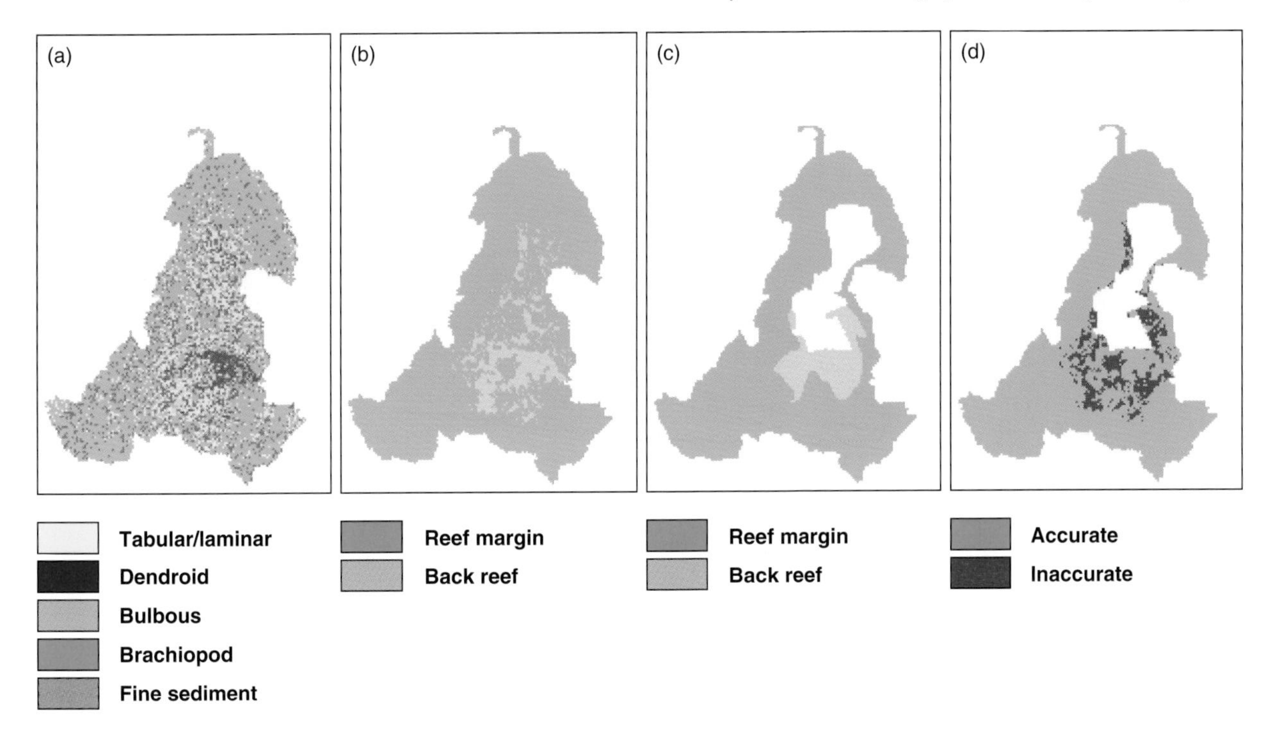

Fig. 14. Comparison of the reef-margin and backreef depositional environments between the digitally mapped outcrop model and fine-scale forward model. (a) Reef-fabric map obtained from the forward model (restored to the surface corresponding with the present-day topography). See Figs 5e and 11 for details. (b) Map showing depositional environments obtained from the forward model (restored to the surface corresponding with the present-day topography). This map is based on the reef-fabric map shown in (a) (see text for details). (c) Map of depositional environments based on the digital mapping. (d) Comparison map evaluating the accuracy of fit between the two models. The prediction is accurate if the depositional environments are identical (green cells) and inaccurate if they differ (red cells). In this case, the accuracy reaches 89.9% (7656/8513 pixels).

SENSITIVITY ANALYSIS

During the progress of this work, we have performed an analysis in order to examine how the predictions made with this model are affected by changes in selected parameters. Two sensitivity tests are presented here: the first one is based on the thickness evolution of the cycles and is used to describe the structure of the reef, the second one is based on the reef fabric and sediment distribution within the reef stages and is used to describe the internal decimetre-scale structure of the reef. The tests have been performed on the sensitivity related to the following parameters: the subsidence rate, the eustatic conditions (amplitude of the cycles), the production rates, and the hydrodynamic parameters (current intensity and wave amplitude).

Figure 15a plots the relative curves showing the evolution of the thickness of the predicted cycles compared to the thickness observed in the field according to individual parameter variations. In order to fit all curves in the same graphs, we use the relative variation of parameters related to those proposed in Table 1. A strong sensitivity to a specific parameter is recorded with a steep curve. Not surprisingly, the external structure of the reef measured here on the thickness of the cycles composing the different Platform stages is very sensitive to the subsidence and the production rates, since these directly influence the accommodation space that constrains the vertical growth of the atoll. The flattening of the curve related to the subsidence rate changes is due to a drowning of the patch reef. If one passes over a critical rate of subsidence, carbonate production is not important enough to fill the accommodation space. The fact that the thickness of the cycles seems not to be influenced by the amplitude chosen for the eustatic cycle is probably related to the low value chosen in this simulation, compared to the subsidence rate. On the other hand, the internal structure predicted using this model does not seem to be more affected by one specific parameter (Fig. 15b). In this case, the model is equally sensitive to hydrodynamic

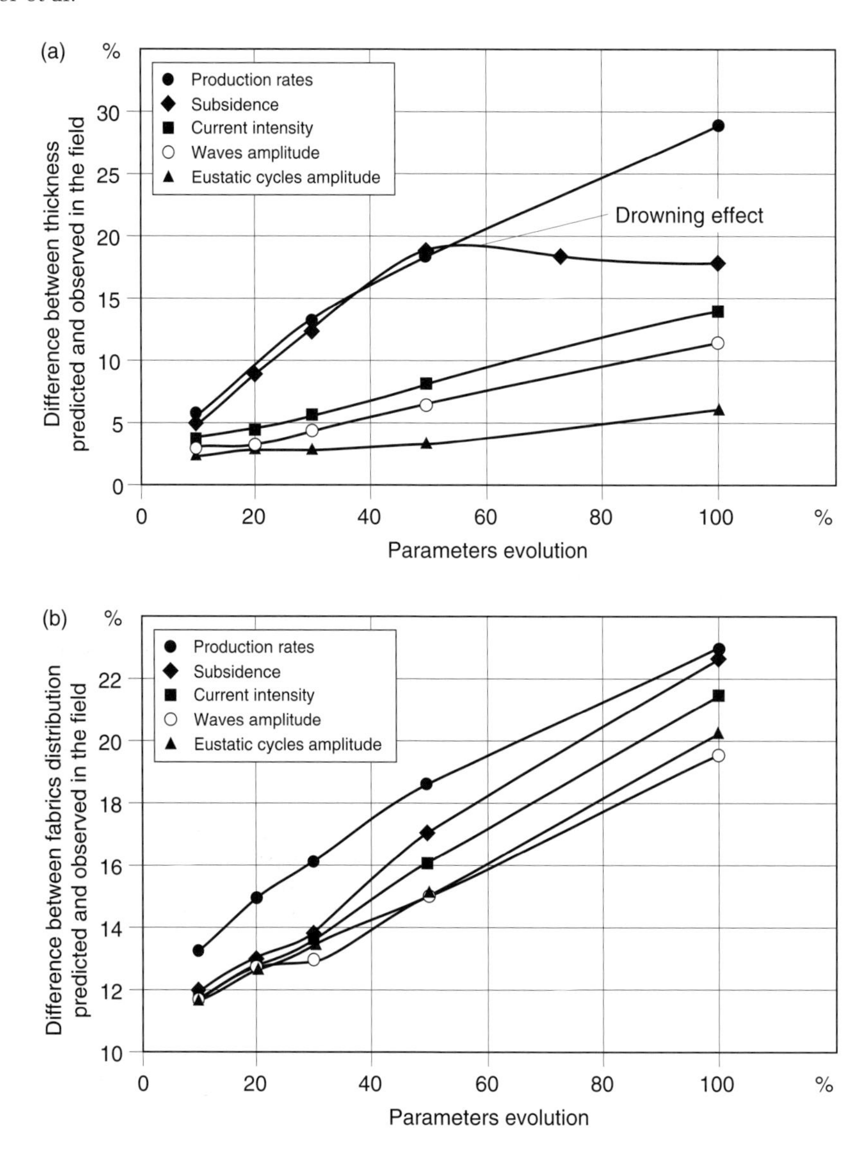

Fig. 15. Evolution of the differences between prediction and outcrop observations based on the thickness of the cycles (a) and on the fabric distribution within the cycles (b), in response to variations of individual parameters. Note that these values would change if different original parameters (see Table 1) were used.

conditions, such as current intensity and wave amplitude, as to subsidence or production rates.

DISCUSSION AND CONCLUSIONS

The excellent quality of exposure of the Devonian outcrops facilitated the choice and helped in defining the magnitude and range of parameters used in the fine decimetre-scale numerical forward model of carbonate reef growth based on the inferred predominant environmental conditions: water depth, energy, turbidity, and light levels on the sea floor. These factors allowed the prediction of the lateral distribution of the major reef fabrics as well as unconsolidated sediments. The vertical evolution of the reefs is constrained by eustatic and subsidence evolution, as well as by the ability of the reef-building organisms to grow, and sediments to be produced and transported.

The fine-scale forward model presented in this paper, mainly based on the simulation of stromatoporoid growth, simulates the development of the major knoll patch reef rather well. The large-scale depositional relationships observed in the field such as backstepping, downlapping, and interfingering are successfully reproduced by the model. The measured thicknesses of platform cycles and the modelled thicknesses are also similar. More than the ability to predict cycle thickness or the three-dimensional extent of stratigraphic layers or reef bodies, it is the scale and the make-up of the modelled cells that are important: the model predicts the distribution of the basic elements that compose the reef. These fundamental elements are characterized by reef-builder species or sediments which posses specific fabrics or textures. Focus at such a fine scale offers an alternative to all carbonate forward models which work at the basin, platform or ramp scale (Aurell *et al.*, 1998; Granjeon & Joseph, 1999; Leyrer *et al.*, 1999; Warrlich *et al.*, 2002; Burgess & Wright, 2003), and to those who use geostatistical rules to work on the individual rock types (Doligez *et al.*, 1999). This fine decimetre-scale may also offer further analytical perspectives: for instance, petrophysical properties such as porosity and permeability could be attributed to each facies in order to simulate, in close detail and in three dimensions, the interaction of fluids with carbonate rocks involved in diagenetic processes (Whitaker *et al.*, 1999) such as early cementation, which is a fundamental process in the deposition of the Devonian reefs in the Canning Basin.

In our example, we used high-frequency eustatic cycles coupled with an almost constant subsidence rate. Even if the predictions fit well with the outcrop data, we cannot favour this hypothesis against a tectonic origin of shallowing-upward cycles. This point is related to one of the major concerns in forward or inverse modelling: the non-uniqueness of predictions due the multiple parameters involved in carbonate systems. Digital field mapping is an ideal tool to obtain precise field data that subsequently can be used to infer modelling parameters. In our case, the hydrodynamic setting and evolution of the accommodation space were particularly problematic. However, with simple and basic settings we were able to reproduce the overall development of the major knoll patch reef.

The comparison between the model predicted by the computing simulation and the field observations is difficult, especially when the goal of the simulation is to work at a very fine scale. However, the high-resolution digitally obtained outcrop data were used to make realistic comparisons at a fine scale between the predictions and outcrop analogues. Both qualitative and quantitative comparisons were made. The quantitative comparisons include cycle thickness, platform-stage boundaries, and depositional environments. In particular, trends and similarity of cycle thicknesses strongly agree. The accuracy of fit of the boundaries of both platform stages and depositional environments is almost 90% between the predicted model and the digitally obtained outcrop model.

The sensitivity tests show that for the forward model, the parameters which have a direct impact on accommodation space such as subsidence and production rate are the most important as far as the prediction of the external structure of the reef is concerned. However, in order to predict the internal facies distribution within reef bodies, other parameters such as hydrodynamic conditions (direction of currents, wave amplitude) are equally important. Unfortunately, if one wants to pass from a model that predicts the stacking of reef bodies to a model focusing on facies distribution within those bodies in order to predict internal heterogeneities, more parameters, other than subsidence, eustasy and production rate, need to be taken into account, thus increasing the problems related to the non-uniqueness of the results.

ACKNOWLEDGEMENTS

We would like to express our gratitude to the Mount Pierre Station and especially to Louis and Marion Dolby for giving us the opportunity to work on their land. We thank Arndt Peterhänsel and Jon Tarasewicz for their help and discussions in the field and Ken McNamara for logistical support. We would like to acknowledge Annette George and Phillip Playford for fruitful discussions and input. We are grateful to Schlumberger Doll Research for supplying surveying equipments. This work is funded by CMI (Cambridge-MIT Institute) Project 039/P-IR Ft. David McCormick is acknowledged for his constructive review of an earlier version of the manuscript. We would like to thank Phillip Playford, Frans van Buchem and Johannes Wendebourg for their highly appreciated reviews.

REFERENCES

Adams, E.W., Morsilli, M., Schlager, W., Keim, L. and **van Hoek, T.** (2002) Quantifying the geometry and sediment fabric of linear slopes: Examples from the tertiary of Italy

(Southern Alps and Gargano Promontory). *Sed. Geol.*, **154**, 11–30.

Adams, E.W., **Schröder, S.**, **Grotzinger, J.P.** and **McCormick, D.S.** (2004) Digital reconstruction and stratigraphic evolution of a microbial-dominated, isolated carbonate platform (terminal Proterozoic, Nama Group, Namibia). *J. Sed. Res.*, **74**, 479–497.

Adams, E.W., **Grotzinger, J.P.**, **Watters W.A.**, **Schröder, S.**, **McCormick D.S.** and **Al-Siyabi, H.A.** (2005) Digital characterization of thrombolite–stromatolite reef distribution in a carbonate ramp system (terminal Proterozoic, Nama Group, Namibia). *AAPG Bull.*, **89**, 1293–1318.

Aigner, T., **Doyle, M.** and **Lawrence, D.T.** (1989) Quantative modelling of carbonate platforms: Some examples. In: *Controls on Carbonate Platform and Basin Development* (Ed. J. Read), Society for Sedimentary Geology (SEPM), Tulsa, OK. *SEPM Spec. Publ.*, **44**, 27–37.

Aurell, M. Badenas, B., **Bosence, D.W.J.** and **Waltham, D.A.** (1998) Carbonate production and offshore transport on a Late Jurassic carbonate ramp (Kimmeridgian, Iberian Basin, NE Spain) – evidence from outcrops and computer modelling. In: *Carbonate Ramps* (Eds V.P. Wright and T.P. Burchette), Geological Society of London, London. *Geol. Soc. London Spec. Publ.*, **137**, 137–161.

Becker, R.T., **House, M.R.** and **Kirchgasser, W.T.** (1993) Devonian goniatite biostratigraphy and timing of facies movements in the Frasnian of the Canning Basin, Western Australia. In: *High Resolution Stratigraphy* (Eds E.A. Hailwood and R.B. Kidd), Geological Society of London, London. *Geol. Soc. London Spec. Publ.*, **70**, 293–321.

Begg, J. (1987) Structuring and controls on Devonian reef development on the north-west Barbwire and adjacent terraces, Canning Basin. *The APEA J.*, **27**, 137–151.

Bellian, J.A., **Kerans, C.** and **Jennette, D.C.** (2005) Digital outcrop models: Application of terrestrial scanning lidar technology in stratigraphic modeling. *J. Sed. Res.*, **75**, 166–176.

Blanchon, P. and **Blackeway, D.** (2003) Are catch-up reefs an artifact of coring? *Sedimentology*, **50**, 1271–1282.

Bosence, D. and **Waltham, D.** (1990) Computer modeling the internal architecture of carbonate platforms. *Geology*, **18**, 26–30.

Bosscher, H. and **Schlager, W.** (1992) Computer simulation of reef growth. *Sedimentology*, **39**, 503–512.

Bosscher, H. and **Schlager, W.** (1993) Accumulation rates of carbonate platforms. *J. Geol.*, **101**, 345–355.

Boylan, A.L., **Waltham, D.A.**, **Bosence, D.W.J.**, **Badenas, B.** and **Aurell, M.** (2002) Digital rocks: Linking forward modelling to carbonate facies. *Basin Res.*, **14**, 401–415.

Burgess, P.M. and **Wright, V.P.** (2003) Numerical forward modelling of carbonate platform dynamics: An evaluation of complexity and completeness in carbonate strata. *J. Sed. Res.*, **73**, 637–652.

Chopard, B. and **Droz, M.** (1998) *Cellular Automata Modeling of Physical Systems.* Cambridge University Press, Cambridge, 353 pp.

Coastal Engineering Research Center (1984) *Shore Protection Manual.* US Government Printing Office, Washington, DC.

Cockbain, A.E. (1984) Stromatoporoids from the Devonian reef complexes, Canning Basin, Western Australia. *Geol. Surv. W. Aust. Bull.*, **129**, 108.

Copp, I.A. (2000) Subsurface facies analysis of Devonian reef complexes, Lennard Shelf, Canning Basin, Western Australia. *W. Aust. Geol. Sur. Rep.*, **58**, 127.

Doligez, B., **Granjeon, D.**, **Joseph, P.**, **Eschard, R.** and **Beucher, H.** (1999) How can stratigraphic modelling help to constrain geostatistical reservoir simulations. In: *Numerical Experiments in Stratigraphy* (Eds J.W. Harbaugh, W.L. Watney, E. Rankey, R. Slingerland, R. Goldstein and E. Franseen), Society for Sedimentary Geology (SEPM), Tulsa, OK. *SEPM Spec. Publ.*, **62**, 239–244.

Druce, E.C. (1976) Conodont biostratigraphy of the Upper Devonian reef complexes of the Canning Basin, Western Australia. *Bur. Mineral Resour. Australia, Bull.*, **158**, 303.

Drummond, B.J., **Sexton, M.J.**, **Barton, T.J.** and **Shaw, R.D.** (1988) The nature of faulting along the margins of the Fitzroy Trough, Canning Basin, and implications for the tectonic development of the trough. *Explor. Geophys.*, **22**, 111–115.

Drummond, C.N. and **Dugan, P.J.** (1999) Self-organizing models of shallow-water carbonate accumulation. *J. Sed. Res.*, **69**, 939–946.

Eberli, G.P., **Kendal, C.G.ST.C.**, **Moore, P.**, **Whittle, G.L.** and **Cannon, R.** (1994) Testing a seismic interpretation of Great Bahama Bank with a computer simulation. *AAPG Bull.*, **78**, 981–1004.

Forman, D.J. and **Wales, D.W.** (1981) Geological evolution of the Canning Basin, Western Australia. *Bur. Mineral Resour. Bull.*, **210**, 91.

Gaylord, R.J. and **Nishidate, K.** (1996) *Modeling Nature – Cellular Automata Simulations with Mathematica.* Telos, Santa Clara, California, 273 pp.

George, A.D. (1999) Deep-water stromatolites, Canning Basin, Northwestern Australia. *Palaios*, **14**, 493–505.

George, A.D. and **Chow, N.** (2002) The depositional record of the Frasnian/Famennian boundary interval in a fore-reef succession, Canning Basin, Western Australia. *Palaeogeogr. Palaeoclimatol. Palaeoecol.*, **181**, 347–374.

George, A.D., **Chow, N.** and **Trinajstic, K.M.** (2002) Integrated approach to platform-basin correlation and deciphering the evolution of Devonian reefs, northern Canning Basin, Western Australia. In: *The Sedimentary Basins of Western Australia 3 Symposium*, Perth, pp. 817–835.

Glenister, B.F. (1958) Upper Devonian ammonoids from the Manticoceras Zone, Fitzroy Basin, Western Australia. *J. Paleontol.*, **32**, 58–96.

Glenister, B.F. and **Klapper, G.** (1966) Upper Devonian conodonts from the Caning Basin, Western Australia. *J. Paleontol.*, **40**, 777–842.

Granjeon, D. and **Joseph, P.** (1999) Concepts and applications of a 3D multiple lithology, diffusive model in stratigraphic modelling. In: Numerical Experiments in stratigraphy (Eds J.W. Harbaugh, W.L. Watney, E. Rankey, R. Slingerland, R. Goldstein and E. Franseen), Society for Sedimentary Geology (SEPM), Tulsa, OK. *SEPM Spec. Publ.*, **62**, 197–210.

Grey, K. (1978) Devonian atrypid brachiopods from the reef complexes of the Canning Basin. *Geol. Sur. W. Aust. Rep.*, **5**, 70 pp.

Harbaugh, J.W. and **Bonham-Carter, G.** (1970) *Computer Simulation in Geology*. Wiley Interscience, New York, 575 pp.

Hardie, L.A., Dunn, P.A. and **Goldhammer, R.K.** (1991) Field and modelling studies of Cambrian carbonate cycles, Virginia Appalachians – discussion. *J. Sed. Petrol.*, **61**, 636–646.

Hocking, R.M. and **Playford, P.E.** (2003) Cycle types in carbonate platform facies, Devonian reef complexes, Canning Basin, Western Australia. *Geol. Surv. W. Aust. Annu. Rev.*, **2001–02**, 74–80.

Isaaks, E.H. and **Srivastava, R.M.** (1990) *An Introduction to Applied Geostatistics*. Oxford University Press, New York, 592 pp.

Kennard, J.M., Southgate, P.N., Jackson, M.J., O'Brien, P.E., Christie-Blick, N., Holmes, A.E. and **Sarg, J.F.** (1992) New sequence perspective on the Devonian reef complex and the Frasnian-Famennian boundary, Canning Basin, Australia. *Palaios*, **1**, 492–503.

Kenter, J.A.M. (1990) Carbonate platform flanks: slope angle and sediment fabric. *Sedimentology*, **72**, 777–794.

Kershaw, S. (1998) The applications of stromatoporoid palaeobiology in palaeoenvironmental analysis. *Palaeontology*, **41**, 509–544.

Le Roux, J.P. (1997) An excel program for computing the dynamic properties of particles in Newtonian fluids. *Comput. Geosci.*, **23**, 671–675.

Leyrer, K., Strohmenger, C., Rockenbauch, K. and **Bechstaedt, T.** (1999) High-resolution forward stratigraphic modeling of Ca_2-carbonate platforms and off-platform highs (Upper Permian, northern Germany). *J. Computerized Modeling of Sedimentary Systems* (Eds Harff, W. Lemke, and K. Stattegger), Springer, Berlin, 307–339.

Long, J.A. (1988) Late Devonian fishes from Gogo, Western Australia. *Natl. Geogr. Res.*, **4**, 436–450.

Long, J.A. (1995) A new plourdosteid arthrodire from the Upper Devonian Gogo formation of Western Australia. *Palaeontology*, **38**, 39–62.

Madsen, O.S. (1994) Spectral wave-current bottom boundary layer flows. *Proceedings of the 24th ICCE/ASCE*, **1**, 384–398.

McCaffrey, K.J.W., Jones, R.R., Holdsworth, R.E., Wilson, R.W., Clegg, P., Imber, J., Holliman, N. and **Trinks, I.** (2005) Unlocking the spatial dimension: Digital technologies and the future of geoscience fieldwork. *J. Geol. Soc. London*, **162**, 927–938.

Meyer, F.O. (1981) Stromatoporoid growth rhythms and rates. *Science*, **213**, 894–895.

Paola, C. (2000) Quantitative models of sedimentary basin filling. *Sedimentology*, **47**, (Suppl. 1), 121–178.

Playford, P.E. (1980) Devonian "Great Barrier Reef" of Canning Basin, Western Australia. *AAPG Bull.*, **64**, 814–840.

Playford, P.E. (2001) The Permo-Carboniferous glaciation of Gondwana: its legacy in Western Australia. *Geol. Surv. W. Aust. Rec.*, **201**, 15–16.

Playford, P.E. (2002) Palaeokarst, pseudokarst, and sequence stratigraphy in Devonian reef complexes of the Canning Basin, Western Australia. In: *The Sedimentary Basins of Western Australia 3 Symposium* (Eds M. Keep and S.J. Moss), pp. 763–793, Perth.

Playford, P.E. and **Hocking, R.M.** (1998) 1:50,000 Geological map of Devonian reef complexes of the Canning Basin. *Geol. Surv. Western Aust. Bull.*, 145, Plate 5: Bugle Gap area.

Playford, P.E. and **Lowry, D.C.** (1966) Devonian reef complexes of the Canning basin, Western Australia. *Geol. Surv. W. Aust. Bull.*, **118**, 150 pp.

Playford, P.E., Cockbain, A.E., Druce, E.C. and **Wray, J.L.** (1976) Devonian stromatolites from the Canning Basin, Western Australia. In: *Stromatolites* (Ed. M.R. Walter), *Dev. Sedimentol.*, **20**, 543–563.

Playford, P.E., Hurley, N.F., Kerans, C. and **Middleton, M.F.** (1989) Reefal platform development, Devonian of the Canning Basin, Western Australia. In: *Controls on Carbonate Platform and Basin Development* (Eds P.D. Crevello, J.L. Wilson, J.F. Sarg and J.F. Read), *SEPM Spec. Publ.*, **44**, 187–202.

Read, J.F. (1973) Carbonate cycles, Pillara Formation (Devonian), Canning Basin, Western Australia. *Bull. Can. Petrol. Geol.*, **21**, 38–51.

Read, J.F. (1982) Carbonate platforms of passive (extensinal) continental margins: types, characteristics and evolution. *Tectonophysics*, **181**, 95–212.

Read, J.F. (1995). Overview of carbonate platform sequences, cycle stratigraphy and reservoirs on carbonate platforms in greenhouse and icehouse worlds. In: *Milankovich Sea-Level Changes, Cycles, and Reservoirs on Carbonate Platforms in Greenhouse and Icehouse* (Eds J.F. Read, C. Kerans, L.J. Weber, J.F. Sarg and F.M. Wright), Society of Economic Paleontologists and Mineralogists, Tulsa, OK. *SEPM Short Course Notes*, **35**, 1–102.

Read, J.F., Koershner, W.F., Oseger, D.A., Bollinger, G.A. and **Coruh, C.** (1991) Field and modelling studies of Cambrian carbonate cycles, Virginia Appalachians – reply. *J. Sed. Petrol.*, **61**, 647–652.

Seddon, G. (1970) Frasnian conodonts from Sadler Ridge-Bugle Gap area, Caning Basin, Western Australia. *J. Geol. Soc. Aust.*, **16**, 723–742.

Silvertown, J., Holtier, S., Johnson, J. and **Dale, P.** (1992) Cellular automaton models of interspecific competition for space – the effect of pattern on process. *J. Ecol.*, **80**, 527–534.

Stephens, N.P. and **Sumner, D.Y.** (2003) Famennian microbial reef facies, Napier and Oscar Ranges, Canning Basin, western Australia. *Sedimentology*, **50**, 1283–1302.

Tipper, J.C. (1997) Modelling carbonate platform sedimentation – lag comes naturally. *Geology*, **25**, 495–498.

Tipper, J.C. (2000) Patterns of stratigraphic cyclicity. *J. Sed. Res.*, **70**, 1262–1279.

Verwer, K., Kenter, J.A.M., Maathuis, B. and **Della Porta, G.** (2004) Stratal patterns and lithofacies of an intact seismic-scale Carboniferous carbonate platform (Asturias, northwestern Spain): A virtual outcrop model. In: *Geological Prior Information: Informing Science and Engineering* (Eds A. Curtis and R. Wood), Geological Society of London, London. *Geol. Soc. London Spec. Publ.*, **239**, 29–41.

Ward, W.B. (1999) Tectonic control on backstepping sequences revealed by mapping of Frasnian backstepped platfroms, Devonian reef complexes, Napier Range, Canning Basin, Western Australia. In: *Advances in Carbonate Sequence Stratigraphy: Application to Reservoirs, Outcrops, and Models* (Eds P.M. Harris, A.H. Saller and J.A. Simo), Society for Sedimentary

Geology (SEPM), Tulsa, OK. *SEPM Spec. Publ.*, **63,** 47–74.

Warrlich, G.M.D., **Waltham, D.A.** and **Bosence, D.W.J.** (2002) Quantifying the sequence stratigraphy and drowning mechanisms of atolls using a new 3-D forward stratigraphic modelling program (CARBONATE 3D). *Basin Res.*, **14,** 379–400.

White, R. and **Engelen, G.** (1993) Cellular automata and fractal urban form: A cellular modelling approach to the evolution of urban land-use patterns. *Environ. Planning A*, **25,** 1175–1199.

Whitaker, F., **Haque, Y. Smart, P. Waltham, D.** and **Boscence, D.** (1999) Structure and fonction of a coupled two-dimensional diagenetic and sedimentological model of carbonate platform evolution. In: *Numerical Experiments in Stratigraphy* (Eds J.W. Harbaugh, W.L. Watney, E. Rankey, R. Slingerland, R. Goldstein and E. Franseen), Society for Sedimentary Geology (SEPM), Tulsa, OK. *SEPM Spec. Publ.*, **62,** 339–356.

Wolfram, S. (2002) *A New Kind of Science*. Wolfram Media, Champaign, Illinois, 1211 pp.

Wood, R. (1999) *Reef Evolution*. Oxford University Press, Oxford, 414 pp.

Wood, R. (2000a) Novel paleoecology of a postextinction reef: Famennian (Late Devonian) of the Canning Basin, northwestern Australia. *Geology*, **28,** 987–990.

Wood, R. (2000b) Paleoecology of a Late Devonian back reef: Canning Basin, Western Australia. *Palaeontology*, **43,** 671–703.

Spec. Publ. Int. Assoc. Sedimentol. (2008) **40**, 65–96

Structural, reverse-basin and forward stratigraphic modelling of the Southern Cantabrian Basin, northwest Spain

ZBYNEK VESELOVSKÝ*, THILO BECHSTÄDT† and RAINER ZÜHLKE†

**Eriksfiord AS, ipark, Postboks 8034, 4068, Stavanger, Norway (E-mail: zv@eriksfiord.com)*

†Department of Earth Sciences, University of Heidelberg, Im Neuenheimer Feld 234, 69120 Heidelberg, Germany (E-mail: zuehlke@uni-hd.de)

ABSTRACT

The evolution of the Palaeozoic Southern Cantabrian Basin is quantified by a multidisciplinary approach of structural balancing, reverse-basin and forward stratigraphic modelling. These methods were applied to the N–S trending, 54 km long Bernesga Transect in the Cantabrian Mountains. Total tectonic shortening of the Southern Cantabrian Basin amounted to a minimum of 54%. These data were derived from two-dimensional structural balancing of the deformed basin infill. Two-dimensional reverse-basin modelling was used for analysing the evolution of the basin architecture, considering lithofacies, incremental compaction, eustatic sea-level changes and flexural loading of the crust. The model comprises the entire basin infill between the top of the Neoproterozoic basement and the time of maximal burial (Cenozoic). Six major subsidence trends, with time spans between 9 and 65 Myr, subdivide the long-term evolution of accommodation space in time. These trends reflect different lithospheric configurations prior to, during and after the Variscan Orogeny, having been primarily triggered by major changes in thermo-tectonic and flexure-induced subsidence, as well as by regional and extraregional tectonics. Basin modelling permits the identification of two encroachment subcycles during the Silurian and Devonian, as well as the approach of the Variscan orogenic front in the Early Carboniferous. The latter is reflected by migrating, strongly subsiding depocentres (orogenic foredeep), controlled by increased flexural-induced and thermo-tectonic subsidence. Two-dimensional stratigraphic forward modelling was used to simulate the evolution of the Southern Cantabrian Basin, and to quantify internal and external parameters governing deposition (e.g. sediment transport, sedimentation rates, *in-situ* carbonate production, erosion and compaction). The model includes data from the Neoproterozoic to the Lower Carboniferous with a focus on the Devonian. Carbonate factories in the Devonian were controlled by fluctuating siliciclastic input and differential thermo-tectonic subsidence, and subordinately influenced by eustatic sea-level changes. Decompacted carbonate production rates reach up to 780 m Myr^{-1}. The changes in carbonate production rates mirror the drift of Iberia from cooler to subtropical/tropical conditions during Devonian times.

Keywords Basin development, numerical modelling, Cantabrian Basin, Palaeozoic, Spain.

INTRODUCTION

Models of sedimentary basins are mostly qualitative in nature, creating difficulties in evaluating and quantifying primary control factors and their interactions. Many sedimentary environments comprise complex, small-scale geometries with abrupt lateral and vertical changes of depositional parameters, which cannot be analysed by static, qualitative models. In recent years, the development of numerical basin-modelling software has provided a powerful tool for improving the comprehension of long-term basin evolution, and for determining the suitability of modelling for exploration purposes (Lawrence *et al.*, 1990; Kendall *et al.*, 1991; Flemings & Grotzinger, 1996; Kendall & Sen, 1998; Whitaker *et al.*, 1999; Bowman & Vail, 1999; Paola, 2000).

In order to carry out basin modelling in a tectonically complex region such as the Cantabrian Zone (Fig. 1), a multidisciplinary approach of detailed

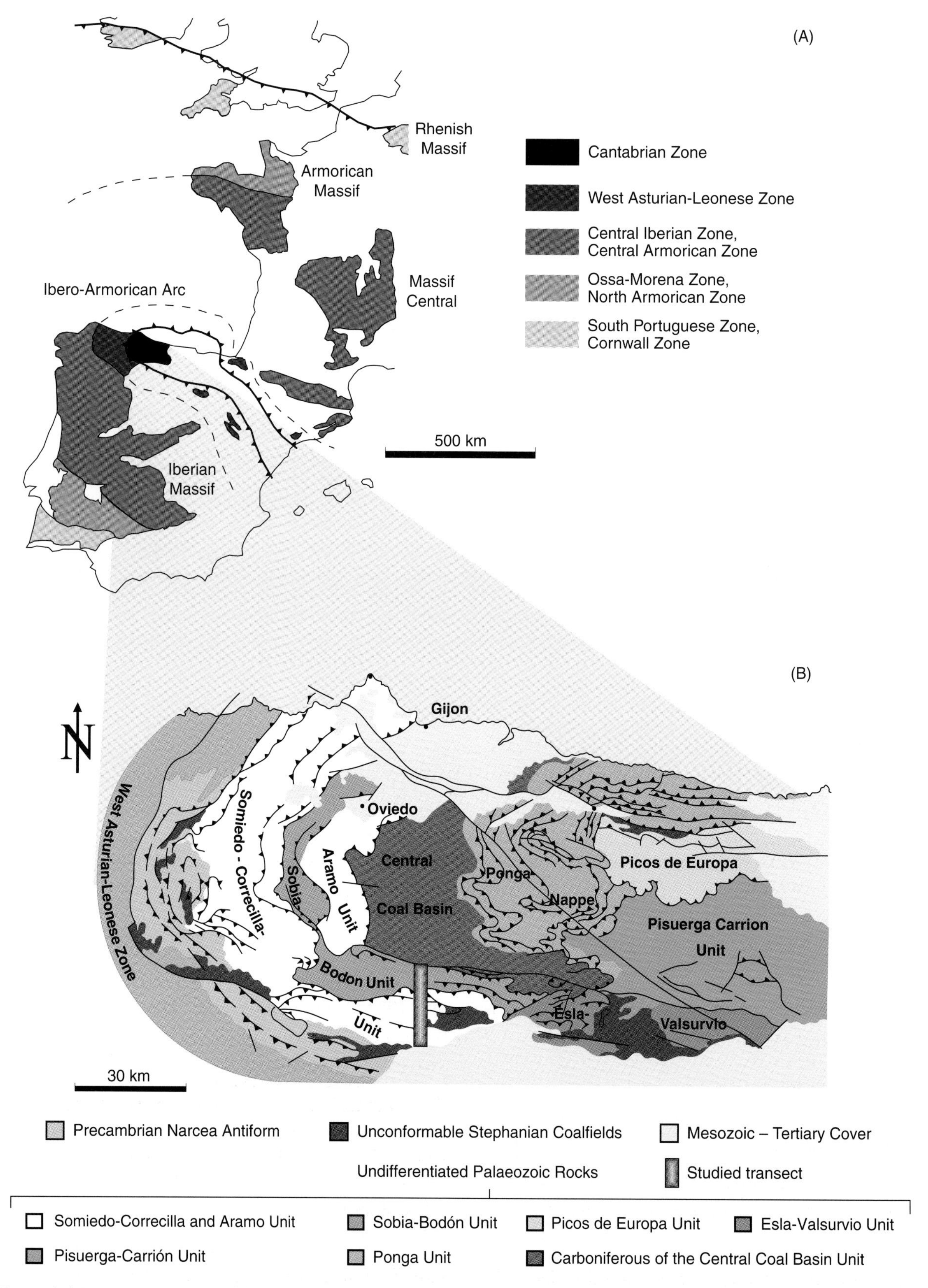

Fig. 1. (A) Tectonic zonation of the outcropping Variscan Belt in western Europe (after Dallmeyer *et al.*, 1997). The Ibero-Armorican Arc is schematically indicated in the west. (B) Tectonic units of the Cantabrian Zone (after Julivert, 1971; Pérez Estaún *et al.*, 1988). The Bernesga Transect is marked by the shaded bar. See Fig. 4 for detailed position of studied cross-sections.

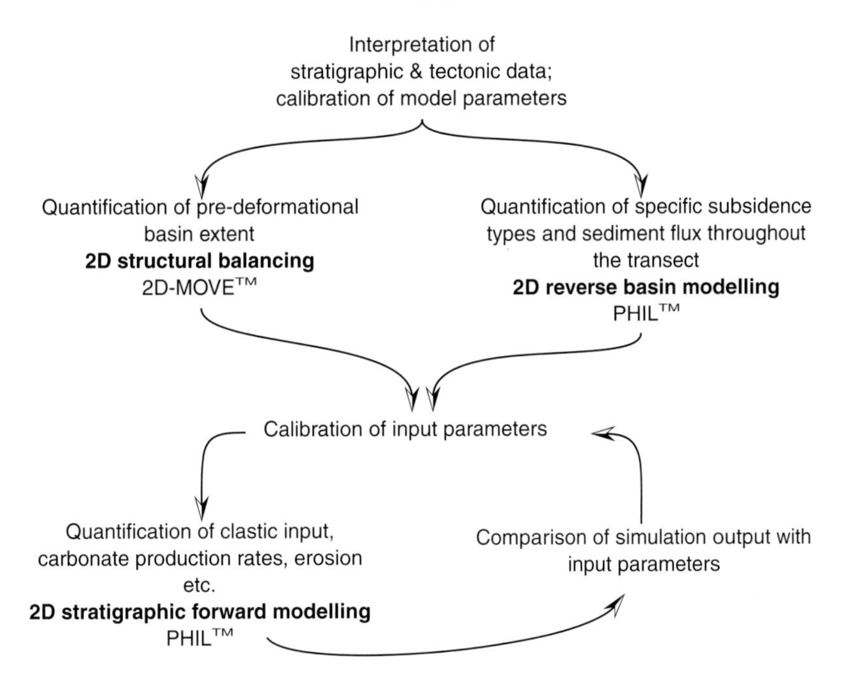

Fig. 2. Workflow of the modelling study, comprising structural balancing, two-dimensional reverse-basin modelling and 2D stratigraphic forward modelling. The circle indicates an iterative approach of simulation runs, which are repeatedly compared with the present-day geological information in order to recalibrate input parameters and achieve the best fit to nature.

fieldwork, structural balancing, subsidence and stratigraphic modelling is required (Fig. 2). Structural balancing has to be carried out before any modelling and reconstruction of the basin architecture prior to the late Palaeozoic/Variscan deformation. The combined dynamic approach of two-dimensional reverse-basin and subsequent stratigraphic forward modelling is an effective tool to analyse geological processes in time. The present study allows the quantification of the basin architecture, the shelf–basin development and the sedimentation history of this highly tectonized foreland thrust and fold belt. The temporal and spatial developments of petrophysical and sedimentary parameters are estimated from the deposition of the oldest sediments in the basin to the time of maximum burial.

Geological setting

The European Variscan Belt is a continental-scale oroclinal bend (Weil *et al.*, 2003), and extends for almost 3000 km in western Europe (Fig. 1A). Resulting from the collision of Laurentia–Baltica and Gondwana, it shows a remarkable bend in the west, called the Ibero-Armorican or Asturian Arc (Matte, 1991; Bastida & Aller, 1992; Ábalos *et al.*, 2002). The Iberian Massif represents the southwestern part of the Variscan Belt, which is subdivided into five tectonostratigraphic zones (Lotze, 1945; Julivert *et al.*, 1972). From northeast to southwest these are the Cantabrian, West Asturian-Leonese, Central Iberian (including the Galicia Tras-Os-Montes Zone), Ossa-Morena and South Portuguese Zones (Fig. 1A). These zones mirror the bilateral symmetry of the Variscan Orogen, allowing the comparison of the two external zones (Cantabrian and South Portuguese zones) and the internal zones (West Asturian-Leonese, Central Iberian and Ossa Morena; Julivert, 1981; Warr, 2000). High-grade metamorphic rocks and granitoids are restricted to the internal zones (Fernández-Suárez *et al.*, 2000). Precambrian rocks of the Narcea antiformal stack separate the Cantabrian Zone from the West Asturian-Leonese Zone (Fig. 1B). The latter shows a much thicker Lower Palaeozoic sedimentary succession and a westward increase of plutonism and metamorphism (Julivert, 1981; Bastida & Aller, 1992). The Cantabrian Zone represents the core of the Ibero-Armorican Arc, being composed of a foreland fold-and-thrust belt, which was deformed at shallow crustal levels without significant metamorphism or penetrative cleavage (García López *et al.*, 1997; Bastida *et al.*, 1999). It can be subdivided into several domains and units (Fig. 1B): the Somiedo-Correcilla, Sobia-Bodón, Aramo, Central Coalfield, Ponga, Picos de Europa and Pisuerga-Carrión (Pérez-Estaún & Bastida, 1990). The basin architecture is characterized by thin-skinned tectonics, with its main detachment horizon located at the base of the Láncara Fm (Lower Cambrian) and some minor décollements within higher stratigraphic horizons (Julivert, 1971; Pérez Estaún *et al.*, 1988). The Palaeozoic succession was strongly deformed during Variscan times by a set of imbricate thrusts and cogenetic folds, as well as by late, high-angle faults (Ábalos *et al.*, 2002). The Cantabrian Zone was described as an orocline (Carey, 1955), indicating oroclinal bending around a vertical axis in the late Stephanian to Early Permian (Van der Voo *et al.*, 1997; Weil *et al.*, 2000, 2001, 2003). Alpine deformation reactivated many Variscan structures, resulting in further compression with steepened thrusts and faults and increased structural shortening. Thus, the present-day structural setting gives only a partial and distorted picture of the Variscan Belt prior to the oroclinal bending (Ábalos *et al.*, 2002; Robardet, 2002). However, the structural situation with subvertical dipping strata in every single thrust sheet, together with magnificent outcrop conditions, provide one of the best insights into the European Variscan Orogen.

During Neoproterozoic times, northwest Iberia was part of a terrane assemblage that developed on an active Gondwana margin (Fernández-Suárez *et al.*, 1999, 2000, 2002a). Cadomian tectonism caused the formation of an angular unconformity (Ábalos *et al.*, 2002). The tectono-sedimentary evolution of the Cantabrian Zone is marked by active tectonism (rifting?) during the Cambrian (Álvaro *et al.*, 2000a, 2000b), followed by a passive margin stage (post-rift), which lasted until the Devonian. The Neoproterozoic to Silurian sedimentary succession (Fig. 3) is dominated by siliciclastic sediments (Herrería, Oville, Barrios, Formigoso and San Pedro Fms) and includes several long-lasting hiatus. From the Early Ordovician (Arenigian), a structural high developed in the northern part of the Iberian Peninsula (Cantabrian High or Umbral Cántabrico-Ibérico: *sensu* Aramburu *et al.*, 1992; Aramburu & García Ramos, 1993; Aramburu, 1995). The Cantabrian High was a location of non-deposition and erosion, being effective from the Middle Silurian until the Early Devonian in the area studied (Veselovský, 2004). Sedimentation occurred in the southern, western and northern basinal areas (according to present directions). During Devonian times, an alternating deposition of carbonate (Abelgas, Santa Lucía and Portilla Fms) and siliciclastic sediments (Esla, Huergas, Nocedo and Fueyo Fms) took place. Each episode of carbonate growth, producing major reefs, was followed by a succession of siliciclastic material and a new phase of carbonate deposition. However, the end of the Givetian marked the demise of Devonian reefs, manifested in the Frasnian Event (García-Alcalde, 1998; García-Alcalde *et al.*, 2002). Close to the Frasnian–Famennian boundary, the first appearance of conglomerates with metamorphic clasts indicates the presence of a new source area in the Central Iberian Zone (Frankenfeld, 1982; Rodríguez Fernández *et al.*, 1985). This points to the onset of crustal thickening of the Iberian Variscan orogen situated to the west of the Cantabrian Zone (Ábalos *et al.*, 2002). Subsequently, the whole Cantabrian Zone was covered by the thin, condensed carbonates of the Alba Fm until the early Namurian. Thick synorogenic turbidites (Olleros Fm), initiated in the Serpukhovian, mark the onset of Variscan orogeny in the Cantabrian Zone. The further evolution of the Variscan deformation was characterized by changes in basin geometry caused by the multiple emplacement of thrust nappes and by the sedimentation of thick synorogenic deposits. Large carbonate platforms developed (Barcaliente and Valdeteja Fms), which were subsequently covered by terrigenous sediments from the approaching Variscan Orogen (San Emiliano Fm). Isolated intra-montane basins with mostly continental deposits formed in the Stephanian (Fig. 1B) due to renewed tensional tectonics. These have been interpreted as pull-apart basins (Nijman & Savage, 1989; Villegas, 1996). The younger Permian and Mesozoic–Cenozoic succession in the Cantabrian Zone was extensively eroded and cannot be reconstructed. The thickness of the mainly shallow-marine Palaeozoic succession ranges between 3800 and 5000 m.

Despite a large amount of published data, there is still a considerable lack of information on the Cantabrian Zone. The following data and models are not yet available:

- a structurally balanced cross-section of the Southern Cantabrian Basin;
- a pre-deformation architectural model of the Southern Cantabrian Basin development including long-term changes in accommodation space between the top of the Neoproterozoic

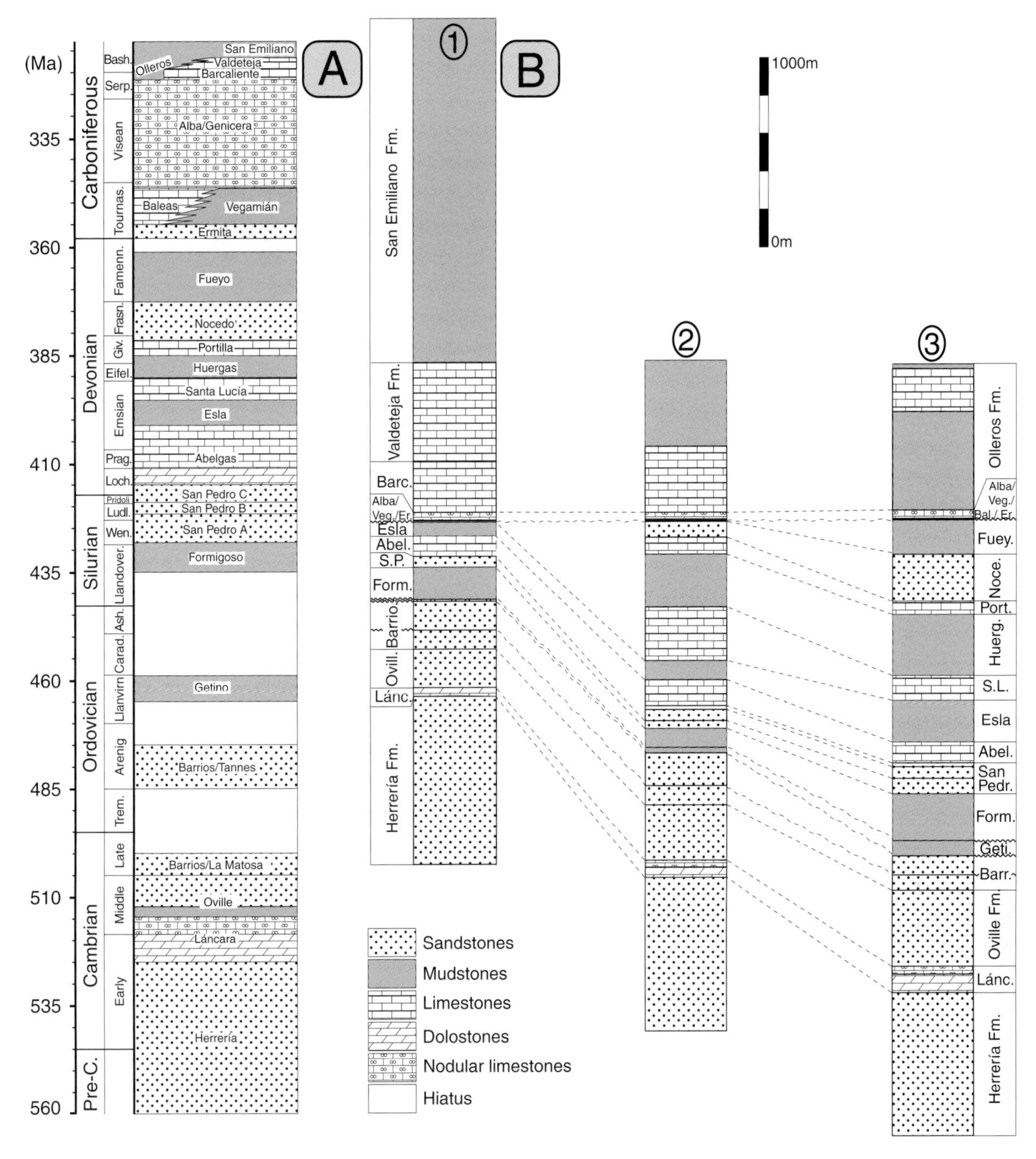

Fig. 3. (A) Stratigraphic chart of the Southern Cantabrian Mountains showing formation names in the stratigraphic columns. Absolute ages are taken from the German Stratigraphic Commission (2002). (B) Three selected synthetic stratigraphic sections along the Bernesga Transect: 1, proximal section at km 7 (northern part of the transect); 2, section at km 42; 3, distal section at km 54 (southern part of the transect). Note the wedge-shaped pattern of the Ordovician to Devonian formation thicknesses, increasing basinwards to the south. Due to the depositional/erosional hiatus in the northern section, most of the Devonian formations are missing. Hiatus are marked by wavy lines.

and the time of maximum burial (flexural reverse-basin modelling);

- a process-oriented model of depositional controls (stratigraphic forward modelling);
- a genetic model of the basin development.

Structural subdivision of the Bernesga Transect

The Bernesga Transect is situated along the N–S trending valleys of the Torio and Bernesga Rivers (Fig. 4). Its orientation follows the

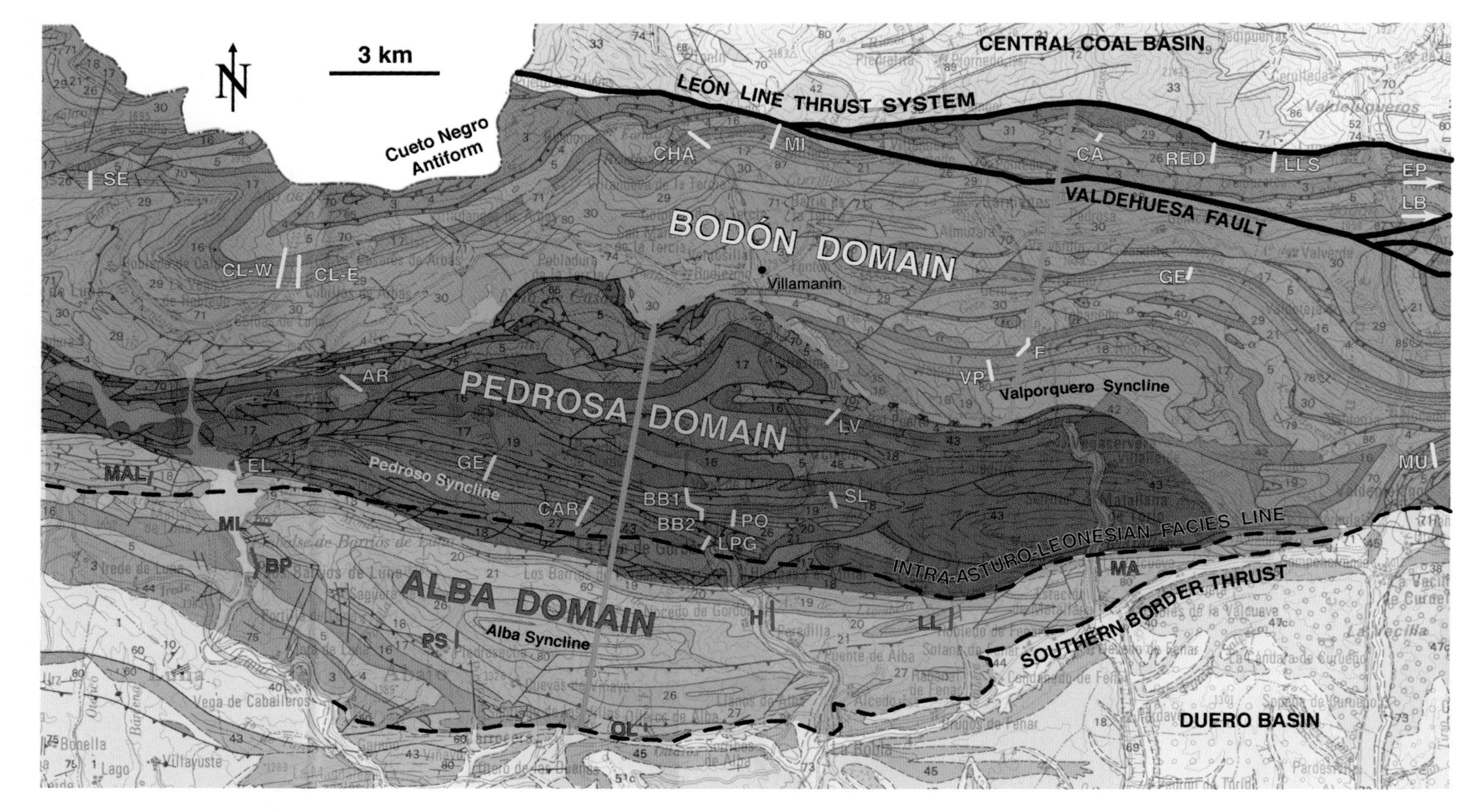

Fig. 4. Subdivision of the Southern Cantabrian Basin into three structural domains: the Bodón, Pedrosa and Alba Domains. Note the position of the León Line Thrust System and the Southern Border Thrust. Measured and interpreted cross-sections are indicated with specific keys according to the locality name (e.g. SL, Santa Lucía cross-section). All sections were plotted and described in Veselovský (2004). Geological map based on Suárez *et al.* (1994). The position of the Bernesga Transect is marked by the orange NNE–SSW trending line. The partitioning of the transect is shown by the dashed line, following the thrust between the Corecillas and Aralla-Rozo Nappes (see Alonso *et al.*, 1991 for detailed tectonic subdivision).

overall tectonic transport direction in the region (parallel to the direction of the displacement vectors/perpendicular to the thrust planes), intersecting the basin from its proximal (north) to distal (south) parts (Fig. 1B). Considering the overall structural framework, this part of the Cantabrian Zone can be divided into three domains, designated from north to south as the Bodón, Pedrosa and Alba Domains (Fig. 4). These domains as well as most of the thrust sheets are bordered by steeply dipping or overturned thrusts.

1. The Bodón Domain in the north comprises the Bodón, Gayo and Correcilla Nappes (see Alonso *et al.*, 1991 for a detailed tectonic subdivision of the area). To the north it is bound by the León Line thrust system and shows three kilometre-scale open folds (Figs 4 and 5). This unit displays the basal detachment at the base of the Láncara Fm, and a second detachment in the Herrería Fm. Alonso & Suárez Rodríguez (1991) assume a staircase geometry of this detachment. The northernmost part of the constructed transect (Fig. 5) suggests in the subsurface an overlapping Herrería ramp-anticline cut by the León Line. This assumption is corroborated by the presence of a huge tectonic window ca. 20 km west of the transect (Cueto Negro Antiform, Fig. 4). The Herrería Fm can be traced along the main strike direction (W–E) to the position of the Bernesga Transect, where it represents a higher tectonic floor of the Herrería antiformal stack. North of the Gayo Thrust the sediments of the San Emiliano Fm show a strong internal deformation. This internal folding has not been included in the present tectonic model, as the complex deformation style requires high-resolution structural studies.
2. The Pedrosa Domain is characterized by multiple overthrust sheets in the form of an imbricated thrust system. The unit is composed of the Aralla-Rozo Nappe and the Pedroso Syncline, which represents the northern part of the Abelgas-Bregón Nappe. The basal detachment is situated at the base of the Láncara Fm (Figs 4 and 5).
3. The Alba Domain, representing the southern part of the Abelgas-Bregón Nappe, is composed of large synformal structures. In the southernmost part of the Bernesga Transect, the Southern Border Thrust marks the contact with the Mesozoic Duero Basin. Palaeozoic sediments of the Southern Cantabrian Basin have been thrusted over the Mesozoic basin infill of the Duero Basin (Fig. 1). The southward displacement was estimated to be about 25 km (Alonso *et al.*, 1996). In comparison to the northern parts, the large open Alba Syncline has suffered the least deformation within the transect.

STRUCTURAL BALANCING

Methods

Balanced cross-sections are the most commonly utilized geometric constraint for overthrust structures (Jamison, 1987). They offer reliable quantitative predictions about the subsurface architecture, even if a seismic data base, borehole information or sufficient surface data are not available. All tectonically affected domains have to be readjusted into their depositional, predeformational position in order to quantify basin shortening caused by tectonic movements and to be able to recognize the former basin configuration and architecture. The balancing procedure follows the strain-compatibility principle during deformation postulated by Ramsay & Huber (1987). The viable and admissible balanced cross-section is the primary test for the consistence of bed lengths, areas or volumes (Jamison, 1987). The principal techniques and restrictions of balanced cross-sections have been discussed by Dahlstrom (1969) and Woodward *et al.* (1985, 1989).

For the balancing of the Bernesga Transect, the fundamental rule of equal bed lengths has been applied (Dahlstrom, 1969). Cross-section balancing starts by constructing a deformed section and then by measuring the bed lengths to relocate the faults in the undeformed state (Woodward *et al.*, 1989). The amount of shortening (ε) in percent can be calculated using the following equation:

$$\varepsilon = (x - x_0)/x_0 \times 100$$

where x_0 represents the restored and x the deformed stage.

In order to avoid problems of Stephanian sediments covering the eastern part of the working area and a prominent lateral ramp (south of Villamanín de la Tercia, Fig. 4), which moved material out of the plane of section (see 'Constraints'), the transect was split into two parts. The thrust plane delimiting the Correcillas Nappe from the Aralla-Rozo Nappe was selected as the boundary between the northern and southern parts of the transect (Fig. 4).

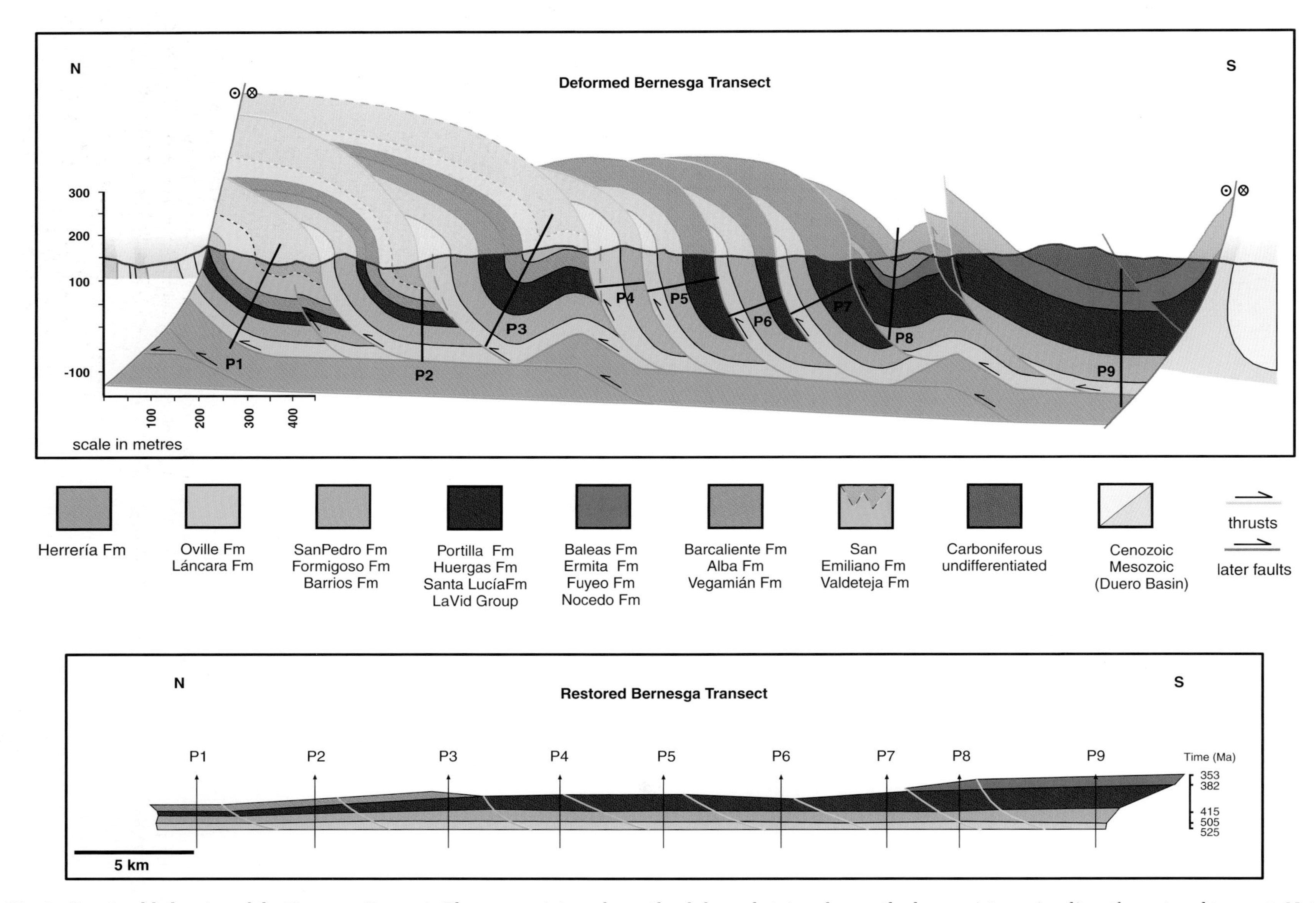

Fig. 5. Structural balancing of the Bernesga Transect. The upper picture shows the deformed state, whereas the lower picture visualizes the restored transect. Note different scales of upper and lower figures! For graphical reasons, different formations were grouped into units with identical colours. The result of structural balancing is an undeformed, largely horizontal stack of sedimentary formations of varying thickness. The topographic feature of the Cantabrian High is only characterized by the diminishing thickness of Devonian formations (red brown colour) towards the north.

Restorable cross-sections normally should contain a completely restorable structure, which comprises the undeformed hinterland and foreland. For this study this is not the case. The undeformed hinterland in the south is covered by Cretaceous and Tertiary rocks of the Duero Basin (Fig. 1), whereas in the north the out-of-sequence León Line fault system cross-cuts the transect (Fig. 4). Consequently pin lines had to be set perpendicular to bedding in both the deformed and the undeformed state (Woodward *et al.*, 1985). Pin lines (Fig. 5) are the starting point for measuring bed length for balancing in every discrete unit.

Constraints

While constructing cross-sections, there are generally a number of geometrical plausible solutions. The number of solutions depends greatly on the quality of the database. The following factors place additional restrictions:

- using the presented simplifications, only plane strain and simple shear can be used;
- parallel folding with flexural slip between beds is assumed;
- cross-sections have to be placed parallel to the direction of strain, analogous to the displacement vectors or perpendicular to the thrust planes;
- no movement into or out of the 2D plane of the transect can be calculated;
- presumption of minimal volume loss or tectonic compaction, for example by means of pressure solution and preservation of line-length during deformation, is necessary;
- possible volume reductions due to compaction by sediment load during and after compression are not considered – these factors might be of importance especially in the successions underneath the Stephanian basins (which were not modelled);
- transects have to contain completely restorable structures.

Results

Models of the pre-orogenic, Late Devonian basin geometry establish a basis for the following, dynamic 2D reverse and stratigraphic forward modelling. The deformed and restored Bernesga Transect is shown in Fig. 5.

Several deformation mechanisms are clearly visible in the deformed transect. The 'piggy-back thrust' – mechanism shows the forward-breaking succession of individual thrust sheets, running chronologically from south to north. The basal accretion within the Herrería Fm is exposed in the Cueto Negro antiformal stack (Fig. 4) and two independent ramp anticlines south of pin line P3 and P8 in the subsurface (Fig. 5). These structures in the subsurface have been interpreted to account for the formation of the Valporquero Syncline (marked by pin line P3 in Fig. 5) and Pedroso Syncline (pin line P8 and Fig. 4), as this independent ramp anticline forms in a more or less symmetrical pattern. The southern part experienced very differential internal deformation from north to south. The imbricated thrust system in the north shows the highest amount of shortening within the whole transect, whereas the wide open fold of the Alba Syncline experienced lower deformation.

Basal accretion occurred after frontal accretion, and resulted in the deformation of the basal detachment at the base of the Láncara Fm (Potent & Reuther, 2000). However, there are several indications for further deformation phases such as out of sequence faults (León Line, Alba Syncline Backthrust), overthrusting of the Palaeozoic sediments on the Mesozoic Duero Basin (Southern Border Thrust) and the steepening and overturning of many Variscan thrusts.

No structural balancing could be accomplished north of the León Line and south of the Southern Border thrust, as these structures represent out-of-sequence faults with unknown displacement factors and characteristics. In Fig. 5 balanced areas (between pin lines P1 and P9) are marked by saturated colours. Consequently the Gayo, Corecillas and Aralla-Rozo-Nappes have been restored completely, whereas the Bodón and the Abelgas-Bregón Nappes have been balanced partially. Small-scale faulting and folding is common, but has not been considered in the balancing process. This represents a valid simplification at basin scale. The deformed section was constructed and balanced simultaneously in an iterative way. Because of the different structural styles, the three domains (see 'Structural subdivision of the Bernesga Transect') have been balanced separately and stitched along their mutual ramps. Each nappe was balanced both manually and with the structural analysis and modelling program 2D-Move^{TM} (Midland Valley Ltd, Glasgow).

Ramp geometries and angles result from the balanced bed lengths. Fault-bend folds are a typical feature in thrust belts and develop as the hangingwall of a thrust is transported through a ramp region on the thrust surface (Jamison, 1987). They are determined by the position of the hangingwall truncation foreland of the ramp and show ramp-angles between 10° and 40° (Boyer & Elliott, 1982; Suppe, 1983; Jamison, 1987; McClay, 1992; Nieuwland *et al.*, 2000). In the present model, ramp angles between the discrete thrust nappes range between 17° and 25° (Fig. 5, lower part). Ramps break through from south to north with an increasing angle towards the top. This listric trend could have been caused by diminished overburden to the top of the succession (Potent & Reuther, 2000) or by high basal friction at the décollement (Nieuwland *et al.*, 2000). Competent beds, which have suffered least ductile deformation, were chosen for bed length balancing. The lowermost horizon of the Herrería Fm was deformed by basal accretion and was not considered within the balancing procedure. Due to an unknown structure and geometry (no high resolution seismics), the duplexes and the antiformal stack in the north were only displayed schematically. The Valdeteja and San Emiliano Fms were not balanced because of thickness variations and interfingering of the uppermost units. Additionally the top of the San Emiliano Fm does not crop out in the region, as it is always cut by thrusts. Internal deformation of the specific thrust sheets and the displacement widths between two neighbouring nappes were calculated independently for every single nappe by measuring the distance between the footwall and hangingwall cut-off points along the different thrust surfaces.

The balancing procedure applied provides minimum shortening values for the Bernesga Transect. The result of the balancing process is an undeformed, horizontal stack of beds with varying thicknesses, formed during deposition. However, consideration of out-of-plane deformation and small-scale folding and faulting may increase shortening values by 10–15%. The length of the constructed, deformed transect amounts to 17.91 km, whereas in its restored state, the length increases to 38.58 km (both distances measured between pin lines P1 and P9) resulting in a shortening distance of 20.67 km. Total tectonic shortening of the deformed basin infill amounts to a minimum of 54%. This result ranges between the values suggested by Julivert & Arboleya (1986), who proposed 40–50% shortening for the Cantabrian Zone and Oczlon (1992) and Ábalos *et al.* (2002), presuming 70–80%. However, Oczlon (1992) and Ábalos *et al.* (2002) considered the whole Cantabrian Zone (Narcea Antiform to the Pisuerga-Carrión Domain, Fig. 1B), so that these values may not be comparable to the southern part of the Cantabrian Basin. Potent & Reuther (2000) assume 60% shortening, balancing a 6.3 km long cross-section of the highly deformed Montuerto Syncline (Curueño Valley).

Table 1. Values for the minimal shortening and displacement widths along transect. See 'Structural subdivision of the Bernesga Transect' for explanation and Fig. 5 for the constructed and balanced transect.

	Bodón Domain	Pedrosa Domain	Alba Domain
Minimal shortening (%)	41	65	25
Displacement widths (km)	1.61–1.76	1.02–1.40	0.31–0.63

In terms of minimal shortening, the Pedrosa Domain displays the highest values followed by the Bodón and Alba Domains (Table 1). The slightly deformed Alba Syncline (20%) shows lower values in proportion to the total minimal shortening of the transect. Nevertheless, because of the simplification of the Alba and highly tectonized Pedroso Synclines, the minimal shortening values may underestimate the true extent of shortening. As expected, the highest shortening values are displayed by the imbricated thrust system north of the Pedroso Syncline (up to 78%) showing minimal shortening to be 23% to 45% higher than the total minimal shortening of the transect. Regarding the Forcada Nappe in the north of the Bernesga Transect, which has not been balanced, values of around 70% can be estimated, as the León Line Thrust system and other residual faults point to a significantly higher level of internal deformation. It can be reasonably assumed that the area between pin line P9 and the Southern Border Thrust will range from around 20% to 25%, considering the backthrust on the southern limb of the Alba Syncline (Fig. 5).

Displacement widths reflect the structural differences between the three domains. With regard to the timing of the Variscan thrusting, the youngest nappes show the lowest displacement widths along the transect (see Table 1 for detailed values). The overall shortening caused by the displacement of the nappes along thrusts alone sums up to 9.21 km. This value corresponds to 52% of the total minimal shortening (internal folding plus thrusting). In other words, thrusting and internal folding of the nappes caused approximately the same amount of total minimal shortening.

BASIN MODELLING

The combination of reverse-basin and stratigraphic forward modelling allows for developing more rigorous genetic models of basin development. Numerical modelling permits the quantitative analysis of key controls in basin development including their temporal and spatial variations. The key controls comprise, e.g.:

- subsidence/uplift rates (overall thermo-tectonic-, compaction- and flexural-induced)
- porosity, density
- palaeobathymetry
- sedimentation rates (input of siliciclastics and redeposited carbonate sediment as well as *in situ* carbonate production)
- erosion rates
- facies distribution
- initial bathymetry
- crustal parameters (effective elastic thickness, mantle density, etc.)

A sedimentation model considers major tectonic, water-level and sediment-supply conditions including short and long-term processes. The 2D reverse-basin and stratigraphic forward modelling for this study were accomplished using the PHIL/BASIM™ program package (PetroDynamics Inc., Houston). Bowman & Vail (1999) published the fundamental algorithms implemented in the stratigraphic simulator. The modelling process applies a procedural loop depicted in Fig. 2.

Two-dimensional reverse-basin modelling

Two-dimensional numerical reverse-basin modelling provides overall thermo-tectonic-, flexural- and compaction-induced subsidence rates as well as sediment flux, which constrain 2D forward stratigraphic models. It additionally offers architectural models of the basin in time (Figs 6–8).

Input data

Field data from cross-sections and outcrops were projected into the modelled transect, including biostratigraphic and lithostratigraphic information, thicknesses, large-scale geometries, characteristic surfaces and structural measurements. Due to the structural character of the investigation area, thrust planes were used as projection lines. The distances between the cross-sections prior to the Variscan orogeny have been obtained by means of the structural balancing procedure discussed above.

Time

The basin fill was subdivided in 35 time layers, based on the available biostratigraphic framework. The youngest time line (Cenozoic, 34 Ma) marks the period of maximum burial depth of the basin. From this point of time, thicknesses of subsequently older time layers were incrementally added up to the top of the basement (Precambrian, 560 Ma). Absolute ages for stage boundaries have been taken from the German Stratigraphic Commission (2002), based on Gradstein *et al.* (1994; Cretaceous), Berggren *et al.* (1995; Tertiary), Tucker & McKerrow (1995; Ordovician, Cambrian), Menning (1995, 2001; Permian), Menning *et al.* (2000; Carboniferous), Pálfy *et al.* (2000; Jurassic) and Weddige (2001; Devonian). Depending on the presence of indicative fossils, time resolution reaches stage level.

For each time line at every location along the transect data, absolute age, burial depth and palaeobathymetry are required. In addition, sea-level history, initial textural porosities for the specific lithologies, as well as flexural parameters were included in the model.

Sea level

Only 2nd order changes of eustatic sea level were considered, because sea-level changes of higher magnitude are poorly constrained throughout the Palaeozoic. The published sea-level curves for the entire Phanerozoic (Vail *et al.*, 1977; Hallam, 1984) show significant discrepancies and do not match the detailed curves published for single series. Therefore a synthetic curve was established for this study based on (see Fig. 12):

0–70 Ma:	Mitchum & van Wagoner (1991) (Haq *et al.*, 1987)
70–160 Ma:	Kendall *et al.* (1992) (Haq *et al.*, 1987)
160–256 Ma:	Haq *et al.* (1987)
255–355 Ma:	Ross & Ross (1988)
355–410 Ma:	Johnson *et al.* (1985) and Dennison (1985)
410–438 Ma:	Ross & Ross (1996)
435–540 Ma:	Ross & Ross (1988)
540–560 Ma:	Vail *et al.* (1977)

Palaeowater depths were estimated from facies analysis and faunal assemblages.

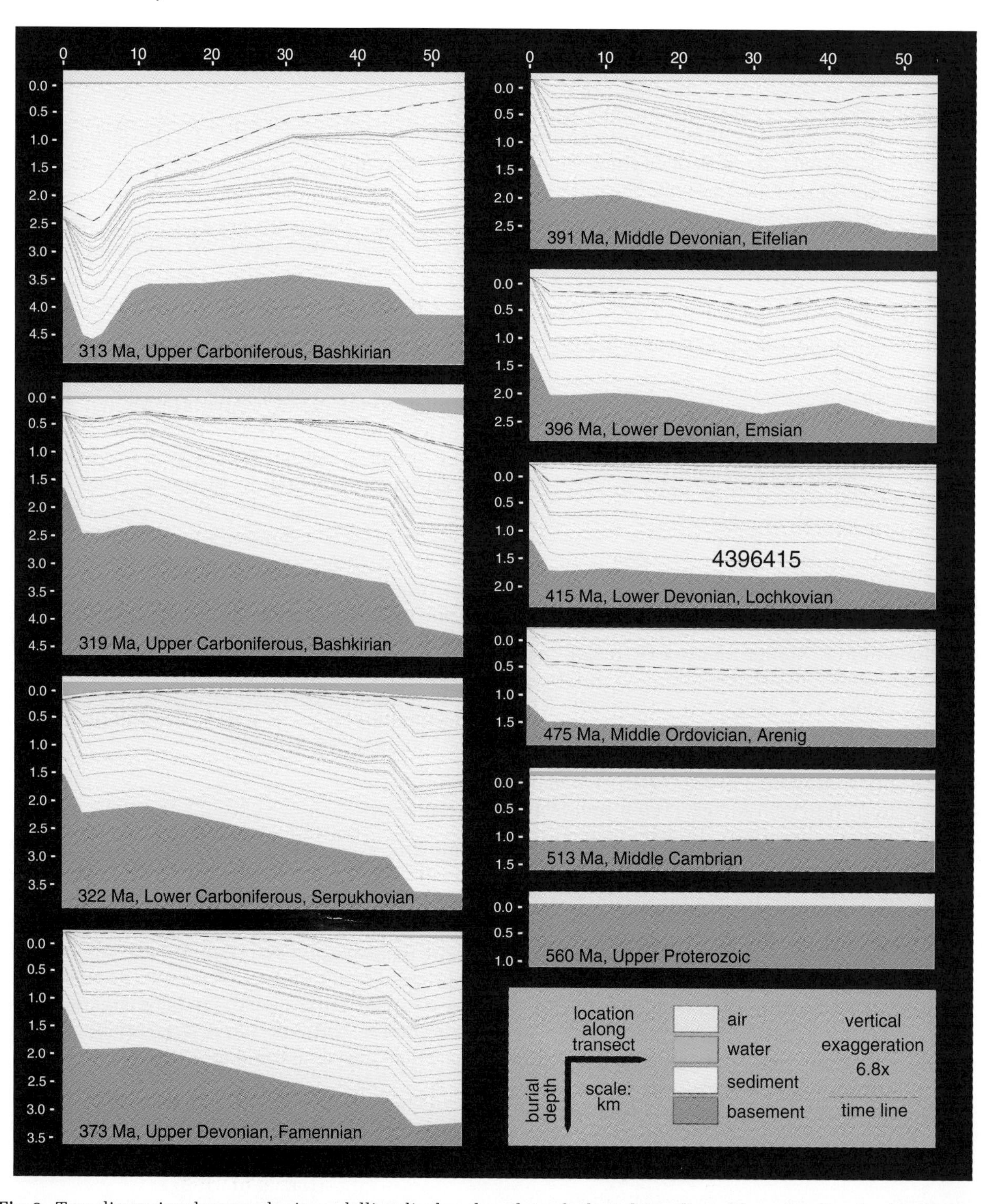

Fig. 6. Two-dimensional reverse-basin modelling displayed as plots of selected time lines. They visualize the development of depositional and structural architecture from Late Proterozoic to Late Carboniferous. The top of an increment is marked by a dashed line in the plot that follows. The position of the Bernesga Transect is shown in Figs 1B and 4.

Compaction and decompaction

The numerical modelling includes incremental decompaction of the succession in each time layer. Decompaction is calculated using empirical porosity/depth relations for specific lithologies. PHIL^{TM} utilizes 12 siliciclastic and eight carbonate/evaporite lithologies with definable

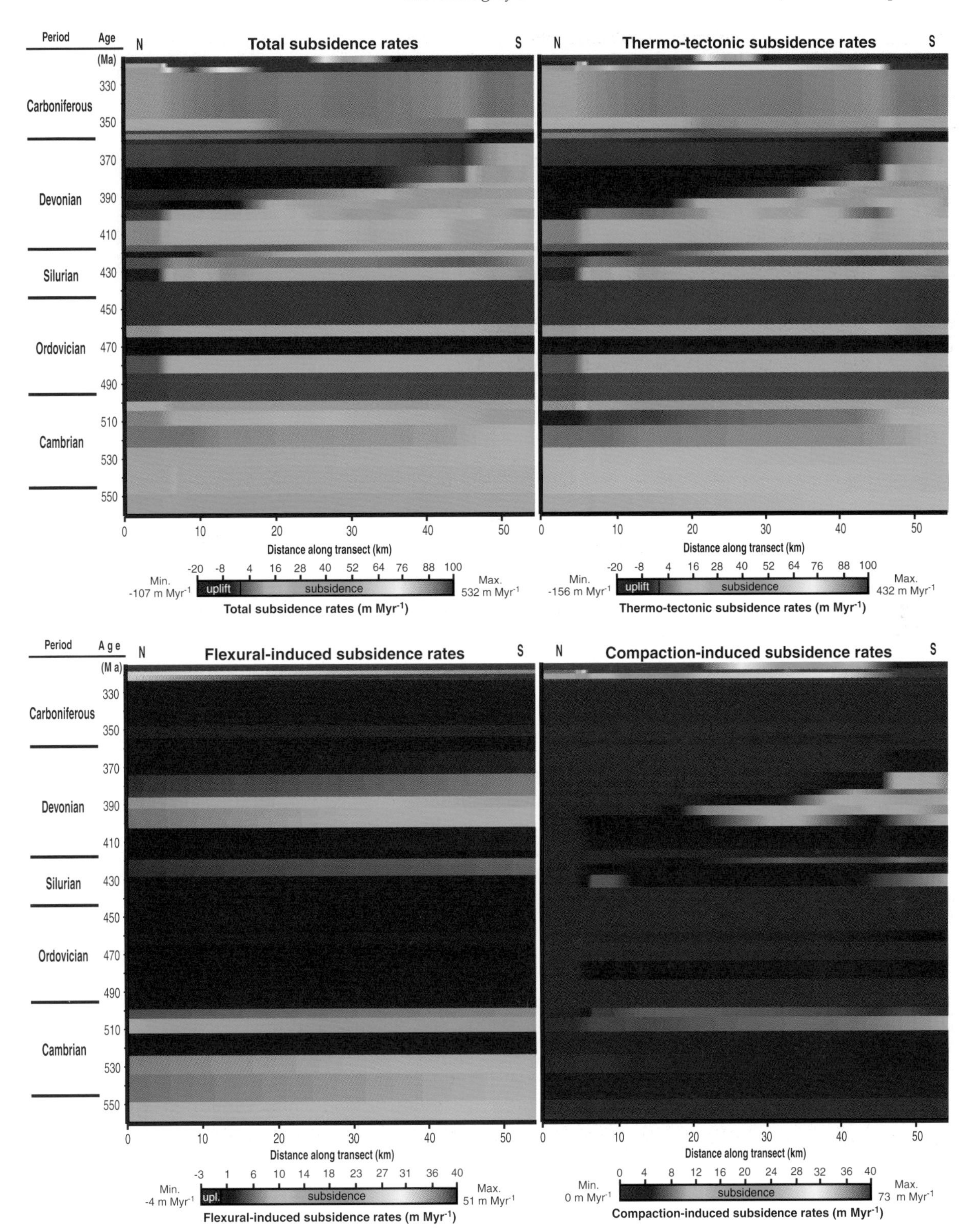

Fig. 7. Spectrogram plots of total subsidence rates and their specific components (thermo-tectonic, flexural-induced and compaction-induced). The y axis of each figure represents the time-development of the basin in Myr, whereas the distance along transect is plotted on the x axis. See Figs 1B and 4 for the position of the Bernesga Transect. Differential subsidence rates are visualized with specific colour codes, cut in order to highlight medium-range values (colour bars at the bottom of each graph). Therefore maximum and minimum values are given for each plot. Detailed maximum, minimum and mean values for all trends are stated in Table 2.

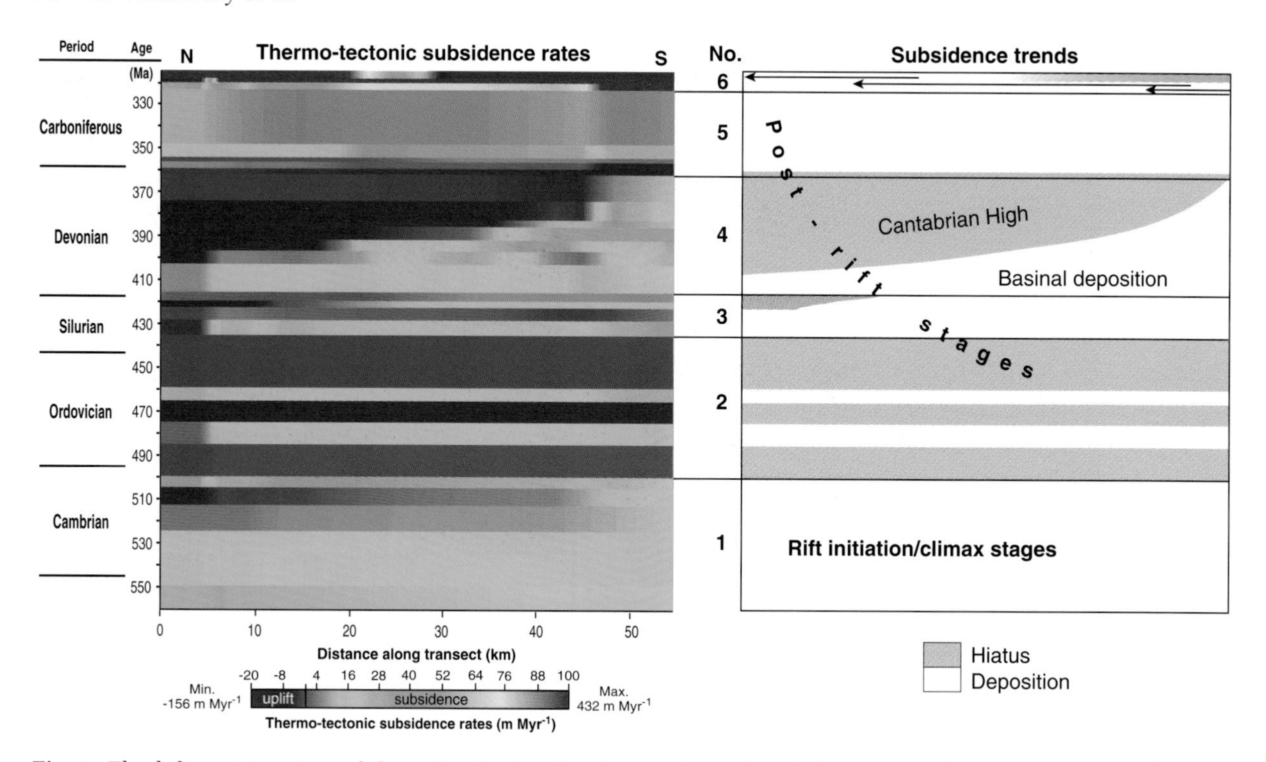

Fig. 8. The left part is extracted from Fig. 7, showing thermo-tectonic subsidence rates between 560 Ma and 313 Ma. The right side indicates the specific numbers of the subsidence trends (see '2D reverse basin modelling') as well as the development stages of the Southern Cantabrian Basin. Trend 6 in the uppermost part of the figure represents the Variscan foredeep stage, with arrows indicating movements of highly subsiding depocentres in time.

bulk rock density, initial porosity, compaction-rate parameter and maximum depth of compaction. Together with burial depth, these values control the compaction of all time layers. Values have been taken from Goldhammer (1997), Welte *et al.* (1997) and Bowman & Vail (1999) who published algorithms for the calculation of burial-depth-related porosity dependent on the lithology.

During each timestep PHIL™ calculates the porosity within each cell according to its depth and lithology from top to bottom of the series. According to the loss of porosity for a time interval, the thickness is decreased equally and the horizon depths are adjusted. Sediment compaction is assumed under normal pressure, with the option of overpressured basin conditions (where compaction is delayed) resulting in thicker sections with lower densities.

Flexural response of the crust

Flexural response of the crust can become very prominent, especially for greater loads and extended duration of loading (Einsele, 2000; Stüwe, 2000). Flexure results from vertical forces, horizontal forces and torques in any combination and affects the elastic upper part of the lithosphere, which retains elastic stresses over geological timescales (Allen & Allen, 1990). Flexural reverse modelling computes the flexure due to total sediment load and subtracts the resulting flexure from the observed thickness in order to obtain tectonic subsidence and uplift (Watts, 2001). The effects of palaeobathymetry and sea-level changes are taken into account. The parameter that characterizes the apparent strength of the lithosphere is the flexural rigidity D, which is commonly expressed by the effective or equivalent elastic thickness (T_e) of the lithosphere (Burov & Diament, 1995, 1996; Cloetingh & Burov, 1996; Einsele, 2000; Stüwe, 2000). García-Castellanos (2002) classifies T_e values of most of the modern foreland basins in a relatively narrow range of 15–40 km. For this model the absence of a significant difference in mechanical behaviour of very old Precambrian lithosphere and middle-aged Variscan lithosphere (Cloetingh & Burov, 1996) is very important. These authors assume that after approximately 400 Ma the temperature has only minor influence on T_e variations. For the 2D reverse modelling a mean T_e value of 35 km was chosen. For detailed discussion see Veselovský (2004).

Methods

Reverse-basin modelling determines the burial history of a basin (e.g. Turcotte & Schubert, 1982). Leeder (1999) distinguishes between tectonic, thermal and load-induced bending subsidence types contributing to total subsidence. In this concept the backstripping procedure calculates and removes the effect of sediment loading and compaction from the basement subsidence, as well as changing palaeobathymetry and eustatic sea-level variations, enabling the quantification of tectonic subsidence (Sachsenhofer *et al.*, 1997; Einsele, 2000; Watts, 2001). By restoring sediment thickness at the time of deposition, taking into account compaction and water-depth changes, and then isostatically unloading it, it is possible to determine the depth at which the basement would be in the absence of water and sediment load (Watts, 2001).

Two-dimensional reverse-basin modelling follows a sequence-stratigraphic concept, which contemplates the creation/destruction of accommodation space and its infill as the principal determinant on sedimentary systems and basins (Zühlke *et al.*, 2004). Accommodation space development is constrained by subsidence/uplift, eustatic sea-level changes and compaction in the buried sedimentary infill. Subsidence/uplift can be caused by a number of forces, such as flexural response to loading/unloading (by sediment, water or any additional load) or thermal cooling/heating. Within this study, total subsidence comprises three genetic components: the thermo-tectonic subsidence, flexural-induced subsidence and compaction-induced subsidence. Bowman & Vail (1999) published algorithms for calculation of the flexural response of the crust to loading/unloading by deposition/erosion of sediment and sea-level changes.

During the procedure, all time layers are incrementally removed, starting at the time of maximum burial depth (Cenozoic, 34 Ma) and continuing until the top of the basement in the basin (Precambrian, 560 Ma). After a specific time layer has been removed, the underlying layer is adjusted according to the predefined palaeobathymetric depth. All older layers are decompacted with predefined compaction parameters. This procedure is repeated for each of the 35 time lines, calculating the sediment flux and different subsidence rates, removing the effects of palaeobathymetric changes, flexural loading, compaction and eustatic sea-level changes. Numerical results are quantitative approximations for thermo-tectonic, flexural and compaction-induced, and total subsidence rates, as well as sediment flux throughout the basin in time (Figs 7 and 8). They represent the fundamental basis for the stratigraphic forward modelling. For details see Bowman & Vail (1999) and Turcotte & Schubert (2002).

Constraints

- Partially the present geochronologic information is not sufficiently precise. During sedimentation in the Southern Cantabrian Mountains only a few volcanoclastics (within the Oville, Barrios, San Pedro Fms) were deposited in the area, and these have not yet been radiometrically dated (Gutiérrez-Alonso, personal communication, 2003). Starting the model in the Precambrian causes uncertainties due to missing biostratigraphic data.
- Linear space interpolation is a basic method in 2D numerical reverse modelling. For all time increments, both subsidence/uplift rates and sediment supply are interpolated between the locations along the transect.
- Flexural behaviour of the crust depends on the effective elastic thickness (T_e) of the lithosphere. Average values for T_e during the reverse-basin modelling cause uncertainties within the flexural-induced subsidence results.
- In this paper, initial porosities and compaction parameters for standard lithologies have been used. They are based on particle size, matrix/grain-support and mud/shale content. Each time increment incorporates information for one representative lithology.

Numerical and graphical output

Figure 6 shows the basin architecture for 10 selected time lines. Rates of subsidence components are shown in the spectrograms of Figs 7 and 8 and in Table 2.

Results

All values are given as subsidence/uplift rates in metres per million years (m Myr^{-1}). Negative values stand for uplift, whereas positive indicate subsidence. Between 560 Ma and 313 Ma six major

Table 2. Trends of total subsidence rates and their specific components (thermo-tectonic, flexural-induced and compaction-induced) for the time interval 560–313 Ma. Spectral plots are shown in Figs 7 and 8. See text for further explanation.

Component		Subsidence rates (m Myr^{-1}) Trend 1 (560–500 Ma)	Trend 2 (500–435 Ma)	Trend 3 (435–415 Ma)	Trend 4 (415–361 Ma)	Trend 5 (361–322 Ma)	Trend 6 (322–313 Ma)
Total	Minimum	7	−6	−4	−10	−9	−107
	Maximum	50	33	45	70	25	532
	Mean	28	6	13	15	4	175
Thermo-tectonic	Minimum	−6	−6	−8	−13	−10	−156
	Maximum	34	30	35	40	28	432
	Mean	16	5	7	5	4	98
Flexural-induced	Minimum	1	−1	−1	2	−4	28
	Maximum	20	0	6	14	1	51
	Mean	8	0	2	6	0	41
Compaction-induced	Minimum	0	0	0	0	0	0
	Maximum	10	5	10	16	2	73
	Mean	4	1	3	4	0	36

subsidence trends can be determined (Fig. 8). Values for all trends and all subsidence components, as well as their absolute ages, are summarized in Table 2.

Thermo-tectonic subsidence. The thermo-tectonic subsidence rates range from −156 to 432 m Myr^{-1} over the entire modelling interval (Fig. 7).

The first trend (560–500 Ma, Fig. 8 and Table 2) is indicated by slightly elevated values with mean rates of 16 m Myr^{-1}. The spatial pattern is uniform, with higher values (up to 34 m Myr^{-1}) in the northern part of the transect (km 6). Within the first four time increments a trend can be observed showing thermo-tectonic subsidence values diminishing in time by 40%.

The second trend (500–435 Ma) is marked by alternating increments of slight subsidence (up to 30 m Myr^{-1}) and no subsidence or very slight uplift (maximum uplift −6 m Myr^{-1}). The distribution of values is uniform, with decreasing values to the north and to the south. Three 10 to 24 Myr long hiatus subdivide the sedimentary record (see also Fig. 3).

The third trend (435–415 Ma) indicates the first inhomogeneity in spatial distribution of subsidence rates. Its first time increment (Formigoso Fm; 435–428 Ma) shows two local subsidence maxima (km 7 and km 54 with values up to 35 m Myr^{-1}), whereas the subsequent increments (San Pedro Fm; 428–415 Ma) are characterized by the development of an uplifted area (Cantabrian Block) of no deposition (up to −8 m Ma^{-1}) in the north. This structural high propagated in time to the south (Members B and C of the San Pedro Fm; 422–415 Ma), restricting the areas of deposition to the south.

The pattern vanishes at the beginning of the fourth trend (415–361 Ma). The Esla Fm (402–396 Ma) shows a marked subsidence peak at km 32 along the transect. However, the subsiding areas start to become restricted once again towards the south from the deposition of the Santa Lucía Fm onwards (396–391 Ma). The uplifted areas (up to −13 m Ma^{-1}) extend to km 47 at the time of the Fuyeo Fm (373–361 Ma). Maximum subsidence rates (40 m Myr^{-1}) are reached during deposition of the Santa Lucía Fm. This wedge-shaped pattern is visible in the spectral plot in Fig. 7.

The fifth trend (361–322 Ma) commences with a short hiatus (3 Ma) and an alternation of thin carbonate and siliciclastic formations. The wedge-shaped pattern vanishes in the uppermost Devonian, switching to fairly uniform subsidence rates (up to 28 m Myr^{-1}). The most important feature of this trend is the deposition of the condensed Alba Fm (347–322 Ma), representing 25 Myr of uniform subsidence and low sedimentation rates along the transect.

In contrast to these uniform conditions, the sixth trend (322–313 Ma) is characterized by highly subsiding areas (up to 432 m Myr^{-1}), with neighbouring regions of strong uplift (up to −156 m Myr^{-1}). During time increment 317 Ma, highest values of 241 m Myr^{-1} are displayed at km 43 along the transect. Throughout time increment 313 Ma, the pattern shows a maximum subsidence restricted to the northernmost part of the Bernesga Transect (up to 432 m Myr^{-1} at km 0), whereas the southern part has been strongly uplifted (−156 m Myr^{-1}). Thus, within the three time increments of this trend, the subsiding depocentres migrate from the southernmost to the northernmost part of the transect (Figs 7 and 8).

Flexural-induced subsidence. This subsidence component does not mirror all six subsidence trends already presented. It has a comparatively minor impact on the total subsidence development with mean proportions of 10–25% of the total subsidence. The flexural-induced values depend to a great extent on the sediment thickness and the palaeobathymetry. However, due to the low specific gravity of water in comparison to sediment, only greater water depth causes major flexural response of the crust.

Compaction-induced subsidence. In comparison with trends one, two and five, trends three (435–415 Ma) and four (415–361 Ma) are fairly well developed, displaying the wedge-shaped pattern of the thermo-tectonic subsidence results. Higher compaction-induced subsidence rates are also due to the highly compactable black shales of the Formigoso Fm and the sediment deposited south of the Cantabrian High. The sixth trend (322–313 Ma) fully corresponds to the thermo-tectonic results, showing highly subsiding regions (up to 73 m Myr^{-1}) migrating northwards in time. The proportion of the compaction-induced subsidence is slightly lower but in the same order as the flexural-induced subsidence. Compaction-induced subsidence is very prominent during times of high sedimentation rates, as for example during deposition of the San Emiliano and Olleros Fms.

Total subsidence. Total subsidence represents the sum of thermo-tectonic, flexural-induced and compaction-induced subsidence. The subsidence values over the whole model period range from −107 to 532 m Myr^{-1}. As the thermo-tectonic subsidence is the dominating component, ranging between 35% and 100%, it shows a fairly identical pattern to total subsidence. During times of high sedimentation rates (e.g. subsidence trends 1 and 6), compaction of deposited sediment and its load cause considerable subsidence, reaching 25–40% of total subsidence rates.

Two-dimensional stratigraphic forward modelling

In order to quantitatively model the depositional history of a basin, only forward models can be applied, as basin evolution consists mostly of irreversible processes, which do not lend themselves to inverse modelling (Cross & Harbaugh, 1990). Stratigraphic forward modelling was applied to simulate basin development, and to quantify the physical factors determining deposition. This study focuses on the evaluation and quantification of internal and external parameters governing deposition, e.g. sediment transport, *in situ* carbonate production, erosion and compaction. Within the stratigraphic forward simulations, these parameters can be calculated and visualized independently in order to determine the primary factors. The modelled, mixed carbonate–clastic depositional system includes subsidence and uplift, flexural loading and eustatic sea-level changes. This method offers geometrical minimum/maximum models of sedimentary geometries and lithofacies distribution in time and space, predicting lithostratigraphic information in areas between outcrops. Consequently, it provides an opportunity to visualize seismic-scale geometries in a tectonically highly disturbed basin (after structural modelling has been applied), where only slices of the basin fill crop out. Further results comprise predictions of data between measured outcrops, such as:

- distribution of lithofacies and depositional environments
- distribution of lithophysical parameters (density, porosity)
- distribution of palaeowater depth
- information about depositional systems

In a procedural loop the data from forward modelling can be used to control and modify reverse-basin modelling results, as it considers sedimentary parameters and has a superior spatial and temporal resolution (Fig. 2). PHIL/BASIM™ comprises a comprehensive set of sedimentation and deformation algorithms that model:

- tectonic history (subsidence and uplift)
- sea-level history
- palaeoenvironmental parameters/proxies
- crustal response due to flexural loading of sediment and water loads
- compaction of sediment

- traction of siliciclastic sediment in fluvial and coastal settings
- dispersion of suspension load in marine settings
- gravity-flow sedimentation (slope failure, turbidites)
- production and redistribution of carbonate sediment
- erosion (subaerial, shoreline).

Input data

For this study, 34 sections were measured in the field (Fig. 4), in terms of lithofacies, lateral thickness variations, large-scale geometries, biostratigraphy and characteristic surfaces important for the sequence-stratigraphic interpretation. Allostratigraphic and sequence stratigraphic correlation of the sections allowed the reconstruction of basin architecture. Field data were interpolated into the synthetic Bernesga Transect according to the tectonic restrictions (see 'Structural balancing') in order to obtain sufficient information for reconstructing the basin development on a seismic scale.

Time and space

All variables are set at the start of the model and can be varied for each timestep. Because of the long simulation time, a constant timestep of 500 kyr was chosen. For the simulation time from 560 Ma to 322 Ma the model comprises 477 layers. Carbonate deposition uses an independent time increment of 21 kyr in order to adequately reflect the sensitivity of carbonate production systems to, for example, climate cycles and relative sea-level changes. The Bernesga Transect is divided into user-defined evenly spaced cells, with a cell spacing of 250 m.

Tectonic subsidence, sea level and compaction

Tectonic subsidence rates vary spatially and temporally. The numerical data have been calculated by 2D reverse-basin analysis (see above). The composed eustatic sea-level curve is identical to the curve used in the 2D reverse-basin modelling (see Fig. 12). The distribution of porosity in the cross-section is determined by the lithofacies and maximum burial depth of the interval. The distribution of physical properties, such as permeability and density is derived from porosity and lithology (Bowman & Vail, 1999).

Flexural response of the crust

The effective elastic lithosphere thickness (T_e) in the region studied changes through time, as the lithospheric plate experiences a broad variety of settings. The value of T_e for the stretched crust and newly created oceanic crust has been estimated to be 5–15 km according to the mid-values for rifting areas (Watts, 2001). This timespan lasted from the Precambrian up to the Lower Cambrian. However, the magmatic event in the Ordovician indicates renewed extension of the continental margin (Ollo de Sapo Event; Dallmeyer *et al.*, 1997; Fernández-Suárez *et al.*, 1998). For the further development of T_e on the passive margin of Gondwana only estimates can be made. The T_e values for passive margins vary from 5 to 60 km with a peak between 10 and 30 km (Watts, 2001; and references therein). A linear increase of T_e from 10 to 60 km was applied for the Ordovician to Early Carboniferous (for detail see Veselovský, 2004).

Simulation workflow

The simulation workflow and key algorithms in PHIL^{TM} have been described by Bowman & Vail (1999). The simulation considers tectonic and sea-level specifications, flexural behaviour of the crust, traction/suspension of siliciclastics, production and redistribution of carbonates, compaction, erosion, slumping, turbidites, etc. Restrictions include no out-of-plane sediment transport, limited bio- and chronostratigraphic resolution in the early Palaeozoic, no hydrodynamic and/or climate simulation. Different combinations of input parameters can lead to the same geological results, reflecting the non-uniqueness of results.

Numerical and graphical output

The evolution of the basin architecture can be visualized by means of a time-series animation (Veselovský, 2004), whereas the numerical results have been plotted as cross-sections, stratigraphic columns and chronostratigraphic plots.

- A series of lithostratigraphic plots illustrates the distribution of facies along the transect as well as the development of sediment geometries (Fig. 9).
- A palaeobathymetry figure plots the bathymetry at the time of deposition (Fig. 10A).
- A chronostratigraphic diagram plots the thickness of each horizontal interval according to its

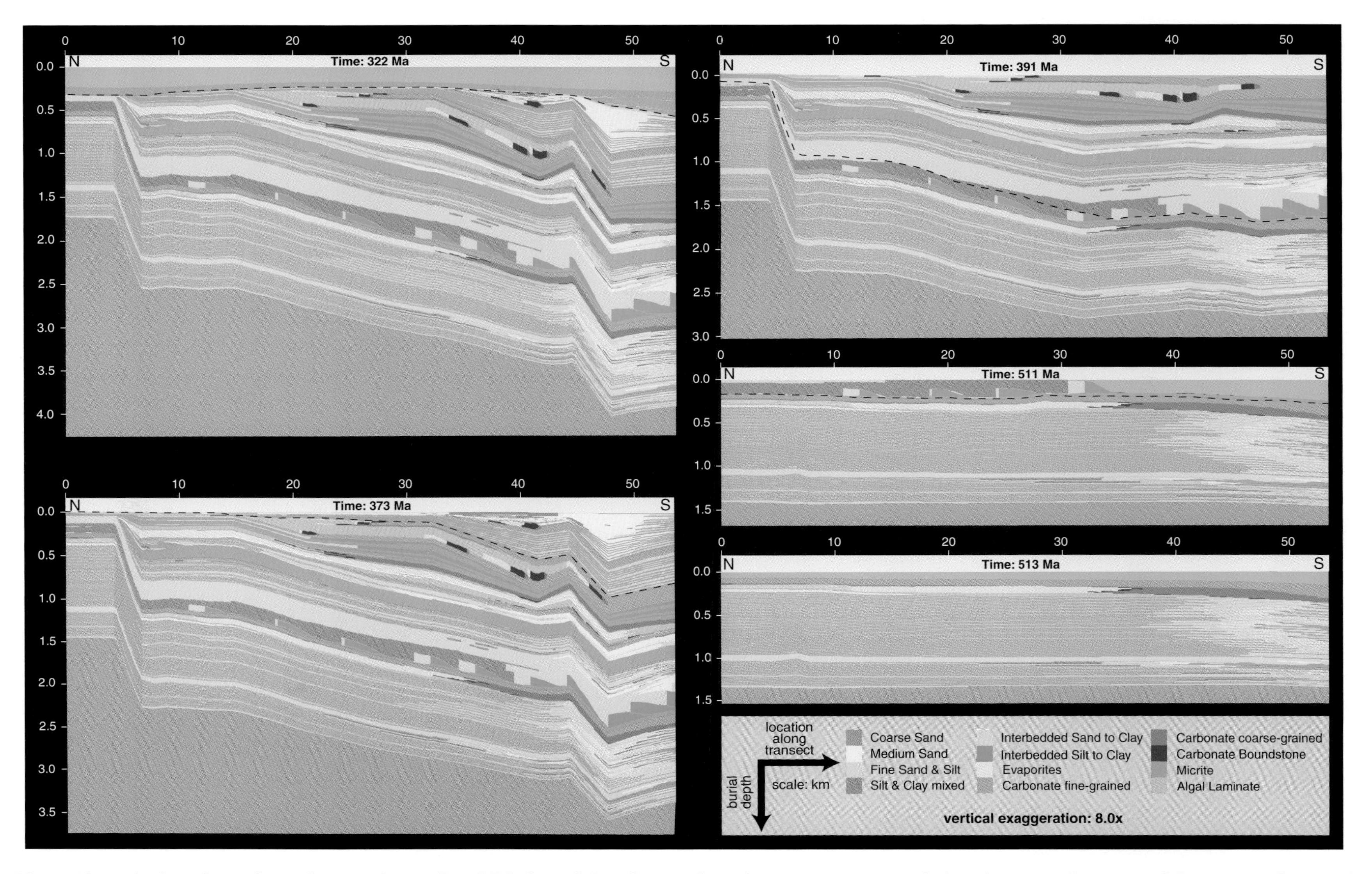

Fig. 9. Plots of selected time lines showing the predicted lithological distribution along the Bernesga Transect before the onset of Variscan deformation. The top of an increment is marked by a dashed line in the plot that follows. 513 Ma: top Láncara Fm; 511 Ma: Oville Fm; 391 Ma: top Santa Lucía Fm; 373 Ma: top Nocedo Fm; 322 Ma: top Alba Fm. See Figs 1B and 4 for the position of the Bernesga Transect.

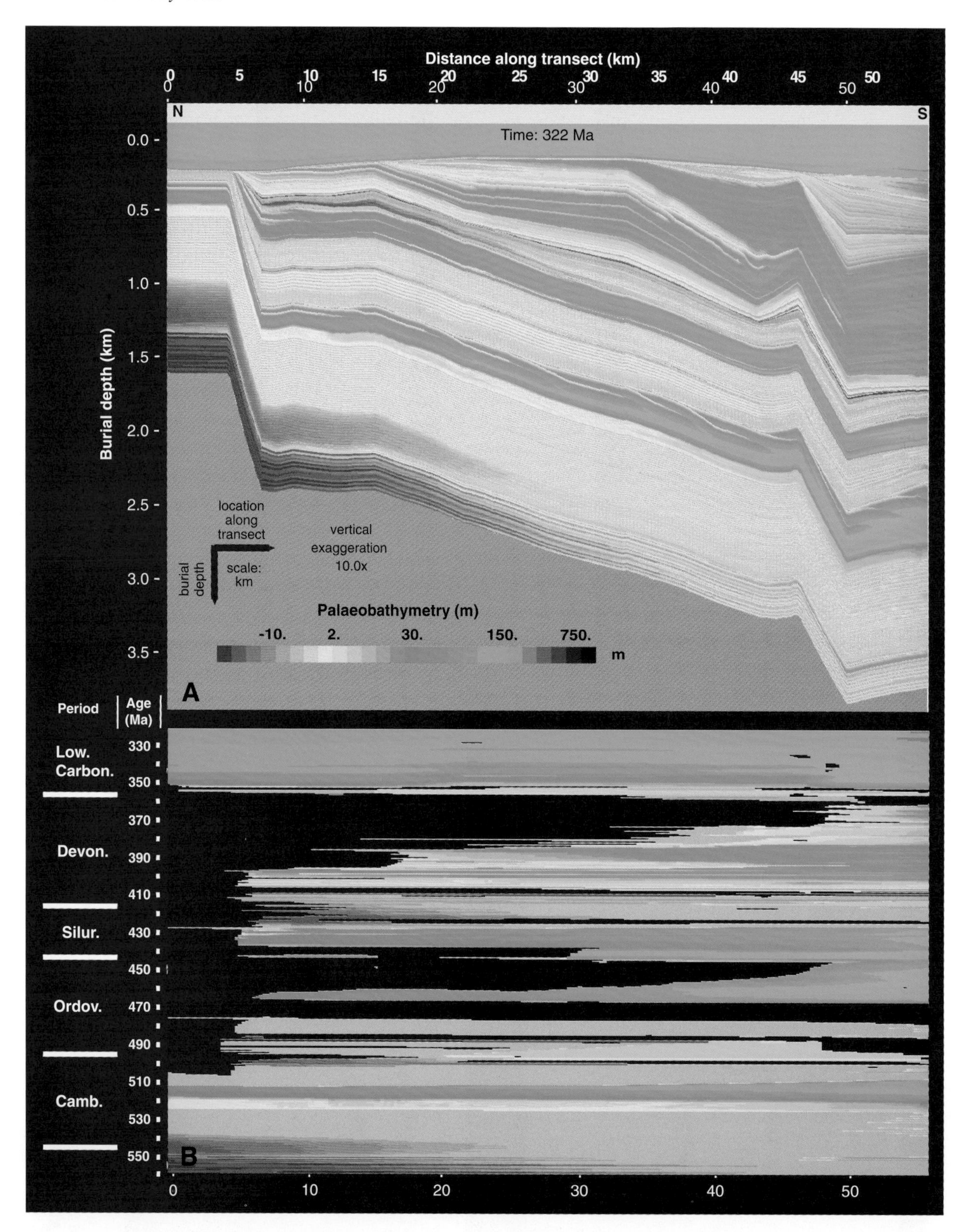

Fig. 10. (A) Depositional depths (palaeobathymetry) of sediments along the Bernesga Transect. The youngest sediments represent the top of the Alba Formation (322 Ma). Note the predominance of shallow marine settings throughout the basin development. (B) Chronostratigraphic plot for the time 560 Ma to 322 Ma. Depositional depths (palaeobathymetry) are plotted in relation to time. Areas of non-deposition (hiatus) are coded in black. See the pronounced backstepping of depositional areas and expansion of the uplifted region to the south in the Devonian.

position along the cross-section. The thickness of the interval is represented by the height of the interval for each cell. Depositional depths (palaeobathymetry) are plotted in relation to time (Fig. 10B).

- Stratigraphic columns can be drawn for each cell and compared with cross-sections in order to calibrate the forward simulation to real-world basin fill data (Fig. 11).

Results

The geometry of a basin is mainly controlled by basin-forming tectonics, but the morphology of a basin, defined by the sediment surface, is the product of the interplay between tectonic movements and sedimentation. Basin-fill development is mainly controlled by two major parameters, accommodation and sediment supply/production. Accommodation space is a function of eustatic sea-level fluctuations and subsidence. The importance of sediment supply/production has been emphasized by Thorne & Swift (1991) for clastic and by Schlager (1993) for carbonate depositional systems.

The siliciclastic sediment supply is given in $m^2\,kyr^{-1}$ as it represents the filling of a two-dimensional area in a two-dimensional model. Changes in sediment input are plotted in Fig. 12. From the latest Neoproterozoic to the Early Ordovician the area was dominated by a high flux of siliciclastic sediments with sedimentation rates ranging between 2600 and 4600 $m^2\,kyr^{-1}$. Only the deposition of a thin carbonate succession (Láncara Fm, 525–513 Ma) and several long-persisting hiatus interrupt the clastic period (hiatus/non-deposition are coded in black in Fig. 10B). During the Silurian only siliciclastic material was deposited. Flux rates diminished by 60% (600 to 1800 $m^2\,kyr^{-1}$), however. The Devonian is characterized by deposition of alternating siliciclastics and carbonates, with highly variable clastic sediment supply (770 $m^2\,kyr^{-1}$ up to 4000 $m^2\,kyr^{-1}$; see Fig. 12). Siliciclastic influence decreased towards the beginning of the Carboniferous, diminishing towards the end of the simulation (322 Ma). Figure 9 shows the development of facies patterns predicted along the Bernesga Transect. The periods of alternating siliciclastic and carbonate deposition reflect the complex interplay between sediment input and eustatic sea-level fluctuations.

The Devonian is one of the most significant maxima of reefal carbonate production in the Phanerozoic (Kiessling *et al.*, 2000). Carbonate accumulation rates are the net result of sediment input and *in situ* production (for carbonate environments), without consideration of sediment export by means of bypass or removal by erosion. They depend on depositional environment, basinal asymmetry, climate, tectonic setting and the time increment being modelled (Enos, 1991). Accumulation rates are mostly calculated in 1D at a specific locality and normally given as compacted values. If the sediment supply/production of a system is greater than the 1D accommodation space for this locality, the accumulation rates are governed by the development of accommodation space and not by the production potential of the organisms.

Accumulation rates should not be confused with the growth potential of carbonate systems, representing the maximum production potential for a specific sedimentary environment. Production rates calculated in the present model are decompacted values in 2D, representing the vertical accumulation together with sediment export and bypass.

Figure 13 shows a compilation of compacted accumulation rates of carbonate systems for the Devonian, Carboniferous and Cenozoic. In the Southern Cantabrian Basin three important carbonate producing intervals occur within the Devonian. These correspond to the Abelgas (415–402 Ma), Santa Lucía (396–391 Ma) and Portilla Fms (385–382 Ma). For comparison of accumulation rates with data from the literature, compacted rates were used, with decompacted values for information. Decompacted rates for the three Devonian carbonate formations reach values 20–39% higher than compacted values (Fig. 13). Only rates with a timespan of observation between 2 and 20 Myr were plotted, corresponding to the timespans of the Cantabrian carbonate formations. Schlager (2000) stressed that the average sedimentation rates of all carbonate systems decrease with an increasing timespan of observation. All values are given in $m\,Myr^{-1}$ (comparable to $\mu m\,yr^{-1}$, Bubnoff unit). Compacted accumulation rates of the Santa Lucía Fm (up to 56 $m\,Myr^{-1}$) reach values slightly lower than values given for comparison (Fig. 13). The Portilla (up to 30 $m\,Myr^{-1}$) and Abelgas (up to 18 $m\,Myr^{-1}$) Formations show considerably lower rates. All three formations have maximum accumulation rates

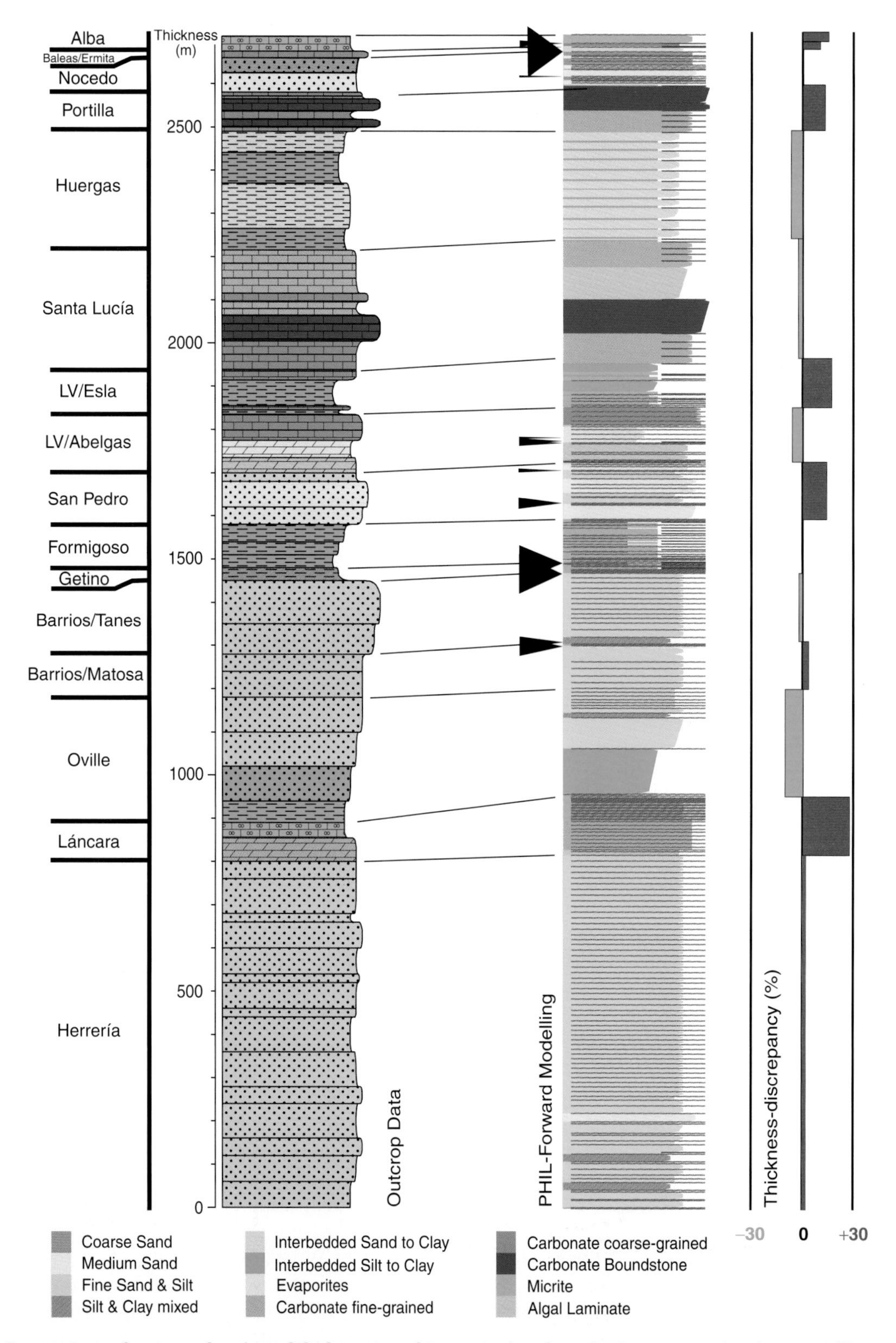

Fig. 11. Comparison of outcrop data (simplified stratigraphic section) and synthetic cross-section (output of forward stratigraphic modelling) at an identical position along the transect (km 42). See Figs 1B and 4 for the position of the Bernesga Transect. Mean values for thickness discrepancy range between 5% and 12%. The study focuses mainly on the timespan between the Ordovician and the Carboniferous. For older strata, the simulation tries to match the depositional facies and thickness and does not claim to be exhaustive. See Fig. 3 for absolute ages of the formations.

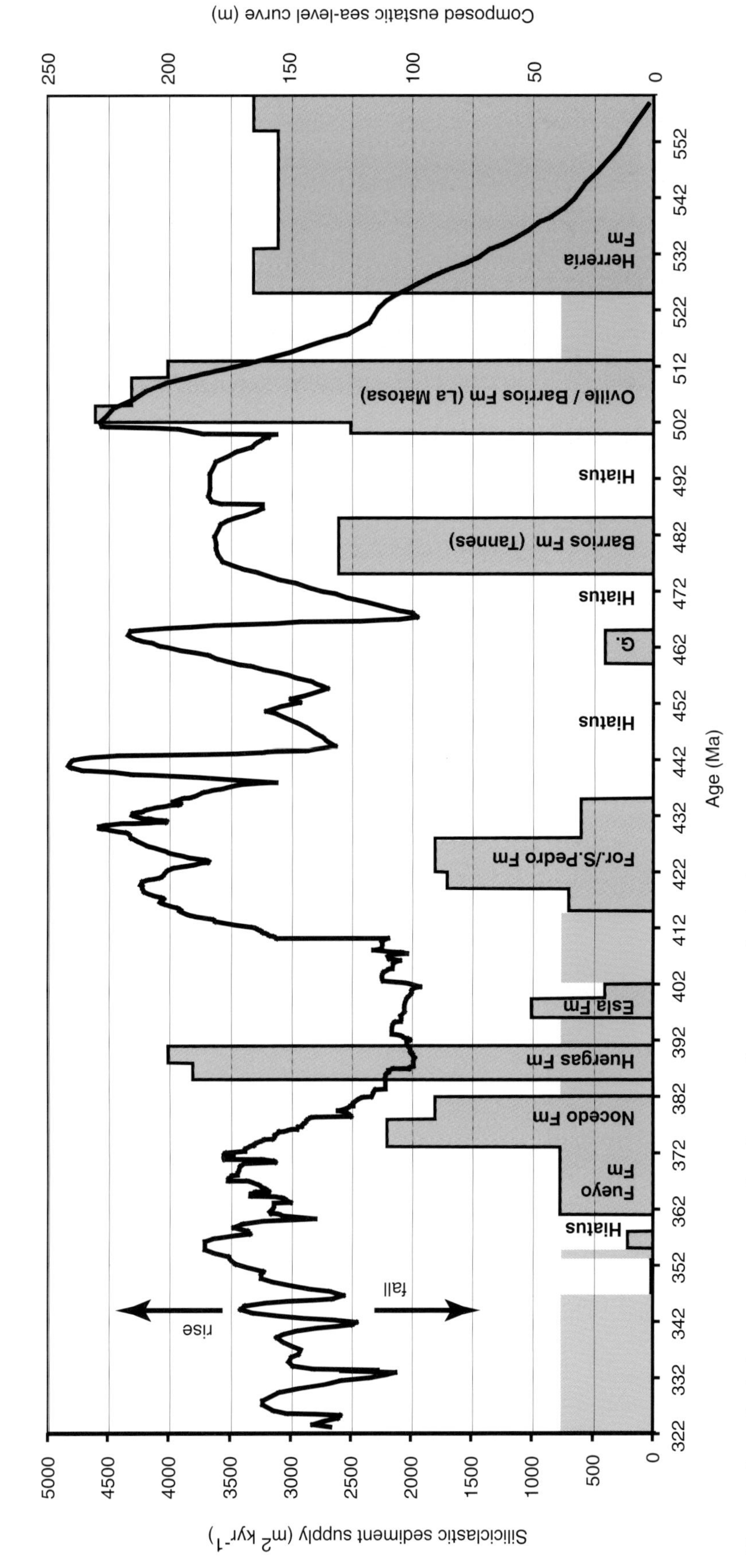

Fig. 12. Grey-shaded areas represent siliciclastic sediment supply for the Bernesga Transect in time. Values are given in square metres per 10,000 years, as they represent the supply for a two-dimensional transect and do not fill a three-dimensional space. Eustatic sea-level fluctuations are plotted as a solid black line. See text for references. Blue-shaded areas correspond to times of carbonate deposition (see Fig. 13 for carbonate accumulation rates). Note the good fit between major eustatic sea-level falls and hiatus of the following timespans: 500–485 Ma, 475–465 Ma, 459–435 Ma and 361–358 Ma. The positions of the hiatus within the stratigraphic record are given in Fig. 3.

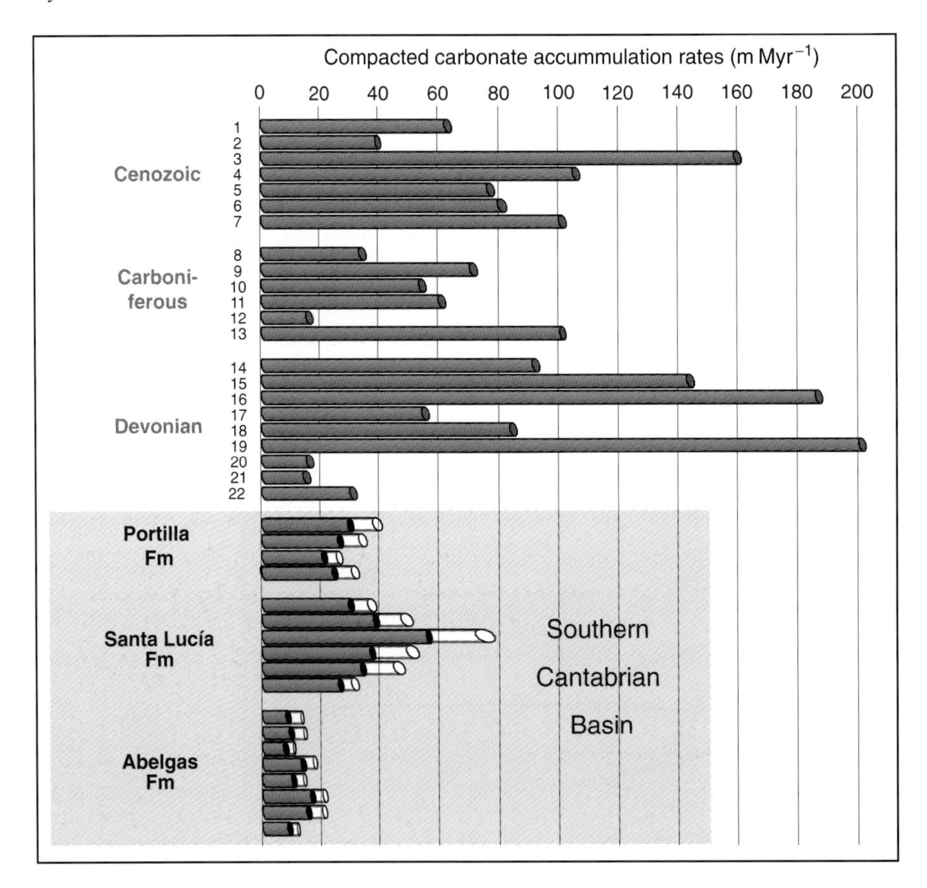

Fig. 13. Compacted carbonate accumulation rates for the Devonian, Carboniferous and Cenozoic, as well as for the three Devonian carbonate formations of the Southern Cantabrian Basin (Abelgas, Santa Lucía and Portilla Formations). Decompacted carbonate accumulation rates of the Cantabrian formations are plotted as white bars. Data outside the Cantabrian Mountains were assembled from the following sources: (1) Chevalier 1973; (2) Frost 1981; (3) Terry & Williams 1969; (4–7) Enos 1991; (8) Parkinson 1957; (9) Lees 1961; (10) Rose 1976; (11–13) Enos 1991; (14) Walls *et al.* 1979; (15) Playford 1980; (16) Krebs 1974; (17) Muir *et al.* 1985; (18, 19) Burchette 1981; (20–22) Enos 1991.

between kilometre 35 and 41 along the Bernesga Transect, with diminishing values towards the south and towards the north. The distribution of reefal build-ups is plotted in Fig. 9.

The three Cantabrian carbonate factories show very different production rates. The oldest one, the Abelgas Fm (415–402 Ma) has the lowest carbonate production rates (90–230 m Myr^{-1}), increasing towards the top of the succession. Following the shaly deposits of the Esla Fm (402–396 Ma), remarkable reefs of the Santa Lucía Fm (396–391 Ma) developed. They show the highest production rates (100–780 m Myr^{-1}) for the whole simulation interval. The carbonate sedimentation is once again followed in time by high siliciclastic sediment influx (4000 m^2 kyr^{-1}) during the deposition of the Huergas Fm (391–385 Ma) and the renewed establishment of carbonate conditions (Portilla Fm, 385–382 Ma). This carbonate factory displays rather low carbonate production rates of 130–150 m Myr^{-1}.

DISCUSSION OF BASIN EVOLUTION

The Southern Cantabrian Basin can be referred to as a 'polyhistory basin' (Klein, 1987) or 'polyphase basin' (Einsele, 2000), having experienced major changes in tectonical style.

Subsidence and stage development

Two-dimensional reverse-basin modelling results reflect a complex evolution from a rift initiation–climax stage (classification after Prosser, 1993) to a post-rift stage (passive continental margin) and finally to a foreland basin, governed by the approach of the Variscan Orogen in the Early Carboniferous. This evolution (560 Ma to 313 Ma) is visualized in Figs 7 and 8, showing six major subsidence trends. Numerical results are summarized in Table 2. The overall pattern is in accordance with the qualitative, palaeogeographical reconstructions of Paris & Robardet (1990), Tait *et al.* (1997), Fernández-Suárez *et al.* (1998, 2000, 2002a, 2002b), Robardet (2000, 2002) and others.

Rifting led to rapid initial subsidence, which was followed by persistent, exponentially decaying thermal subsidence (Einsele, 2000). The pattern of the first subsidence trend (560–500 Ma; mean value 16 m Myr^{-1}) fits this assumption of the development of the highest subsidence rates during the early stages of the model (rifting) and declining subsidence rates towards the post-rift stage. Dating by Sdzuy (1967, 1968) suggests a diachronous, tectonic breakdown of the Láncara carbonate platform. The subsequently deposited uppermost nodular limestones (Láncara Fm) and the offshore shales of the Oville Formation (Genestosa Member) can be interpreted as corresponding to the rift climax stage (according to the classification of Prosser, 1993).

The second trend (500–435 Ma) indicates a phase of low subsidence rates (mean value 5 m Myr^{-1}) with minor spatial differences in subsidence along the N–S trending Bernesga Transect and several long persisting hiatus (10–24 Myr). An Early Ordovician magmatic event (Ollo de Sapo Event; Dallmeyer *et al.*, 1997, 460–490 Ma; Fernández-Suárez *et al.*, 1998, 465–520 Ma) together with regional unconformities indicate renewed thinning and further extension of the crust up to the Ordovician. During this subsidence trend, terrigenous sedimentation rates (2500–2600 m^2 kyr^{-1}), derived from the 2D stratigraphic forward modelling declined by 46% in comparison with the end of the first trend (4600 m^2 kyr^{-1}). According to Einsele (2000) this feature is typical for the shift from rift to drift stage (immediate post-rift stage according to Prosser, 1993). The absence of carbonates on the shallow-marine platform points to a position at higher latitudes during this time in contrast to the Lower Cambrian, which bears evaporitic remains in the lower member of the Láncara Fm.

Differential subsidence with an uplifted area in the north (−8 m Myr^{-1}; Cantabrian High) and higher subsidence to the south (up to 35 m Myr^{-1}) becomes visible in the third trend (435–415 Ma), and is reinforced in the fourth trend (415–361 Ma; −13 to 40 m Myr^{-1}). The Cantabrian High was a structural high from Ludlovian times, as expressed by erosional unconformities and diminishing sediment thicknesses towards the elevated region. The northernmost areas were subsequently eroded down to the Ordovician Barrios Fm. Despite a rising eustatic sea level (Emsian to early Famennian; Fig. 12) and a worldwide transgressive trend, the Devonian sediments of the Asturo-Leonese Domain were deposited in a regressive pattern (García-Alcalde *et al.*, 2002). This pattern was caused by the regional tectonic movements of the Cantabrian High, which was a southward tilting block until the Frasnian. Since that time, the synsedimentary Intra-Asturo-Leonesian Facies Line (Raven, 1983; see also Fig. 4) was a sharp facies and thickness boundary (approximately at km 45 along the Bernesga Transect), creating accommodation space in the southernmost part of the basin (deposition of the Nocedo and Fueyo Fms). Thus in time, the Cantabrian High extended in two pulses (428–415 Ma and 415–361 Ma) to the south, displacing the regions of main subsidence and deposition basinwards. Consequently the non-depositional/erosional hiatus of the Cantabrian High extended spatially to the south (Figs 8 and 10B). Compaction-induced subsidence rates highlight this movement, as compactable sediment was only deposited south of the shifting topographic high. Figure 10A shows the depositional depths (palaeobathymetry) along the Bernesga Transect. As most of the sediments were deposited under shallow-marine conditions, flexure caused by water burden can be excluded.

During the fifth trend (361–322 Ma) the region was flooded, depositing condensed, deeper marine sediments throughout the entire basin (nodular limestones of the Alba Fm). During this interval, the inner parts of the orogen (Central Iberian Zone) were already deformed due to the Variscan collision, causing the change from a passive continental to an active margin (Colmenero *et al.*, 2002). The terrigenous input from the evolving mountain range was trapped in an orogenic foredeep, initially situated in the West Asturian-Leonese Zone (Fig. 1). This is, however, not visible in the field, as the sediments of this foredeep became uplifted, eroded and later filled the younger foredeep of the Cantabrian Zone (Ábalos *et al.*, 2002). This might be the reason that the Cantabrian Carboniferous clastics are quite poor in terrigenous provenance indicator minerals.

The sixth trend (Variscan foredeep stage; 322–313 Ma) marks a sharp change in subsidence rate distribution in the uppermost Lower Carboniferous and a marked basin reorganization. High thermo-tectonic subsidence rates in the southernmost part and moderate rates in the north indicate strong spatial differences along the Bernesga Transect. The southern, highly subsiding depocentres migrated in time to the north (according to the present coordinate system), indicating the approaching orogen and its foredeep (Fig. 8). Considering the chronology established by Truyols & Sánchez de Posada (1983), first evidence for the advance of the Variscan Orogen in the examined area became visible in the middle Namurian A (approximately 322 Ma). Nevertheless the area investigated might be too small to provide information about the migration of a forebulge.

In a global context the subsidence configuration of the Southern Cantabrian Basin points to the onset of a long-term, first-order continental encroachment cycle, with a basinward shift of regional onlap. This concept was proposed by Duval *et al.* (1998). Two continental encroachment cycles with a duration of more than 50 Myr (1st order) were recognized during the Phanerozoic, caused by the changes in ocean-basin volume induced by the break-up and subsequent assembly of supercontinents (Duval *et al.*, 1998). During the older encroachment cycle (Neoproterozoic to the end of the Permian) the change from a backstepping/transgressive phase to a forestepping/regressive phase, marked by a major downlap surface, took place at ca. 500 Ma. After a eustatic high during the Ordovician–Silurian and a maximum marine transgression around the Ordovician–Silurian boundary, there was a gradual restriction of the marine domain from the Silurian to the Permian. Cycles of higher orders are superimposed (2nd to 4th order). Regression–transgression cycles (2nd order; duration 3–50 Myr) bound by hiatus and associated with 2nd order eustatic cycles, are assumed to be the result of changes in the rate of regional tectonic subsidence and/or changes in the rate of sea-floor spreading (Vail *et al.*, 1984). Not every aspect of this concept is reflected in the present model. The change from the transgressive to the regressive phase is not visible in the Cantabrian Zone as this time is characterized by numerous hiatus (Figs 3 and 10B) and the tectonic overprint does not allow the recognition of seismic-scale surfaces. The maximum marine flooding is marked by the black shales of the Formigoso Fm (Figs 3 and 12), with a slightly delayed onset (early Llandoverian) in comparison with the data of Duval *et al.* (1998). It is also the start of a second-order encroachment subcycle, bounded by a significant hiatus (top 435 Ma). The San Pedro Fm shows a basinward shift of regional onlap from member A to member C. The subcycle has a duration of 20 Ma and ends at time line 415 Ma. The second and more pronounced basinward displacement of regional onlap (Fig. 10B) commences with the deposition of the Esla Fm and lasts until the Upper Devonian hiatus (361 Ma). The duration of this encroachment subcycle amounts to 41 Myr. There is no marked hiatus separating the two continental encroachment subcycles in the Southern Cantabrian Basin (see above).

Sedimentary evolution

Two-dimensional stratigraphic forward modelling creates a high-resolution synthetic shelf–basin transect (Figs 9 and 10). The comparison of outcrop data with modelling results in terms of lithofacies distribution, thicknesses, palaeowater depth and depositional systems, shows a good fit (Fig. 11). The dominance of siliciclastic deposition from the Neoproterozoic to the Ordovician (only interrupted by Early Cambrian, often microbial carbonate sedimentation) is reflected by very high sediment flux rates throughout the basin, ranging between 2500 and 4600 $m^2\ kyr^{-1}$ (Fig. 12). The siliciclastic succession is interrupted by numerous levels of non-deposition (hiatus, Fig. 10B), which have been correlated to major falls of eustatic sea level (Fig. 12). Reef growth is chiefly controlled by numerous, closely related extrinsic factors, such as sea level, palaeoclimate, oceanography, plate tectonics, nutrients and tectono-sedimentary setting. The Early Devonian carbonates of the Abelgas Fm developed because the siliciclastic input dropped to 700 $m^2\ kyr^{-1}$ at the end of the Silurian. Nevertheless, during later times strong fluctuations in siliciclastic input were one of the major limiting factors for carbonate growth. The carbonate Santa Lucía (Eifelian) and Portilla (Upper Givetian) Formations were drowned by sandy siliciclastics with mean flux rates of 2000 to 3950 $m^2\ kyr^{-1}$ (Fig. 12). Additionally, carbonate production rates were governed by differential thermo-tectonic and flexural-induced subsidence and their strong influence on the creation and destruction of accommodation space. During the deposition of the Abelgas Fm, low thermo-tectonic (7–24 $m\ Myr^{-1}$) and flexural-induced (1–3 $m\ Myr^{-1}$) subsidence rates together with a falling eustatic sea level limited the accommodation space available. In the case of the Santa Lucía Fm, high subsidence (up to

40 m Myr^{-1}) and a slowly rising eustatic sea level (Johnson *et al.*, 1985; Dennison, 1985; Fig. 12) offered sufficient accommodation space for the development of a thick carbonate succession (up to 356 m – decompacted value). Throughout the modelled period the amplitudes of the synthetic eustatic sea-level curve show excessive values for this basin and had to be reduced by an average of 9% (Fig. 12).

Based on their great number and dimensions, the Devonian reefs probably produced the largest amount of reefal carbonate in the Phanerozoic (Kiessling *et al.*, 2000). Decompacted carbonate production rates show evidence of a significant increase in production during the Devonian. The values rise from 90 m Myr^{-1} in the lowermost Gedinnian (Abelgas Fm) up to 780 m Myr^{-1} in the Eifelian (Santa Lucía Fm). This formation marks the first appearance of highly productive reef organisms in the Cantabrian Zone. The quantitative results point to a shift of the sedimentary environment towards tropical conditions adequate for reef growth. This is consistent with the qualitative palaeogeographical models of Paris & Robardet (1990) proposing a northward drift of Iberia during this time interval. Carbonate production rates of the Portilla Fm (upper Givetian) dropped to 150 m Myr^{-1}, which is 81% less than in the Eifelian. This drop is caused by an increased terrigenous input suppressing the carbonate productivity and possibly diminishing accommodation space during this time. Spatially, the results of this study point to an earlier development of a syn-sedimentary fault between km 45 and 47 (Intra-Asturo-Leonesian Facies Line, see above and Fig. 4), starting at the latest at the beginning of the Late Devonian (see also discussion in Veselovský, 2004). The identical position of reefal development during Santa Lucía and Portilla times at km 42, only a few kilometres north of the Intra-Asturo-Leonesian Facies Line, indicates the influence of this line on carbonate palaeogeography (Fig. 9).

The carbonate production rates in the Southern Cantabrian Mountains reached their peak close to the Emsian–Eifelian boundary. This is in slight disagreement with the global data published by Kiessling *et al.* (2000), who claim that the maximum of total production rates was in the Givetian–Frasnian. The Lower Carboniferous Alba Fm is represented by condensed sediments with very low accumulation rates (1.2–1.8 m Myr^{-1}).

The Southern Cantabrian Basin experienced very low bathymetric fluctuations throughout its history. Apart from the uppermost part of the Láncara Fm and the Alba Fm (red nodular limestone) and the black shales of the Formigoso and Vegamián Fms, the depositional environment was constantly at shallow-marine depths (Fig. 10B). Long-persisting morphological basin configuration and maintenance of shallow-marine facies was described as a 'balanced basin' (Einsele, 2000), where sediment supply and vertical build-up approximately compensated for subsidence. On the other hand, due to the Mesozoic and Cenozoic cover of the Duero Basin and the structural situation, no data are available for the estimation of the true former extent of the Southern Cantabrian Basin at this time. Consequently, long-persisting shallow-marine conditions probably signify that the basin was oversupplied with sediment at some time. In this case the sediment flux throughout the basin may have been underestimated, as an uncertain amount of sediment was transported to deeper regions further south, not visible in the outcropping geological record.

Open questions

Stratigraphic modelling depends to a great extent on an adequate chronostratigraphical and /or biostratigraphical framework. Even though numerous authors have contributed to the present knowledge of the Cantabrian stratigraphy, the precise dating of formations across the Cantabrian Mountains still remains the largest uncertainty in this study. Thus the model focuses on Ordovician to Carboniferous strata, as the time-resolution diminishes significantly towards the older strata.

Although structural balancing forms part of the present study (see also Veselovský, 2004), it represents only a first approximation of a highly complex topic. A detailed structurally balanced cross-section along the Bernesga Valley, with a strong focus on outcrop data would offer a better insight into the spatial extent of the basin before the onset of tectonic deformation. A study along these lines would also need to touch on the difficult task of distinguishing between Variscan and Alpidic deformation.

SUMMARY

This multidisciplinary approach provides new quantified data for the evolution of the Palaeozoic Southern Cantabrian Basin.

Structural balancing offers approximations of pre-deformational, spatial relationships between measured cross-sections, as well as minimal basin shortening and Late Devonian basin geometry. At the basin scale, the amount of total tectonic shortening of the deformed basin infill was 54% at minimum. The Pedrosa Domain displays the highest values reaching 65%, followed by the Bodón (41%) and Alba (25%) Domains.

Using 2D reverse-basin modelling, six major subsidence trends within the total subsidence values and their components can be distinguished between 560 Ma and 313 Ma. These trends reflect a complex evolution from a rift initiation–climax stage, to a post-rift stage (passive continental margin) and finally to a foreland basin, governed by the approach of the Variscan Orogen in the Early Carboniferous. The strongly subsiding foredeep moved from south to north in time (present-day coordinates), reflecting movements of the Variscan Orogen. Two second-order encroachment subcycles with durations of 20 Myr and 41 Myr were identified, based on subsidence patterns and basinward shifts of regional onlap. Maximum marine flooding was reached during the early Llandovery.

Two-dimensional stratigraphic forward modelling visualizes the predicted depositional history along the synthetic Bernesga Transect. Differential thermo-tectonic and flexural-induced subsidence, along with the fluctuating eustatic sea level controlled the production and destruction of accommodation space. Siliciclastic flux rates between 2500 and 4600 $m^2 kyr^{-1}$ impeded carbonate production, which was able to initiate after a significant drop of clastic input (700 $m^2 kyr^{-1}$) in the Early Devonian. Three Devonian carbonate factories display decompacted carbonate production rates from 90 to 780 $m Myr^{-1}$, each depending on (i) the amount of siliciclastic input, (ii) the accommodation space available and (iii) palaeogeographical position of the depositional area. The present study shows that carbonate production rates are often underestimated in the literature.

ACKNOWLEDGEMENTS

Thanks go to J. Adam for highly stimulating discussions about structural balancing of the Bernesga Transect and S. Bowman for his assistance and updates of PHIL^{TM}. We also thank the reviewers, J.J. Alvaro and M. Wagreich, for their helpful comments. We are especially grateful to the inhabitants of Villamanín, León for their friendship and limitless help during our field seasons in Spain. We also thank N. Wilson, who kindly read over the English of this manuscript. Financial support was provided by the German Research Fund (Be 641/36-1).

REFERENCES

Ábalos, B., Carreras, J., Druguet, E., Escuder Viruete, J., Gómez Pugnaire, T., Lorenzo Álvarez, S., Quesada, C., Rodríguez Fernández, L.R. and **Gil-Ibarguchi, J.I.** (2002) Variscan and Pre-Variscan tectonics. In: *The Geology of Spain* (Eds W. Gibbons and T. Moreno), Geological Society of London, London, pp. 155–183.

Allen, P.A. and **Allen, J.R.** (1990) *Basin Analysis: Principles and Applications.* Blackwell Science, Oxford, pp. 451.

Alonso, J.L. and **Suárez Rodríguez, A.** (1991) Tectónica. In: *Mapa geológico de España, Escala 1:50.000, La Pola de Gordón (103).* Instituto Tecnológico y Geominero de España, Madrid, 79–88.

Alonso, J.L., Suárez Rodríguez, A., Rodríguez Fernández, L.R., Farias, P. and **Villegas, F** (1991) *Mapa geológico de España, Escala 1:50.000, La Pola de Gordón (103),* Instituto Tecnológico y Geominero de España, Madrid.

Alonso, J.L., Pulgar, J.A., García-Ramos, J.C. and **Barba, P.** (1996) Tertiary basins and Alpine tectonics in the Cantabrian Mountains (NWSpain). In: *Tertiary Basins of Spain* (Eds P.F. Friend and C. Dabrío), Cambridge University Press, Cambridge, pp. 214–227.

Álvaro, J.J., Rouchy, J.M., Bechstädt, T., Boucout, A., Boyer, F., Debrenne, F., Moreno-Eiris, E., Perejón, A. and **Vennin, E.** (2000a) Evaporitic constraints on the southward drifting of the western Gondwana margin during Early Cambrian times. *Palaeogeogr. Palaeoclimatol. Palaeoecol.*, **160,** 105–122.

Álvaro, J.J., Vennin, E., Moreno-Eiris, E., Perejón, A. and **Bechstädt, T.** (2000b) Sedimentary patterns across the Lower-Middle Cambrian transition in the Esla nappe (Cantabrian Mountains, northern Spain). *Sed. Geol.*, **137,** 43–61.

Aramburu, C. (1995) El precámbrico y el paleozoico inférior. In: *Geología de Asturias* (Eds C. Aramburu and F. Bastida), Ediciones Trea, S.L., pp. 35–50.

Aramburu, C. and **García-Ramos, J.C.** (1993) La sedimentación cambroordovícica en la Zona Cantábrica (NO de España). *Trabajos de Geologia*, **19,** 45–73.

Aramburu, C., Truyols, J., Arbizu, M., Méndez-Bedía, I., Zamarreño, I., García-Ramos, J.C., Suarez de Centi, C. and **Valenzuela, M.** (1992) El Paleozoico Inferior de la Zona Cantábrica. In: *Paleozoico Inferior de Ibero-America* (Eds J.C. Gutiérrez Marco, J. Saavedra and I. Rábano), Unex Press, pp. 397–422.

Bastida, F. and **Aller, J.** (1992) Rasgos geológicos generales. In: *Geología de Asturias* (Eds C. Aramburu and F. Bastida). Ediciones Trea, S.L., Somonte, Spain, pp. 27–34.

Bastida, F., Brime, C., García López, S. and **Sarmiento, G.N.** (1999) Tectono-thermal evolution in a region with thin-skinned tectonics: The western nappes in the Cantabrian Zone (Variscan belt of NW Spain). *Int. J. Earth Sci.*, **88,** 115–130.

Berggren, W.A., Kent, D.V., Swisher III, C.C. and **Aubry, P.P.** (1995) A revised Cenozoic geochronology and chronostratigraphy. In: *Geochronology, Time Scales and Global Stratigraphic Correlation* (Eds W.A. Berggren, D.V. Kent, M.P. Aubry and J. Hardenbol). SEPM Spec. Publ., Tulsa, USA, Vol. 54, 386pp.

Bowman, S.A. and **Vail, P.R.** (1999) Interpreting the stratigraphy of the Baltimore Canyon section, offshore New Jersey with PHIL, a stratigraphic simulator. In: *Numerical Experiments in Stratigraphy: Recent Advances in Stratigraphic and Sedimentologic Computer Simulations* (Eds. J.W. Harbaugh, W.L. Watney, E. Rankey, R. Slingerland, R. Goldstein and E. Franseen), SEPM Spec. Publ., Tulsa, USA, Vol. 62, pp. 117–138.

Boyer, S.E. and **Elliott, D.** (1982) Thrust systems. *AAPG Bull.*, **66,** 1196–1230.

Burchette, T.P. (1981) European Devonian reefs; a review of current concepts and models. In: *European Fossil Reef Models* (Ed. D.F. Toomey), SEPM Spec. Publ., Tulsa, USA, Vol. 30, pp. 85–142.

Burov, E.B. and **Diament, M.** (1995) The effective elastic thickness (T_e) of continental lithosphere; what does it really mean? *J. Geophys. Res.*, B, **100(3),** 3905–3927.

Burov, E.B. and **Diament, M.** (1996) Isostasy, equivalent elastic thickness, and inelastic rheology of continents and oceans. *Geology*, **24,** 419–422.

Carey, S.W. (1955) The orocline concept in geotectonics. *R. Soc. Tasmania Proc.*, **89,** 255–288.

Chevalier, J.P. (1973) Geomorphology and Geology of Coral Reefs in French Polynesia. Biology and Geology of Coral Reefs (Eds O.A. Jones and R. Endean), Vol. 1, Geology 1. Academic Press, New York.

Cloetingh, S. and **Burov, E.B.** (1996) Thermomechanical structure of European continental lithosphere: Constraints from rheological profiles and ETT estimates. *Geophys. J. Int.*, **124,** 695–723.

Colmenero, J.R., Fernández, L.P., Moreno, C., Bahamonde, J.R., Barba, P., Heredia, N. and **González, F.** (2002) Carboniferous. In: *The Geology of Spain* (Eds W. Gibbons, and T. Moreno), *Geol. Soc. London*, 93–116.

Cross, T.A. and **Harbaugh, J.W.** (1990) Quantitative dynamic stratigraphy: A workshop, a philosophy, a methodology. In: *Quantitative Dynamic Stratigraphy* (Ed. T.A. Cross), Prentice-Hall, Englewood Cliffs, pp. 3–20.

Dahlstrom, C.D.A. (1969) Balanced cross sections. *Can. J. Earth Sci.*, **6,** 743–757.

Dallmeyer, R.D., Martínez-Catalán, J.R., Arenas, R., Gil Ibarguchi, J.I., Gutiérrez-Alonso, G., Farias, P., Bastida, F. and **Aller, J.** (1997) Diachronous Variscan tectonothermal activity in the NW Iberian Massif: Evidence from 40Ar/39Ar dating of regional fabrics. *Tectonophysics*, **277,** 307–337.

Dennison, J.M. (1985) Devonian eustatic fluctuations in Euramerica: Discussion. *GSA Bull.*, **96,** 1595–1597.

Duval, B.C., Cramez, C. and **Vail, P.R.** (1998) Stratigraphic cycles and major marine source rocks. In: *Mesozoic and Cenozoic Sequence Stratigraphy of European Basins* (Eds P.C. De Graciansky, J. Hardenbol, J. Thierry and P.R. Vail), Soc. Sediment. Geol. SEPM, Tulsa, USA, Vol. 60, pp. 43–52.

Einsele, G. (2000) *Sedimentary Basins. Evolution, Facies, and Sedimentary Budget.* Springer, Heidelberg, Germany, pp. 792.

Enos, P. (1991) Sedimentary parameters for computer modeling. In: *Sedimentary Modeling; Computer Simulations and Methods for Improved Parameter Definition* (Eds E.K. Franseen, W.L. Watney, C.G.St.C. Kendall and W. Ross), *Bulletin Kansas State Geol. Surv.*, Lawrence, USA, Vol. 233, pp. 63–99.

Fernández-Suárez, J., Gutiérrez-Alonso, G., Jenner, G.A. and **Jackson, S.E.** (1998) Geochronology and geochemistry of the Pola de Allande granitoids (northern Spain): Their bearing on the Cadomian-Avalonian evolution of northwest Iberia. *Can. J. Earth Sci.*, **35,** 1–15.

Fernández-Suárez, J., Gutiérrez-Alonso, G., Jenner, G.A. and **Tubrett, M.N.** (1999) Crustal sources in Lower Palaeozoic rocks from NW Iberia: Insights from laser ablation U-Pb ages of detrital zircons. *J. Geol. Soc. London*, **156,** 1065–1068.

Fernández-Suárez, J., Gutiérrez-Alonso, G., Jenner, G.A. and **Tubrett, M.N.** (2000) New ideas on the Proterozioc-Early Paleozoic evolution of NW Iberia: Insights from U-Pb detrital zircon ages. *Precambrian Res.*, **102,** 185–206.

Fernández-Suárez, J., Gutiérrez-Alonso, G., Cox, R. and **Jenner, G.A.** (2002a) Assembly of the Armorica microplate: A strike-slip terrane delivery? Evidence from U-Pb ages of detrital Zircons. *J. Geol.*, **110,** 619–626.

Fernández-Suárez, J., Gutiérrez-Alonso, G. and **Jeffries, T.E.** (2002b) The importance of along-margin terrane transport in northern Gondwana: Insights from detrital zircon parentage in Neoproterozoic rocks from Iberia and Britany. *Earth Planet Sci. Lett.*, **204,** 75–88.

Flemings, P.B. and **Grotzinger, J.P.** (1996) Strata: Frameware for analysing classic stratigraphic problems. *GSA Today*, **6,** 1–7.

Frankenfeld, H. (1982) Das Ende der devonischen Riff-Fazies im nordspanischen Variszikum. *Neues Jb. Geol. Paläontol. Abh.*, **163,** 238–241.

Frost, S.H. (1981) Oligocene reef coral biofacies of the Vicentin, Northeast Italy. In: *European Fossil Reef Models* (Ed. D.F. Toomey), SEPM Spec. Publ., Tulsa, USA, Vol. 30, 483–539.

García López, S., Brime, C., Bastida, F. and **Sarmiento, G.N.** (1997) Simultaneous use of thermal indicators to analyse the transition from diagenesis to metamorphism: An example from the Variscan Belt of northwest Spain. *Geol. Mag.*, **134,** 323–334.

García-Alcalde, J.L. (1998) Devonian events in northern Spain. *Newslett. Stratigr.*, **36,** 157–175.

García-Alcalde, J.L., Carls, P., Pardo Alonso, M.V., Sanz López, J., Soto, F., Truyols-Massoni, M. and **Valenzuela-Ríos, J.I.** (2002) Devonian. In: *The Geology of Spain* (Eds W. Gibbons and T. Moreno). Geological Society of London, London, pp. 67–91.

García-Castellanos, D. (2002) Interplay between lithospheric flexure and river transport in foreland basins. *Basin Res.*, **14,** 89–104.

German Stratigraphic Commision (Ed.) (2002) *Stratigraphic Table of Germany 2002.* Stein, Potsdam.

Goldhammer, R.K. (1997) Compaction and decompaction algorithms for sedimentary carbonates. *J. Sed. Res.*, **67,** 26–35.

Gradstein, F.M., Agterberg, F.P., Ogg, J.G., Hardenbol, J., van Veen, P., Thierry, J. and **Huang, Z.** (1994) A Mesozoic time scale. *J. Geophys. Res*, **99,** 24,051–24,074.

Hallam, A. (1984) Pre-Quaternary sea-level changes. *Annu. Rev. Earth Planet. Sci.*, **12,** 205–243.

Haq, B.U., Hardenbol, J. and **Vail, P.R.** (1987) Chronology of fluctuating sea levels since the Triassic. *Science*, **235,** 1156–1167.

Jamison, W.R. (1987) Geometric analysis of fold development in overthrust terranes. *J. Struct. Geol.*, **9,** 207–219.

Johnson, J.G., Klapper, G. and **Sandberg, C.A.** (1985) Devonian eustatic fluctuations in Euramerica. *GSA Bull.*, **96,** 567–587.

Julivert, M. (1971) Decollement tectonics in the Hercynian Cordillera of Northwest Spain. *Am. J. Sci.*, **270,** 1–29.

Julivert, M. (1981) A cross-section through the northern part of the Iberian Massif. *Geol. Mijnbouw*, **60,** 107–128.

Julivert, M. and **Arboleya, M.L.** (1986) Areal balancing and estimate of areal reduction in a thin-skinned fold and thrust belt (Cantabarian Zone, NW Spain); constraints on its emplacement mechanism. In: *Thrusting and Deformation* (Eds J. Platt, M.P. Coward, J. Deramond, and J. Hossack), Pergamon, Oxford, pp. 407–414.

Julivert, M., Fontboté, J.M., Ribeiro, A. and **Conde, L.N.** (1972) *Mapa tectónico de la Península Ibérica y Baleares.* Instituto Geológico y Minero de España, Madrid.

Kendall, C.G.S.C. and **Sen, A.** (1998) Use of sedimentary simulations and measuring the size of eustatic sea-level changes: An example from the Neogene of the Bahamas. In: *Computerized Modelling of Sedimentary Systems* (Eds J. Harff, W. Lemke and K. Stattegger), Springer, Heidelberg, Germany, pp. 291–306.

Kendall, C.G.S.C., Moore, P., Strobel, J., Cannon, R., Perlmutter, M., Bezdek, J. and **Biswas, G.** (1991) Simulation of the sedimentary fill of basins. *Kansas Geol. Surv.*, **293,** 9–30.

Kendall, C.G.S.C., Moore, P., Whittle, G. and **Cannon, R.** (1992) A challenge; is it possible to determine eustasy and does it matter? In: *Eustasy; the Historical Ups and Downs of a Major Geological Concept* (Ed. R.H. Jr. Dott), *Geol. Soc. Am. Mem.*, **180,** 93–107.

Kiessling, W., Flügel, E. and **Golonka, J.** (2000) Fluctuations in the carbonate production of Phanerozoic reefs. In: *Carbonate Platform Systems: Components and Interaction* (Eds E. Insalaco, P.W. Skelton and T.J. Palmer), *Geol. Soc. London, Spec. Publ.*, **178,** 191–215.

Klein, G. de V. (1987) Current aspects of basin analysis. *Sed. Geol.*, **50,** 95–118.

Krebs, W. (1974) Devonian carbonate complexes of central Europe. In: *Reefs in Time and Space; Selected Examples from the Recent and Ancient*, SEPM Spec. Publ., Tulsa, USA, Vol. 18, pp. 155–208.

Lawrence, D.T., Doyle, M. and **Aigner, T.** (1990) Stratigraphic simulation of sedimentary basins: Concepts and calibration. *AAPG Bull.*, **74,** 273–295.

Leeder, M. (1999) Sedimentology and sedimentary basins. *From Turbulence to Tectonics.* Blackwell Science, Oxford, 592pp.

Lees, A. (1961) The Waulsortian 'reefs' of Eire; a carbonate mudbank complex of lower Carboniferous age. *J. Geol.*, **69,** 101–109.

Lotze, F. (1945) Zur Gliederung der Varisziden der Iberischen Meseta (Spanien). *Geotekt. Forsch.*, **6,** 78–92.

Matte, P. (1991) Accretionary history and crustal evolution of the Variscan belt in Western Europe. *Tectonophysics*, **196,** 309–337.

McClay, K.R. (1992) Glossary of thrust tectonics terms. In: *Thrust Tectonics* (Ed. K.R. Mc Clay,), Chapman & Hall, London, pp. 419–433.

Menning, M. (1995) A numerical time scale for the Permian and Triassic Periods. An integrated time analysis. In: *Permian of the Northern Continents* (Eds P. Scholle, T.M. Peryt and D.S. Ulmer-Scholle), Springer, Heidelberg, Germany, Vol. 1, pp. 77–97.

Menning, M. (2001) A Permian time scale 2000 and correlation of marine and continental sequences using the Illawarra reversal (265 Ma). *Natura Bresciana, Ann. Mus. Civ. Sc. Nat.*, Monografia, Brescia, **25,** 355–362.

Menning, M., Weyer, D., Drozdzewski, G., Amerom, H.W.J. and **Wendt, I.** (2000) A Carboniferous time scale 2000: Discussion and use of geological parameters as time indicators from Central and Western Europe. *Geol. Jb.*, **A 156,** 3–44.

Mitchum, R.M. Jr. and **Van Wagoner, J.C.** (1991) High-frequency sequences and their stacking patterns: Sequence-stratigraphic evidence of high-frequency eustatic cycles. *Sed. Geol.*, **70,** 131–160.

Muir, I., Wong, P. and **Wendte, J.** (1985) Devonian Hare Indian-Ramparts (Kee Scarp) evolution, MacKenzie Mountains and subsurface Norman Wells, N.W.T.; basin-fill and platform reef development. In: *Rocky Mountain Carbonate Reservoirs; a Core Workshop* (Eds M.W. Longman, K.W. Shanley, R.F. Lindsay and D.E. Eby), *SEPM Core Workshop*, **7,** pp. 311–341.

Nieuwland, D.A., Leutscher, J.H. and **Gast, J.** (2000) Wedge equilibrium in fold-and-thrusts belts; prediction of out-of-sequence thrusting based on sandbox experiments and natural examples. *Geol. Mijnbouw*, **79,** 81–91.

Nijman, W. and **Savage, J.F.** (1989) Persistent basement wrenching as controlling mechanism of Variscan thin-skinned thrusting and sedimentation, Cantabrian Mountains, Spain. *Tectonophysics*, **169,** 281–302.

Oczlon, M.S. (1992) Gondwana and Laurussia before and during the Variscan Orogeny in Europe and related areas. *Heidelb. Geowiss. Abh.*, **53,** 1–56.

Pálfy, J., Smith, P.L. and **Mortensen, J.K.** (2000) U-Pb and ^{40}Ar / ^{39}Ar time scale for the Jurassic. *Can. J. Earth Sci.*, **37,** 923–944.

Paola, C. (2000) Quantitative models of sedimentary basin filling. *Sedimentology*, **47,** 121–179.

Paris, F. and **Robardet, M.** (1990) Early Paleozoic palaeobiogeography of the Variscan regions. *Tectonophysics*, **177,** 193–213.

Parkinson, D. (1957) Lower carboniferous reefs of northern England. *AAPG Bull.*, **41,** 511–537.

Pérez-Estaún A. and **Bastida, F.** (1990) Structure: Cantabrian Zone. In: *Pre-Mesozoic Geology of Iberia* (Eds R.D. Dallmeyer and E. Martínez García), Springer, Heidelberg, Germany, pp. 55–69.

Pérez-Estaún, A., Bastida, F., Alonso, J.L., Marquínez, J., Aller, J., Álvarez-Marrón, J., Marcos, A. and **Pulgar, J.A.** (1988) A thin-skinned tectonics model for an arcuate fold and thrust belt; the Cantabrian Zone (Variscan Ibero-Armorican Arc). *Tectonics*, **7,** 517–537.

Playford, P.E. (1980) Devonian 'great barrier reef' of Canning Basin, Western Australia. *AAPG Bull.*, **64,** 814–840.

Potent, S. and **Reuther, C.D.** (2000) Kinematik der Faltungs- und Überschiebungsprozesse der variszisch angelegten Montuerto-Struktur im südlichen Kantabrischen Gebirge, Nord-Spanien. *Mitt. Geol.-Paläont. Inst. Univ. Hamburg*, **84,** 83–110.

Prosser, S. (1993) Rift-related linked depositional systems and their seismic expression. In: *Tectonics and Seismic Sequence Stratigraphy* (Eds G.D. Williams and A. Dobb), *Geol. Soc. Spec. Publ.*, **71,** 35–66.

Ramsay, J.G. and **Huber, M.I.** (1987) *The Techniques of Modern Structural Geology; Volume 2; Folds and Fractures*, Academic Press, London, pp. 462.

Raven, J.G.M. (1983) Conodont biostratigraphy and depositional history of the Middle Devonian to Lower Carboniferous in the Cantabrian Zone (Cantabrian Mountains, Spain). *Leidse Geol. Meded.*, **52,** 265–339.

Robardet, M. (2000) An alternative approach to consider the Variscan belt in SW Europe: The pre-orogenic paleogeographical constraints. *Basement Tectonics, A Coruña*, **15,** 23–26.

Robardet, M. (2002) Alternative approach to the Variscan Belt in southwestern Europe: Preorogenic palaeobiogeographical constraints. In: *Variscan-Apalachian Dynamics: The Building of the Late Paleozoic Basement* (Eds J.R. Martínez Catalán, R.D. Hatcher, R. Arenas and F. Díaz García), *Geol. Soc. Am. Spec. Paper*, **364,** 1–15.

Rodríguez Fernández, L.R., García-Alcade, J.L. and **Menéndez-Álvarez, J.R.** (1985) La sucesión del Devónico superior y Carbonífero inferior en el sinclinal de Alba (León, N.O. de España). *Dixième Congrès Int. de Stratigraphie et de Géologie du Carbonifère*, Madrid, 133–144.

Rose, P.R. (1976) Mississippian carbonate shelf margins, western United States. *J. Res. USGS*, **4,** 449–466.

Ross, C.A. and **Ross, J.R.P.** (1988) Late Paleozoic transgressive–regressive deposition. In: *Sea-Level Changes; an Integrated Approach* (Eds C.L. Wilgus, B.S. Hastings, C.A. Ross, H.W. Posamentier, J. Van Wagoner and C.G.St.C. Kendall), Soc. Sediment. Geol. SEPM, Tulsa, USA, Vol. 42, pp. 227–247.

Ross, C.A. and **Ross, J.R.P.** (1996) Silurian sea-level fluctuations. In: *Paleozoic Sequence Stratigraphy: Views from the North American Continent* (Eds B.J. Witzke, G.A. Ludvigsen and J. Day), *Geol. Soc. Am. Spec. Paper*, Boulder, USA, Vol. 187, pp. 203–211.

Sachsenhofer, R.F., Lankreijer, A., Cloetingh, S. and **Ebner, F.** (1997) Subsidence analysis and quantitative basin modelling in the Styrian Basin (Pannonian Basin System, Austria). *Tectonophysics*, **272,** 175–196.

Schlager, W. (1993) Accommodation and supply – A dual control on stratigraphic sequences. *Sed. Geol.*, **86,** 111–136.

Schlager, W. (2000) Sedimentation rates and growth potential of tropical, cool-water and mud-mound carbonate systems. In: *Carbonate Platform Systems: Components and Interaction* (Eds E. Insalaco, P.W. Skelton, and T.J. Palmer), *Geol. Soc. London Spec. Publ.*, **178,** 217–227.

Sdzuy, K. (1967) Trilobites del Cámbrico medio de Asturias. *Trabajos de Geología*, **1,** 77–133.

Sdzuy, K. (1968) Biostratigrafía de la griotte cámbrica de los Barrios de Luna (Léon) y de otras sucesiones comparables. *Trabajos de Geología*, **2,** 45–57.

Stüwe, K. (2000) *Geodynamics of the Lithosphere*. Springer, Heidelberg, Germany, pp. 449.

Suárez, A., Barba, P., Heredía, N. and **Rodríguez Fernández** (1994) *Mapa geológico de la provincia de León, Escala 1:200,000*, Instituto Tecnológico y Geominero de España, Madrid.

Suppe, J. (1983) Geometry and kinematics of fault-bend folding. *Am. J. Sci.*, **283,** 684–721.

Tait, J., Bachtadse, V., Franke, W. and **Soffel, H.C.** (1997) Geodynamic evolution of the European Variscan fold belt: Palaeomagnetic and geological constraints. *Geol. Rundsch.*, **86,** 585–598.

Terry, C.E. and **Williams, J.J.** (1969) The Idris 'A' bioherm and oilfield, Sirte Basin, Libya – its commercial development, regional Palaeocene geologic setting and stratigraphy. In: *The Exploration for Petroleum in Europe and North Africa* (Ed. P. Hepple), The Institute of Petroleum, London, pp. 31–48.

Thorne, J.A. and **Swift, D.J.P.** (1991) Sedimentation on continental margins, VI: A regime model for depositional sequences, their component systems tracts, and bounding surfaces. In: *Shelf Sand and Sandstone Bodies* (Eds D.J.P. Swift, G.F. Oertel, R.W. Tillman, and J.A. Thorne), *Int. Assoc. Sedimentol. Spec. Publ.*, **14,** 189–255.

Truyols, J. and **Sánchez de Posada, L.C.** (1983) El Carbonífero inferior y medio de la Región de Pliegues y Mantos. In: *Carbonífero y Pérmico de España* (Ed. J. Truyols), IGME, Madrid, pp. 39–59.

Tucker, R.D. and **McKerrow, W.S.** (1995) Early Paleozoic chronology: A review in light of new U-Pb zircon ages from Newfoundland and Britain. *Can. J. Earth Sci.*, **32,** 68–379.

Turcotte, D.L. and **Schubert, G.** (1982) *Geodynamics-Applications of Continuum Physics to Geological Problems*, Wiley, Cambridge, London, pp. 450.

Turcotte, D.L. and **Schubert, G.** (2002) *Geodynamics*. Cambridge University Press, Cambridge, London, pp. 456.

Vail, P.R., Mitchum, R.M. Jr., Todd, R.G., Widmier, J.M., Thompson, S., Sangree, J.B., Bubb, J.N. and **Hatlelid, W.G.** (1977) Seismic stratigraphy and global changes of sea level. In: *Seismic Stratigraphy – Applications to Hydrocarbon Exploration* (Ed. C.E. Payton), *AAPG Mem.*, **26,** 49–205.

Vail, P.R., Hardenbol, J. and **Todd, R.G.** (1984) Jurassic unconformities, chronostratigraphy, and sea level changes from seismic stratigraphy and biostratigraphy. In: *Interregional Unconformities and Hydrocarbon Accumulation* (Ed. J.S. Schlee), *AAPG Mem.*, **36,** 129–144.

Van der Voo, R., Stakamatos, J.A. and **Parés, J.M.** (1997) Kinematic constraints on thrust-belt curvature from syndeformational magnetizations in the Lagos del Valle syncline in the Cantabrian Arc, Spain. *J. Geophys. Res.*, **102,** 10,105–10,119.

Veselovský, Z. (2004) Integrated numerical modelling of a polyhistory basin, Southern Cantabrian Basin (Palaeozoic, NW-Spain). *(CD-ROM) – Gaea heidelbergensis*, **13,** Heidelberg, pp. 225.

Villegas, H.F.J. (1996) *Exploración e investigación de un nuevo yacimiento de carbón en la cuenca minera Cinera-Matallana (León).* Unpublished Doctoral Thesis, Facultad Ciencias Geológicas, Universidad Complutense, Madrid, pp. 417.

Walls, R.A., Mountjoy, E. and **Fritz, P.** (1979) Isotopic composition and diagenetic history of carbonate cements in Devonian Golden Spike Reef, Alberta, Canada. *Geol. Soc. Am. Bull.*, **90,** I 963–I 982.

Warr, L.N. (2000) The Variscan Orogeny: the welding of Pangea. In: *Geological History of Britain and Ireland* (Eds N. Woodcock and R. Strachan), Blackwell Science, Oxford, London, pp. 271–294.

Watts, A.B. (2001) *Isostasy and Flexure of the Lithosphere,* Cambridge University Press, Cambridge, London, pp. 458.

Weddige, K. (Ed.) (2001) Devon-Bibliographie. http://www.senckenberg.de/publ/Dev-Bibl.htm.

Weil, A.B., Van der Voo, R., Van der Pluijm, B.A. and **Parés, J.M.** (2000) The formation of an orocline by multiphased deformation: A paleomagnetic investigation of the Cantabria-Asturias arc hinge-zone (northern Spain). *J. Struct. Geol.*, **22,** 735–756.

Weil, A.B., Van der Voo, R. and **Van der Pluijm, B.A.** (2001) Oroclinal bending and evidence against the Pangea megashear: The Cantabria-Asturias arc (northern Spain). *Geology*, **29,** 991–994.

Weil, A.B., Gutiérrez-Alonso, G. and **Fernández-Suárez, J.** (2003) Orocline triggered lithospheric delamination. GSA Abst. with Prog., **35,** no. 6, Annual meeting 2003, Seattle, p. 346.

Welte, D.H., Horsfield, B. and **Baker, D.R.** (1997) *Petroleum and Basin Evolution,* Springer, Heidelberg, Germany, pp. 535.

Whitaker, F., Smart, P., Hague, Y., Waltham, D. and **Bosence, D.** (1999) Structure and function of a coupled two-dimensional diagenetic and sedimentological model of carbonate platform evolution. In: *Numerical Experiments in Stratigraphy; Recent Advances in Stratigraphic and Sedimentologic Computer Simulations* (Eds J.W. Harbaugh, W.L. Watney, E.C. Rankey, R. Slingerland, R.H. Goldstein and E.K. Franseen), *Soc. Sedim. Geol., Spec. Publ.*, **62,** 337–355.

Woodward, N.B., Boyer, S.E. and **Suppe, J.** (1985) An Outline of Balanced Cross-Sections. *Stud. Geol., Univ. Tennessee*, **11,** pp. 166.

Woodward, N.B., Boyer, S.E. and **Suppe, J.** (1989) Balanced geological cross-sections: An essential technique in geological research and exploration. In: *Short Course in Geology* (Eds M.L. Crawford and E. Padovani), American Geophysical Union, Washington, DC, USA, Vol. 6, pp. 132.

Zühlke, R., Bouaouda, M.S., Ouajhain, B., Bechstädt, T. and **Leinfelder, R.** (2004) Quantitative Meso-/Cenozoic development of the Eastern Central Atlantic Continental Shelf, onshore Agadir Basin, Morocco. *Mar. Petrol. Geol.*, **21,** 225–276.

Spec. Publ. Int. Assoc. Sedimentol. (2008) **40**, 97–138

Numerical modelling of alluvial deposits: recent developments

JOHN BRIDGE

Department of Geological Sciences, Binghamton University, Binghamton, NY 13902-6000, USA
(E-mail: jbridge@binghamton.edu)

ABSTRACT

Alluvial deposits occur over a range of superimposed scales, dependent on the scales of associated topographic features (such as ripples, dunes, bars, channels, floodplains, valleys, river systems), the time and spatial extent over which deposition occurred, and the degree of preservation. Understanding and prediction of alluvial deposits is aided by numerical (forward) modelling of depositional forms and processes. The most desirable approach to such numerical modelling is through solution of the fundamental equations of motion for the fluid and the sediment (e.g. conservation of mass, momentum and energy) for all of the scales of deposits. Construction of the equations of motion requires an understanding of the interactions among unsteady, non-uniform, turbulent water flow, the supplied and transported sediment, and the topography of the sediment bed. This understanding is very incomplete.

Simplified numerical forward models have been applied to relatively short-term, small-scale processes such as bed degradation and armouring downstream of dams, reservoir sedimentation, and sorting of sediment (e.g. downstream fining) during deposition in spatially decelerating flows. However, such models are rarely applied over long time periods, over large spatial scales, and where there are complicated temporal and spatial variations in the geometry, water and sediment supply. This is because of limitations to computing facilities, and because of lack of understanding of the workings of the alluvial system. As a result, long-term, large-scale alluvial processes are commonly treated using 'process-imitating' models. Process-imitating models do not necessarily represent processes accurately or completely. This review of process-based models demonstrates that they are generally undeveloped, and linkages between models for different scales are lacking. Process-based modelling is in its infancy. As a result, structure-imitating stochastic models are widely used for simulating alluvial hydrocarbon reservoirs and aquifers, given some initial data. However, such models cannot help understanding of alluvial deposits, nor can they predict their nature outside the data region.

Keywords Alluvial deposits, numerical models, depositional scales, process-based versus stochastic models.

INTRODUCTION

Understanding of fluvial sedimentary processes and deposits has come from: (1) field studies of modern fluvial environments; (2) laboratory flume studies using physical models; and (3) construction of idealized models based on these studies. Understanding and prediction of ancient alluvial deposits is based on direct modern analogues or idealized models.

Field studies of modern fluvial environments are absolutely essential for understanding fluvial sedimentary processes and deposits, but such studies are difficult to undertake during floods, over large areas, and for a long time, and it is difficult to describe deposits below the water table. Recently, some of these problems have been overcome by: using remote-sensing images and digital elevation models (DEMs) for studying changes in channel and floodplain geometry; measurement of water flow and sediment transport during floods using new types of equipment (e.g. acoustic Doppler current profilers); description of deposits using ground-penetrating radar

(GPR) in combination with coring, and optically stimulated luminescence (OSL) dating.

Laboratory flumes have been used to study fluvial processes and deposits over a wide range of physical scales. These physical models may be full scale, reduced scale, or unscaled (analogue). However, scaling problems can limit the applicability of these studies to the real world, and these problems increase as the scale of the physical model decreases relative to the real-world prototype. In particular, all superimposed scales of bedform and associated strata cannot be produced in flumes, and rates of sedimentary processes are unrealistically high.

Idealized models may be qualitative (graphic), quantitative (numerical), static, dynamic (forward), stochastic, deterministic. As these models are ultimately based on studies of modern sedimentary processes and deposits, models that are relatively short-term and small-scale are well developed and most easy to test. As the spatial and temporal scales increase, useful models are more difficult to construct and test. As a result, it is difficult to link models of different scales of deposition. Numerical models of fluvial deposits are discussed here, with emphasis on process-based (forward) models. The models are discussed in order of increasing scale, following definition of the different scales of fluvial forms and associated deposits.

SCALES OF FLUVIAL FORMS AND DEPOSITS

Different scales of fluvial deposits depend on the scale of the associated topographic features, the time and spatial extent over which deposition occurred, and the degree of preservation (Fig. 1). The scales of topographic features and associated deposits discussed here are, in order of increasing scale: (1) relatively small-scale bedforms such as ripples, dunes, antidunes, and bedload sheets on pseudo-planar beds, that give rise to small- and medium-scale cross strata and planar strata; (2) depositional increments in channels and on floodplains formed during distinct floods, referred to here as large-scale strata; (3) channels and bars, forming sets of large-scale inclined strata; (4) channel belts, that are composed of groups of large-scale inclined stratasets; (5) floodplains with levees, crevasse splays, channels and floodplains, that are also composed of groups of large-scale inclined stratasets; (6) alluvial valleys, containing deposits of channel belts and floodplains (or fans or deltas); (7) river systems, that comprise erosional landscapes and basin fills.

The geometry of a particular scale of strataset is related to the geometry and migration of the associated topographic feature, and subsequent erosion. With wavy bedforms such as ripples, dunes, and bars, the length-to-thickness ratio of associated stratasets is similar to the wavelength-to-height ratio of the bedforms (Fig. 2; Bridge & Lunt, 2006). Furthermore, the wavelength and height of dunes and bars are related to channel depth and width. Therefore, the thickness of medium-scale cross stratasets (formed by dunes) and large-scale inclined stratasets (formed by bars) will vary with river dimensions. These ideas need to be tested more fully and applied to larger scales of topographic feature and strataset.

NUMERICAL FORWARD MODELS OF FLUVIAL DEPOSITS

Desirable approaches and compromises

The most desirable approach to numerical modelling of physical deposition, irrespective of scale, is through solution of the fundamental equations of motion for the fluid and the sediment (e.g. conservation of mass, momentum and energy). In the case of fluvial environments, construction of the equations of motion requires an understanding of the interactions between turbulent water flow, the transported sediment, and

Fig. 1. (Opposite page) Scales of fluvial forms and deposits. (a) Alluvial valley of Senguerr River, Argentina with channel belt of a meandering river (100 m across) and floodplain (background). (b) Deposits of channel belts (thick grey sandstones) and floodplains (red siltstones and thin grey sandstones) from the Miocene of the Potwar Plateau, Pakistan. Marked channel-belt deposit is 10 to 15 m thick. (c) Channel belt of Rakaia River, New Zealand showing unit bars and compound braid bars composed of accreted unit bars. View is about 500 m across. (d) Outcrop (about 10 m high) of Carboniferous channel sandstone deposits (from West Virginia), showing large-scale inclined strata formed by bar migration. (e) GPR profile of braided-channel deposits from the Sagavanirktok River, Alaska, showing large-scale inclined strata associated with migration of unit bars and compound braid bars. (f) Curved crested dunes on point bar (Congaree River, South Carolina), with medium-scale trough cross strata visible in trench. Scale is 0.15 m across. (g) Cross-section (1 m across) showing experimental antidune (top) and antidune cross strata.

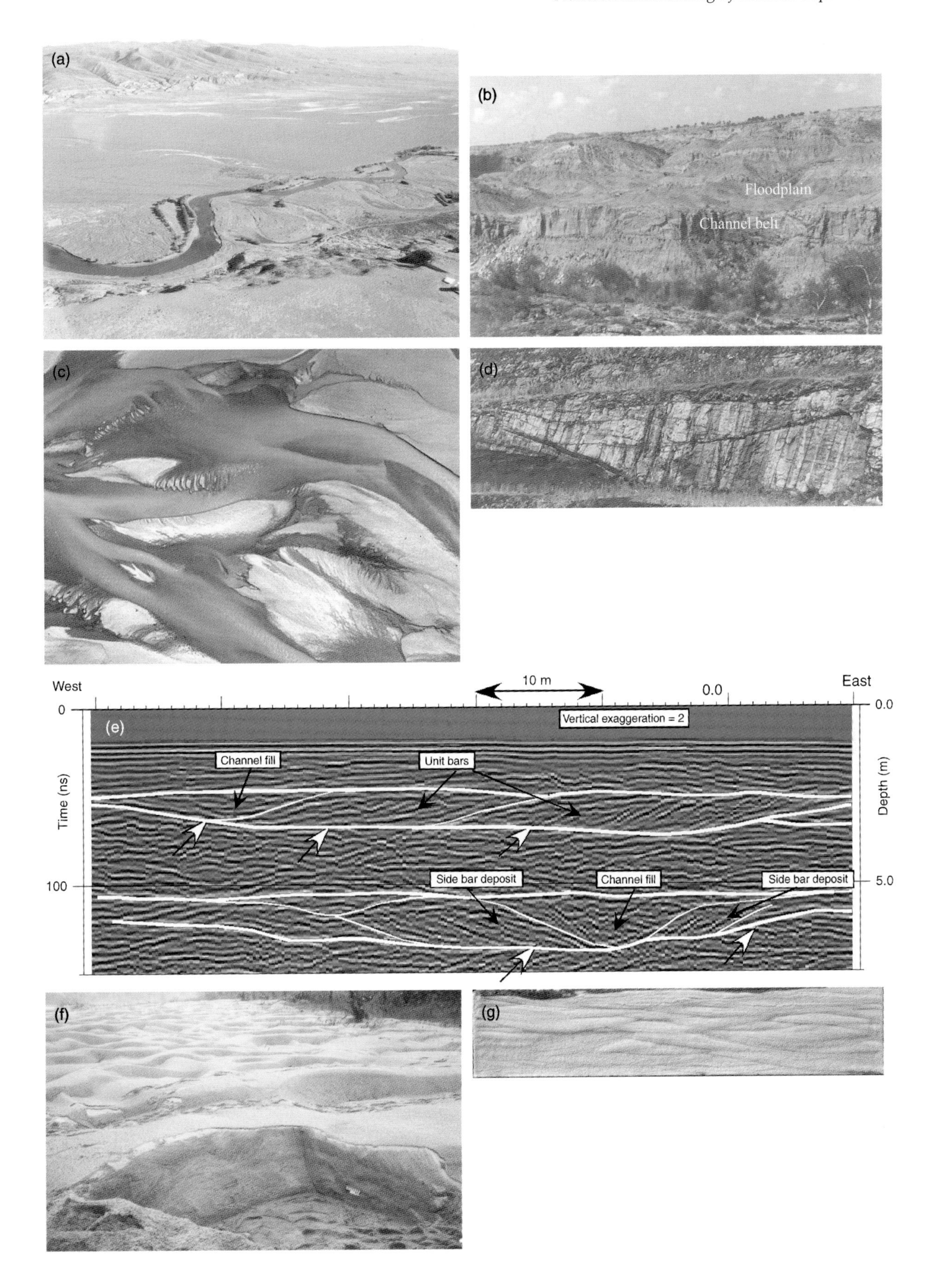
(a)
(b)
Floodplain
Channel belt
(c)
(d)
West
10 m
0.0
East
0
0.0
(e)
Vertical exaggeration = 2
Channel fill
Unit bars
Time (ns)
Depth (m)
100
5.0
Side bar deposit
Channel fill
Side bar deposit
(f)
(g)

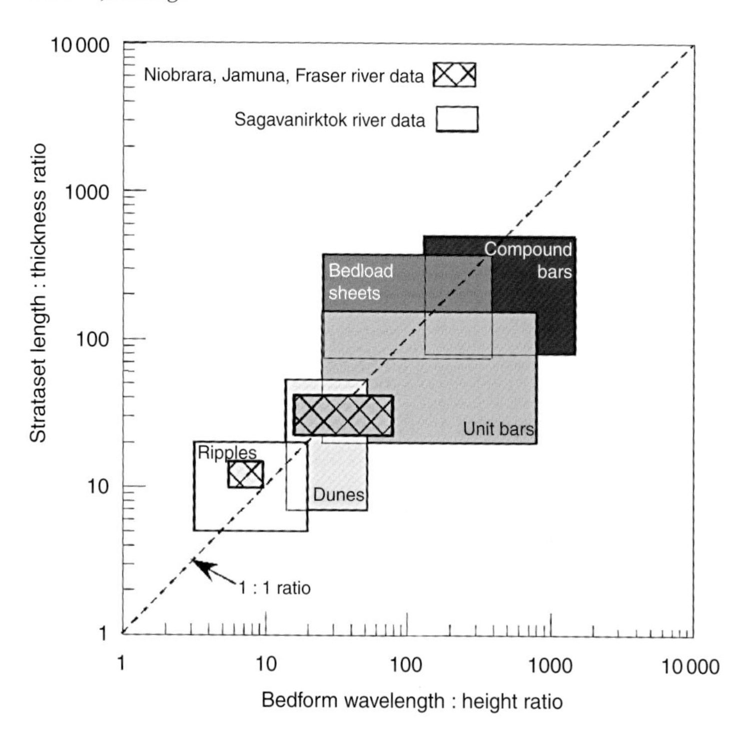

Fig. 2. Bedform length/height plotted against corresponding strataset length/thickness (from Bridge and Lunt, 2006). Bedform geometry was taken from aerial photographs or descriptions, and strataset dimensions were taken from ground-penetrating radar profiles and descriptions. Some bedform dimensions were measured at low flow stage, and may not represent formative bedform dimensions.

the topography of the sediment bed. Models that describe this interaction are sometimes referred to as sediment routing models. Sediment routing models normally involve treatment of unsteady, non-uniform flow acting upon heterogeneous sediment beds. A realistic sediment routing model requires specification of: (1) the original geometry of the system; (2) time varying supply of water and sediment to the system; (3) sediment types (grain size, shape, density) available for transport; (4) the mean and turbulent fluctuating values of fluid force acting upon sediment grains, and how these forces vary in space and time; (4) the interaction between fluid forces and individual sediment fractions, resulting in sediment entrainment and transport as bedload (normally in characteristic bedforms such as ripples, dunes and bars) and suspended load; (5) erosion and deposition (change in bed position) as determined by a sediment continuity equation applied to different grain fractions; (6) the concept of an active layer to which sediment can be added or from which it can be removed; and (7) an accounting scheme to record the nature of deposited sediment. The geometry of the system, and the supply of water and sediment will change in response to tectonic uplift and subsidence, and to base-level change. Such changes inevitably result in erosion and deposition, in turn changing the geometry of the system.

Many extant sediment routing models (review by Bridge, 2003) do not have all of these desirable components, detracting from their usefulness. Nevertheless, simplified sediment routing models have been applied to relatively short-term, small-scale processes such as bed degradation and armouring downstream of dams, reservoir sedimentation, and sorting of sediment (e.g. downstream fining) during deposition in spatially decelerating flows. Sediment routing models have rarely been applied to flow and sediment transport acting over long time periods, over large spatial scales, and where there are complicated temporal and spatial variations in geometry, water and sediment supply. This is because of limitations to computing facilities, and because of lack of understanding of the dynamics of the alluvial system. As a result, long-term, large-scale fluvial processes are commonly treated using 'process-imitating' models (term after Koltermann & Gorelick, 1996). Process-imitating models do not necessarily represent processes accurately or completely.

Small-scale bedforms

Laboratory flume studies (full-scale) have been invaluable for elucidating the water flow and sediment transport, bedform geometry and migration, and deposits associated with small-scale fluvial bedforms such as ripples, dunes, antidunes and bedload sheets on pseudo-planar beds (review by Bridge, 2003). Many field studies of small-scale bedforms have also been undertaken: however, it is difficult to adequately describe the geometry and migration of bedforms in the field, and to relate them to the formative flow and sediment transport. This is because river flows are unsteady and non-uniform, and because of the limited time periods and areas over which detailed and coordinated field observations are possible. As a result, empirical data used for defining equilibrium hydraulic stability fields of small-scale bedforms come almost entirely from laboratory flume experiments, and the behaviour of bedforms under unsteady flows is poorly understood.

Water flow and sediment transport over incipient small-scale bedforms such as ripples, dunes and antidunes have been modelled numerically as part of stability analysis of bedforms (reviewed in Allen, 1982; McLean, 1990; Bridge, 2003). A stability analysis starts by defining the uniform, steady water flow and sediment transport over a plane bed. Fluid motion is defined using the equations for conservation of momentum and mass (continuity equation). Erosion and deposition of sediment are defined using the conservation of mass (sediment continuity equation) and a function that relates sediment transport rate to fluid flow characteristics. Stability analyses vary greatly in the mathematical treatment of water flow and sediment transport. It is then assumed that a sinusoidal disturbance of the planar bed (a bed defect) occurs, and the stability analysis attempts to predict whether a bed defect of a given wavelength will become amplified (giving rise to bedforms) or whether it will be damped (the bed remains plane). All stability analyses require introduction of a spatial lag of sediment transport rate behind a spatial change in flow velocity and bed topography in order to allow amplification of the initial bed defect. The spatial lag has been related to: (1) the interaction between turbulent bed shear stress and bed topography; (2) the inertia of suspended load; and (3) the effects of local bed slope on bed-load transport rate. If a bed defect becomes amplified, those wavelengths with the fastest rate of amplification are considered to become dominant. Thus, a stability analysis will normally predict the wavelength of the resulting bedforms. Stability analyses have been successful in predicting the stability fields of antidunes, plane beds, dunes, and ripples.

Stability analyses only describe flow and sediment transport during the initial stages of bedform growth, and before flow separation occurs on the lee side of growing bedforms. Flow separation downstream of bedform crests has an important effect on turbulent flow structure, bed shear stress and sediment transport both in bedform troughs and on the back of the next bedform downstream. No stability analysis describes these interactions between turbulent flow structure, sediment transport, and bed topography. As a result, there are no reliable models to predict bedform geometry and kinematics based on flow and sediment transport characteristics.

Even the most sophisticated stability analyses do not treat unsteady, non-uniform flow (averaged over several bedforms). Theoretical models are available for predicting changes in the height and length of individual dunes as discharge changes (reviews by Allen, 1982; Bridge, 2003). Unfortunately, these models are very simplistic and there are many unjustified assumptions, such that it is difficult to assess their usefulness. There are no theories for the interaction between water flow, sediment transport, and bed topography over a train of changing bedforms.

In view of the lack of adequate numerical forward models for flow and sediment transport over migrating small-scale bedforms, it is not surprising that numerical forward models for the deposits associated with these bedforms are also poorly developed. In order to model the deposits of bedforms such as ripples, dunes and bedload sheets, it is necessary to account for the sorting of heterogeneous sediment during transport and deposition on the downstream (lee) side, associated with: (1) temporal variation in the texture, composition and transport rate of sediment arriving at the bedform crest; (2) grain sorting during grain flow (avalanching) on the steep downstream face of the bedform; (3) settling of suspended sediment (grain fall) on the downstream face; and (4) modification of the lee-side deposits by the near-bed flow in the vortex beneath the separated boundary layer (Fig. 3). Much more is known about factors (2) and (3) than factors (1) and (4) (reviews by Bridge, 2003; Kleinhans, 2004).

The texture, composition, and transport rate of sediment arriving at the bedform crest vary in time due to turbulence, the passage of smaller bed waves over the back of the larger bedform, and longer-term (e.g. diurnal or seasonal) variations in sediment transport. For example, relatively large amounts of coarse-grained bed load may be associated with the arrival at the depositional lee side of a superimposed bed wave, whereas small amounts of relatively fine-grained sediment may be associated with arrival of the trough region. The flow

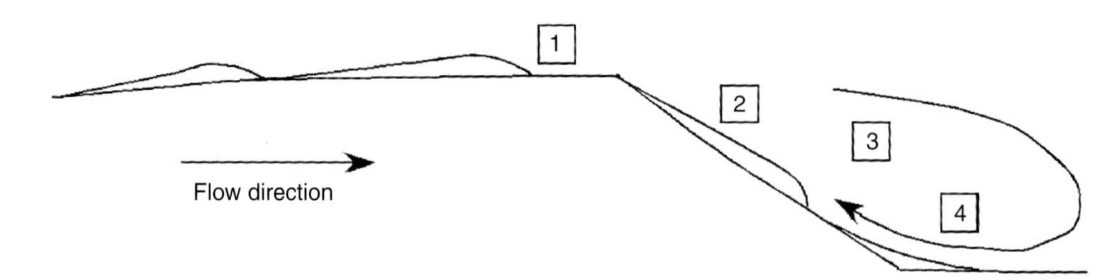

Fig. 3. Controls on the sorting of sediment deposited on the lee side of small-scale bedforms: (1) temporal variation in the texture, composition and transport rate of sediment arriving at the bedform crest; (2) grain sorting during grain flow (avalanching) on the steep downstream face of the bedform; (3) settling of suspended sediment (grain fall) on the downstream face; and (4) modification of the lee-side deposits by the near-bed flow in the vortex beneath the separated boundary layer.

pattern associated with a relatively large superimposed bedform tends to modify the lee slope of the host bedform. Relatively large amounts of coarse-grained bedload may be associated with diurnal snowmelt floods or seasonal floods. At relatively low rates of sediment transport, bedload avalanching on the lee side is intermittent, and discrete grain flows may be distinguishable. The coarsest grains tend to accumulate preferentially at the outer margin and base of such grain flows, because finer grains move downwards through the spaces between the larger grains (kinetic sieving). At high rates of sediment transport, bedload avalanching is more-or-less continuous, and kinetic sieving is not as effective. Also, the associated increased rate of deposition of suspended load on the lower lee slope of the bedform results in a build-up of relatively fine-grained sediment here. This build-up of sediment near the base of the lee slope may also be augmented by the near-bed reverse flow. Numerical models that take into account all of these processes of sediment transport and deposition still need to be developed. This is also true of sedimentary processes associated with pseudo-planar beds and antidunes described below.

Planar strata are formed by migration of low-relief bed waves (bedload sheets) on lower-stage plane beds and upper-stage plane beds (Bridge & Best, 1988, 1997; Paola *et al.*, 1989; Best & Bridge, 1992; Bennett & Bridge, 1995b). Therefore, plane beds are not strictly planar. Low-relief bedforms on lower-stage plane beds are associated with low bedload transport rates of gravels mainly. These bedforms are a few grain diameters high, and their length is proportional to flow depth. The grain size of these bedforms is less than the bed material, and the largest grains in the bed are more-or-less immobile. The gravel on most of the back of these bedforms is imbricated, and may be in the form of pebble or cobble clusters. The steep front of the bedforms may have platy grains dipping in the flow direction at the angle of repose (pseudo-imbrication). Thus gravelly planar strata formed by slow downstream migration of low-relief bed waves typically fine upward, and contain both imbricated and pseudo-imbricated gravel clasts. As bed shear stress increases, the grain size of the bedload becomes more like the bed material, and low-relief bed waves are transformed into dunes.

Bedload sheets on upper-stage plane beds are associated with high bedload transport rates of sands mainly, and there is a substantial suspended sediment load. The bedforms are millimetres high, and their length is about six flow depths. As these bedforms migrate downstream (at rates of millimetres per second), suspended sediment is deposited in the trough of the bedform. As the bedform migrates over this trough, a lamina is produced that has a fine-grained base formed by suspended sediment deposition.

Recent experiments on antidune migration under aggradational conditions (Alexander *et al.*, 2001) have revealed that the dominant internal structure (as seen in along-stream sections) is trough-shaped stratasets containing low-angle cross strata (Fig. 1g). The cross-strata may dip upstream or downstream, or in no preferred direction. These stratasets are formed as antidunes and associated water-surface waves migrate upstream, increase in height and asymmetry, and then break, filling the antidune trough with sediment very rapidly.

The deposits of bedforms described above vary in space because the geometry of an individual bedform varies in space (e.g. sinuous-crested ripples and dunes), and because individual bedforms vary in geometry and migration rate during their lifespans, even under steady, uniform flow conditions (review in Bridge, 2003). Furthermore, the preservation of these deposits at any location depends on the sequence of bedforms of a given geometry arriving at that location, and the net deposition rate. In particular, the thickness of cross-sets observed in cross-sections parallel to mean flow direction depends on: (1) the average rate of deposition relative to the rate of bedform migration; (2) the sequence of bedforms of a given scour depth (related to bedform height) passing a given bed location; and (3) changes in the geometry and migration rate of individual bedforms as they migrate (Allen, 1973, 1982; Paola & Borgman, 1991; Bridge, 1997; Leclair & Bridge, 2001; Leclair, 2002). Quantitative models are available to predict the thickness distribution of cross-sets given a distribution of bedform scour depths (Paola & Borgman, 1991; Bridge, 1997; Leclair and Bridge, 2001) and some other empirical information, but these models are not based on theoretical models of fluid flow, sediment transport and bedform mechanics (as such theoretical models do not exist).

Process-imitating (forward) models of the geometry of cross-strata sets produced by migrating bedforms were proposed by Rubin (1987) and Leclair (2002). In Rubin's (1987) model, bedforms can be straight-crested or curved-crested, can change geometry and migration rate in time, can be superimposed on different types of bedform, and bedforms can migrate in different directions. This versatility in the model requires 75 input variables. The geometry and migration of bedforms is represented by translation of sine curves. It is assumed that bedform height is equally represented above and below mean bed level, and that the cross-strata sets are formed by climbing (net aggradation). Experimental studies discussed above have shown that these last two assumptions are incorrect. Furthermore, interactions among bedforms are not considered. These deficiencies are reflected in the unrealistic geometry of the cross-strata produced by the model. Leclair (2002) modelled the geometry of cross-sets by simulating the dune trough trajectories of a succession of migrating dunes. The trough trajectories were obtained by random sampling of empirical distributions of dune migration rate and changes in dune trough elevation.

Geometrical and kinematical models such as that of Rubin (1987) and Leclair (2002) do not consider how bedform geometry and migration are controlled by flow and sediment transport mechanics. This is a very poorly understood field. Although theoretical models are available for predicting changes in height and length of single dunes as flow velocity, depth and sediment transport rate change in time (review in Bridge, 2003), there are no such theoretical models to account for simultaneous changes in several adjacent bedforms in a train, even under steady uniform flows. Indeed, no experimental studies have been undertaken to address this problem.

The variation of sediment texture that defines stratification has a major influence on spatial variation of porosity and permeability, and therefore on the nature of fluid flow in hydrocarbon reservoirs and aquifers (reviews by Brayshaw *et al.*, 1996; Bridge, 2003). It is therefore desirable to improve our ability to predict sediment texture within different scales of strata using numerical forward models.

Flood deposits

Figure 4 shows some typical sedimentary sequences in which mean grain size and sedimentary structures vary vertically and laterally, associated with changing flow conditions over a typical flood. Figure 4 (from Bridge, 2003) is based on theoretical reasoning, experimental studies of bedforms and sedimentary structures, and numerous field observations of river deposits. Temporal changes in sediment transport rate and grain size (hence bedform) at a point are associated with changes in flow strength (e.g. flow velocity or bed shear stress) over the flood, and these produce vertical variations in grain size and sedimentary structures. Spatial decrease in sediment transport rate is mainly responsible for the deposition and for downstream changes in mean grain size and sedimentary structures. The availability of different sediment sizes and the overall flow strength controls the overall grain size of the deposited sediment. Sequences such as those shown in Fig. 4 must be incorporated into larger-scale sequences associated with channel bars and channel fills, levees and crevasse splays; however, Fig. 4 is qualitative. It is possible to predict such facies sequences quantitatively using existing sediment routing models, but it has not been attempted yet.

Channel and channel bar

Theoretical stability analyses for the formation of unit bars in river channels, and the development of meandering and braided rivers, have been approached in a way similar to smaller bedforms (review by Bridge, 2003). These analyses have resulted in predictions of the geometry and

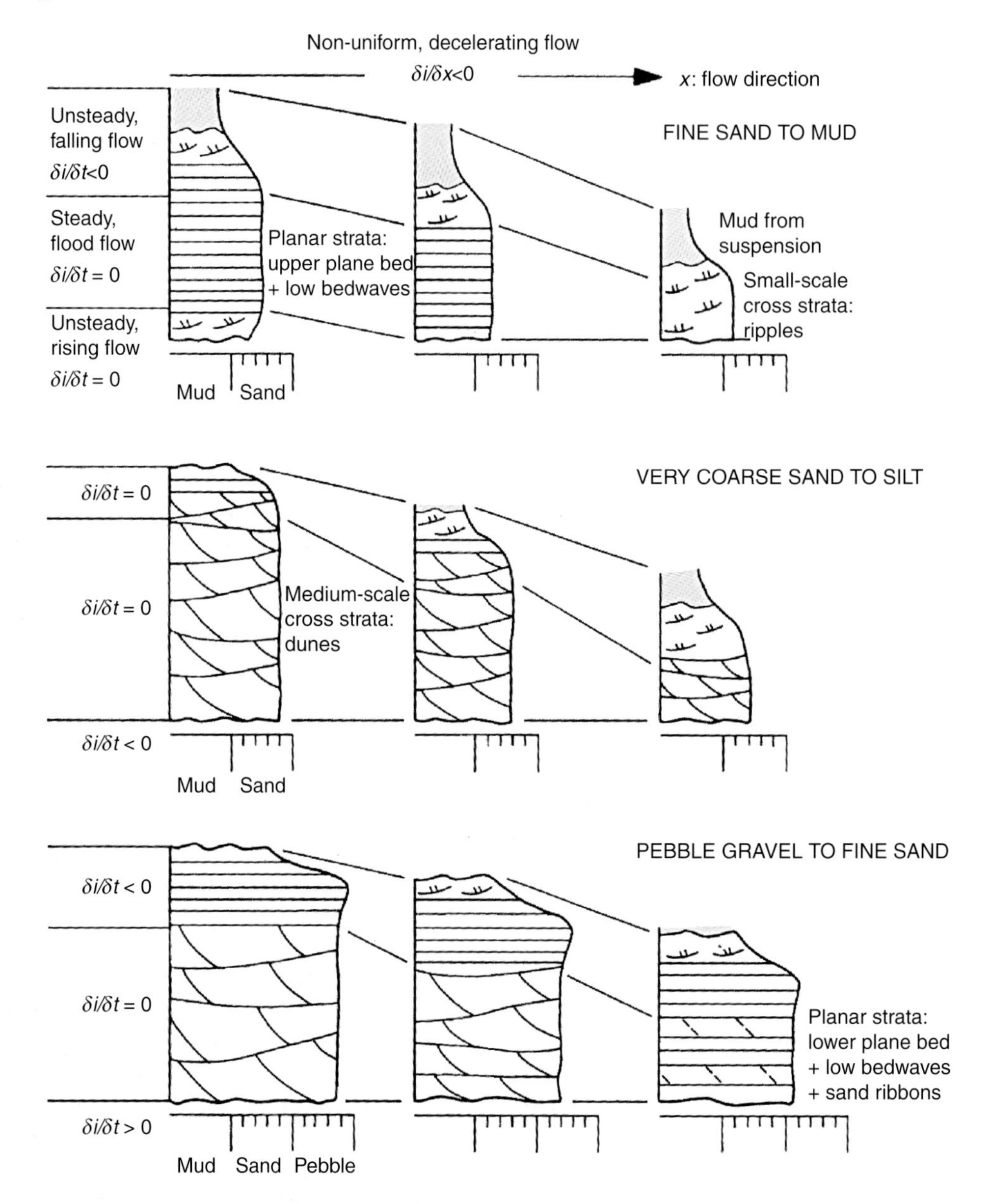

Fig. 4. Flood-generated stratasets for different grain-size ranges (from Bridge, 2003). Sediment transport rate varied in time (t) during a single flood, and in space. Deposition and lateral changes in grain size, bedforms and sedimentary structures, are caused mainly by decrease in sediment transport rate (i) in the flow direction (x). Vertical changes in grain size and sedimentary structures are caused by temporal changes in sediment transport rate. Flood-generated stratasets are typically centimetres to metres thick.

migration rates of bars, and simple patterns of flow and sediment transport. Equilibrium cross-sectional and plan geometry of channels has also been predicted using a range of different theories, including: minimization of energy expenditure (maximum sediment transporting efficiency); channel bank stability with mobile sediment beds dependent on lateral variations in bed shear stress relative to the critical shear stress for bank erosion; and constancy of dimensionless bed shear stress in sand-bed or gravel-bed rivers (discussed by Bridge, 2003). There are many models for the interaction between flow, sediment transport and equilibrium bed topography in channel bends, but most of them do not consider sorting of bed sediment (reviews in Bridge, 1992, 2003). Such interactive theoretical models are not well developed for channel confluences and diffluences. Although there are numerical models of turbulent flow in confluences that agree fairly well with observed flows (e.g. Bradbrook *et al.*, 1998, 2000, 2001; Lane & Richards, 1998; Nicholas & Sambrook-Smith, 1999; Lane *et al.*, 2000; Huang *et al.*, 2002), they do not describe the interaction between flow, sediment transport and bed topography. Theoretical models for flow, sediment transport, and bed topography at channel diffluences are rudimentary (e.g. one-dimensional) at present (e.g. Wang *et al.*, 1995; Sun *et al.*, 2002; Bolla-Pittaluga *et al.*, 2003). They commonly involve assumptions of uniform steady flow, constant slope, and constant channel width. These models have been used to predict whether both diffluent channels are stable, or whether one of them progressively fills with sediment while the other becomes dominant. Theoretical models for the influence of riparian vegetation on the geometry, water flow, sediment transport, and migration of river channels are being increasingly developed (reviews in Bridge, 2003; Bennett and Simon, 2004).

Forward models for the temporal evolution of plan geometry and migration of single meanders are based on models of flow velocity and depth variation in channel bends, and empirical equations for bank erosion rate (e.g. Ikeda *et al.*, 1981; Parker *et al.*, 1982, 1983; Parker & Andrews, 1986; Odgaard, 1989a,b; Johannesson & Parker, 1989; Howard, 1992, 1996; Sun *et al.*, 1996, 2001a, 2001b, 2001c; Nagata *et al.*, 2000; Duan *et al.*, 2001; Seminara *et al.*, 2001; Zolezzi & Seminara, 2001; Lancaster & Bras, 2002). Bank erosion rate is dependent on either near-bank velocity or near-bank depth (each dependency giving different results). Bank erosion models require empirical bank erodibility coefficients that are not known well, and most models do not consider changing modes of bank erosion associated with unsteady flows (but see Nagata *et al.*, 2000). Despite these shortcomings, these models correctly predict both lateral and downstream migration of bends, transformation of symmetrical meander loops to asymmetrical ones as bend sinuousity increases, and reduction in rate of bank erosion as bend amplitude increases. Bend cut-off was predicted by Bridge (1975), Howard (1996), and Sun *et al.* (1996), but not based on an analytical understanding of the process. The model of Sun *et al.* (2001b, 2001c) is the most sophisticated, and can predict the grain size of point-bar surfaces (Fig. 5).

Quantitative, dynamic, 3D depositional models of river channel deposits (e.g. Willis, 1989; Bridge, 1993) are at a rudimentary stage. These types of models require prediction of the interaction between bed topography, water flow, sediment transport rate, mean grain size of bedload, and bedforms within channels of prescribed geometry. The flow conditions are assumed to be steady and bankfull, with the bed topography, water flow and sediment transport in equilibrium. The models apply to either single channel bends with an associated point bar, or two channel bends separated by a braid bar. The plan forms of the channels are sine-generated curves, and features such as unit bars and cross-bar channels are not considered. The channels and associated bars must be put in a dynamic context by allowing them to migrate by bank erosion and bar deposition, and to change geometry in time. Net vertical deposition is not allowed over the timespans considered in the models. Despite being simplified, these models give important insights into the nature of channel-bar deposits that could not come from the previous static 1D and 2D models (review in Bridge, 2003). Examples of these models (shown in Figs 6 and 7) illustrate a number of fundamentally important aspects of river channel deposits.

1. As channels migrate by lateral and downstream migration, the deposits from different parts of channel bars become vertically superimposed (e.g. bar-head deposits overlying bar-tail deposits, and bar-tail deposits overlying confluence scour deposits).
2. Systematic spatial variations in the thickness of channel-bar deposits, and the inclination and orientation of large-scale strata, are due to bed topography and the mode of channel migration. For example, it is common for channel-bar deposits to thicken (by up to a factor of two), and for large-scale strata to steepen, towards a cut bank (channel-belt margin) or confluence scour.
3. Lateral and vertical variation in grain size and sedimentary structures are controlled by the bed topography, flow, sediment transport and bedforms, and by the mode of channel migration. Channel-bar deposits normally fine upwards, but they also commonly show little vertical variation in grain size. Some channel-bar deposits coarsen at the top if bar-head deposits are preserved.

These models of Bridge and Willis predict the geometry, grain size and sedimentary structure of the deposits of single point bars or braid bars. However, they do not consider the somewhat complicated flow structures at channel diffluences

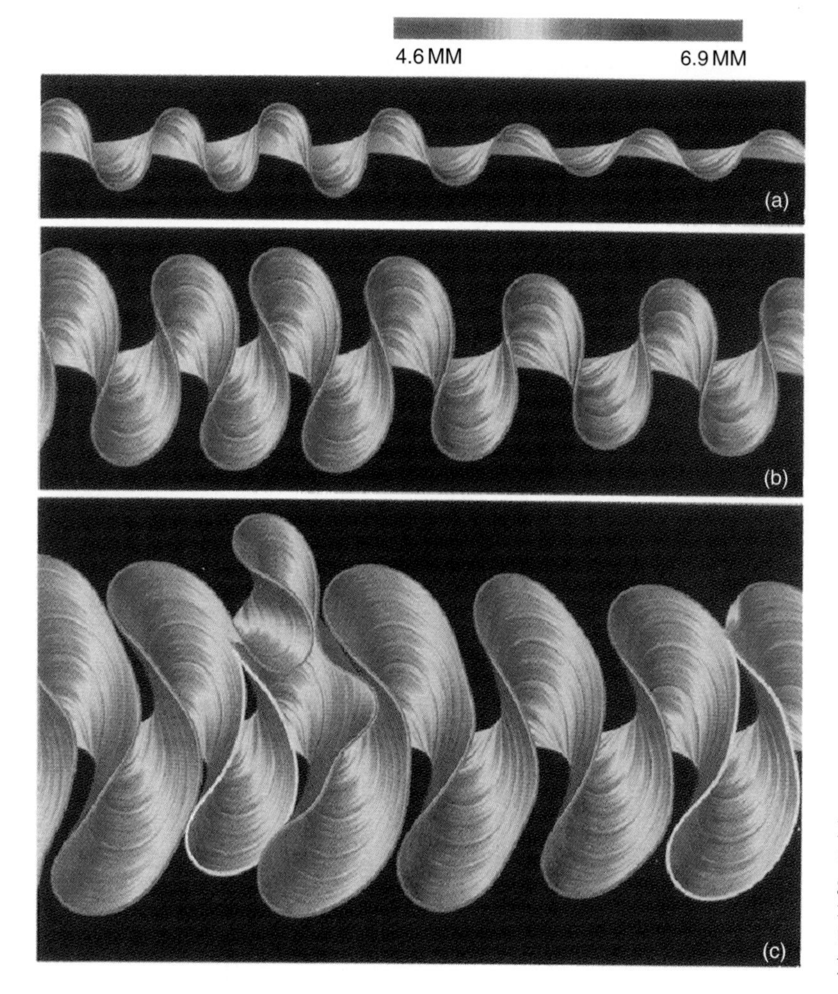

Fig. 5. Meander-belt model of Sun *et al.* (2001b,c). The colour scale is for the grain size of surface sediment. The maps indicate meander belt development after (a) 2762 years, (b) 3263 years, and (c) 4000 years.

and confluences. Finally, there are no quantitative models for the flow, sediment transport and deposition in abandoned-channel fills.

Channel belt

Numerical simulation of the nature of channel deposits within channel belts is in its infancy. The spatial distribution of the deposits of individual channel bars and fills could not be included easily in the bar depositional models of Willis (1989) and Bridge (1993) because it is very difficult to predict how individual channel segments and bars will migrate and become preserved within channel belts. It is necessary to develop models for the deposits of several adjacent bars and channel fills within channel belts. Quantitative models for the simultaneous migration of different curved channels in meander belts (e.g. Howard, 1992, 1996; Sun *et al.*, 1996, 2001a, 2001b, 2001c) cannot yet predict the nature of the deposits. Tetzlaff (1991) proposed a very simple empirical model for the geometry, lateral channel migration and aggradation of meandering and braided channels within channel belts. This model was used (in a way not clearly explained) to predict spatial distributions of channel deposits within floodplain deposits, in which the shapes of channel sandstone bodies look very unrealistic. The predicted deposits were used as training images for a transition probability model that was used to simulate oilfield data.

SEDSIM (Tetzlaff & Harbaugh, 1989) is a sediment routing model of fluvial deposition that is based on solution of the simplified equations of motion of water and sediment. It is claimed to be able to simulate flow, sediment transport, erosion and deposition in river channel bends, braided rivers, alluvial fans, and deltas. However, there are fundamental problems with the basic assumptions and construction of this model, particularly in the treatment of sediment transport, erosion and

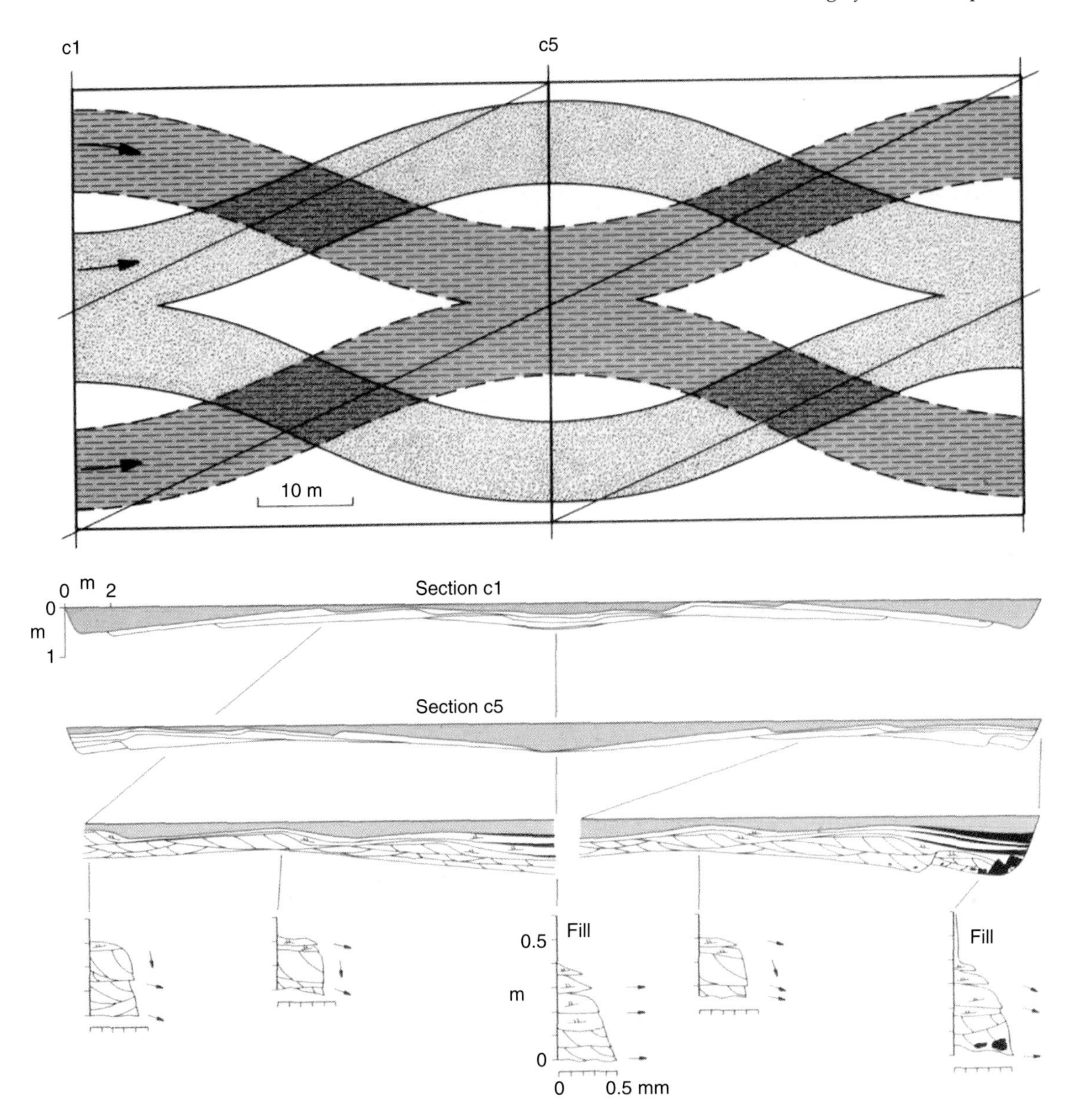

Fig. 6. Example of a quantitative model of braid bar deposits (Bridge, 1993). Upper figure shows plan geometry of initial braided channels (stippled) and migrated channels (dashed). The braid bar migrated downstream in four discrete increments. Cross-sections show basal erosion surface of bar deposits, large-scale inclined strata due to incremental deposition, and details of spatial variation in deposit thickness, grain size, sedimentary structure, and paleocurrents. Deposit thickness and inclination of large-scale inclined strata vary systematically. Bar sequences generally either fine upwards or have little vertical variation in grain size. The dominant internal structure in this example is medium-scale trough cross-strata (formed by dunes), with subordinate small-scale cross-strata (formed by ripples).

deposition (North, 1996). SEDSIM has not been tested by detailed comparison with real-world data (Paola, 2000). Tetzlaff and Priddy (2001) and Griffiths *et al.* (2001) describe the development of SEDSIM into STRATSIM, in order to simulate a range of fluvial and marine sedimentary processes acting over time intervals of hundreds of thousands of years. Unfortunately, few details are given of the workings of most of these models, nor of how or whether they were tested against data from natural sedimentary environments (Paola, 2000). Despite the shortcomings of these models, they have been used to simulate the general character of hydrocarbon reservoirs.

MIDAS is a sediment routing model that predicts the sorting of sediment grains by size and density, and is different from most others in that it considers the effect of turbulent fluctuations

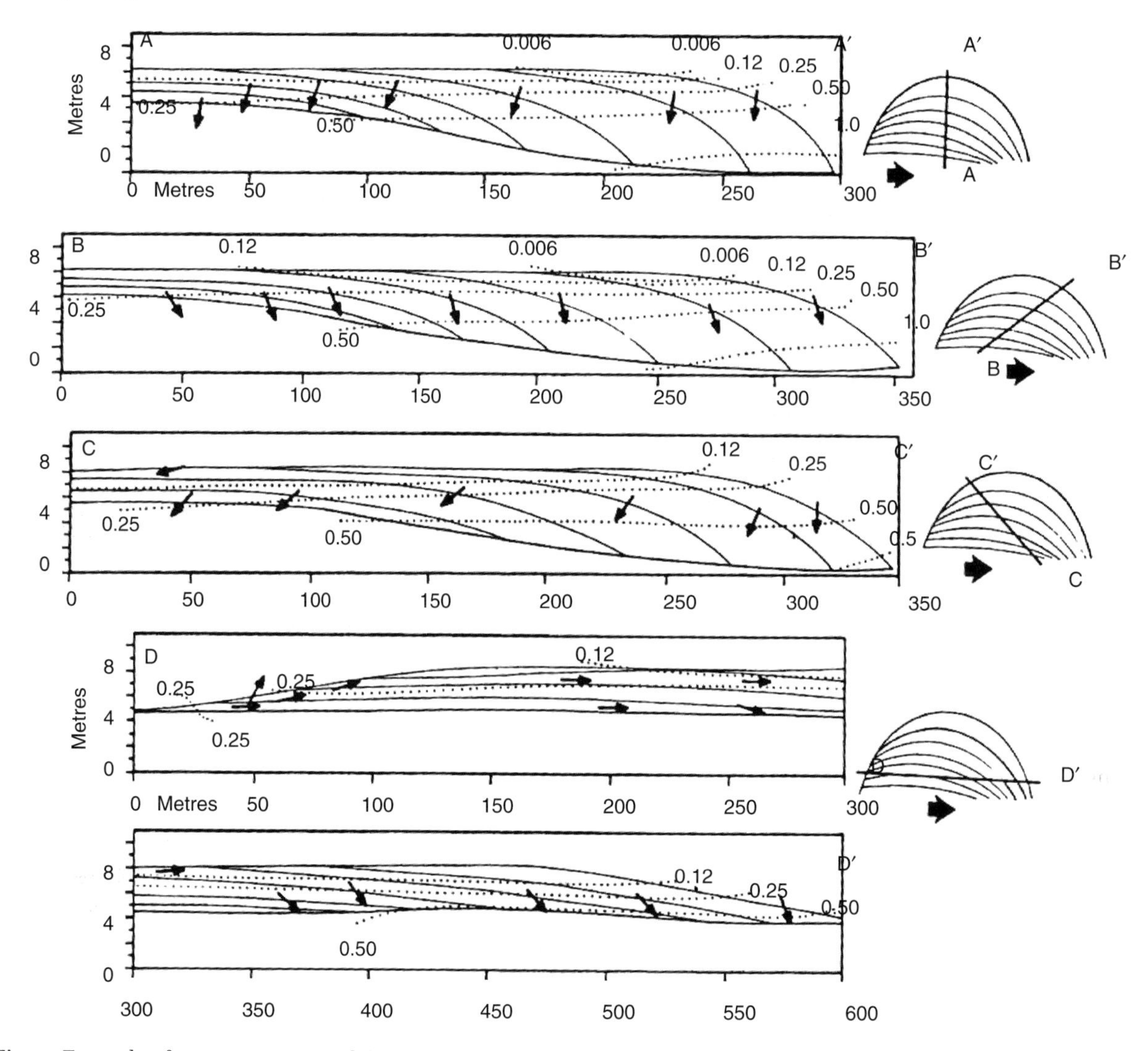

Fig. 7. Example of a quantitative model of point-bar deposits (Willis, 1989). Meander plans to right indicate downstream and lateral growth of a point bar in discrete increments, and position of cross-sections in various orientations. Cross-sections indicate basal erosion surface of point-bar deposits, large-scale inclined strata due to incremental deposition, contours of mean grain size (dotted lines annotated in mm), and current orientations relative to the cross-section (arrows pointing down indicate flow out of plane of cross-section). Point-bar deposits thicken, and inclination of large-scale inclined strata increases, from left to right in sections A, B, and C.

of fluid drag and lift on sediment transport (Van Niekerk *et al.*, 1992). MIDAS considers width-averaged flow and sediment transport in straight channels, and therefore cannot predict the flow and sediment transport associated with point bars and braid bars. Vogel *et al.* (1992), Bennett & Bridge (1995a), and Robinson *et al.* (2001) used MIDAS to predict conditions of erosion and deposition in various flumes and modern rivers. Under erosional conditions, the model predicted well the temporal and spatial variation in sediment transport rate, grain-size distributions of the bed, armour layer, and eroded sediment, and the amount of erosion. In the case of aggradation produced by overloading (sediment supply in excess of transport rate), the grain-size distributions of the bedload and active layer, bedload transport rates, and amounts of deposition were predicted well by MIDAS. The downstream decrease in bed shear stress that occurs under aggradational conditions results in the coarsest sediments being deposited upstream, and the deposit becomes finer grained downstream, as found by many others (review in Bridge, 2003). Robinson & Slingerland (1998a, 1998b) have applied MIDAS to predict downstream fining of bed sediment in ancient channel belts under conditions of active subsidence and aggradation. They concluded that the rate of downstream fining is influenced largely by sediment supply rate and subsidence rate, and to a lesser extent by downstream variations (normally increases) in channel width and depth. For example, if sediment supply rate is large relative to subsidence rate, a relatively large amount of sediment travels past the zone of subsidence rather than being deposited in it and the coarsest grains cannot be effectively separated from the finer grains.

Murray & Paola (1994, 1997) modelled channel belts of braided rivers using a grid of cells, the top surfaces of which formed a uniform downstream slope but with random perturbations in elevation superimposed. Water discharge moving down the grid is distributed between adjacent cells such that more discharge moves down the steeper slopes. Sediment discharge between cells is a power function of either water discharge or a discharge–slope product. The result of the model is that hollows are scoured and high areas receive deposits (due to the non-linear dependence of sediment transport rate on discharge), so that braided channels and bars form. Despite somewhat unrealistic assumptions in the model, Murray & Paola (1994, 1996, 1997) claim that it can produce braided-channel patterns that are similar to those in real rivers. This model cannot predict the nature of deposits. A recent development of this approach is by Thomas and Nicholas (2002).

Vertical superposition of channel-bar and channel-fill deposits in single-channel belts can occur by superposition of a cross-bar channel on a main-channel bar and by migration of one main-channel bar over another. In the latter case, the degree of preservation of the overridden bar depends on the relative elevations of the two superposed basal erosion surfaces. The likelihood of preservation of the lower parts of the eroded bar increases with the vertical deposition rate relative to the lateral migration rate of the superposed bar, and the variability of channel scour depth and bar thickness. The relative importance of deposition rate/lateral migration rate of bars and the variability of channel scour depths (bar heights) in controlling the amount of preservation of truncated bars can be assessed using the method outlined in Bridge and Lunt (2006). In general, the variability of channel scour depths is the main control.

Object-based stochastic models (e.g. MOHERES, FLUVSIM) have been used to distribute channel deposits within channel belts (e.g. Tyler *et al.*, 1994; Webb, 1994, 1995; Webb & Anderson, 1996; Deutsch & Wang, 1996; Holden *et al.*, 1998; Deutsch & Tran, 2002). Webb (1994, 1995) and Webb & Anderson (1996) modelled a braided network of river channels as a random walk. Channel shapes were assigned using hydraulic geometry equations, and sedimentary facies within the channel fills were assigned using a calculated Froude number. A 3D pattern of channel fills was produced by repeated simulation of channel networks with a fixed aggradation rate superimposed. Hydraulic conductivity values were assigned to the various lithofacies types in order to explore the effects of 3D lithofacies heterogeneity on groundwater flow (Webb & Anderson, 1996; Anderson *et al.*, 1999). In the approaches of Tyler *et al.* (1994), Deutsch & Wang (1996), and Holden *et al.* (1998), individual channel deposits within channel belts are represented by a series of sinuous channels superimposed in space (Fig. 8). None of these approaches correctly represents the nature of channel deposits in channel belts, which are in fact composed predominantly of parts of channel bars with relatively minor volumes of channel fills (Fig. 1). It is necessary to define shapes of objects properly, as shown in Fig. 9. The information shown in Fig. 2 will assist in the scaling of objects.

Prediction of the width and thickness of subsurface channel-belt deposits is an important objective. Channel-belt thickness can be determined relatively easily from well logs and cores (e.g. Bridge & Tye, 2000). The most common method for predicting the width of channel-belt deposits is probably lithostratigraphic correlation of channel sandstone bodies between wells. Other methods include use of outcrop analogues, empirical equations relating channel depth to channel-belt width, and, rarely, 3D seismic data. Drawbacks of these methods for predicting channel-belt width were discussed by Bridge & Tye (2000) and Bridge (2001, 2003). A reasonable numerical model for the dynamics of channels and bars within channel belts should be capable of predicting channel-belt width and thickness, given parameters such as bankfull channel width and depth, channel sinuosity and degree of braiding, deposition rate, and lifespan of the channel belt (Bridge, 2003).

Floodplain

The interaction between water flow, sediment transport, and bed topography of floodplains is not known as well as for channels (review by Bridge, 2003). Detailed field studies of water flow

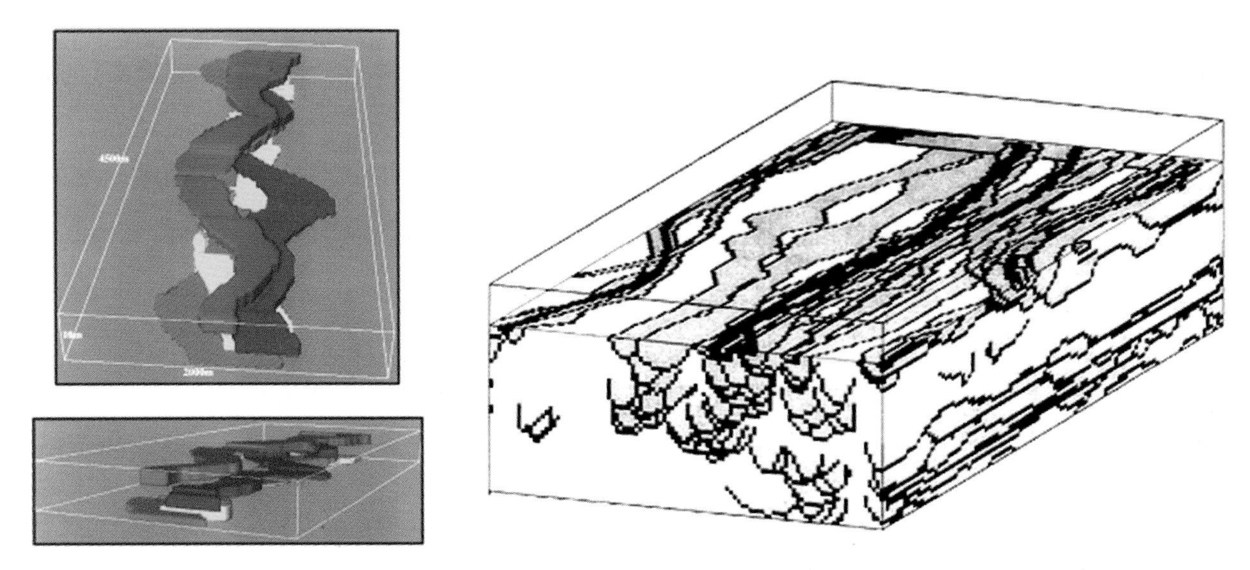

Fig. 8. Stochastic object models for channel belts from Tyler *et al.* (1994) (left) and Deutsch & Wang (1996) (right). Both representations are unrealistic because real channel belts are composed of the deposits of channel bars and channel fills.

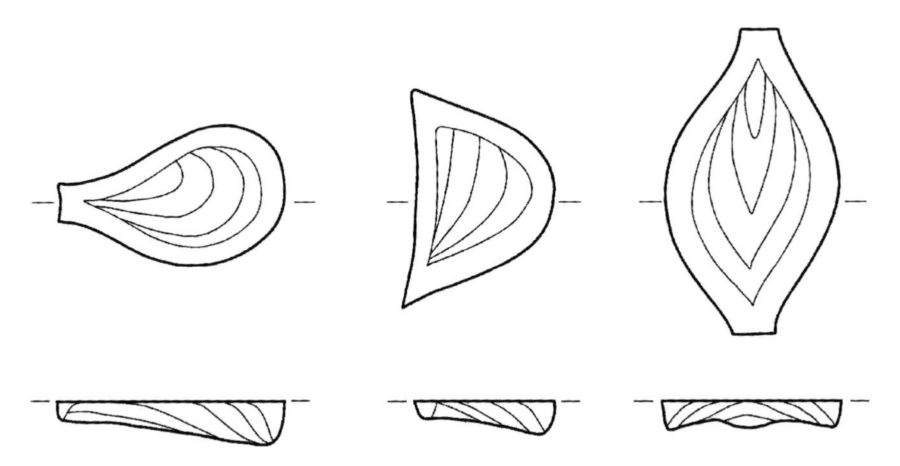

Fig. 9. Realistic shapes of channel bar and fill objects (plans and cross-sections of channel bars and adjacent channel fills) that must be distributed within channel belts.

and sediment transport over floodplains during overbank floods do not exist, mainly because of difficulties of observation and lack of willing students. Water flow and sediment transport on floodplains is complicated by variable floodplain width and surface topography (channels; depressions; mounds of sediment such as levees, crevasse splays, abandoned channel belts, and marginal alluvial fans; vegetation and structures produced by humans and other animals). Many experimental studies of overbank flow adjacent to river channels have been undertaken, but mostly with steady flows over simple channel–floodplain geometry, and with immobile boundaries without sediment movement. Numerical models of floodplain flow and sediment transport are inadequate at present. Therefore, there are no quantitative, 3D depositional models for floodplains.

Several numerical models of suspended-sediment transport on floodplains have been proposed (e.g. James, 1985; Pizzuto, 1987; Nicholas & Walling, 1997b, 1998; Middlekoop & Van Der Perk, 1998; Hardy *et al.*, 2000). The models of James and Pizzuto are based on lateral turbulent diffusion from a zone of high suspended-sediment concentration at the channel margin to a zone of low concentration on the floodplain. However, such a diffusion approach cannot describe the common convective transport of suspended sediment from channels to floodplains, and within floodplains, in sheet flows and channels. Furthermore, bedload is commonly an important component of sediment transport on floodplains. Once again, a sediment routing approach is required.

Empirical equations for the spatial variation in time-averaged deposition rate on floodplains by (Bridge & Mackey, 1993a) are:

$$r/a = (1 + z/z_{\mathrm{m}})^{-b} \tag{1}$$

$$r/a = \mathrm{e}^{-bz/z_{\mathrm{m}}} \tag{2}$$

where r is deposition rate at distance z from the edge of the channel belt, a is mean channel-belt deposition rate, z_{m} is the maximum floodplain distance from the edge of the channel belt, and b is an exponent that determines the rate of decrease of floodplain deposition with distance from the channel belt. The value of b ranges up to about 10, but there is no rational theory to explain its variability. Apparently, b values decrease as the timescale of deposition increases, implying that successions of overbank floods do not maintain the same areal distribution of deposition rate as observed during a single flood. Several others have proposed quantitative models for spatial variation of overbank aggradation rate (e.g. James, 1985; Pizzuto, 1987; Howard, 1992, 1996; Marriott, 1996; Nicholas & Walling, 1997a, 1998; Middelkoop & Van Der Perk, 1998). As stated above, those based on lateral diffusion of suspended sediment cannot be justified in general. Several recently proposed floodplain deposition models (e.g. Howard, 1992, 1996; Walling *et al.*, 1996; Nicholas & Walling, 1997a; Middelkoop & Van Der Perk, 1998; Hardy *et al.*, 2000) have incorporated the role of floodplain topography (or flow depth), which can exert considerable control on short-term overbank aggradation rate.

Alluvial valley

Models of channel-belt movements across floodplains (and fans)

A ubiquitous feature of aggrading floodplains (and fans and deltas) is the periodic shifting of channel belts to new locations on the floodplain, a process called avulsion (reviews in Smith & Rodgers, 1999; Bridge, 2003; Slingerland & Smith, 2004). It has also been suggested that channel belts can migrate across floodplains by more gradual erosion and deposition at the edge of the channel belt (not discussed here, but see Bridge, 2003). Process-based, numerical models of avulsion are rare. Mackey & Bridge (1995) suggested that the probability of avulsion at a given location along a channel belt could be expressed empirically as:

$$P(a) = (Q_{\mathrm{f}}/Q_{\mathrm{a}})^{eq}(cs\ S_{\mathrm{cv}}/S_{\mathrm{dv}})^{es} \tag{3}$$

where Q_{f} is maximum flood discharge for a given year, Q_{a} is a threshold flood discharge, S_{cv} is the cross-valley slope at the edge of the channel belt, S_{dv} is the local down-valley slope of the channel belt, and cs is an empirical coefficient. This expression is based on the assumptions that avulsions are initiated during periods of extreme discharge when erosive power of the stream is greatest, and that a sufficient stream power advantage exists for a new course to be established on the floodplain. In reality, avulsion could also be triggered by a flow blockage downstream. The model produces realistic results if cs is 0.2 to 0.3, and the exponents eq and es are between 1 and 5.

The slope ratio concept was used because the discharge–slope product (QS) is proportional to the sediment transporting power per unit length of the stream. Erosion and enlargement of a crevasse channel will be possible if the sediment transport rate increases from the point at which it bifurcates from the main channel down its course over the floodplain. The water discharge in a growing crevasse channel will initially be less than that in the larger main channel. Therefore, in order for the sediment transport rate (per unit channel length) to increase from the main channel to the crevasse channel, the crevasse channel slope must be greater than the main channel slope. In the limiting case where the water discharges of the main channel and the developing crevasse channel are equal, the slope in the crevasse channel must exceed that of the main channel for the avulsion to proceed. The water level in a floodbasin must be lower than that in the main channel to allow water to flow away from the main channel through a crevasse channel. When the water surface elevations of the main channel and the floodbasin are the same, there can be no such crevasse-channel flow. Therefore, crevasse-channel enlargement can only operate during certain overbank flood stages. Accordingly, it may take a number of overbank flood periods for a crevasse channel to enlarge to the point of avulsion.

Mackey & Bridge (1995) used equation (3) to simulate avulsions where floodplain slopes vary in space and time due to variations in deposition rate, tectonic tilting and faulting within the alluvial valley (Fig. 10). *Down-valley increase in channel-belt deposition rate* produces a down-valley decrease in channel-belt slope but an

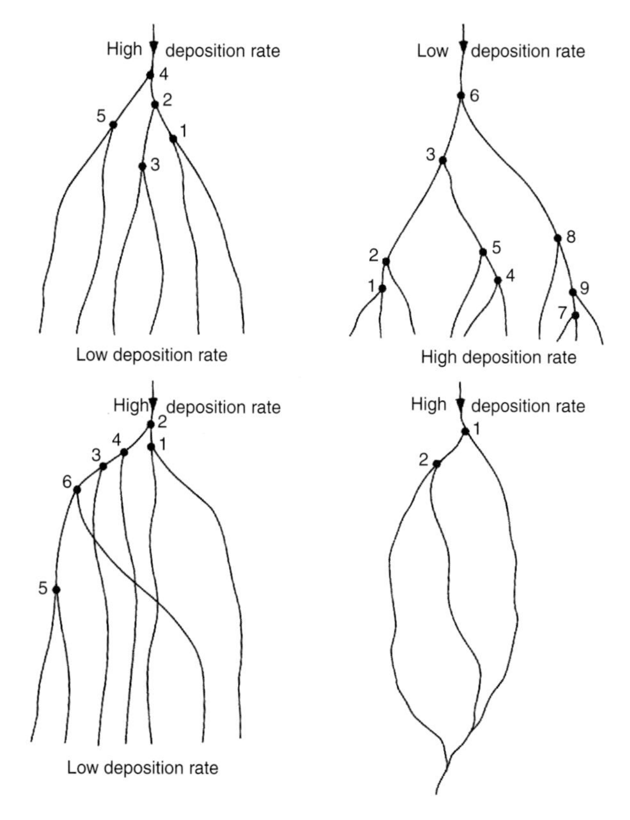

Fig. 10. Typical patterns of channel-belt avulsion, dependent upon spatial variation in deposition rate and pre-existing floodplain topography. Channel belts represented by lines. Numbers represent points of avulsion in chronological order. Based on observations and theoretical models (full explanation in Bridge, 2003).

increase in cross-valley slope as the alluvial ridge grows, as is likely to happen during base-level rise. Under these circumstances, avulsion probability increases through time, and is greatest in the down-valley part of the floodplain where channel-belt slopes are smallest but cross-valley slopes are largest. The model predicts that avulsion is initiated in the down-valley part of the floodplain and successive avulsion points shift up-valley with a progressive decrease in interavulsion period. This is due to gradual increase in avulsion probability up-valley of avulsion points where alluvial ridge growth continues uninterrupted. New channel-belt segments down-valley from avulsion points have not had time to aggrade significantly and develop alluvial ridges, and therefore have low avulsion probabilities. After a finite number of avulsions stepping up-valley, the progressive decrease in channel-belt slopes in the down-valley part of the floodplain causes abrupt shift in the locus of avulsion to this location. Although model results agree broadly with what is observed in nature, the model does not take into account the effects of the increased slope of the new channel as it leaves the old channel. This steep slope would result in channel incision and upstream retreat of a knickpoint in the vicinity of the point of avulsion. Therefore, the probability of avulsion would be greatly reduced immediately upstream of a recent avulsion point.

Down-valley decrease in channel-belt aggradation rate is typical of alluvial fans and where base-level is falling, and produces a down-valley decrease in down-valley slope. Avulsion probability decreases with time because overall down-valley channel-belt slope increases with time. However, avulsion probability is high in the up-valley parts of the floodplain where cross-valley slopes are increased by high aggradation rate. The concentration of avulsions up-valley where deposition rate is high produces nodal avulsions, which are characteristic of alluvial fans (Fig. 10).

The model predicts that avulsion periods vary greatly depending on the stage of growth of alluvial ridges. Sections of alluvial ridges that are well developed are associated with short avulsion periods (say decades or centuries), whereas newly formed channel belts may not experience an avulsion for on the order of 1000 years. Also, the obstruction to flow caused by pre-existing alluvial ridges may cause subsequent channel belts to be clustered preferentially on one side of the floodplain with a distinctive en échelon pattern (Fig. 10). This means that other parts of the floodplain distant from the active channels experience relatively low overbank deposition rates for extended periods of time, allowing soils to develop.

Tectonic tilting and faulting within the floodplain increase avulsion probability locally (Fig. 11). Channel belts shift away from zones of uplift and towards zones of maximum subsidence. However, if channel-belt aggradation keeps pace with fault displacement or tilting, alluvial ridge topography causes channels to shift *away* from areas of maximum subsidence. Although these predictions agree broadly with data from modern rivers, data on the relationship between tectonism and avulsion are insufficient to test model predictions in detail.

The Mackey–Bridge model was a first attempt, and there is ample room for improvement. The model does not simulate changes in channel pattern and channel-belt width, floodplain lakes, and channel-belt and floodplain deposition and

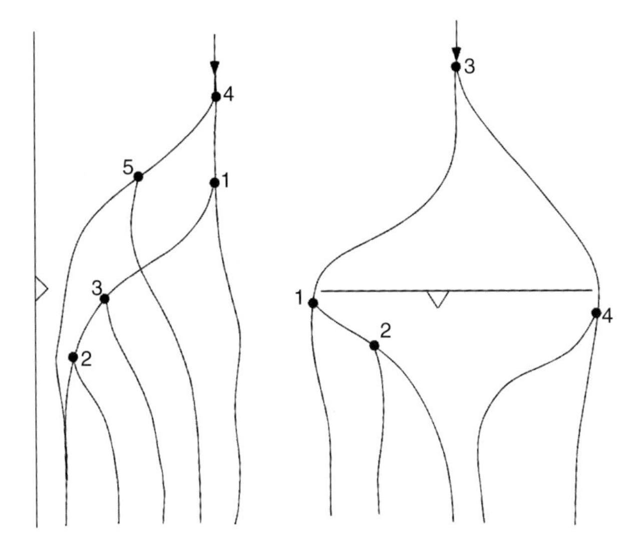

Fig. 11. Typical effect of faults and tilting on channel-belt avulsion. Channel belts indicated by lines. Numbers represent points of avulsion in chronological order. Downthrown side of faults indicated by triangles. Based on observations and theoretical models (full explanation in Bridge, 2003).

erosion associated with tectonism. Tributaries and downstream increases in channel belt size are not simulated, nor are the influences of alluvial fans at valley margins. The model does not consider widespread erosion of channel–floodplain systems, and it is not capable of simulating the diversion of a channel belt into a pre-existing channel. Model development is, however, under way, and has been greatly aided by new data from Holocene fluvial and deltaic settings (reviewed by Bridge, 2003). The component models for aggradation, avulsion, and the development of channel belts following avulsion have been greatly improved (details in Karssenberg *et al.*, 2003, 2004). New channel belts are formed by channel bifurcation, and multiple channel belts may develop and coexist (forming anastomosing rivers). Anastomosing rivers occur in the new model where the channel belt bifurcates but there are insufficient gradient advantages to cause avulsion. Avulsions occur when a gradient advantage causes the discharge in one of the channel belts to increase. The new model can simulate the occupation of pre-existing floodplain channels by an avulsing channel. Degradation of the channel and floodplain is treated using a diffusion–advection approach, such that upstream migration of knickpoints and formation of incised channels and terraces can be simulated. If a channel belt is incised up-valley, it becomes fixed in space, such that in depositional areas downstream the channel belt can be associated with abnormally thick and wide deposits. The effect of cyclic variation in degradation and aggradation on avulsion and alluvial architecture can be simulated (allowing a link to sequence-stratigraphy models).

Slingerland & Smith (1998) have made the only analytical approach to the cause of avulsion. Their model is based on simplified equations of motion for fluid and sediment applied to simple channel geometry. The crux of the model is that the suspended sediment concentration at the entrance to a crevasse channel leading from a deeper main channel is different from the equilibrium concentration that should exist in the crevasse channel. Then, depending on local hydraulic conditions, the crevasse channel would deepen or fill with sediment until the equilibrium suspended sediment concentration is reached. A condition where the channel progressively deepens is taken as a criterion for avulsion. Avulsions are predicted to occur wherever the slope ratio exceeds 5. One drawback of this approach is that bedload sediment transport is not treated, so that the model cannot explain avulsions in the many rivers that transport mainly bedload during floods. Slingerland & Smith (2004) mention that theories for the stability of channel bifurcations (including treatment of bedload) in braided rivers (e.g. Bolla Pittaluga *et al.*, 2003) might be applied to prediction of avulsion.

Sun *et al.* (2002) made a numerical model of channel flow and sediment transport dynamics linked to channel avulsion for the case of fluvial fan deltas. They used the models of Parker *et al.* (1998a, 1998b) for predicting channel geometry, flow, and sediment transport, assuming steady uniform flow conditions. They used a cellular modelling approach similar to Murray and Paola (1994) to simulate channel bifurcations, and assumed that the water discharge to each bifurcation channel is proportional to the square root of the channel slope for that channel. Of course, the discharge in any channel should really also be related to its depth and width. A channel avulsion occurs if the channel elevation at the bifurcation is greater than the elevation at some reference point on the fan. Therefore, a degree of channel superelevation is required for an avulsion to occur. If an avulsion is predicted, the channel follows the line of steepest descent, but is allowed some statistical variation in its course. These fan channels do not form channel belts *per se*. This avulsion model is not compared with those of Mackey & Bridge (1995) and Slingerland & Smith (1998): indeed, no reference is made to this previous work.

Deposits of alluvial valleys (alluvial architecture)

The term *alluvial architecture* was used by Allen (1978) to refer to variations in the mean grain size, geometry, proportion and spatial distribution of channel-belt and floodplain deposits within alluvial valleys. Definition of the architecture of ancient deposits requires extensive exposures, and/or high-resolution (preferably three-dimensional) seismic data, and/or many closely spaced cores or borehole logs, and accurate age dating. As such data are commonly lacking or incomplete, it is necessary to 'fill in' three-dimensional space in order to produce a complete (and hypothetical) representation of alluvial architecture. Furthermore, as it is not possible to observe directly the processes of development of alluvial architecture, it is necessary to use models to interpret and predict alluvial architecture. Most quantitative models of alluvial architecture (reviewed by Bryant & Flint, 1993; Koltermann & Gorelick, 1996; North, 1996; Anderson, 1997; Bridge, 2003) are either process-based (process-imitating) or stochastic (structure-imitating), as discussed below.

Alluvial architecture is primarily controlled by: the geometry and sediment type of channel belts and floodplains; the rate of deposition or erosion in channel belts and on floodplains; local tectonic deformation within the alluvial valley, and; the nature of channel-belt movements (mainly avulsions) over floodplains. These *intrinsic* (intrabasinal) controlling factors are in turn controlled by *extrinsic* (extrabasinal) factors such as tectonism, climate, and eustatic sea-level changes. For example, the geometry and sediment type of channel belts and floodplains are determined by water and sediment supply (rate and type), which are in turn controlled by the source rocks, topographic relief, climate, and vegetation of the drainage basin. Floodplain width is also influenced by its tectonic setting and development history. Long-term deposition in alluvial valleys is due mainly to long-term decrease in sediment transport rate in the down-flow direction (by virtue of conservation of mass), which can be accomplished by either increasing sediment supply up-valley (by tectonic uplift, climate change, or river diversions) and/or by decreasing sediment transport rate down valley (by flow expansion associated with tectonic subsidence or base-level rise). Subsequent erosion depends upon increasing sediment transport in the down-flow direction, such as caused by basin uplift or base-level fall. Local tectonic deformation in alluvial valleys can cause local changes in channel and floodplain geometry and location, deposition and erosion. Channel-belt movements across floodplains are influenced by the severity of floods and the development of cross-floodplain slopes associated with alluvial ridge deposition and local tectonic deformation. Most process-based models of alluvial architecture consider intrinsic processes.

Process-based (process-imitating) models of alluvial architecture

Leeder (1978) developed the first quantitative, process-imitating model of alluvial architecture. A channel belt of specified cross-sectional area was allowed to migrate by avulsion within a floodplain of finite width as aggradation (balancing subsidence) continued. The period of time separating avulsions and the location of avulsing channels on the floodplain were chosen randomly within defined limits. The deposits were modelled within a single cross-valley section. Leeder defined 2D measures of the proportion of channel-belt deposits and their degree of connectedness, that depend upon aggradation rate, avulsion frequency, channel-belt cross-sectional area, and floodplain width. Allen's (1978, 1979) later approach followed Leeder's closely, but Allen added a function to allow diverted channel belts to avoid high floodplain areas underlain by pre-existing channel belts. The 2D approach was extended (Bridge & Leeder, 1979; Bridge & Mackey, 1993a, 1993b) by considering also the effects on alluvial architecture of compaction, tectonic tilting of the floodplain, and variation of aggradation rate with distance from channel belts, and predicting the width and thickness of channel sandstone/conglomerate bodies comprising single or connected channel belts (Fig. 12). Channel-deposit proportion, sandstone-body width and thickness increase as bankfull channel depth and channel-belt width increase, and as floodplain width, aggradation rate and inter-avulsion period decrease. Channel-deposit proportion and connectedness also increase in locally subsided areas of floodplain. These quantitative 2D models have been used (and misused) widely to interpret and model alluvial architecture, despite the fact that they have not been tested comprehensively against field data (but see Leeder *et al.*, 1996). However, 2D models are unable to realistically simulate down-valley variations in the location

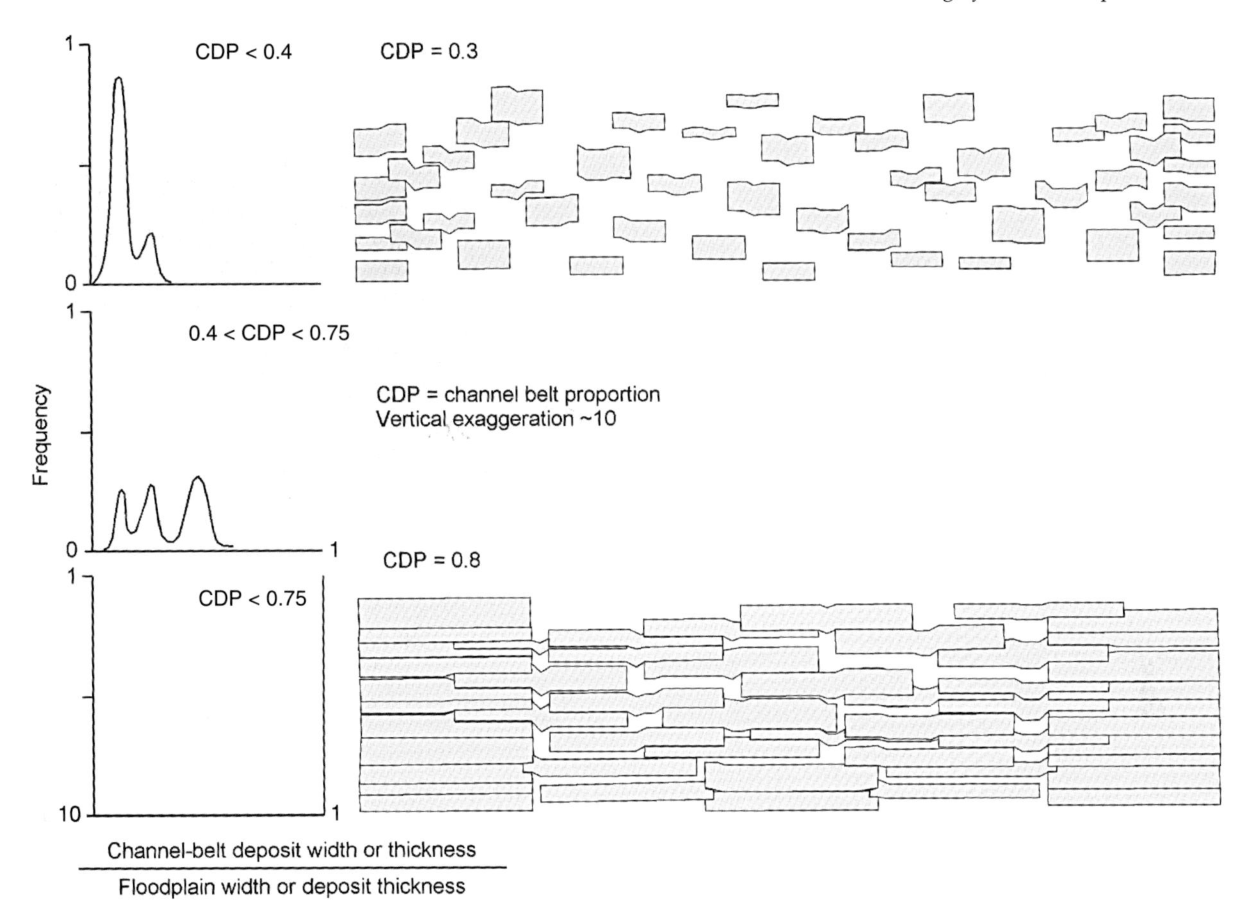

Fig. 12. Channel belt connectedness increases with channel deposit proportion (CDP). For CDP < 0.4, most channel belts (shown as stippled boxes in cross-floodplain sections) are unconnected, such that frequency distributions of channel deposit width or thickness are bimodal with a large mode equivalent to the width or thickness of an unconnected channel belt. As CDP increases, more channel belts are connected, channel deposits become larger, and the frequency distributions of channel deposit width or thickness become polymodal. For CDP > 0.75, all channel belts are connected and the channel deposit is as wide as the floodplain. Based on 2D alluvial architecture model of Bridge & Mackey (1993a).

and orientation of individual channel belts. This is only possible with 3D models.

Mackey & Bridge (1995) developed the first 3D process-based model of alluvial architecture. The floodplain contains a single active channel belt (Fig. 13). Changes in floodplain topography are produced by spatial and temporal variation of channel-belt and floodplain deposition rates, by compaction, and by tectonism. The location and timing of avulsions are determined by local changes in floodplain slope relative to channel belt slope, and by flood magnitude and frequency (equation 3). The diverted channel follows the locus of maximum floodplain slope. Major differences between this model and the 2D models are the treatment of avulsion location and period as dependent variables, and constraints on the location of avulsing channels by the points of avulsion and topographic highs on the floodplain. The behaviour of the avulsion model was discussed above. Avulsions occur preferentially where there is a decrease in channel-belt slope and/or an increase in cross-valley slope that may be related to spatial variations in deposition rate and/or tectonism and/or base-level change. Evolution of alluvial ridges over time in different parts of the floodplain greatly influences the timing and location of avulsions. This may result in sedimentary sequences that increase upwards in channel-deposit proportion and connectedness, capped with overbank deposits with well developed soils. Such sequences may take on the order of 10^3–10^5 years to form, comparable to cycles attributed to tectonism or climate change. Predictions of the Mackey–Bridge model agree with the somewhat limited data from modern rivers, and the model has been applied to interpreting and predicting the alluvial architecture of ancient deposits. However,

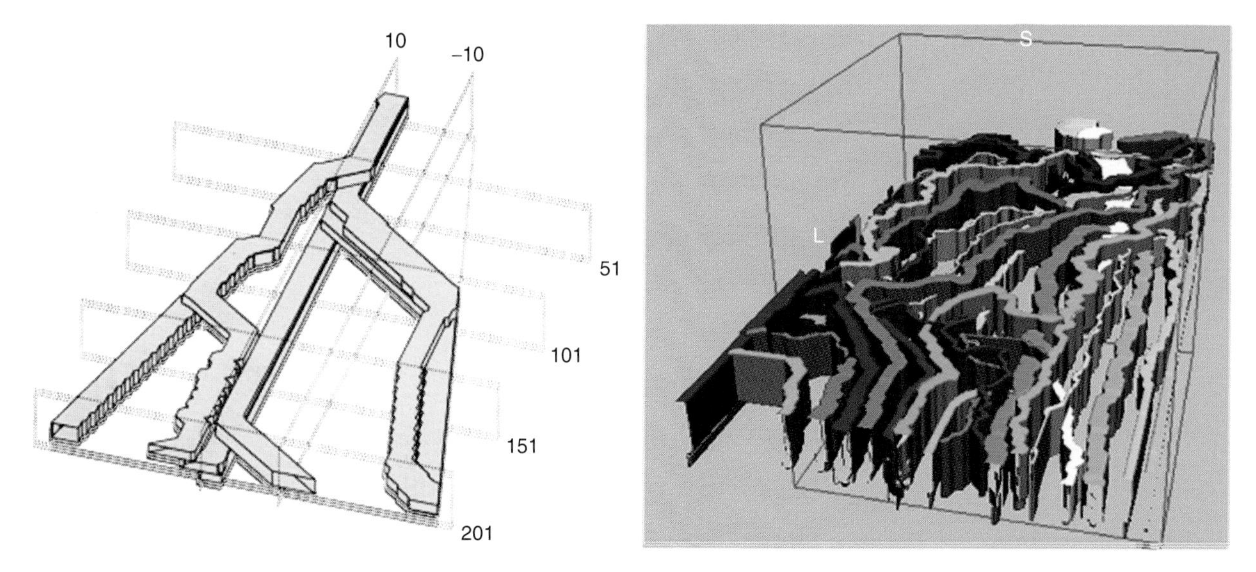

Fig. 13. Quantitative, 3D alluvial architecture models from Mackey and Bridge (1995) (left) and Karssenberg *et al.* (2004) (right). Coloured objects are channel belts.

the model is very simplified, and further development of the model is underway, as discussed above.

Even though desirable, sediment routing models have not been used to simulate the complicated spatial and temporal variations in sediment transport and deposition in channels and floodplains that occur over the long term in alluvial valleys. However, one very simple sediment routing approach applied to alluvial architecture (Koltermann & Gorelick, 1992) is discussed below.

PREDICTION OF ALLUVIAL ARCHITECTURE OF SUBSURFACE DEPOSITS

The most common approach to predicting the architecture of subsurface fluvial reservoirs and aquifers (discussed in Bridge & Tye, 2000) is to: (1) determine the geometry, proportion and location of different types of sediment bodies (e.g. sandstones, shales) from well logs, cores, seismic or GPR; (2) interpret the origin of the sediment bodies; (3) use outcrop analogues to predict more sediment-body characteristics; and (4) use stochastic (structure-imitating) models to simulate the alluvial architecture between wells, and the rock properties with sediment bodies such as channel-belt sandstones. Stochastic (structure-imitating) models are either object-based (also known as discrete or Boolean) or continuous, or both (reviewed by Haldorsen & Damsleth, 1990; Bryant & Flint, 1993; Srivastava, 1994; North, 1996; Koltermann & Gorelick, 1996; Deutsch, 2002). A common combined approach is to use object-based models to simulate the distribution of channel-belt sandstone bodies and floodplain shales, and then use continuous models for simulating 'continuous' variables such as porosity and permeability within the objects.

With object-based models, the geometry and orientation of specified objects (e.g. channel-belt sandstone bodies or discrete shales) are determined by Monte-Carlo sampling from empirical distribution functions derived mainly from outcrop analogues. 'Conditioned simulations' begin by placing objects such that their thickness and position correspond with the available well data. Then, objects are placed in the space between wells until the required volumetric proportion is reached. Objects are placed more-or-less randomly, although arbitrary overlap/repulsion rules may be employed to produce 'realistic' spatial distributions of objects (Figs 14 and 15).

Continuous stochastic models have been used mainly to simulate the spatial distribution of continuous data such as permeability, porosity, and grain size. With these models, a parameter value predicted to occur at any point in space depends on its value at a neighbouring site. The conditional probabilities of occurrence are commonly based on an empirical semivariogram. These approaches have been modified to predict the distribution of discrete facies by using indicator semivariograms and simulated annealing (e.g. Johnson &

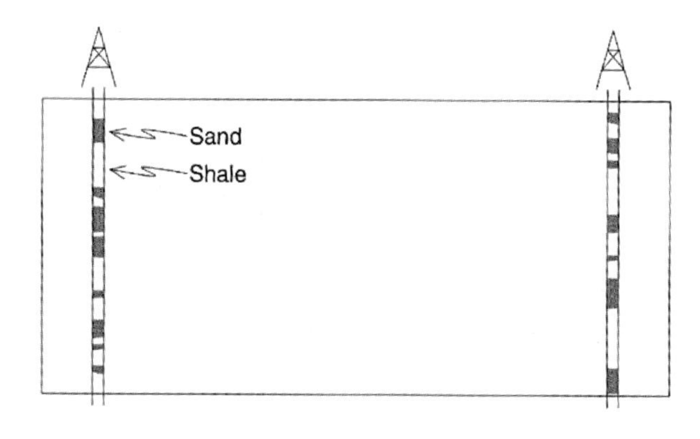

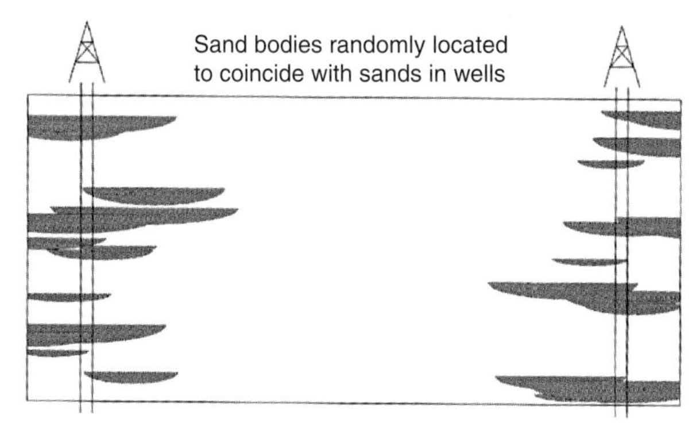

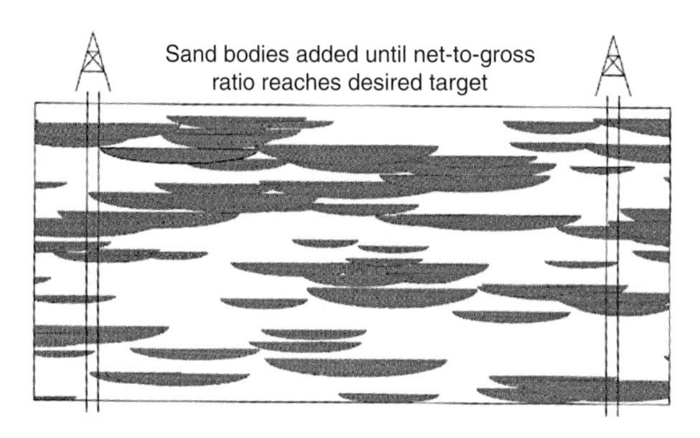

Fig. 14. Example of 2D object-based stochastic modelling of alluvial architecture (modified from Srivastava, 1994).

Driess, 1989; Bierkens & Weerts, 1994; Deutsch & Cockerham, 1994; Seifert & Jensen, 1999, 2000). A variant of the indicator semivariogram approach is transition probability (Markov) models in which the spatial change from a particular sediment type (e.g. channel sandstone) to another (e.g. floodplain mudstone) is based on the probability of the transition. The probability of spatial transition to a particular sediment type depends on the existing sediment type, and this dependence is called a Markov property. The matrix of probabilities of transition from one sediment type to another can be used to simulate sedimentary sequences in one, two or three dimensions (e.g. Tyler *et al.*, 1994; Doveton, 1994; Carle *et al.*, 1998; Elfeki & Dekking, 2001; Fig. 15). Another variant on the variogram approach is multiple-point geostatistics, which can simulate complex spatial heterogeneity given training images (Caers *et al.*, 2000; Caers and Zhang, 2004; Strebelle, 2002).

It is commonly difficult to define the input parameters for stochastic models, especially the semivariograms and transition probability matrices in lateral directions, and spatial heterogeneity is difficult to simulate realistically. The shapes, dimensions and locations of objects in object-based models are also difficult to define realistically (see the models of Tyler *et al.*, 1994; Deutsch & Wang, 1996; Holden *et al.*, 1998; Seifert & Jensen, 1999, 2000; Fig. 15). If definition of the dimensions of objects relies upon use of outcrop analogues and determination of palaeochannel patterns from subsurface data, there may be serious problems (Bridge & Tye, 2000). Process-based models and sequence-stratigraphy models demonstrate that the spatial distribution of channel-belt sandstones is not random. Unrealistic shapes, dimensions and spatial distributions of sediment types means that it is difficult to get the model to fit observed data and predict reservoir/aquifer behaviour. Furthermore, as stochastic models do not simulate processes of deposition, they cannot give any insight into the origin of the alluvial architecture, and they have no predictive value outside the data region.

If there are so many problems with stochastic models, why are they used? Commercial software is available. Simulations can easily be conditioned using well data, cores, seismic, GPR, and other types of geological information. Detailed understanding of the origin of the subsurface strata is not necessary in order to use stochastic models, even though it is desirable. Numerical forward (process-imitating) models are considered difficult to fit to subsurface data, and the models and software have not yet been well developed. Therefore, process-imitating models have had limited application in quantitative simulation of the architecture of hydrocarbon reservoirs or aquifers. However, process-imitating models provide genetic interpretations of deposits, and can predict more realistic sedimentary architecture than structure-imitating stochastic models. Karssenberg *et al.* (2001) have demonstrated that fitting of process-based models to well data using an essentially trial-and-error approach is possible in principle. Such an approach involves multiple

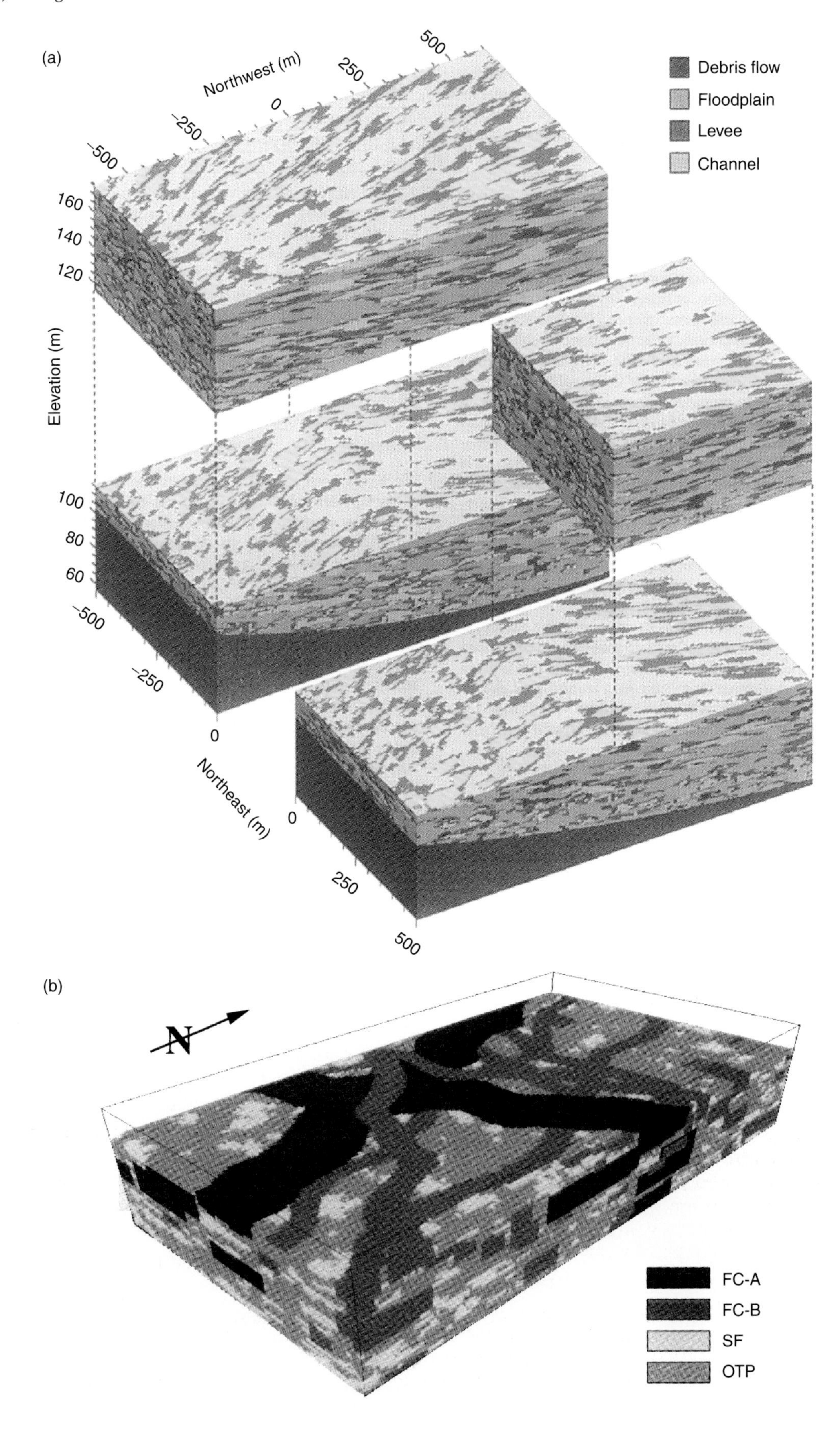
(a)
Northwest (m)
-500
-250
0
250
500
Debris flow
Floodplain
Levee
Channel
160
140
120
Elevation (m)
100
80
60
-500
-250
0
Northeast (m)
0
250
500
(b)
N
FC-A
FC-B
SF
OTP

runs of a process-based model under different input conditions, and optimization of the fitting of output data to observed data. Process-based models are being developed by Karssenberg *et al.* (2003), including development of software so that models can be fitted to subsurface data (inversion approach). Another approach is to use output from process-imitating forward models to provide input (training images) for stochastic models that can be more easily conditioned with subsurface data.

RIVER SYSTEM

Process-imitating models of river systems consider interactions between tectonic uplift and subsidence, climate, base-level change, vegetation development, water and sediment supply, erosion and deposition, and evolution of drainage systems. However, it is uncommon for all of these interactions to be considered simultaneously. Modelling of the influence of vegetation on water and sediment supply, erosion and deposition is at a rudimentary stage.

Landscape erosion models and evolution of drainage systems

Many models now exist for the evolution of drainage systems in upland regions dominated by erosion (e.g. Kirkby, 1986, 1987; Willgoose *et al.*, 1991a, 1991b, 1991c, 1992, 1994; Chase, 1992; Beaumont *et al.*, 1992, 2000; Kooi & Beaumont, 1994, 1996; Howard, 1994; Tucker and Slingerland, 1994, 1996, 1997; Izumi & Parker, 1995; Johnson & Beaumont, 1995; Moglen & Bras, 1995; Braun & Sambridge, 1997; Tucker & Bras, 1998; Coulthard *et al.*, 1999, 2002; Ellis *et al.*, 1999; Allen & Densmore, 2000; Crave & Davy, 2001; Tucker *et al.*, 2001, 2002; Bogaart *et al.*, 2003a; Clevis *et al.*, 2003, 2004a, 2004b; Tucker, 2004). Although these various approaches differ in the way erosion and deposition are modelled, there are many similarities. Sediment transport on hillslopes associated with mass wasting processes such as soil creep is commonly modelled using a linear diffusion approach, and such transport may be limited by production of sediment by weathering. Landslides have been modelled by considering the shear strength of rock or soil and whether or not this material exceeds some critical hillslope angle. Erosion rate of bedrock and cohesive sediments by streams is commonly taken as proportional to kA^mS^n in excess of some threshold value, where k is bedrock erodibility, A is drainage basin area, and S is surface slope; A is a surrogate measure for stream water discharge, such that bedrock erosion rate becomes a function of stream power per unit channel length. The exponents m and n are not known well, although they are commonly assumed equal to unity. There has been extensive discussion of the various models for bedrock erosion (e.g. Seidl & Dietrich, 1992; Howard *et al.*, 1994; Sklar & Dietrich, 1998, 2001; Stock & Montgomery, 1999; Whipple & Tucker, 1999, 2002; Nieman *et al.*, 2001; Tucker & Whipple, 2002; Baldwin *et al.*, 2003). Sediment transport in channels and by overland flow is also commonly taken as proportional to a function of stream power in excess of some limiting value. Sediment transport by flowing water may also be limited by sediment production by weathering. Grain-size sorting during sediment transport is generally only treated at a rudimentary level (e.g. separate treatment of sand and gravel: Gasparini *et al.*, 1999, 2004; Tucker *et al.*, 2002; Clevis *et al.*, 2003) if at all. Gasparini *et al.* (2004) examined theoretically the interaction between the degree of stream-profile concavity and bed grain-size sorting.

Hillslope processes are not explicitly distinguished from channel processes in all of these models. In some models, the formation of channels is dependent on a channel initiation function (proportional to functions of water discharge (drainage area) and slope) exceeding some specified threshold. Headward extension of channels and development of tributaries continues until the channel initiation function decreases below the threshold because the drainage areas contributing water to the tributary channels progressively decrease.

The development of drainage systems has also been approached from the point of view of the energy required to transport water and sediment

Fig. 15. (Opposite page) Three-dimensional stochastic model simulations. (a) Alluvial fan deposits simulated using a Markov model (Carle *et al.*, 1998). Note the unrealistic depiction of channel deposits. (b) Fluvial deposits simulated using an object model for channel belts (FC-A and FC-B) and a sequential indicator simulator for the 'background' sheetflood (SF) and lacustrine (OTP) deposits (Seifert & Jensen, 2000). Note the unrealistic distribution of channel-belt orientations.

from the land as efficiently as possible. Drainage systems are most efficient (require the least energy expenditure) when flow resistance due to boundary friction is at a minimum. The best way of minimizing flow resistance is to have channel flow instead of overland (sheet) flow, and to have a relatively small number of large channels. However, channel initiation requires a certain amount of overland and/or shallow subsurface flow, which is associated with high flow resistance. According to Rodriguez-Iturbe *et al.* (1992) and Rinaldo *et al.* (1992, 1998), drainage networks are developed such that the energy expenditure in any stream segment and in the whole system are at a minimum, and the energy expenditure per unit bed area is constant. Although these assumptions regarding distribution of energy expenditure can be argued with, this theory agrees with the well known morphometric laws for drainage networks.

The influence of climate on stream erosion and sediment transport has been modelled by taking water discharge proportional to the product of effective precipitation rate per unit area and drainage basin area (e.g. Rinaldo *et al.*, 1995; Tucker & Slingerland, 1997; Moglen *et al.*, 1998; Coulthard *et al.*, 1999, 2002; Tucker & Bras, 2000; Tucker *et al.*, 2002; Bogaart *et al.*, 2003a; Snyder *et al.*, 2003; Tucker, 2004). River erosion through bedrock and development of drainage systems is critically dependent on the temporal and spatial variation of rainfall events, the form of the bedrock erosion law, and the magnitude of the erosion thresholds relative to that of the erosive forces. Bogaart *et al.* (2003a) examined theoretically the effects of alternating permafrost and temperate regimes on channel-network geometry and sediment dynamics. In a different type of approach, Bogaart *et al.* (2003b, 2003c) modelled the effect of rapid climate change on fluvial systems by using models for: weather generation; precipitation linked to runoff; hillslope erosion; channel geometry; and sediment transport.

The behaviour of these landscape-evolution models depends on the nature of discretization of space and time. For example, some models are based on rectangular grid cells and others are based on irregular networks and node points (discussed by Braun & Sambridge, 1997; Tucker *et al.*, 2001). The type of discretization determines how water and sediment are moved downslope. The dimensions of grid cells determine whether or not discrete channels can be modelled. The time steps determine the way in which surface processes can be averaged over time. These issues are of course relevant to all numerical forward models.

General models for tectonic subsidence, fluvial sediment supply and deposition

Paola *et al.* (1992) developed a theoretical model of large-scale variations in mean grain size in alluvial basins, as seen in a 2D section parallel to the sediment transport direction (see also Paola, 1988, 1990). All sediment transport was modelled using a linear diffusion approach, with the diffusivity of sediment being controlled mainly by water discharge and channel pattern (braided or single channel). Grain-size partitioning in the model was based on the crude assumption that gravel will dominate a deposit until all gravel in transport is exhausted, at which point deposition of sand will begin (referred to here as the perfect sorting model). They examined the response of an alluvial basin to sinusoidal variation of four extrinsic controlling variables: rate of sediment supply, diffusivity of sediment, tectonic subsidence rate, and proportion of gravel in the sediment supply (Fig. 16). The basin response depends strongly on the time scale over which variation in the controlling variables occurs. 'Slow' and 'rapid' variations are defined as those that vary with periods that are respectively longer or shorter than a so-called basin equilibrium time, defined as the square of basin length divided by sediment diffusivity. Changes in the rates of uplift of uplands, erosion, sediment supply, and subsidence are not linked (i.e., uncoupled) in this model, as they must be in nature. Also, tectonic and sedimentary processes in sedimentary basins are distinctly 3D. Notwithstanding the simplifications in this model, Heller & Paola (1992) applied it to three alluvial basins to help determine whether conglomerate progradation was coincident with tectonic uplift and increase in erosion and sediment supply (i.e., syntectonic) or not (antitectonic).

Paola *et al.* (1999) calibrated the sediment diffusion coefficient for a braided river on a fan using field data and a simple model for spatially averaged sediment transport rate and bed topography. Marr *et al.* (2000) further developed the Paola *et al.* (1992) model. They used different diffusion models for sand and gravel transport, constant values of dimensionless bed shear stress for sand-bed rivers and gravel-bed rivers, and a single constant friction coefficient for sand-bed and gravel-bed rivers. These are dubious assumptions (e.g. Robinson & Slingerland, 1998a). Results of the

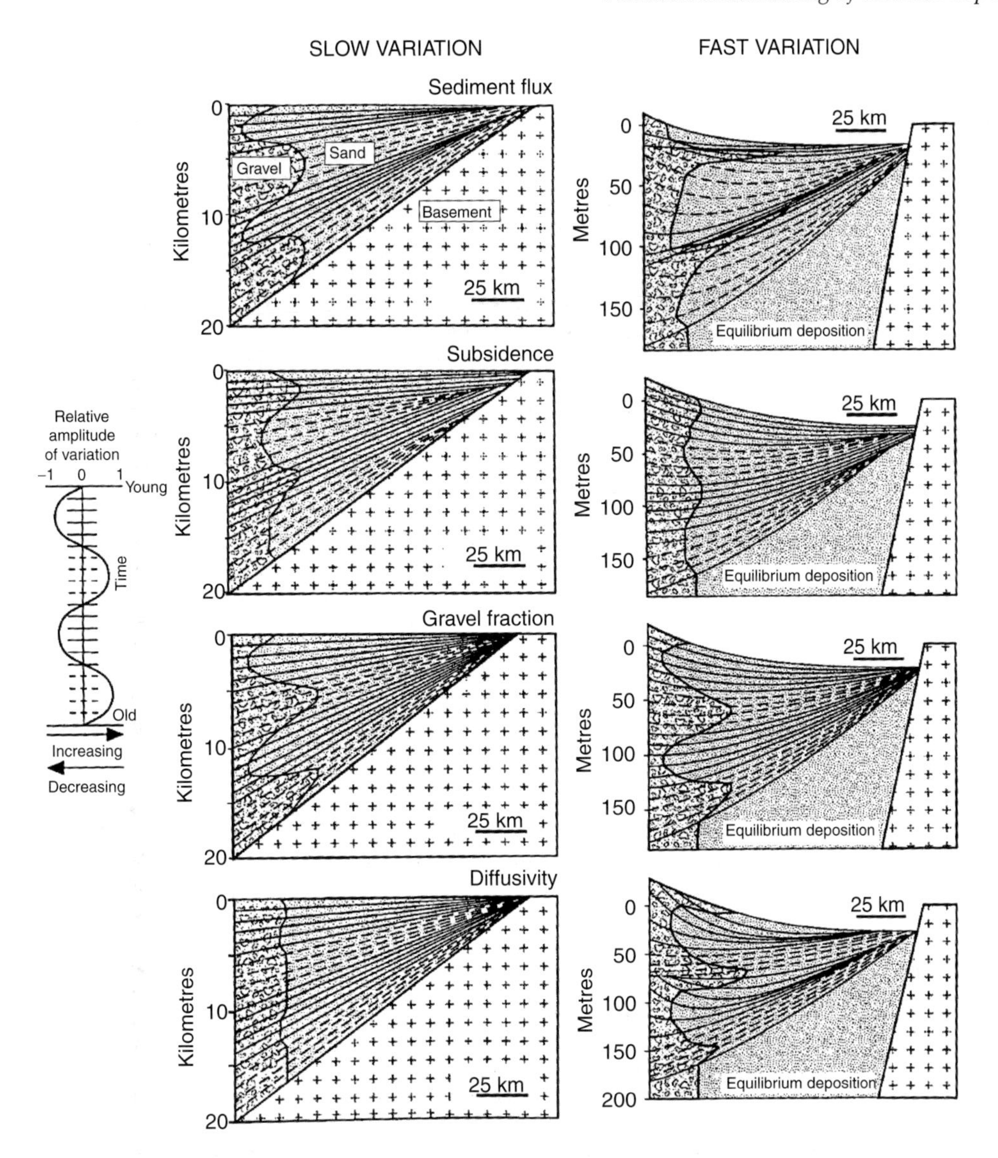

Fig. 16. Hypothetical variation in distribution of sand and gravel in a sedimentary basin subjected to periodic variation in sediment flux, subsidence, gravel fraction, and sediment diffusivity (from Paola *et al.*, 1992). Figures on left are for slow variation, and lines are isochrons drawn at 1 million year intervals. Figures on right are for fast variation and isochrons are drawn every 10,000 years.

model are similar to the original model with a few minor differences. Rivenæs (1992, 1997) modelled fluvial sediment transport using the diffusion equation applied separately to sand and mud with separate, depth-dependent diffusivities. These types of diffusion models give no information on alluvial architecture.

Slingerland *et al.* (1994) applied a simple sediment routing model for river channels to predict changes in the depth and longitudinal profiles of rivers as a result of downstream changes in water discharge, sediment supply and channel width. This model was also coupled with models for tectonic subsidence and uplift, varying sea level, and variable sediment supply related to climate change, in order to explore their effects on long profiles of rivers (hence erosion and deposition). These simple sediment routing models (Fig. 17) are potentially more realistic than diffusion-based fluvial models, but they still require development before they are capable of simulating the nature of sediment deposits in detail. As described above, Robinson & Slingerland (1998a, 1998b) were the first to combine a sophisticated sediment routing model for river channels (MIDAS) with basin subsidence and aggradation.

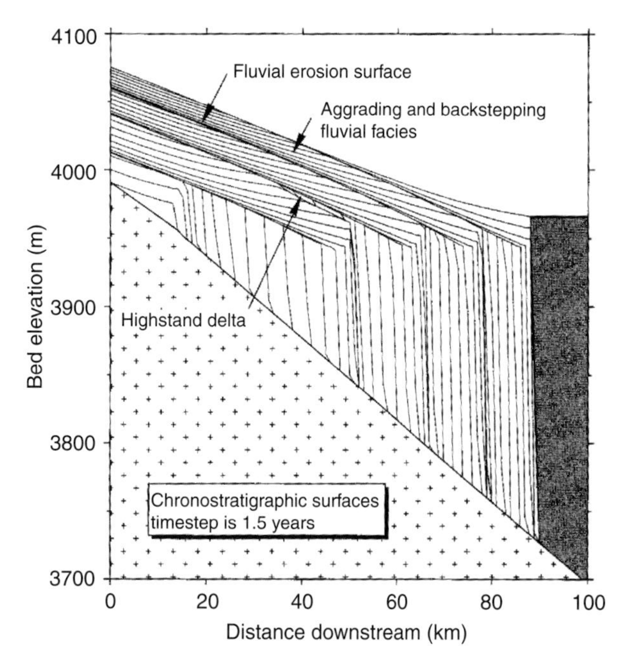

Fig. 17. Two-dimensional fluvial–deltaic model of Slingerland *et al.* (1994) that links a simple sediment routing model to an elastic flexure model and sinusoidal variation in sea level.

Models for deposition in compressional basins

Compressional basins are formed by lithospheric thickening arising from thrusting and folding. The overthrusted lithosphere is uplifted, but the underthrusted lithosphere is flexed downward due to gravitational loading, and the adjacent downflexed regions become sedimentary basins. An important aspect of the lithospheric flexure is a flexural bulge at the periphery of the basin. If the crust has high flexural rigidity and viscosity, the basin is relatively shallow and wide: otherwise it is deep and narrow. Thus, temporal changes in the rheological properties of the lithosphere result in changes in the shape of the basin.

An episode of lithospheric thickening and loading results in uplift and increases in topographic slopes, erosion rate and sediment supply in the vicinity of the uplift. It also results in subsidence and deposition in the basin, and growth and migration of the peripheral bulge towards the basin. Erosion of uplifted lithosphere and deposition in the basin causes further isostatic uplift of the highlands and subsidence in the basin. The erosional and depositional response to lithospheric thickening, uplift and subsidence depend upon the relative timing, positions and rates of these events, and these are very difficult to ascertain. For example, the nature of lithospheric subsidence in response to loading depends on lithospheric rheology, specifically whether the crust is elastic or viscoelastic, and how rheological properties change in time with temperature and pressure. The response of weathering, erosion and sediment supply to changes in source-rock type, elevation, slope and vegetation are difficult to predict.

The early quantitative models for compressional basins considered flexural isostatic response to loads created by tectonic thrusting (critically tapered wedges) and sediment deposition, as represented in 2D sections parallel to the direction of thrusting (e.g. Beaumont, 1981; Jordan, 1981; Quinlan & Beaumont, 1984; Beaumont *et al.*, 1988). Erosion and deposition were not modelled explicitly. Subsequently, erosion and deposition were modelled explicity by Flemings & Jordan (1989) using a linear diffusion approach. This model was further discussed by Flemings & Jordan (1990) and Jordan & Flemings (1989, 1991), and the effects of sea-level change on basin stratigraphy were examined. Sinclair *et al.* (1991) had a similar model. The model of Paola *et al.* (1992) discussed above is also essentially a model for stratigraphy in compressional basins, but it does not consider the mechanics of uplift and subsidence. Paola (2000) compared the stratigraphy predicted by these various 2D, diffusion-based models (Fig. 18). The treatment of surface processes in these models is very simplistic, and the 2D treatment of tectonics and sedimentation can be misleading.

A major step forward in the modelling of deposition at the river system and basin scale was the treatment of surface processes and drainage basin evolution in 3D, and linking them to climate, tectonic activity, and base-level change (e.g. Beaumont *et al.*, 1992; Kooi & Beaumont, 1994, 1996; Johnson & Beaumont, 1995; Coulthard *et al.*, 1999, 2002; Ellis *et al.*, 1999; Allen & Densmore, 2000; Garcia-Castellanos, 2002; Tucker *et al.*, 2002; Clevis *et al.*, 2003, 2004a, 2004b; Fig. 19). The compressional basin model of Johnson & Beaumont (1995) implicitly suggests deposited sediment size by associating it with stream power (Fig. 19). Explicit modelling of the grain size of deposited sediment was attempted by Coulthard *et al.* (1999, 2002), Tucker *et al.* (2002), and Clevis *et al.* (2003). However, Coulthard *et al.* (2002) showed no examples of how their model predicts grain size of deposits, and Tucker *et al.* (2002) and Clevis *et al.* (2003) use the simplistic 'perfect sorting' model of Paola *et al.* (1992).

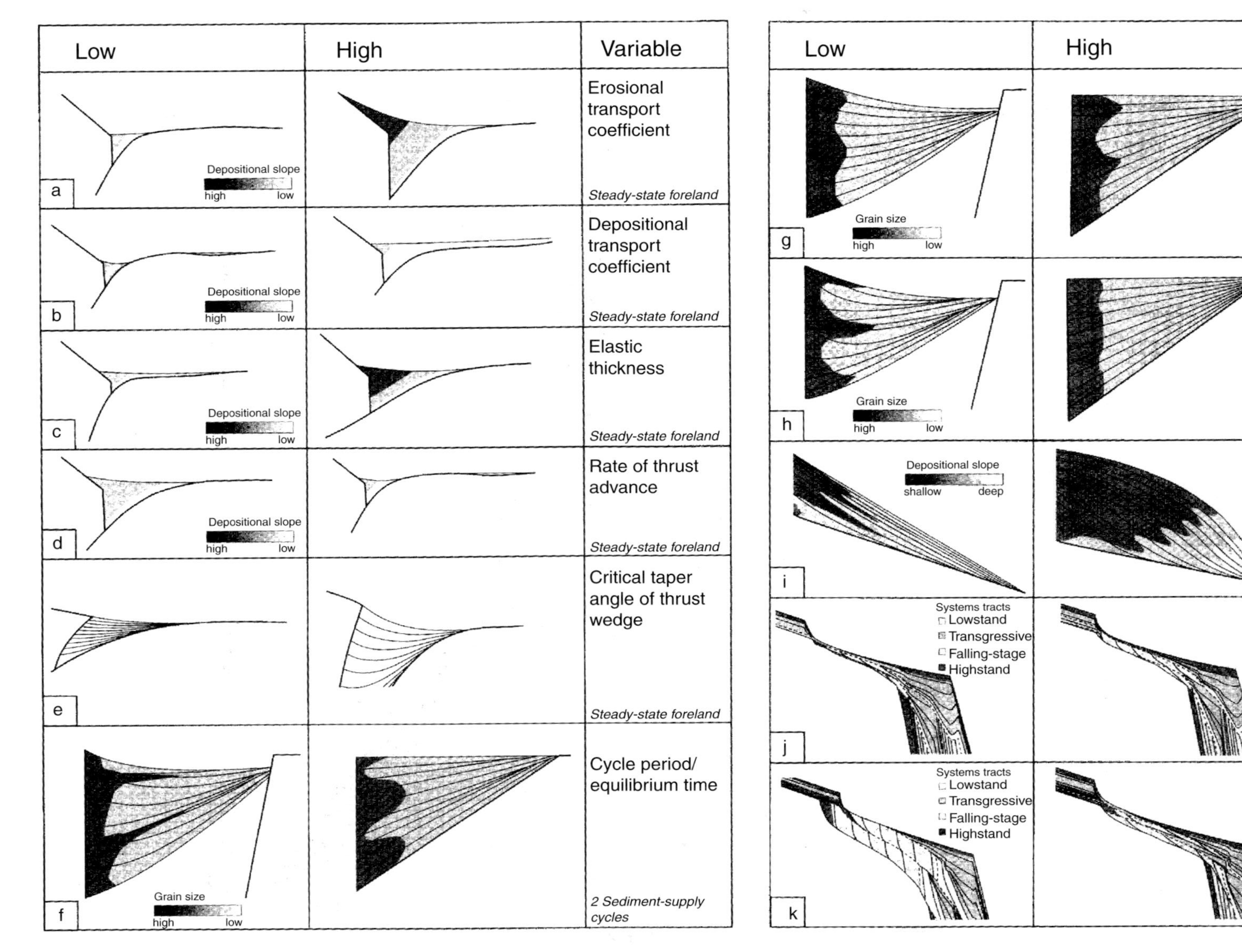

Fig. 18. Comparison of large-scale alluvial–marine stratigraphy predicted by various 2D, diffusional basin-filling models (from Paola, 2000).

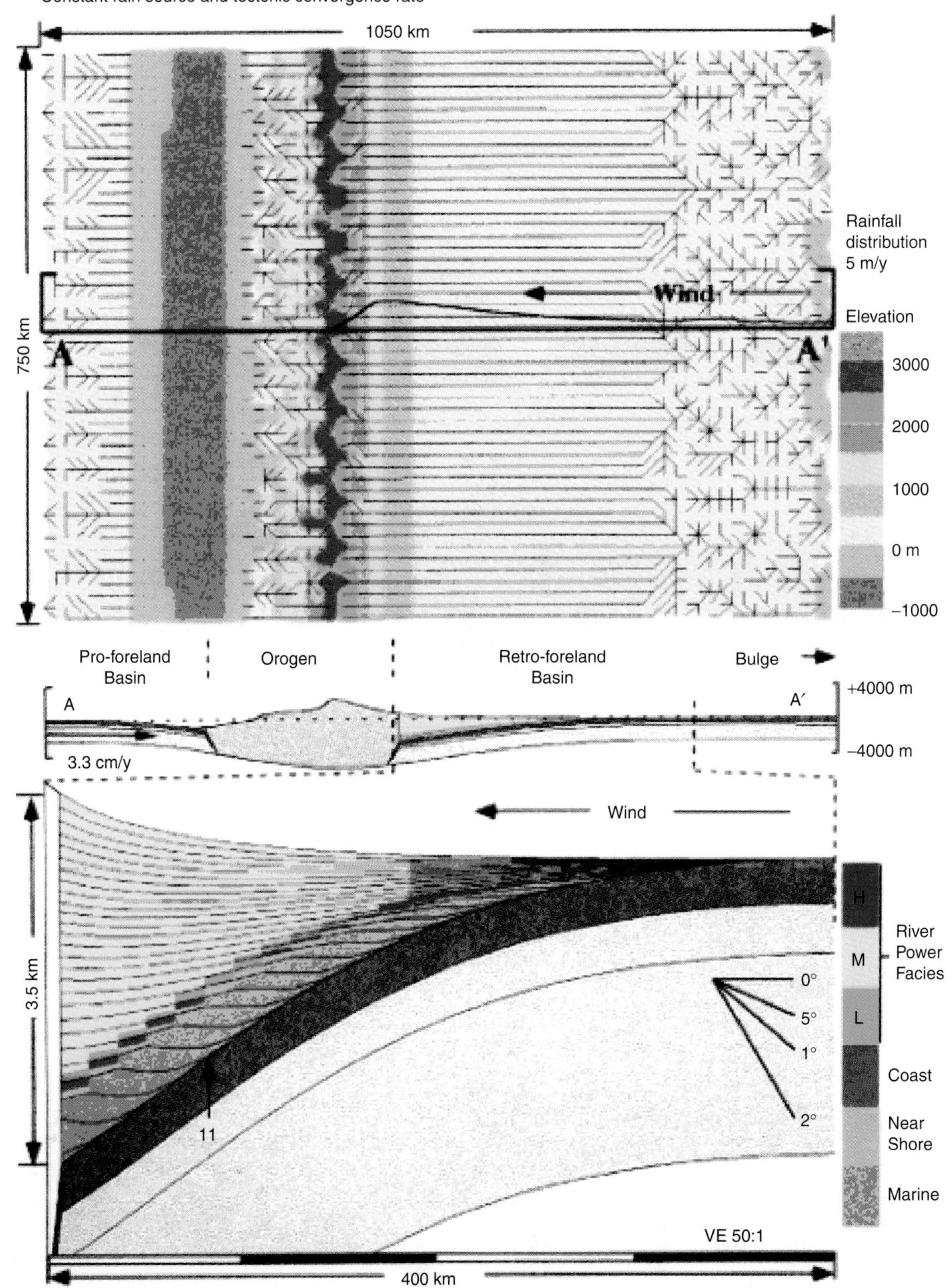

Fig. 19. (A) Three-dimensional compressional-basin model with surface processes (Johnson & Beaumont, 1995). (B) Three-dimensional compressional basin model with surface processes and grain-size sorting (Clevis *et al.*, 2003). Perspective views of successive stages of landscape subjected to tectonic pulsations. Phases of tectonic activity (A) reflected by retreat of coastline and gravel front. Quiescent periods (Q) associated with progradation of coastline and gravel front. (C) Three-dimensional compressional basin model with surface processes and grain-size sorting (Clevis *et al.*, 2003). Cross-sections (c,d) showing distribution of gravel during tectonic pulsation (200,000 period). Gravel progrades during tectonic quiescence. Sinusoidal sea-level fluctuation (period 100,000 years, amplitude 20 m) superimposed on tectonic pulsation shown in (e).

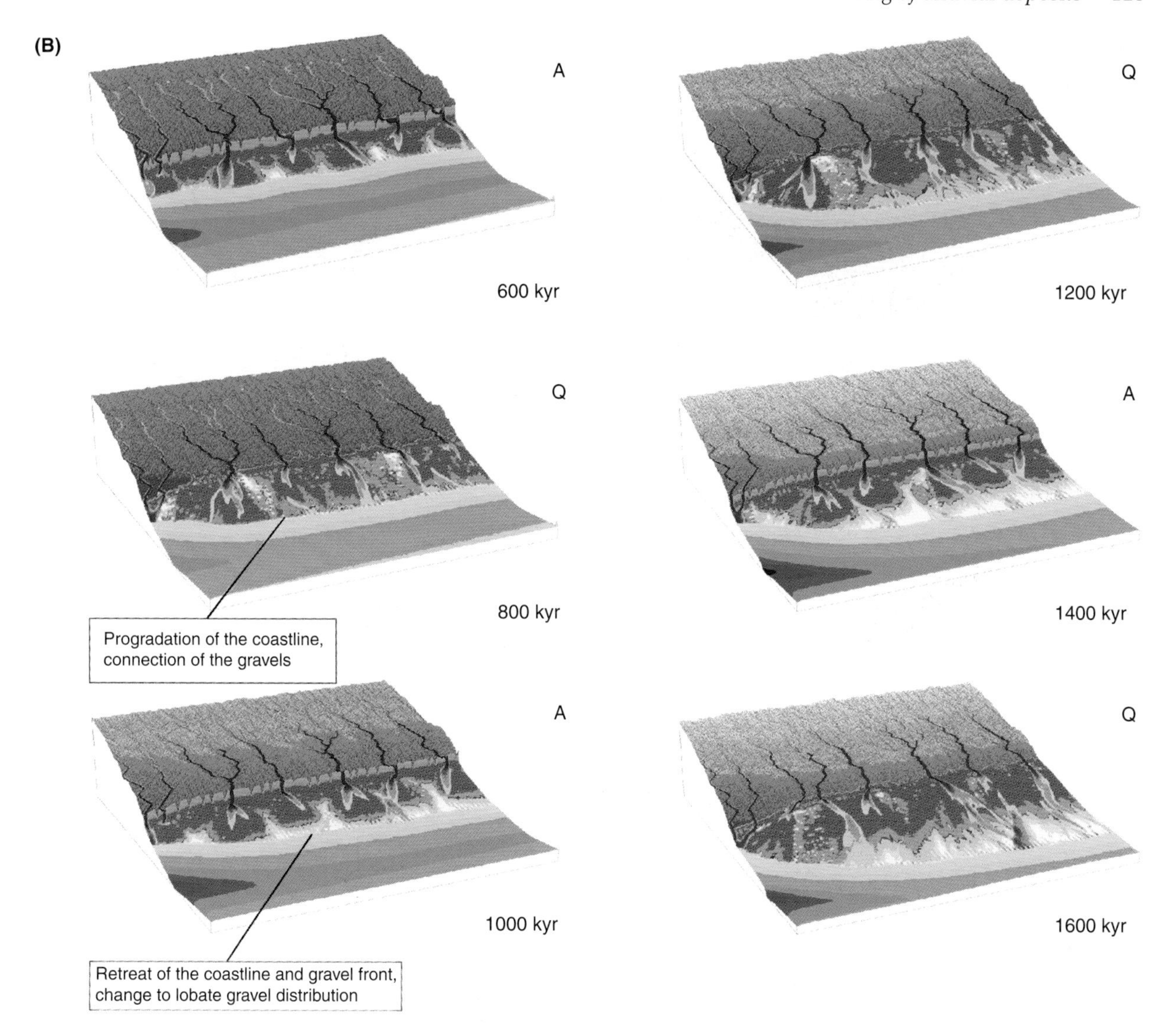

Fig. 19. Continued.

These 3D basin models indicate that the availability of sediment, related to weathering rate and bedrock erodibility, exerts a strong control on sediment transport rate and basin deposition. Changes in sediment transport rate to basins lag behind episodes of tectonic uplift, because of limits to sediment availability (weathering and bedrock erodibility) and the time it takes for sediment to move downslope through the drainage network. Sediment may be stored temporarily in an orogen because of the development of intermontane basins related to local thrusting and folding (e.g. Tucker & Slingerland, 1996). The lag time may be on the order of 10^4 to 10^6 years.

Episodic uplift and subsidence leads to episodic progradation and retrogradation of fluvial gravels and coastlines, but the relative timing of these events is equivocal (review in Bridge, 2003). In models that assume an elastic lithosphere, tectonic subsidence in the basin is an immediate response to thrusting, crustal thickening and loading. Thus, subsidence rate may exceed sediment supply and basin deposition rate during crustal loading. Relatively coarse sediment produced as a result of uplift is deposited close to the source, and marine transgression may occur. Such basins may be called 'under-filled'. Subsequent to uplift (time lag on the order of 10^4 to 10^6 years), the rate of sediment supply may begin to exceed subsidence rate, and as deposition proceeds there is progradation of relatively coarse sediment across the basin, and possibly marine regression. Such basins may be

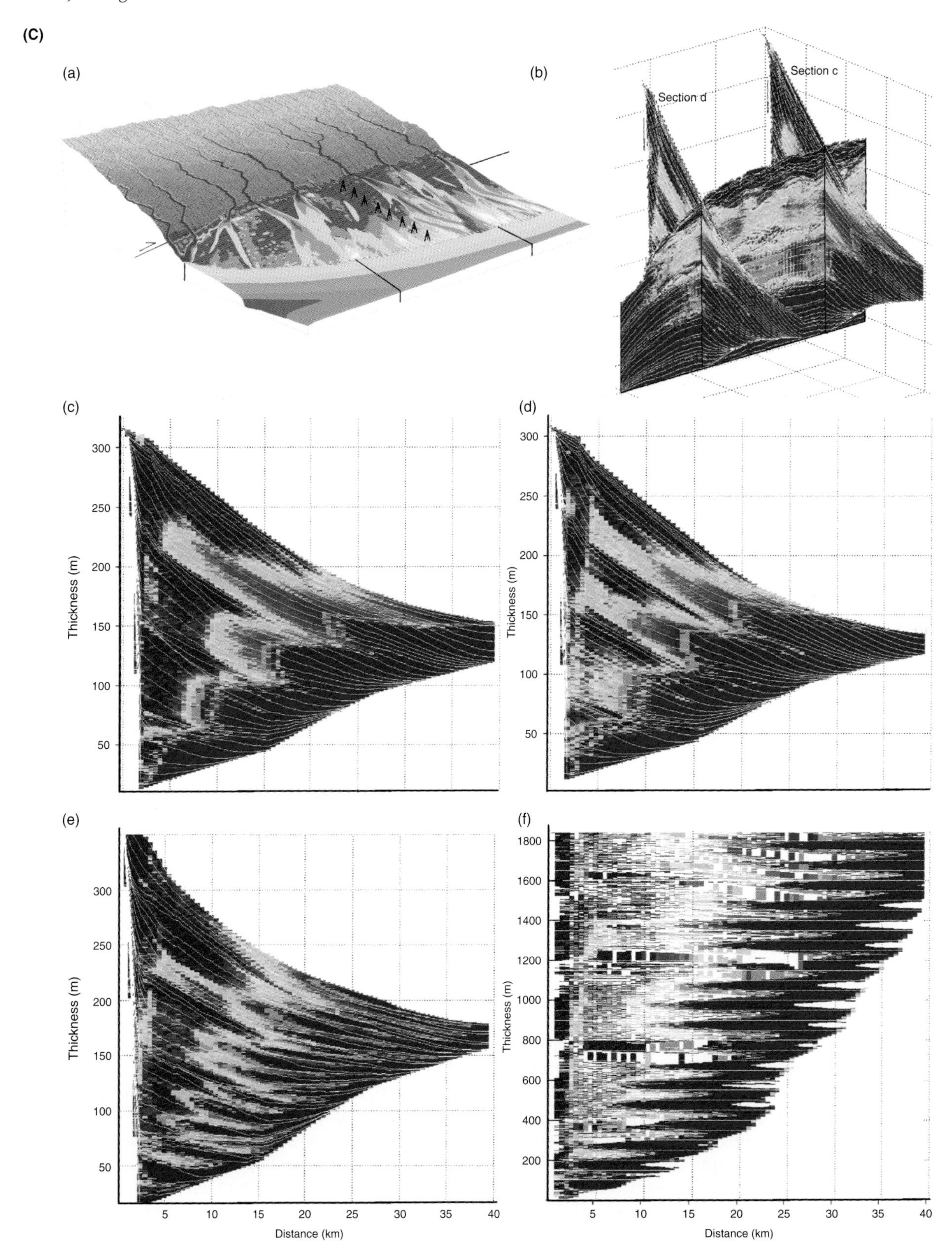

Fig. 19. Continued.

called 'overfilled'. Prograding coarse material is called 'antitectonic' in this case because it is not coincident with the tectonic uplift. In models that assume viscoelastic lithosphere, sediment supply and deposition rate may exceed subsidence rate during thrusting and uplift, such that relatively coarse sediment fills the basin and progrades basinwards (i.e., 'syntectonic'), possibly resulting in marine regression. Subsequently, subsidence rate exceeds deposition rate, the coarsest sediment is limited to areas near the uplift, and marine transgression may occur. The predictions of these models can be changed dramatically by different assumptions about the response to uplift of erosion and transport of sediment.

Eustatic sea-level fluctuations add more complexity to basin stratigraphy, especially in near-coastal fluvial deposits (see below). Basin stratigraphy is strongly linked to climate (e.g. rainfall) changes, mainly because of the strong link between rainfall, water discharge, bedrock erosion rate, and sediment transport rate in rivers. Climate also indirectly affects basin stratigraphy through its effect on vegetation and weathering rate, which influence both effective precipitation and sediment production. Changes in sediment transport rate to basins may lag behind changes in rainfall by on the order of hundreds to thousands of years. Soil creep is not affected by climate change according to the diffusive transport models.

Uplift is likely to be associated with climate change in the mountain belt and surrounding basins, and this climate change will affect the nature of weathering, erosion and sediment transport. Furthermore, climate change independent of uplift can influence rates of erosion in mountain belts. Burbank (1992) suggested that periods of accelerated isostatic uplift associated with climatically induced increase in erosion rate should result in deposits that do not vary greatly in thickness across the basin, because uplift is not associated with downwarping of the foreland basin. Ettensohn (1985) proposed a qualitative model of foreland basin deposition in which uplift is related to climate change. This model can now be tested using the new generation of quantitative models. Johnson & Beaumont's (1995) model specifically considers the effects on surface processes of orographic rainfall related to tectonic uplift.

The spatial and temporal scales considered in these basin models preclude consideration of individual channels and floodplains, and important controls on alluvial architecture such as channel-belt avulsion and local tectonics. Therefore, these models do not predict alluvial architecture. However, it is possible to link changes in alluvial architecture qualitatively with changes in subsidence rate, deposition rate and grain size. High deposition rate of relatively coarse sediment in basins that are back-tilting should result in relatively high avulsion frequency close to the uplands, producing fans with nodal avulsion, as observed in the Himalayan foreland (e.g. Kosi fan). Overlapping channel belts on such fans result in sandstone-conglomerate bodies with large width/thickness, with channel-deposit proportion and connectedness increasing towards the mountain belt (Mackey & Bridge, 1995). Palaeoslopes and river courses are approximately normal to the edge of the mountain belt (Fig. 20). Periods of relatively low deposition rate of relatively finer sediment should be associated with relatively low avulsion frequency, and the possibility of a relative rise in base level may lead to the highest avulsion frequencies distant from the edge of the mountain belt. In this case, rivers may be flowing parallel or oblique to the axis of the foreland basin (Fig. 20). Garcia-Castellanos (2002) discusses the tectonic influences on the orientation of rivers flowing across foreland basins.

Tectonic uplift occurs at different rates at different times in different parts of a mountain belt, which may result in diversions of rivers within mountain belts (e.g. Tucker & Slingerland, 1996; Gupta, 1997). Thus, the supply of sediment and water to rivers entering the basin, and the positions of the entry points of rivers, may vary in space as well as time. Variations in water and sediment supply in different rivers may be congruent or incongruent. Furthermore, changes in climate in different parts of the mountain belt (especially during glaciations) could also result in both congruent and incongruent changes in the discharges and sizes of different rivers flowing from the mountain belt. Figure 20 illustrates a scenario where a river is diverted by a thrust-related anticline near the edge of a compressional basin. Diversion of the river results in a reduction in the supply of water and sediment to a basin-marginal fan. The size and slope of this fan may then become more influenced by tectonic subsidence than by sediment progradation. The river that receives the diverted flow will experience an increased discharge of water and sediment. Its basin-marginal fan would experience an increase in deposition rate, and the size and slope of the fan would

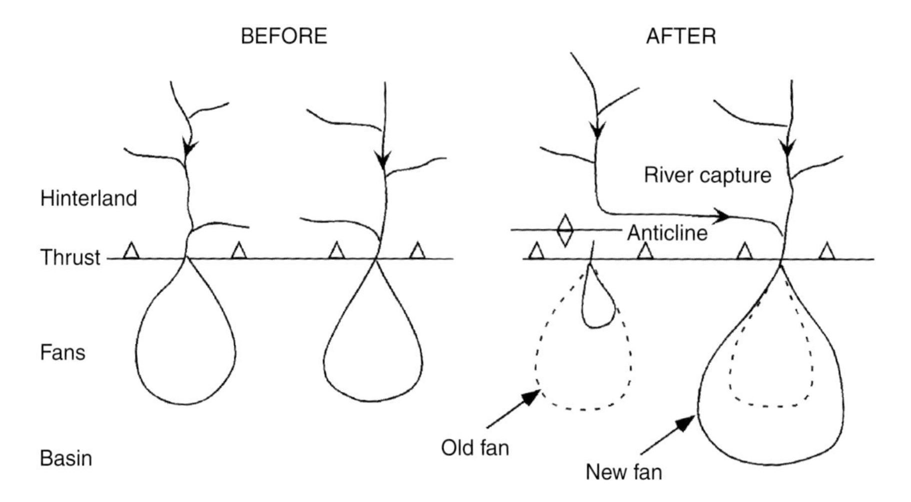

Fig. 20. Hypothetical changes in alluvial-fan size in a foreland (compressional) basin due to hinterland tectonics and river capture (from Bridge, 2003). Increase in water and sediment supply to growing fan should result in increased channel size and frequent avulsion, hence increased channel-deposit proportion. Furthermore, river channels on the growing fan tend to be oriented normal to the edge of the basin, whereas those on the shrinking fan may be oblique to the basin margin.

become dominated by this sediment progradation. According to the MIDAS sediment routing model (Van Niekerk *et al.*, 1992), downstream fining of river bed material would not be as effective on the growing fan as it would be on the shrinking fan (cf., Robinson & Slingerland, 1998a, 1998b). Increase in the discharges and sizes of rivers, and in deposition rate and avulsion frequency, on the growing fan would probably result in increasing channel-deposit proportion (cf. Fig. 14) and connectedness. Such increases in deposition rate in a cross-section oriented normal to the thrust belt and basin axis may not be related to a change in uplift rate or subsidence rate in the same cross-section. Thus, although tectonism (with or without climate change) may occur over a broad region over a long period of time, the depositional responses may not be the same in different parts of the basin. This illustrates the limitations of modelling tectonics and sedimentation in two dimensions.

Models for deposition in extensional basins

Extensional basins are caused by lithospheric stretching and by thermal subsidence (reviews in Allen & Allen, 1990; Leeder, 1999). Lithospheric stretching causes brittle fracture and normal faulting in relatively shallow parts of the lithosphere, but thinning by plastic deformation in deeper parts. Upwelling of hot asthenosphere beneath the area of thinned lithosphere increases the thermal gradient and causes decreasing density and thermal expansion of the lithosphere. This results in both isostatic and expansional uplift at the margins of the thinned lithosphere. Sediment is eroded from these peripheral uplifts (and other local uplifts) and deposited in the extensional basin. As the lithosphere cools, density increases and subsidence occurs as a result of isostasy and thermal contraction. The rate of thermal subsidence decreases with the square root of time. Loading of sediment deposited in the basin causes downward flexure of the lithosphere and the onlap of sediment at the basin margins. As flexural rigidity increases as the lithosphere cools, the zone of onlap increases in width with time, and the basin becomes wider and shallower. Such lapping of marine sediments onto fluvial sediments is commonly ascribed to eustatic sea-level rise, but this is clearly not necessarily so.

The qualitative model of Gawthorpe & Leeder (2000) relates erosion and deposition in extensional basins to the initiation, growth, propagation and death of arrays of normal faults (Fig. 21). There are no quantitative numerical models that include all of the processes discussed by Gawthorpe & Leeder (2000). However, Allen & Densmore (2000) have linked a 3D surface processes model to models for tectonic uplift, subsidence and climate change in order to explore these controls on fan deposition at the margins of rift basins. Major changes in climate affecting catchments tend to be reflected relatively rapidly (order of thousands of years) in downstream alluvial

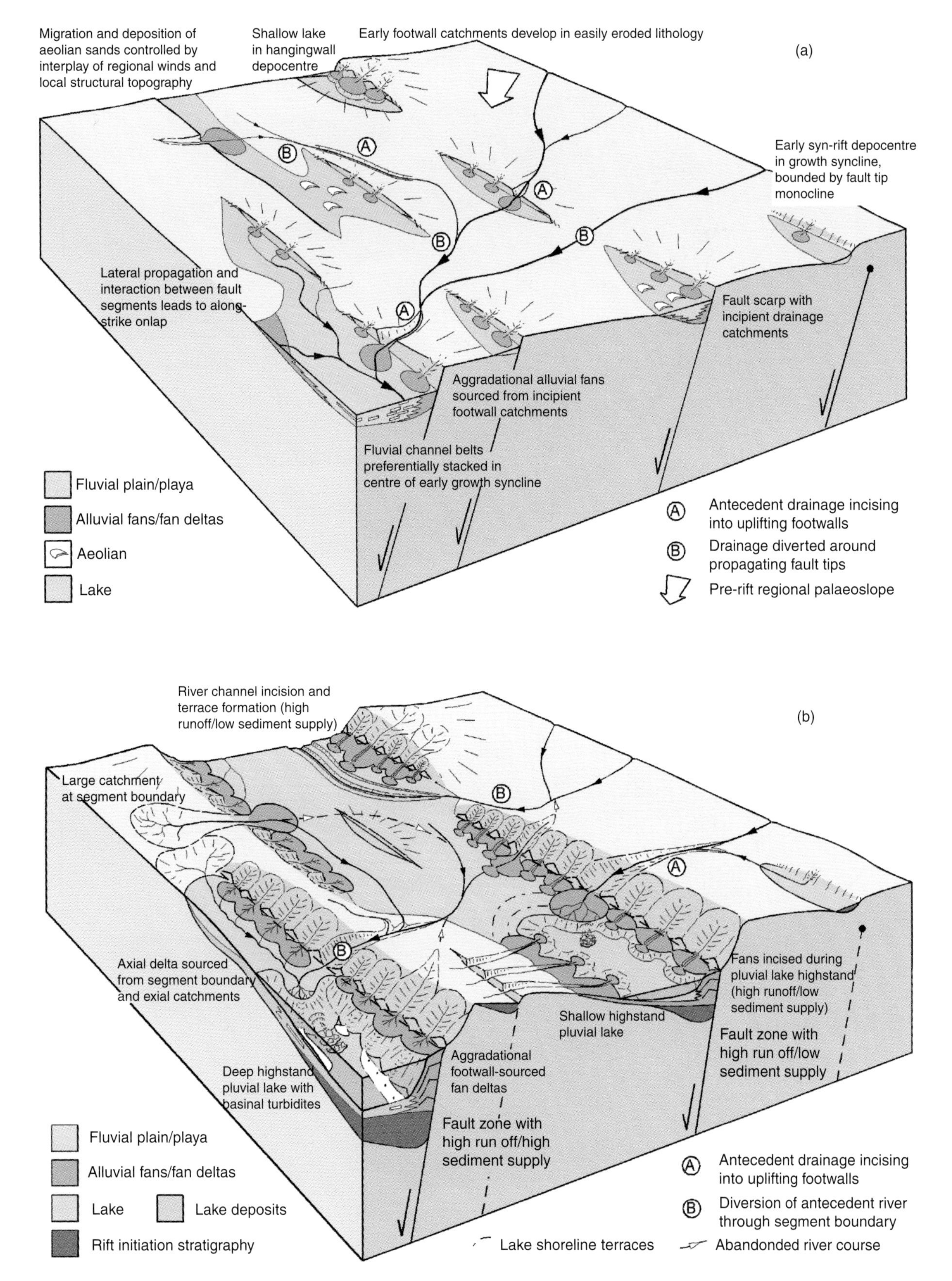

Fig. 21. Model of alluvial architecture in evolving extensional basins (from Gawthorpe & Leeder, 2000). (a) Initiation stage. (b) Interaction and linkage stage. (c) Through-going fault stage. (d) 'Fault death' stage.

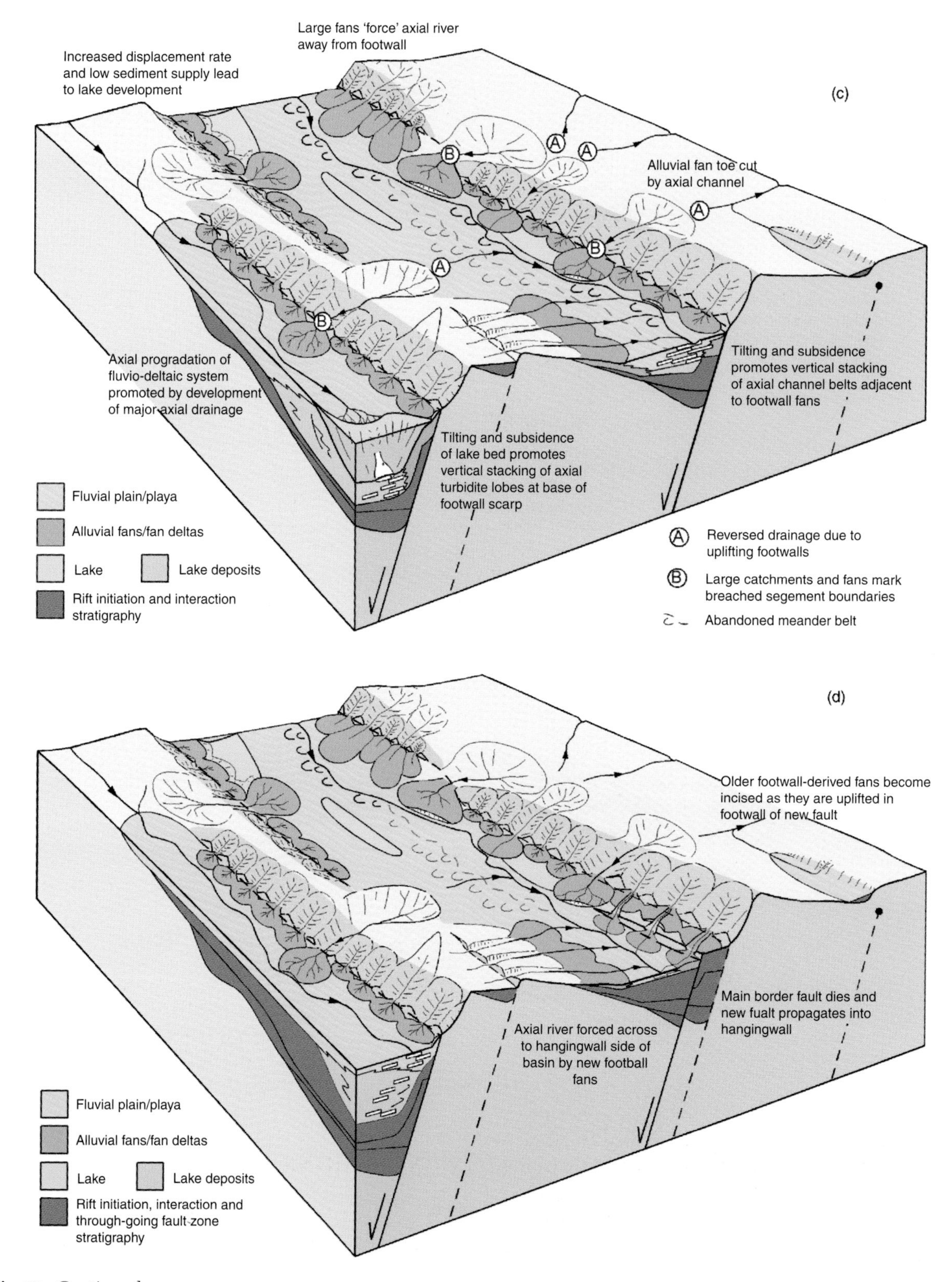

Fig. 21. Continued.

deposition. Episodic uplift of footwalls causes a wave of incision to move up the catchment, providing increased sediment supply to alluvial fans. However, it may take many tens of thousands of years for this to be reflected in increased fan growth, if at all (Allen & Densmore, 2000). Incision of fan channels and local growth of fans can also be associated with more or less random increases in sediment supply and river avulsions. Episodes of fan progradation into the basin result in upward-coarsening sequences on the order of tens of metres thick, whereas fan recession results in fining upward sequences.

Low channel-deposit proportion and connectedness are expected from low avulsion frequency and larger floodplain widths, but high values are expected if deposition rate is very low, allowing for extensive reworking of the deposits. High avulsion frequency is expected in local areas affected by base-level rise. Thus, patterns of deposition in extensional basins change in space and time over a range of different scales. Determination of whether an observed facies change is related to intrinsic (intrabasinal) processes, climate change, base-level change, or tectonic activity will not be straightforward.

Koltermann & Gorelick (1992) explained changes in the architecture of alluvial fan deposits due to climate change using a process-based model based on simplified equations for fluid flow and sediment transport (similar to SEDSIM). Floods were generated using a stochastic simulator. Variations in water discharge and sediment transport during large floods were crudely linked to Quaternary palaeoclimate change. The effects of a basin-bounding fault, compaction and base-level change were also considered in this model. Periods of high deposition rate of relatively coarse sediment (fan progradation) were associated with the periods of wet, cool climate when flood discharge of water and sediment were high. Fine sediment was deposited in warmer, drier climate. Transcurrent movement on the basin-margin fault caused the fan to move horizontally relative to the feeder stream, such that successive progradations of the fan were offset relative to each other, producing a 100 m thick sequence where the thickness of successive fan deposits either increased or decreased upwards.

The only way to test the various basin-filling models that relate uplift, subsidence, erosion and deposition is to have independent estimates of the timing and magnitude of these processes. These are not normally available.

Models for effect of base-level change on alluvial architecture

There are now many different (sequence stratigraphic) models for the effects of relative sea-level change on deposition rate and alluvial architecture (reviewed in Bridge, 2003). Most of them are qualitative and only 2D, and do not adequately represent all of the controls on alluvial architecture. Miall (1996) has criticized some of the earlier models of the effects of sea-level change on near-coastal alluvial deposition. His main point is that a relative fall in sea level is not normally associated with alluvial aggradation, except for the newly exposed part of the sea bed, and even then only under special circumstances. Whether or not a river valley is incised or aggraded during sea-level fall depends, among other things, on the slope of the exposed shelf relative to that of the river valley. In general, effects of sea-level change are expected to decrease up-valley.

The quantitative, basin-scale forward models discussed above that have been used to examine the effects of base-level change on basin deposition (e.g. Jordan & Flemings, 1991; Rivenæs, 1992, 1997; Slingerland *et al.*, 1994; Johnson & Beaumont, 1995; Clevis *et al.*, 2003, 2004a; see also Leeder and Stewart, 1996) can only describe the stratigraphic response to extrinsic factors in a general way. The 3D alluvial stratigraphy model of Mackey and Bridge (1995) has been used to predict the effects of base-level changes on alluvial architecture (e.g. Bridge, 1999). However, this model does not adequately consider the effects of base-level fall and stream incision (but see Karssenberg *et al.*, 2003, 2004), and it is not coupled to river-system-scale surface process models and tectonic models. There is a clear need for 3D river-system-scale numerical forward models that predict details of alluvial architecture under conditions of tectonic activity, climate change, and eustatic sea-level change.

CONCLUSIONS

Process-based (forward) models of most scales of alluvial deposits are generally undeveloped, and linkages between models for different scales are lacking. As a result, structure-imitating stochastic models are widely used for simulating alluvial hydrocarbon reservoirs and aquifers, given some initial data. However, such models cannot help understand alluvial deposits, nor do they have any predictive value outside the data region. Development of process-based (forward) models, and improvement in our understanding of alluvial deposits, requires much more comprehensive and detailed study of modern depositional processes in the field and in the laboratory. Better collaboration between Earth scientists, engineers and software developers is also required. None of these developments are likely to happen without serious financial investment in a broad range of research groups.

ACKNOWLEDGEMENTS

This paper benefited from the comments of Mike Leeder, Ulf Nordlund, and Dan Tetzlaff.

REFERENCES

Alexander, J., **Bridge, J.S.**, **Cheel, R.J.** and **Leclair, S.F.** (2001) Bed forms and associated sedimentary structures formed under supercritical water flows over aggrading sand beds. *Sedimentology*, **48**, 133–152.

Allen, J.R.L. (1973) Features of cross-stratified units due to random and other changes in bedforms. *Sedimentology*, **20**, 189–202.

Allen, J.R.L. (1978) Studies in fluviatile sedimentation: An exploratory quantitative model for architecture of avulsion-controlled alluvial suites. *Sed. Geol.*, **21**, 129–147.

Allen, J.R.L. (1979) Studies in fluviatile sedimentation: An elementary geometrical model for the connectedness of avulsion-related channel sand bodies. *Sed. Geol.*, **24**, 253–267.

Allen, J.R.L. (1982) *Sedimentary Structures: Their Character and Physical Basis*. Volume I. *Developments in Sedimentology* 30, Elsevier Science Publishers, Amsterdam, 593 p.

Allen, P.A. and **Allen, J.R.** (1990) *Basin Analysis*. Blackwell, Oxford.

Allen, P.A. and **Densmore, A.L.** (2000) Sediment flux from an uplifting fault block. *Basin. Res.*, **12**, 367–380.

Anderson, M.P. (1997) Characterization of geological heterogeneity. In: *Subsurface Flow and Transport: A Stochastic Approach* (Eds G. Dagan and S.P. Neuman), Cambridge University Press, Cambridge, UK, pp. 23–43.

Anderson, M.P., **Aiken, J.S.**, **Webb, E.K.** and **Mickelson, D.M.** (1999) Sedimentology and hydrogeology of two braided stream systems. *Sed. Geol.*, **129**, 187–200.

Baldwin, J.A., **Whipple, K.X.** and **Tucker, G.E.** (2003) Implications of the shear stress river incision model for the timescale of postorogenic decay of topography. *J. Geophys. Res.*, **108**, 2158 p.

Beaumont, C. (1981) Foreland basins. *Geophys. J. R. Astro. Soc.*, **65**, 291–329.

Beaumont, C. Quinlan, G. and **Hamilton, J.** (1988) Orogeny and stratigraphy: Numerical models of the Paleozoic in the eastern interior of North America. *Tectonics*, **7**, 389–416.

Beaumont, C., **Fullsack, P.** and **Hamilton, J.** (1992) Erosional control of active compressional orogens. In: *Thrust Tectonics* (Ed. K.R. McClay), Chapman & Hall, London, pp. 1–18.

Beaumont, C., **Kooi, H.** and **Willet, S.** (2000) Coupled tectonic-surface process models with applications to rifted margins and collisional orogens. In: *Geomorphology and Global Tectonics* (Ed. M.A. Summerfield), Wiley, Chichester, pp. 29–35.

Bennett, S.J. and **Bridge, J.S.** (1995a) An experimental study of flow, bedload transport and bed topography under conditions of erosion and deposition and comparison with theoretical models. *Sedimentology*, **42**, 117–146.

Bennett, S.J. and **Bridge, J.S.** (1995b) The geometry and dynamics of low-relief bed forms in heterogeneous sediment in a laboratory channel, and their relationship to water flow and sediment transport. *J. Sed. Res.*, **A65**, 29–39.

Bennett, S.J. and **Simon, A.** (Eds) (2004) *Riparian Vegetation and Fluvial Geomorphology*. American Geophysical Union, Washington, D.C.

Best, J.L. and **Bridge, J.S.** (1992) The morphology and dynamics of low amplitude bedwaves upon upper stage plane beds and the preservation of planar laminae. *Sedimentology*, **39**, 737–752.

Bierkens, M.F.P. and **Weerts, H.J.T.** (1994) Application of indicator simulation to modeling the lithological properties of a complex confining layer. *Geoderma*, **62**, 265–284.

Bogaart, P.W., **Tucker, G.E.** and **de Vries, J.J.** (2003a) Channel network morphology and sediment dynamics under alternating periglacial and temperate regimes: A numerical simulation study. *Geomorphology*, **54**, 257–277.

Bogaart, P.W., **Van Balen, R.T.**, **Kasse, C.** and **Vandenberghe, J.** (2003b) Process-based modeling of fluvial system response to rapid climate change – I: Model formulation and generic applications. *Quat. Sci. Rev.*, **22**, 2077–2095.

Bogaart, P.W., **Van Balen, R.T.**, **Kasse, C.** and **Vandenberghe, J.** (2003c) Process-based modeling of fluvial system response to rapid climate change – II: Application to the River Maas (The Netherlands) during the Last Glacial–Interglacial Transition. *Quat. Sci. Rev.*, **22**, 2097–2110.

Bolla-Pittaluga, M., **Repetto, R**, and **Tubino, M.** (2003) Channel bifurcations in braided rivers: Equilibrium configurations and stability. *Water Resour. Res.*, **39**, 1046–1059.

Bradbrook, K.F., **Biron, P.M.**, **Lane, S.N.**, **Richards, K.S.** and **Roy, A.G.** (1998) Investigation of controls on secondary circulation in a simple confluence using a three-dimensional numerical model. *Hydrol. Proc.*, **12**, 1371–1396.

Bradbrook, K.F., **Lane, S.N.** and **Richards, K.S.** (2000) Numerical simulation of three-dimensional, time averaged flow structure at river channel confluences. *Water Resour. Res.*, **36**, 2731–2746.

Bradbrook, K.F., **Lane, S.N.**, **Richards, K.S.**, **Biron, P.M.** and **Roy, A.G.** (2001) Role of bed discordance at asymmetrical river confluences. *J. Hydraul. Eng., ASCE*, **127**, 351–368.

Braun, J. and **Sambridge, M.** (1997) Modelling landscape evolution on geological time scales: A new method based on irregular spatial discretization. *Basin Res.*, **9**, 27–52.

Brayshaw, A.C., Davies, G.W. and **Corbett, P.W.M.** (1996) Depositional controls on primary permeability and porosity at the bedform scale in fluvial reservoir sandstones. In: *Advances in Fluvial Dynamics and Stratigraphy* (Eds P.A. Carling and M.R. Dawson), Wiley, Chichester, pp. 374–394.

Bridge, J.S. (1975) Computer simulation of sedimentation in meandering streams. *Sedimentology*, **22**, 3–43.

Bridge, J.S. (1992) A revised model for water flow, sediment transport, bed topography and grain size sorting in natural river bends. *Water Resour. Res.*, **28**, 999–1023.

Bridge, J.S. (1993) The interaction between channel geometry, water flow, sediment transport and deposition in braided rivers. In: *Braided Rivers* (Eds J.L. Best and C.S. Bristow), *Geol. Soc. London, Spec. Pub.* **75**, 13–72.

Bridge, J.S. (1997) Thickness of sets of cross strata and planar strata as a function of formative bed-wave geometry and migration, and aggradation rate. *Geology*, **25**, 971–974.

Bridge, J.S. (1999) Alluvial architecture of the Mississippi Valley: Predictions using a 3D simulation model. In: *Floodplains: Interdisciplinary Approaches* (Eds S.B. Marriott and J. Alexander), *Geol. Soc. London, Spec. Publ.*, **163**, 269–278.

Bridge, J.S. (2001) Characterization of fluvial hydrocarbon reservoirs and aquifers: Problems and solutions. *AAS Rev., J. Argentinian Assoc. Sedimentol.*, **8**, 87–114.

Bridge, J.S. (2003) *Rivers and Floodplains.* Blackwells, Oxford.

Bridge, J.S. and **Best, J.L.** (1997) Preservation of planar laminae arising from low-relief bed waves migrating over aggrading plane beds: Comparison of experimental data with theory. *Sedimentology*, **44**, 253–262.

Bridge, J.S. and **Best, J.L.** (1988) Flow, sediment transport and bedform dynamics over the transition from upper-stage plane beds: Implications for the formation of planar laminae. *Sedimentology*, **35**, 753–763.

Bridge, J.S. and **Leeder, M.R.** (1979) A simulation model of alluvial stratigraphy. *Sedimentology*, **26**, 617–644.

Bridge, J.S. and **Lunt, I.A.** (2006) Depositional models of braided rivers. In: *Braided Rivers II* (Eds G.H. Sambrook Smith, J.L. Best, C.S. Bristow and G. Petts), *Int. Assoc. Sediment. Spec. Publ.*, **36**, 11–50.

Bridge, J.S. and **Mackey, S.D.** (1993a) A revised alluvial stratigraphy model. In: *Alluvial Sedimentation* (Eds M. Marzo and C. Puidefabregas), *Int. Assoc. Sediment. Spec. Publ.*, **17**, 319–337.

Bridge, J.S. and **Mackey, S.D.** (1993b) A theoretical study of fluvial sandstone body dimensions. In: *Geological Modeling of Hydrocarbon Reservoirs* (Eds S.S. Flint and I.D. Bryant), Blackwells, Oxford, UK. *Int. Ass. Sediment. Spec. Pub.*, **15**, 213–236.

Bridge, J.S. and **Tye, R.S.** (2000) Interpreting the dimensions of ancient fluvial channel bars, channels, and channel belts from wireline-logs and cores. *Am. Assoc. Petrol. Geol. Bull.*, **84**, 1205–1228.

Bryant, I.D. and **Flint, S.** (1993) Quantitative clastic reservoir geological modelling: Problems and perspectives. In: *Geological Modelling of Hydrocarbon Reservoirs* (Eds S. Flint and I.D. Bryant), *Int. Ass. Sediment. Spec. Pub.*, **15**, 3–20.

Burbank, D.W. (1992) Causes of recent Himalayan uplift deduced from deposited patterns in the Ganges basin. *Nature*, **357**, 680–683.

Caers, J., Srinivasan, S. and **Journel, A.G.** (2000) Geostatistical quantification of geological information for a fluvial-type North Sea reservoir. *Soc. Petrol. Eng., Reservoir Eval. Eng.*, Richardson, TX, **3**, 457–467.

Caers, J. and **Zhang, T.** (2004) Multiple-point geostatistics: A quantitative vehicle for integrating geological analogs into multiple reservoir models. In: *Integration of Outcrop and Modern Analogs in Reservoir Modeling* (Eds G.M. Grammer, P.M. Harris and G.P. Eberli), Richardson, TX. *AAPG Memoir*, **80**, 383–394.

Carle, S.F., Labolle, E.M., Weissmann, G.S., Van Brocklin, D. and **Fogg, G.E.** (1998) Conditional simulation of hydrofacies architecture: A transition probability/Markov approach. In: *Hydrogeologic Models of Sedimentary Aquifers* (Eds G.S. Fraser and J.M. Davis), *SEPM Concepts Hydrogeol. Environ. Geol.*, **1**, 147–170.

Chase, C.G. (1992) Fluvial landsculpting and the fractal dimension of topography. *Geomorphology*, **5**, 39–57.

Clevis, Q., de Boer, P.L. and **Wachter, M.** (2003) Numerical modelling of drainage basin evolution and three-dimensional alluvial fan stratigraphy. *Sed. Geol.*, **163**, 85–110.

Clevis, Q., de Boer, P.L. and **Nijman, W.** (2004a) Differentiating the effect of episodic tectonism and eustatic sea-level fluctuations in foreland basins filled by alluvial fans and axial deltaic systems: Insights from a three-dimensional stratigraphic forward model. *Sedimentology*, **51**, 809–835.

Clevis, Q., de Jager, G., Nijman, W. and **de Boer, P.L.** (2004b) Stratigraphic signatures of translation of thrust-sheet top basins over low-angle detachment faults. *Basin Res.*, **16**, 145–163.

Coulthard, T.J., Kirkby, M.J. and **Macklin, M.G.** (1999) Modelling the impacts of Holocene environmental change on the fluvial and hillslope morphology of an upland landscape, using a cellular automaton approach. In: *Fluvial Processes and Environmental Change* (Eds A.G. Brown and T.M. Quine), Wiley and Sons, Chichester, pp. 31–47.

Coulthard, T.J., Macklin, M.G. and **Kirkby, M.J.** (2002) A cellular model of Holocene upland river basin and alluvial fan evolution. *Earth Surf. Proc. Land.*, **27**, 269–288.

Crave, A. and **Davy, P.** (2001) A stochastic 'precipiton' model for simulating erosion/sedimentation dynamics. *Comp. Geosci.*, **27**, 815–827.

Deutsch, C.V. (2002) *Geostatistical Reservoir Modeling.* Oxford University Press, New York.

Deutsch, C.V. and **Cockerham, P.** (1994) Practical considerations in the application of simulated annealing to stochastic simulation. *Math. Geol.*, **26**, 67–82.

Deutsch, C.V. and **Tran, T.T.** (2002) FLUVSIM: A program for object-based modeling of fluvial depositional systems. *Comp. Geosci.*, **28**, 525–535.

Deutsch, C.V. and **Wang, L.** (1996) Hierarchical object-based stochastic modeling of fluvial reservoirs. *Math. Geol.*, **28**, 857–880.

Doveton, J.H. (1994) Theory and applications of vertical variability measures from Markov Chain analysis. In: *Stochastic Modeling and Geostatistics* (Eds J.M. Yarus and R.L. Chambers), *Am. Assoc. Petrol. Geol. Computer Applications in Geology*, Tulsa, OK, **3**, 55–64.

Duan, J.G., **Wang, S.S.Y.** and **Jia, Y.** (2001) The applications of the enhanced CCHE2D model to study the alluvial migration process. *J. Hydraul. Res.*, **39**, 469–480.

Elfeki, A. and **Dekking, M.** (2001) A Markov chain model for subsurface characterization: Theory and application. *Math. Geol.*, **33**, 569–589.

Ellis, M.A., **Densmore, A.L.** and **Anderson, R.S.** (1999) Development of mountainous topography in the Basin Ranges, USA. *Basin Res.*, **11**, 21–41.

Ettensohn, F.R. (1985) Controls on development of Catskill Delta complex basin-facies. In: *The Catskill Delta* (Eds D.L. Woodrow and W.D. Sevon), *Geol. Soc. Am. Spec. Paper*, Boulder, CO, **201**, 65–75.

Flemings, P.B. and **Jordan, T.E.** (1989) A synthetic stratigraphic model of foreland basin development. *J. Geophys. Res.*, **94**, 3851–3866.

Flemings, P.B. and **Jordan, T.E.** (1990) Stratigraphic modeling of foreland basins: Interpreting thrust deformation and lithospheric rheology. *Geology*, **18**, 430–434.

Garcia-Castellanos, D. (2002) Interplay between lithospheric flexure and river transport in foreland basins. *Basin Res.*, **14**, 89–104.

Gasparini, N.M., **Tucker, G.E.** and **Bras, R.L.** (1999) Downstream fining through selective particle sorting in an equilibrium drainage network. *Geology*, **27**, 1079–1082.

Gasparini, N.M., **Tucker, G.E.** and **Bras, R.L.** (2004) Network-scale dynamics of grain-size sorting: Implications for downstream fining, stream-profile concavity, and drainage basin morphology. *Earth Surf. Proc. Land.*, **29**, 401–421.

Gawthorpe, R.L. and **Leeder, M.R.** (2000) Tectono-sedimentary evolution of active extensional basins. *Basin Res.*, **12**, 195–218.

Griffiths, C.M., **Dyt, C.**, **Paraschivoiu, E.** and **Liu, K.** (2001) SEDSIM in hydrocarbon exploration. In: *Geologic Modeling and Simulation: Sedimentary Systems* (Eds D.F. Merriam and J.C. Davis), Kluwer Academic/Plenum Publishers, New York, pp. 71–97.

Gupta, S. (1997) Himalayan drainage patterns and the origin of fluvial megafans in the Ganges foreland basin. *Geology*, **25**, 11–14.

Haldorsen, H.H. and **Damsleth, E.** (1990). Stochastic modeling. *J. Petrol. Technol.*, **42**, 404–412.

Hardy, R.J., **Bates, P.D.** and **Anderson, M.G.** (2000) Modelling suspended sediment deposition on a fluvial floodplain using a two-dimensional dynamic finite element model. *J. Hydrol.*, **229**, 202–218.

Heller, P.L. and **Paola, C.** (1992) The large scale dynamics of grain-size variation in alluvial basins, 2. Application to syntectonic conglomerate. *Basin Res.*, **4**, 91–102.

Holden, L., **Hauge, R.**, **Skare, Ø.**, and **Skorstad, A.** (1998) Modeling of fluvial reservoirs with object models. *Math. Geol.*, **30**, 473–496.

Howard, A.D. (1992) Modelling channel migration and floodplain development in meandering streams. In: *Lowland Floodplain Rivers* (Eds P.A. Carling and G.E. Petts), Wiley, Chichester, pp. 1–42.

Howard, A.D. (1994) A detachment-limited model of drainage basin evolution. *Water Resour. Res.*, **30**, 2261–2285.

Howard, A.D. (1996) Modelling channel evolution and floodplain morphology. In: *Floodplain Processes* (Eds M.G. Anderson, D.E. Walling and P.D. Bates), Wiley, Chichester, pp. 15–62.

Howard, A.D., **Dietrich, W.E.** and **Seidl, M.A.** (1994) Modeling fluvial erosion on regional to continental scales. *J. Geophys. Res.*, **99**, 13971–13986.

Huang, J., **Weber, L.J.** and **Lai, Y.G.** (2002) Three-dimensional numerical study of flows in open-channel junctions. *J. Hydraul. Eng., ASCE*, **128**, 268–280.

Ikeda, S., **Parker, G.** and **Sawai, K.** (1981) Bend theory of river meanders. 1. Linear development. *J. Fluid Mech.*, **112**, 363–377.

Izumi, N. and **Parker, G.** (1995) Inception of channelization and drainage basin formation: Upstream-driven theory. *J. Fluid Mech.*, **283**, 341–363.

James, C.S. (1985) Sediment transfer to overbank sections. *J. Hydraul. Res.*, **23**, 435–452.

Johannesson, H. and **Parker, G.** (1989) Linear theory of river meanders. In: *River Meandering* (Eds S. Ikeda and G. Parker), *AGU Water Resour. Mon.*, **12**, 181–213.

Johnson, D.D. and **Beaumont, C.** (1995) Preliminary results from a planform kinematic model of Orogen evolution, surface processes and the development of clastic foreland basin stratigraphy. In: *Stratigraphic Evolution of Foreland Basins* (Eds S.L. Dorobek and G.M. Ross), *SEPM Spec. Publ.*, Tulsa, OK, **52**, 3–24.

Johnson, N.M. and **Driess, S.J.** (1989) Hydrostratigraphic interpretation using indicator geostatistics. *Water Resour. Res.*, **25**, 2501–2510.

Jordan, T.E. (1981) Thrust loads and foreland basin evolution, Cretaceous western United States. *Am. Assoc. Petrol. Geol. Bull.*, **65**, 2506–2620.

Jordan, T.E. and **Flemings, P.B.** (1989) From geodynamical models to basin fill – A stratigraphic perspective. In: *Quantitative Dynamic Stratigraphy* (Ed. T.A. Cross), Prentice-Hall, Englewood Cliffs, New Jersey, pp. 149–163.

Jordan, T.E. and **Flemings, P.B.** (1991) Large-scale stratigraphic architecture, eustatic variation, and unsteady tectonism: A theoretical evaluation. *J. Geophys. Res.*, **96**, 6681–6699.

Karssenberg, D., **Tornqvist, T.E.** and **Bridge, J.S.** (2001) Conditioning a process-based model of sedimentary architecture to well data. *J. Sed. Res.*, **71**, 868–879.

Karssenberg, D., **Bridge, J.S.**, **Stouthamer, E.**, **Kleinhans, M.G.** and **Berendsen, H.J.A.** (2003) Modelling cycles of fluvial aggradation and degradation using a process-based alluvial stratigraphy model. *Proceedings of 'Analogue and Numerical Modelling of Sedimentary Systems; from Understanding to Prediction'*, 9–11 October 2003, Utrecht, The Netherlands.

Karssenberg, D., **Dalman, R.**, **Weltje, G.J.**, **Postma, G.** and **Bridge, J.S.** (2004) Numerical modelling of delta evolution by nesting high and low resolution process-based models of sedimentary basin filling. *Abstract for Joint EURODELTA/EUROSTRATAFORM meeting*, Venice, Italy, October 20–23, 2004.

Kirkby, M.J. (1986) A two-dimensional simulation model for slope and stream evolution. In: *Hillsope Processes* (Ed. A.D. Abrahams), Allen and Unwin, Boston, pp. 203–222.

Kirkby, M.J. (1987) Modelling some influences of soil erosion, landslides and valley gradient on drainage density and hollow development. *Catena Suppl.*, **10**, 1–11.

Kleinhans, M.G. (2004) Sorting in grains flows at the lee side of dunes. *Earth-Sci. Rev.*, **65**, 75–102.

Koltermann, C.E. and **Gorelick, S.M.** (1992) Paleoclimatic signature in terrestrial flood deposits. *Science*, **256**, 1775–1782.

Koltermann, C.E. and **Gorelick, S.M.** (1996) Heterogeneity in sedimentary deposits: A review of structure-imitating, process-imitating and descriptive approaches. *Water Resour. Res.*, **32**, 2617–2658.

Kooi, H. and **Beaumont, C.** (1994) Escarpment evolution on high-elevation rifted margins: Insights derived from a surface processes model that combines diffusion, advection, and reaction. *J. Geophys. Res.*, **99**, 12191–12209.

Kooi, H. and **Beaumont, C.** (1996) Large-scale geomorphology: Classical concepts reconciled and integrated with contemporary ideas using a surface processes model. *J. Geophys. Res.*, **101**, 3361–3386.

Lancaster, S.T. and **Bras, R.L.** (2002) A simple model of river meandering and its comparison to natural channels. *Hydrol. Proc.*, **16**, 1–26.

Lane, S.N. and **Richards, K.S.** (1998) High resolution, two-dimensional spatial modeling of flow processes in a multi-thread channel. *Hydrol. Proc.*, **12**, 1279–1298.

Lane, S.N., Bradbrook, K.F., Richards, K.S., Biron, P.M. and **Roy, A.G.** (2000) Secondary circulation cells in river channel confluences: Measurement artifacts or coherent flow structures? *Hydrol. Proc.*, **14**, 2047–2071.

Leclair, S.F. (2002) Preservation of cross-strata due to the migration of subaqueous dunes: An experimental investigation. *Sedimentology*, **49**, 1157–1180.

Leclair, S.F. and **Bridge, J.S.** (2001) Quantitative interpretation of sedimentary structures formed by river dunes. *J. Sed. Res.*, **71**, 713–716.

Leeder, M.R. (1978) A quantitative stratigraphic model for alluvium with special reference to channel deposit density and interconnectedness. In: *Fluvial Sedimentology* (Ed. A.D. Miall), *Can. Soc. Petrol. Geol. Mem.*, Calgary, Canada, **5**, 587–596.

Leeder, M.R. (1999) *Sedimentology and Sedimentary Basins*. Blackwell, Oxford, 592 pp.

Leeder, M.R. and **Stewart, M.** (1996) Fluvial incision and sequence stratigraphy: Alluvial responses to relative base level fall and their detection in the geological record. In: *Sequence Stratigraphy in British Geology* (Eds S.P. Hesselbo and D.N. Parkinson), *Geol. Soc., Spec. Publ.*, London, **103**, 47–61.

Leeder, M.R., Mack, G.H., Peakall, J. and **Salyards, S.L.** (1996) First quantitative test of alluvial stratigraphy models: Southern Rio Grande rift, New Mexico. *Geology*, **24**, 87–90.

Mackey, S.D. and **Bridge, J.S.** (1995) Three dimensional model of alluvial stratigraphy: Theory and application. *J. Sed. Res.*, **B65**, 7–31.

Marr, J.G., Swenson, J.B., Paola, C. and **Voller, V.R.** (2000) A two-diffusion model of fluvial stratigraphy in closed depositional basins. *Basin Res.*, **12**, 381–398.

Marriott, S.B. (1996) Analysis and modelling overbank deposits. In: *Floodplain Processes* (Eds M.G. Anderson, D.E. Walling and P.D. Bates), Wiley, Chichester, pp. 63–94.

McLean, S.R. (1990) The stability of ripples and dunes. *Earth-Sci. Rev.*, **29**, 131–144.

Miall, A.D. (1996) *The Geology of Fluvial Deposits*. Springer-Verlag, New York, 582 p.

Middlekoop, H. and **Van Der Perk, M.** (1998) Modelling spatial patterns of overbank sedimentation on embanked floodplains. *Geogr. Ann.*, **80A**, 95–109.

Moglen, G.E. and **Bras, R.L.** (1995) The effect of spatial heterogeneities on geomorphic expression in a model of basin evolution. *Water Resour. Res.*, **31**, 2613–2623.

Moglen, G.E., Eltahir, E.A.B. and **Bras, R.L.** (1998) On the sensitivity of drainage density to climate change. *Water Resour. Res.*, **34**, 855–862.

Murray, A.B. and **Paola, C.** (1994) A cellular model of braided rivers. *Nature*, **371**, 54–57.

Murray, A.B. and **Paola, C.** (1996) A new quantitative test of geomorphic models, applied to a model of braided streams. *Water Resour. Res.*, **32**, 2579–2587.

Murray, A.B. and **Paola, C.** (1997) Properties of a cellular braided stream model. *Earth Surf. Proc. Land.*, **22**, 1001–1025.

Nagata, N., Hosoda, T. and **Muramoto, Y.** (2000) Numerical analysis of river channel processes and bank erosion. *J. Hydraul. Eng., ASCE*, **126**, 243–252.

Nicholas, A.P. and **Sambrook-Smith, G.H.** (1999) Numerical simulation of three-dimensional flow hydraulics in a braided channel. *Hydrol. Proc.*, **13**, 913–929.

Nicholas, A.P. and **Walling, D.E.** (1997a) Investigating patterns of medium-term overbank sedimentation on floodplains: A combined numerical modelling and radiocaesium-based approach. *Geomorphology*, **19**, 133–150.

Nicholas, A.P. and **Walling, D.E.** (1997b) Modelling flood hydraulics and overbank deposition on river floodplains. *Earth Surf. Proc. Land.*, **22**, 59–77.

Nicholas, A.P. and **Walling, D.E.** (1998) Numerical modelling of floodplain hydraulics and suspended sediment transport and deposition. *Hydrol. Proc.*, **12**, 1339–1355.

Niemann, J.D., Gasparini, N.M., Tucker, G.E. and **Bras, R.L.** (2001) A quantitative evaluation of Playfair's law and its use in testing long-term stream erosion models. *Earth Surf. Proc. Land.*, **26**, 1317–1332.

North, C.P. (1996). The prediction and modelling of subsurface fluvial stratigraphy. In: *Advances in Fluvial Dynamics and Stratigraphy* (Eds P.A. Carling and M.R. Dawson), Wiley, Chichester, pp. 395–508.

Odgaard, A.J. (1989a) River-meander model, I, Development. *J. Hydraul. Eng. ASCE.*, **115**, 1433–1450.

Odgaard, A.J. (1989b) River-meander model, II, Applications. *J. Hydraul. Eng. ASCE*, **115**, 1451–1464.

Paola, C. (1988) Subsidence and gravel transport in alluvial basins. In: *New Perspectives in Basin Analysis* (Eds K.L. Kleinspehn and C. Paola), 231–243, Springer-Verlag, New York.

Paola, C. (1990) A simple basin-filling model for coarse-grained alluvial systems. In: *Quantitative Dynamic Stratigraphy* (Eds T.A. Cross), Prentice-Hall, Englewood Cliffs, pp. 363–374.

Paola, C. (2000) Quantitative models of sedimentary basin filling. *Sedimentology*, **47** (Suppl. 1), 121–178.

Paola, C. and **Borgman, L.** (1991) Reconstructing random topography from preserved stratification. *Sedimentology*, **38**, 553–565.

Paola, C., Wiele, S. M. and **Reinhart, M.A.** (1989) Upper-regime parallel lamination as the result of turbulent sediment transport and low-amplitude bedforms. *Sedimentology*, **36**, 47–60.

Paola, C., Heller, P. L. and **Angevine, C.L.** (1992) The large-scale dynamics of grain-size variation in alluvial basins. 1 – Theory. *Basin Res.*, **4**, 73–90.

Paola, C., Parker, G., Mohrig, D.C. and **Whipple, K.X.** (1999). The influence of transport fluctuations on spatially averaged topography on a sandy, braided fluvial plain. In: *Numerical Experiments in Stratigraphy: Recent Advances in Stratigraphic and Sedimentologic Computer Simulations* (Eds J.W. Harbaugh, W.L. Watney, E. Rankey, R. Slingerland, R. Goldstein and E. Franseen), *SEPM Spec. Pub.* Tulsa, OK, **62**, 211–218.

Parker, G. and **Andrews, E.A.** (1986) On the time development of meander bends. *J. Fluid Mech.*, **162**, 139–156.

Parker, G., Diplas, P. and **Akiyama, J.** (1983) Meander bends of high amplitude. *J. Hydraul. Eng., ASCE*, **109**, 1323–1337.

Parker, G., Sawai, K. and **Ikeda, S.** (1982) Bend theory of river meanders. 2. Nonlinear deformation of finite amplitude bends. *J. Fluid Mech.*, **112**, 303–314.

Parker, G., Paola, C., Whipple, K.X. and **Mohrig, D.C.** (1998a) Alluvial fans formed by channelized fluvial and sheet flow, I, Theory. *J. Hydraul. Eng., ASCE*, **124**, 985–995.

Parker, G., Paola, C., Whipple, K.X., Mohrig, D.C., Toro-Escobar, C.M., Halverson, M. and **Skoglund, T.W.** (1998b) Alluvial fans formed by channelized fluvial and sheet flow, II, Application. *J. Hydraul. Eng., ASCE*, **124**, 996–1004.

Pizzuto, J.E. (1987) Sediment diffusion during overbank flows. *Sedimentology*, **34**, 301–317.

Quinlan, G.M. and **Beaumont, C.** (1984) Appalachian thrusting, lithospheric flexure, and the Paleozoic stratigraphy of the eastern interior of North America. *Can. J. Earth Sci.*, **21**, 973–996.

Rinaldo, A., Dietrich, W.E., Rigon, R., Vogel, G. and **Rodriguez-Iturbe, I.** (1995) Geomorphological signatures of varying climate. *Nature*, **374**, 632–634.

Rinaldo, A., Rodriguez-Iturbe, I., Rigon, R., Bras, R.L., Ijjasz-Vasquez, E.J. and **Marani, A.** (1992) Minimum energy and fractal structures of drainage networks. *Water Resour. Res.*, **28**, 2183–2195.

Rinaldo, A., Rodriguez-Iturbe, I. and **Rigon, R.** (1998) Channel networks. *Ann. Rev. Earth Planet. Sci.*, **26**, 289–327.

Rivenæs, J.C. (1992) Application of a dual-lithology, depth-dependent diffusion equation in stratigraphic simulation. *Basin Res.*, **4**, 133–146.

Rivenæs, J.C. (1997) Impact of sediment transport efficiency on large-scale sequence architecture: Results from stratigraphic computer simulation. *Basin Res.*, **9**, 91–105.

Robinson, R.L. and **Slingerland, R.L.** (1998a) Origin of fluvial grain-size trends in a foreland basin: The Pocono formation of the central Appalachian basin. *J. Sed. Res.*, **A68**, 473–486.

Robinson, R.L. and **Slingerland, R.L.** (1998b) Grain-size trends and basin subsidence in the Campanian Castlegate Sandstone and equivalent conglomerates of central Utah. *Basin Res.*, **10**, 109–127.

Robinson, R.A.J., Slingerland, R.L. and **Walsh, J.M.** (2001) Predicting fluvial-deltaic aggradation in Lake Roxburgh, New Zealand: test of a water and sediment routing model. In: *Geologic Modeling and Simulation: Sedimentary Systems* (Eds D.F. Merriam and J.C. Davis), Kluwer Academic/Plenum Publishers, New York, pp. 119–132.

Rodriguez-Iturbe, I., Rinaldo, A., Rigon, R., Bras, R.L., Marani, A. and **Ijjasz-Vasquez, E.J.** (1992) Energy dissipation, runoff production, and the three-dimensional structure of river basins. *Water Resour. Res.*, **28**, 1095–1103.

Rubin, D.M. (1987) *Cross-Bedding, Bedforms and Paleocurrents.* SEPM, Tulsa, Oklahoma, Concepts in Sedimentology and Paleontology, **1**, 187p.

Seidl, M.A. and **Dietrich, W.E.** (1992) The problem of channel erosion into bedrock. *Catena Suppl.*, **23**, 101–124.

Seifert, D. and **Jensen, J.L.** (1999). Using sequential indicator simulation as a tool in reservoir description: Issues and uncertainties. *Math. Geol.*, **31**, 527–550.

Seifert, D. and **Jensen, J.L.** (2000). Object and pixel-based reservoir modeling of a braided fluvial reservoir. *Math. Geol.*, **32**, 581–603.

Seminara, G., Zolezzi, G., Tubino, M. and **Zardi, D.** (2001) Downstream and upstream influence in river meandering. Part 2. Planimetric development. *J. Fluid. Mech.*, **438**, 213–230.

Sinclair, H.D., Coakley, B.J., Allen, P.A. and **Watts, A.B.** (1991) Simulation of foreland basin stratigraphy using a diffusion model of mountain belt uplift and erosion: An example from the central Alps, Switzerland. *Tectonics*, **10**, 599–620.

Sklar, L. and **Dietrich, W.E.** (1998) River longitudinal profiles and bedrock incision models: stream power and the influence of sediment supply. In: *Rivers Over Rock: Fluvial Processes in Bedrock Channels* (Eds E.E. Wohl and K.J. Tinkler), AGU Press, Washington, pp. 237–260.

Sklar, L. and **Dietrich, W.E.** (2001) Sediment and rock strength controls on river incision into bedrock. *Geology*, **29**, 1087–1090.

Slingerland, R.L. and **Smith, N.D.** (1998) Necessary conditions for a meandering-river avulsion. *Geology*, **26**, 435–438.

Slingerland, R.L. and **Smith, N.D.** (2004) River avulsions and their deposits. *Ann. Rev. Earth Planet. Sci.*, **32**, 255–283.

Slingerland, R., Harbaugh, J.W. and **Furlong, K.P.** (1994) *Simulating Clastic Sedimentary Basins.* Prentice-Hall, Englewood Cliffs, New Jersey.

Smith, N.D. and **Rogers, J.** (Eds) (1999) *Fluvial Sedimentology VI. Int. Assoc. Sediment. Spec. Pub.* **28**, Blackwell, Oxford, 478 p.

Snyder, N.P., Whipple, K.X., Tucker, G.E. and **Merritts, D.J.** (2003) Importance of a stochastic distribution of floods and erosion thresholds in the bedrock river incision problem. *J. Geophys. Res.*, **108**, 2117.

Srivastava, R.M. (1994) An overview of stochastic methods for reservoir characterization. In: *Stochastic Modeling and Geostatistics* (Eds J.M. Yarus and R.L. Chambers), *Am. Assoc. Petrol. Geol. Comput. Appl. Geol.*, Tulsa, OK, **3**, pp. 3–16.

Stock, J. and **Montgomery, D.R.** (1999) Geologic constraints on bedrock river incision using the stream power law. *J. Geophys. Res.*, **104**, 4983–4993.

Strebelle, S. (2002) Conditional simulation of complex geological structures using multiple-point statistics. *Math. Geol.*, **28**, 937–950.

Sun, T., Meakin, P., Jossang, T. and **Schwarz, K.** (1996) A simulation model of meandering rivers. *Water Resour. Res.*, **32**, 2937–2954.

Sun, T., Meakin, P. and **Jossang, T.** (2001a) Meander migration and the lateral tilting of floodplains. *Water Resour. Res.*, **37**, 1485–1502.

Sun, T., Meakin, P. and **Jossang, T.** (2001b) A computer model for meandering rivers with multiple bedload sediment sizes 1. Theory. *Water Resour. Res.*, **37**, 2227–2241.

Sun, T., Meakin, P. and **Jossang, T.** (2001c) A computer model for meandering rivers with multiple bedload sediment sizes 2. Computer simulations. *Water Resour. Res.*, **37**, 2243–2258.

Sun, T., Paola, C., Parker, G., and **Meakin, P.** (2002) Fluvial fan deltas: Linking channel processes with large-scale morphodynamics. *Water Resour. Res.*, **38**, 1151, 10.1029/2001WR000284.

Tetzlaff, D. (1991) The combined use of sedimentary process modeling and statistical simulation in reservoir characterization. *SPE Paper 22759*, 937–942.

Tetzlaff, D.M. and **Harbaugh, J.W.** (1989) *Simulating Clastic Sedimentation*, Van Nostrand Reinhold, New York.

Tetzlaff, D. and **Priddy, G.** (2001). Sedimentary process modeling: From academia to industry. In: *Geologic Modeling and Simulation: Sedimentary Systems* (Eds D.F. Merriam, and J.C. Davis), Kluwer Academic/Plenum Publishers, New York, pp. 45–69.

Thomas, R. and **Nicholas, A.P.** (2002) Simulation of braided river flow using a new cellular routing scheme. *Geomorphology*, **43**, 179–195.

Tucker, G.E. (2004) Drainage basin sensitivity to tectonic and climatic forcing: Implications of a stochastic model for the role of entrainment and erosion thresholds. *Earth Surf. Proc. Land.*, **29**, 185–205.

Tucker, G.E. and **Bras, R.L.** (1998) Hillslope processes, drainage density, and landscape morphology. *Water Resour. Res.*, **34**, 2751–2764.

Tucker, G.E. and **Bras, R.L.** (2000) A stochastic approach to modeling the role of rainfall variability in drainage basin evolution. *Water Resour. Res.*, **36**, 1953–1964.

Tucker, G.E. and **Slingerland, R.L.** (1996) Predicting sediment flux from fold and thrust belts. *Basin Res.*, **8**, 329–349.

Tucker, G.E. and **Slingerland, R.L.** (1997) Drainage basin responses to climate change. *Water Resour. Res.*, **33**, 2031–2047.

Tucker, G.E. and **Whipple, K.X.** (2002) Topographic outcomes predicted by stream erosion models: Sensitivity analysis and intermodel comparison. *J. Geophys. Res.*, **107**, 2179, doi:10.1029/2001JB000162.

Tucker, G.E. and **Slingerland, R.L.** (1994) Erosional dynamics, flexural isostasy and long-lived escarpments: A numerical modeling study. *J. Geophys. Res.*, **99**, 12229–12243.

Tucker, G.E., Lancaster, S.T., Gasparini, N.M., Bras, R.L and **Rybarczyk, S.M.** (2001) An object-oriented framework for hydrogeologic and geomorphic modeling using triangulated irregular networks. *Comp. Geosci.*, **27**, 959–973.

Tucker, G.E., Lancaster, S.T., Gasparini, N.M., and **Bras, R.L.** (2002) The Channel-Hillslope Integrated Landscape development Model (CHILD). In: (Eds R.S. Harmon and W.W. III Doe), *Landscape Erosion and Evolution Modeling*. Kluwer Academic Publishing, New York, pp. 349–388.

Tyler, K., Henriquez, A. and **Svanes, T.** (1994) Modeling heterogeneities in fluvial domains: A review of the influence on production profiles. In: *Stochastic Modeling and Geostatistics* (Eds J.M. Yarus and R.L. Chambers), *Am. Assoc. Petrol. Geol. Comput. Appl. Geol.*, Tulsa, OK, **3**, 77–89.

Van Niekerk, Vogel, A.K., Slingerland, R. and **Bridge, J.** (1992) Routing heterogeneous size-density sediments over a moveable bed: Model development. *J. Hydraul. Eng., ASCE*, **118**, 246–262.

Vogel, K., Van Niekerk, A., Slingerland, R. and **Bridge, J.S.** (1992) Routing of heterogeneous size-density sediments over a moveable bed: Model verification and testing. *J. Hydraul. Eng., ASCE*, **118**, 263–279.

Walling, D.E., He, Q. and **Nicholas, A.P.** (1996) Floodplains as suspended sediment sinks. In: *Floodplain Processes* (Eds M.G. Anderson, D.E. Walling and P.D. Bates), Wiley, Chichester, pp. 399–440.

Wang, Z.B., Fokkink, R.J. and **De Vries, M.** (1995) Stability of river bifurcations in 1D morphodynamic models. *J. Hydraul. Res.*, **33**, 739–750.

Webb, E.K. (1994) Simulating the three-dimensional distribution of sediment units in braided stream deposits. *J. Sed. Res.*, **B64**, 219–231.

Webb, E.K. (1995) Simulation of braided channel topology and topography. *Water Resour. Res.*, **31**, 2603–2611.

Webb, E.K. and **Anderson, M.P.** (1996). Simulation of preferential flow in three-dimensional, heterogeneous conductivity fields with realistic internal architecture. *Water Resour. Res.*, **32**, 533–545.

Whipple, K.X. and **Tucker, G.E.** (1999) Dynamics of the stream-power river incision model: Implications for height limits of mountain ranges, landscape response timescales, and research needs. *J. Geophys. Res.*, **104**, 17661–17674.

Whipple, K.X. and **Tucker, G.E.** (2002) Implications of sediment-flux-dependent river incision models for landscape evolution. *J. Geophys. Res.*, **107**, 1029.

Willgoose, G.R., Bras, R.L. and **Rodriguez-Iturbe, I.** (1991a) Results from a new model of river basin evolution. *Earth. Surf. Proc. Land.*, **16**, 237–254.

Willgoose, G.R., Bras, R.L. and **Rodriguez-Iturbe, I.** (1991b) A physically-based coupled network growth and hillslope evolution model: 1. Theory. *Water Resour. Res.*, **27**, 1671–1684.

Willgoose, G.R., Bras, R.L. and **Rodriguez-Iturbe, I.** (1991c) A physically-based coupled network growth and hillslope evolution model: 1. Applications. *Water Resour. Res.*, **27**, 1685–1696.

Willgoose, G.R., Bras, R.L. and **Rodriguez-Iturbe, I.** (1992) The relationship between catchment and hillslope

properties: Implications of a catchment evolution model. *Geomorphology*, **5**, 21–37.
Willgoose, G.R., Bras, R.L. and **Rodriguez-Iturbe, I.** (1994) Hydrogeomorphology modeling with a physically based river basin evolution model. In: *Process Models and Theoretical Geomorphology* (Ed. M.J. Kirby), Wiley, Chichester, pp. 3–22.
Willis, B.J. (1989) Paleochannel reconstructions from point bar deposits: A three-dimensional perspective. *Sedimentology*, **36**, 757–766.
Zolezzi, G. and **Seminara, G.** (2001) Downstream and upstream influence in river meandering. Part 1. General theory and application to overdeepening. *J. Fluid. Mech.*, **438**, 183–211.

Spec. Publ. Int. Assoc. Sedimentol. (2008) **40**, 139–144

Process-based stochastic modelling: meandering channelized reservoirs

SIMON LOPEZ, ISABELLE COJAN, JACQUES RIVOIRARD and ALAIN GALLI

Ecole des Mines de Paris, 35 rue Saint-Honoré, 77305 Fontainebleau Cedex, France
(E-mail: isabelle.cojan@ensmp.fr)

ABSTRACT

A combined stochastic and process-based approach has been developed for modelling fluvial meandering channels at the hydrocarbon-reservoir scale. The model is based on the spatial evolution of the channel in time, and deposition of the associated sediment bodies. The different elements have been implemented taking into account physical results and case study reports. The three-dimensional evolution of the channel stems from equations developed in hydraulic studies and proven to generate realistic two-dimensional shapes. Modifications have been made to account for the vertical evolution of the longitudinal profile. A stochastic algorithm linked with physical parameters allows the simulation of chute cut-off, levee breaching and avulsions. Where appropriate, the model allows for the generation of the different elements of fluvial deposits: point bars, crevasse splays, overbank alluvium, sand and mud plugs as well as organic-rich deposits in low areas.

To be operational, a limited number of key parameters are user-specified. These allow for the building and testing of different architectures, e.g. allowing the sinuosity and connectivity of sand bodies to vary with avulsion frequency, or reproducing external forcing (aggradation or incision). The resulting model is simple but robust and computationally fast. It depends on a limited number of key parameters, but it is also able to represent various architectures, and can be used to produce one or multiple realizations of a reservoir. It produces detailed three-dimensional blocks that makes it easy to test hypotheses on architecture, and to extract training images or virtual wells as input data for other simulation techniques.

Keywords Fluvial system, reservoir, meander, process-based model.

INTRODUCTION

In several depositional environments, reservoirs may have a complex internal architecture and thus complex properties. Since the 1980s, the characterization of these types of heterogeneous reservoirs has received much attention, and research and development have focused on different types of models in order to improve exploration and production (Galli & Beucher, 1997). Although stochastic models, either pixel-based or object-based, are flexible, they may fail to reproduce realistic sediment bodies or consistent stratigraphic arrangements in complex systems. Here a stochastic and process-based approach are combined in the modelling of a meandering fluvial system at the hydrocarbon-reservoir scale. In this type of depositional setting, point bars form the key reservoir targets. This modelling technique focuses on an accurate representation of point-bar shape, connectivity and the distribution of bodies that represent potential flow barriers.

DESCRIPTION OF THE MODEL

Previous numerical models of alluvial architecture have already proven useful in the understanding of ancient deposits. While most of these models rely on a rough modelling of the channel belt (e.g. Mackey & Bridge, 1995), the present methodology is based upon a rather detailed description of the channel evolution through time and of the deposits associated with the various processes linked to fluvial systems (Lopez, 2003). It uses

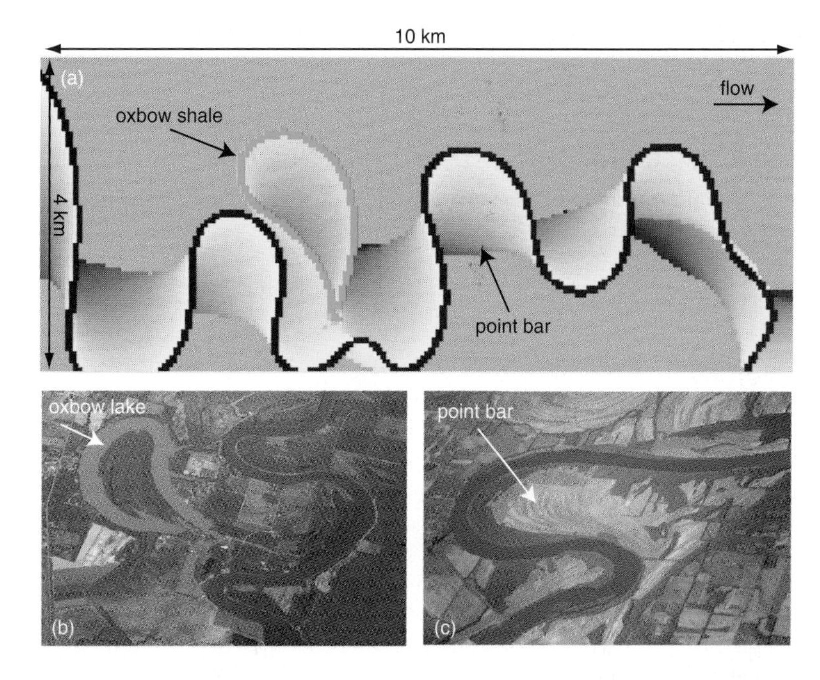

Fig. 1. Meandering channel and associated deposits. (a) Results from the model (red to yellow, sands from older to more recent; green, shale; purple, undefined facies). (b and c) Aerial views of the Mississippi River.

some selected key parameters that represent the major effects of the processes involved at the scale considered, for instance floodplain slope, channel width, channel depth, floods and overbank-flood frequency and intensity, and erodibility. The lateral migration of the channel is implemented taking into account physical equations derived from hydraulic studies (Ikeda *et al.*, 1981; Sun *et al.*, 1996, 2001). While developed in 2D and assuming a constant channel width, these models proved to generate realistic shapes (Howard, 1996; Gross & Small, 1998). They have been modified by taking into account the longitudinal profile to allow smoothing of the local floodplain slope following a cut-off (Lopez, 2003). Each iteration of the model corresponds to the occurrence of bankfull floods (e.g. every year), which are considered to be the major cause of migration. Migration is accompanied by the deposition of the associated sediment bodies: point bars in the inner part of the meander loops, sand and mud plugs in the oxbow lakes (Figs 1 and 2).

At regular or irregular intervals, overbank floods occur. This results in the deposition of levees on the borders of the channel, and of alluvium on the floodplain. The intensity of the flood defines the thickness of the deposited levees, while the thickness as well as the grain size of the material deposited on the floodplain decrease away from the channel as negative exponential functions. In between overbank floodings, the floodplain is stable and organic-rich deposits can accumulate in low areas. From time to time levee breaches may occur. These are preferentially located in areas where flow velocity reaches a local maximum, and they can generate either a chute cut-off, or a crevasse-splay deposit that may lead to an avulsion resulting in a new channel path downslope. Avulsions may also occur upstream of the modelled floodplain, and are then referred to as regional avulsions, as opposed to local avulsions that develop within the modelled domain (Fig. 3).

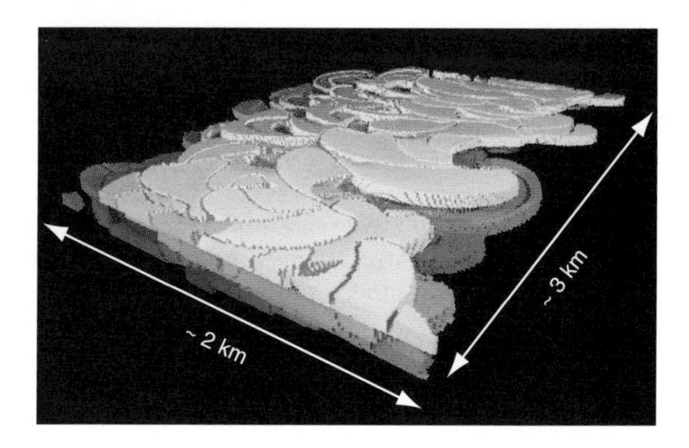

Fig. 2. Filtered view of point bars produced over a 20 kyr period (colours as in Fig. 1a). From Lopez (2003).

CONTROL OF THE MODEL

In addition to the parameters controlling the evolution of the channel and the various deposits, the evolution of the model in terms of aggradation and incision is controlled by the accommodation

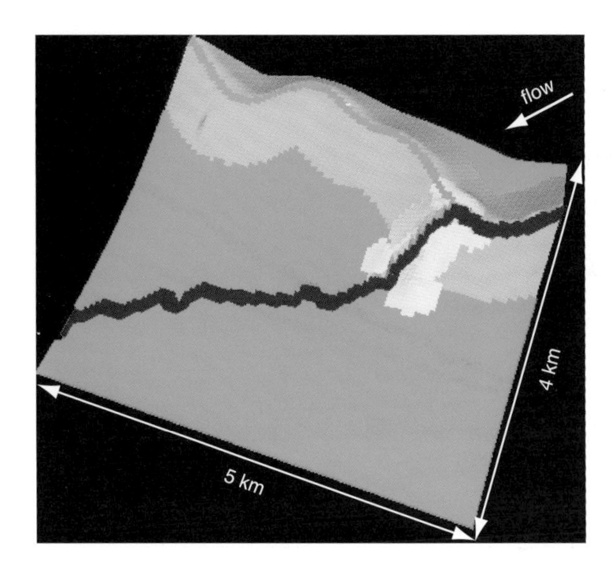

Fig. 3. Perspective view of the modelled floodplain. The present stream (dark blue) after an avulsion initiated by a levee breach. Note the crevasse splay deposits (tan colours) at the levee breach, and the former course of the river bordered by the levees (light green) and by overbank shale (dark green).

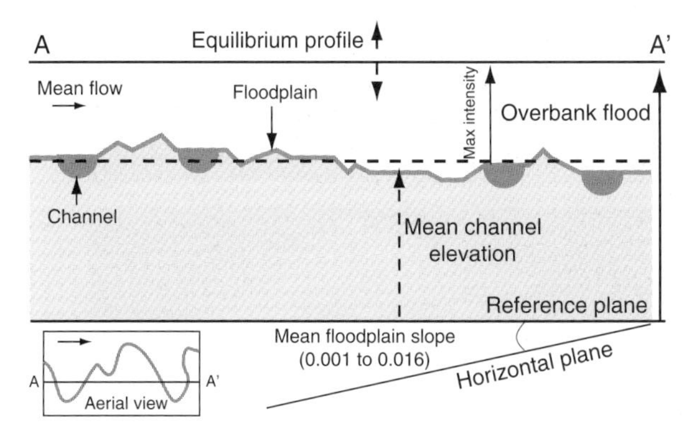

Fig. 4. Representation of the accommodation space by an equilibrium profile parallel to the floodplain.

space. At the reservoir scale that is considered, this can be schematically represented by the elevation of a so-called 'equilibrium profile', a plane parallel to the floodplain (Fig. 4). Its variations through time are input by the user. At any given time, if the floodplain is close to the equilibrium profile, levee aggradation is limited as accommodation space for deposits from overbank floods is restricted. When the equilibrium profile is far above the floodplain, aggradation evolves freely and depends on the occurrence and intensity of overbank floods. Conversely, incision occurs when the equilibrium profile is lower than the floodplain.

The influence of some key parameters, such as frequency of avulsion and overbank floods, and in particular of the equilibrium profile, is illustrated in the following examples, where floodplain slope, and channel depth and width are kept constant. The first set of examples (Fig. 5) presents very different situations: an equilibrium profile moving up slowly and associated with low-frequency avulsions (Fig. 5a) opposed to an equilibrium profile moving up rapidly and associated to high-frequency avulsions (Fig. 5b).

In Fig. 5a, the equilibrium profile increases very slowly at regular time steps. The accommodation space thus created is small and it is quickly filled by deposits during the next overbank floods, so that the floodplain again matches the equilibrium profile. The system is greatly constrained, and there is little aggradation. Drops of mean channel elevation correspond to regional avulsions, when the channel settles lower on the floodplain. However, these avulsions are rare, so that meander point bars and oxbow fills are well developed. The available space is nearly fully filled with sand bodies, except for mud plugs which are also quite frequent in such a setting.

In Fig. 5b, the equilibrium profile increases very rapidly (30 times faster than in Fig. 5a). Overbank floods are extremely frequent, and the mean channel elevation stays well below the equilibrium profile. The system evolves freely, favouring aggradation. Avulsions are very frequent, so that the channel has no time to migrate, and sand bodies form long ribbons. Connectivity is poor as a consequence of the high aggradation rate.

The second set of examples (Fig. 6) illustrates the influence of frequency and intensity of overbank floods on the connectivity of the sand bodies, all the other parameters being equal.

The equilibrium profile is raised with significant jumps at regular time steps. They are concomitant with regional avulsions, so that the mean channel elevation drops at these steps. In Fig. 6a, overbank floods are frequent and intense. The floodplain rapidly reaches the equilibrium profile again after each drop in channel elevation linked to the avulsion. Between two avulsions, channel migration produces sand bodies with a fair lateral extent. These are, however, weakly connected in the vertical sense because of the rapid aggradation. In Fig. 6b, overbank floods are much less intense (half that of the example of Fig. 6a). The floodplain never reaches the equilibrium profile. Sand bodies, controlled by the lateral migration and the frequency of the avulsions, are of a similar size as in the previous case whereas the parameters are

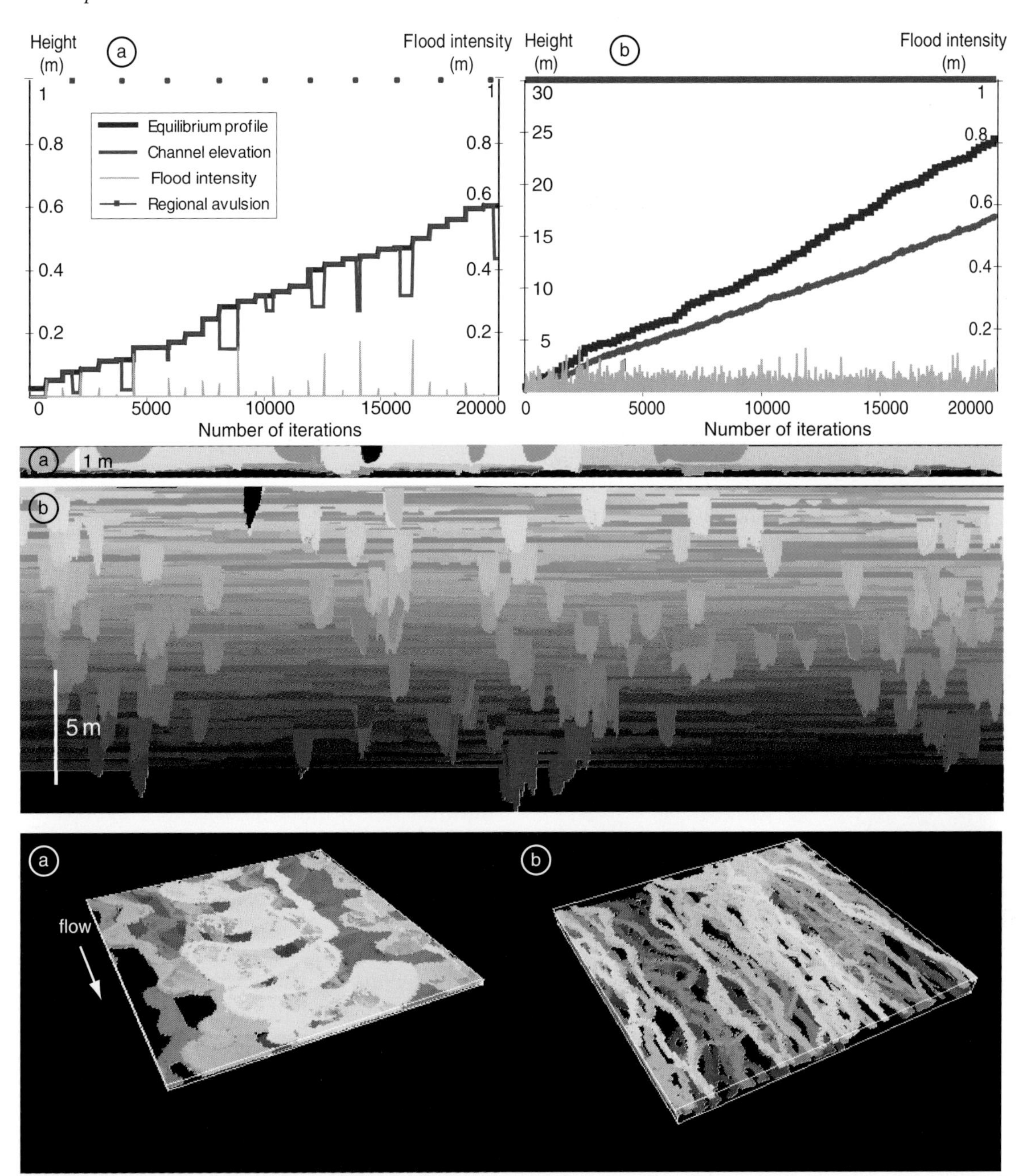

Fig. 5. (a) Channel evolution constrained by a small accommodation space that favours lateral migration. Note the good connectivity of sand bodies (red to yellow, from older to more recent) and the frequent mud plugs (green). Top: Evolution of equilibrium profile, mean channel elevation and overbank flood intensity (all in metres), as well as occurrence of regional avulsion, as a function of time. Middle: Cross-valley vertical section. Bottom: Perspective view of sand deposits (flow from top). From Lopez (2003). (b) Channel evolution in a high aggradation context that is not constrained by the accommodation space, but is subject to frequent regional avulsions. Note the dominance of ribbons and the poor connectivity of the sand bodies (red to yellow, from older to more recent). Overbank shale in green (darker being older). Top, Middle and Bottom as in Fig. 5a. From Lopez (2003).

the same. However, connectivity is fair due to the low aggradation rate of the floodplain.

These few examples show that in a model ruled by a small number of key parameters (channel width, channel depth, floodplain slope, avulsion or overbank-flood frequency, levee aggradation, equilibrium profile), the variations of these parameters can produce clearly distinct

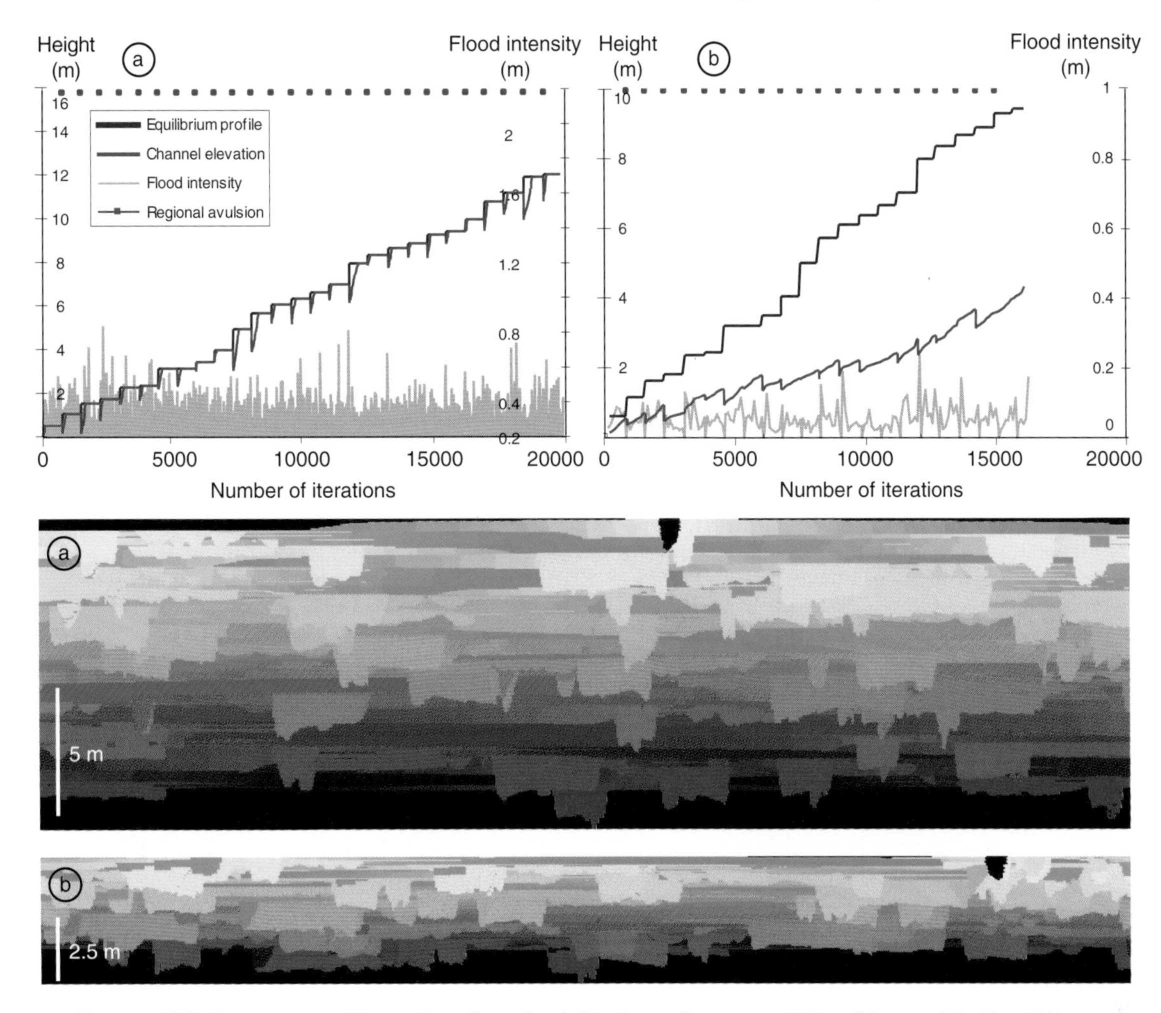

Fig. 6. Influence of the frequency and intensity of overbank floods on the connectivity of the sand bodies. Other parameters are constant (red to yellow, sands from older to more recent; dark green to light green, shale from older to more recent). (a) High frequency and intensity of overbank floods. (b) Less frequent overbank floods at half intensity. From Lopez (2003).

architectures. This offers the opportunity to build and compare 3D blocks, and to extract virtual wells and training images that can be used as input data for other conditioning statistical methods. As it is a stochastic technique, multiple realizations of the model can be produced using the same set of parameters, providing as many representations of the modelled system.

CONCLUSION

Our 3D process-based stochastic model offers a new way to reproduce realistic sediment bodies and their distribution at the reservoir scale. The model is controlled by a limited number of key parameters, which allows the model to be operational while capturing the essential part of the natural processes involved at the reservoir scale. In the present case of meandering fluvial systems, the construction of the 3D blocks relies upon the migration of the channel and the filling of the accommodation space in relation to migration and events associated with floods. Modelled sediment bodies are realistic, their shapes and sizes are directly related to some controlling parameters such as channel width, channel depth, and avulsion frequency. Distinct connectivity values are related to the elevation of the equilibrium profile and to the frequencies of avulsions and overbank floods.

This paper describes the direct construction of the model, starting from chosen parameter values. However, for the operational use for reservoir appraisal, there is a reverse aspect which consists in conditioning the model and valuating the parameters from the data (e.g. wells and seismic). Elements of this aspect of the model are described by Lopez (2003) and Cojan *et al.* (2005), while others are under development.

ACKNOWLEDGEMENTS

SHELL, GDF (Gaz de France), IFP (French Institute for Petroleum), EXXON and PETROBRAS are greatly acknowledged for their support through the Meandering Channelized Reservoirs Consortium. We are grateful to the reviewers for their helpful comments.

REFERENCES

Cojan, I., **Fouché, O.**, **Lopez, S.** and **Rivoirard, J.** (2005) Process-based reservoir modelling in the example of meandering channel. In: *Geostatistics 2004, Quantitative Geology and Geostatistics* (Eds. O. Leuhangthong and C. Deutsch), Springer, Dordrecht, **14/2**, pp. 611–619.

Galli, A. and **Beucher, H.** (1997) Stochastic models for reservoir characterization: A user-friendly review. In: *Fifth Latin American and Caribbean Petroleum Engineering Conference and Exhibition*, Rio de Janeiro, Brazil, 30 August–3 September 1997, 11 pp. (SPE 38999).

Gross, L.J. and **Small, M.J.** (1998) River and floodplain process simulation for subsurface characterization. *Water Resour. Res.*, **34**, 2365–2376.

Howard, A.D. (1996) Modelling channel migration and floodplain morphology. *Floodplain Processes*, John Wiley and Sons, New York, pp. 15–62.

Ikeda, S., **Parker, G.** and **Sawai, K.** (1981) Bend theory of river meanders. Part 1. Linear development. *J. Fluid Mech.*, **112**, 363–377.

Lopez, S. (2003) *Modélisation de réservoirs chenalisés méandriformes, approche génétique et stochastique*. PhD, Ecole des Mines de Paris, 287 pp., URL: http://cg.ensmp.fr/bibliotheque/index.html

Mackey, S.D. and **Bridge, J.S.** (1995) Three-dimensional model of alluvial stratigraphy: Theory and application. *J. Sed. Res.*, **B65**, 7–31.

Sun, T., **Meakin, P.** and **Jossang, T.** (2001) Meander migration and the lateral tilting of floodplains. *Water Resour. Res.*, **37(5)**, 1485–1502.

Sun, T., **Meakin, P.**, **Jossang, T.** and **Schwartz, K.** (1996) A simulation model for meandering rivers. *Water Resour. Res.*, **32**, 2937–2954.

Spec. Publ. Int. Assoc. Sedimentol. (2008) **40**, 145–169

Simulation of tidal flow and circulation patterns in the Early Miocene (Upper Marine Molasse) of the Alpine foreland basin

ULRICH BIEG*, MICHAEL PETER SÜSS† and JOACHIM KUHLEMANN†

**Technical University of Darmstadt, Schnittspahnstrasse 9, D-64287 Darmstadt, Germany (E-mail: ulrich.bieg@rohoel.at)*

†Department of Geosciences, University of Tübingen, Sigwartstrasse 10, D-72076 Tübingen, Germany

ABSTRACT

During the Ottnangian (Early Miocene) the peripheral Alpine foreland basin was flooded by a shallow sea, which linked the Mediterranean in the south with the Paratethys in the northeast. Since this seaway shows a clear evolution and did not exist uniformly during the period investigated, we reconstructed its palaeogeography and bathymetry for the time of its maximum extent in the German part of the Molasse Basin. For this time interval tidal deposits are known from France and Switzerland towards Austria. A fully non-linear three-dimensional hydrodynamic model has been applied to reconstruct tidal currents and amplitudes. Residual currents are interpreted to represent net transport directions of sediments within the seaway. The model is forced at its open boundaries by semi-diurnal M_2 waves. The sensitivity of the tidal model has been tested, using different tidal amplitudes and phases at the open model boundaries. Based on this, the model indicates high mesotidal ranges, similar to those deduced from outcrops in Switzerland, Germany and Austria. The model suggests a complex system of convergent and divergent currents in the seaway, mainly along circular pathways.

Keywords Tidal simulation, ancient sea, North Alpine Foreland Basin, mesoscale circulation pattern.

INTRODUCTION

The Upper Marine Molasse deposits are part of the North Alpine Foreland Basin (NAFB) that was formed during the Oligocene and Miocene as a mechanical response to the tectonic load of the evolving Alpine thrust wedge (e.g. Homewood *et al.*, 1986; Sinclair *et al.*, 1991; Schlunegger *et al.*, 1997) (see Fig. 1).

The stratigraphic unit investigated, i.e. the Upper Marine Molasse (OMM), has a preserved thickness of approximately 1000 m adjacent to the Alpine thrust front and thins to 70–100 m near the northern edge of the Swiss part of the Molasse basin (Berger, 1983; Keller, 1989).

The development of the OMM seaway was coeval with a reduction of thrust advance rates of the Alpine orogenic wedge and with strongly reduced sediment discharge in the western and central part of the NAFB (Kuhlemann, 2000; Schlunegger *et al.*, 2001). It also can be correlated to a global eustatic sea-level rise (e.g. Haq *et al.*, 1988; Crumeyrolle *et al.*, 1991; Zweigel *et al.*, 1998).

While marine conditions had prevailed since the Oligocene in the Austrian part of the peripheral foredeep (Kuhlemann & Kempf, 2002), a basin-wide transgression during the Early Miocene (~20 Ma) flooded the whole Alpine foredeep and generated shallow-marine conditions in all other parts of the Molasse basin. Shallow marine conditions existed for about 3 million years mainly in the Rhodanina area, but showed a distinctive evolution of trans- and regressive cycles within the rest of the foredeep (Fig. 2). At the time of maximum transgression (~18 Ma) the so-called Burdigalian Seaway evolved, stretching from southeastern France to eastern Austria (Fig. 1, and see Fig. 4), and it had a length of about 1200 km

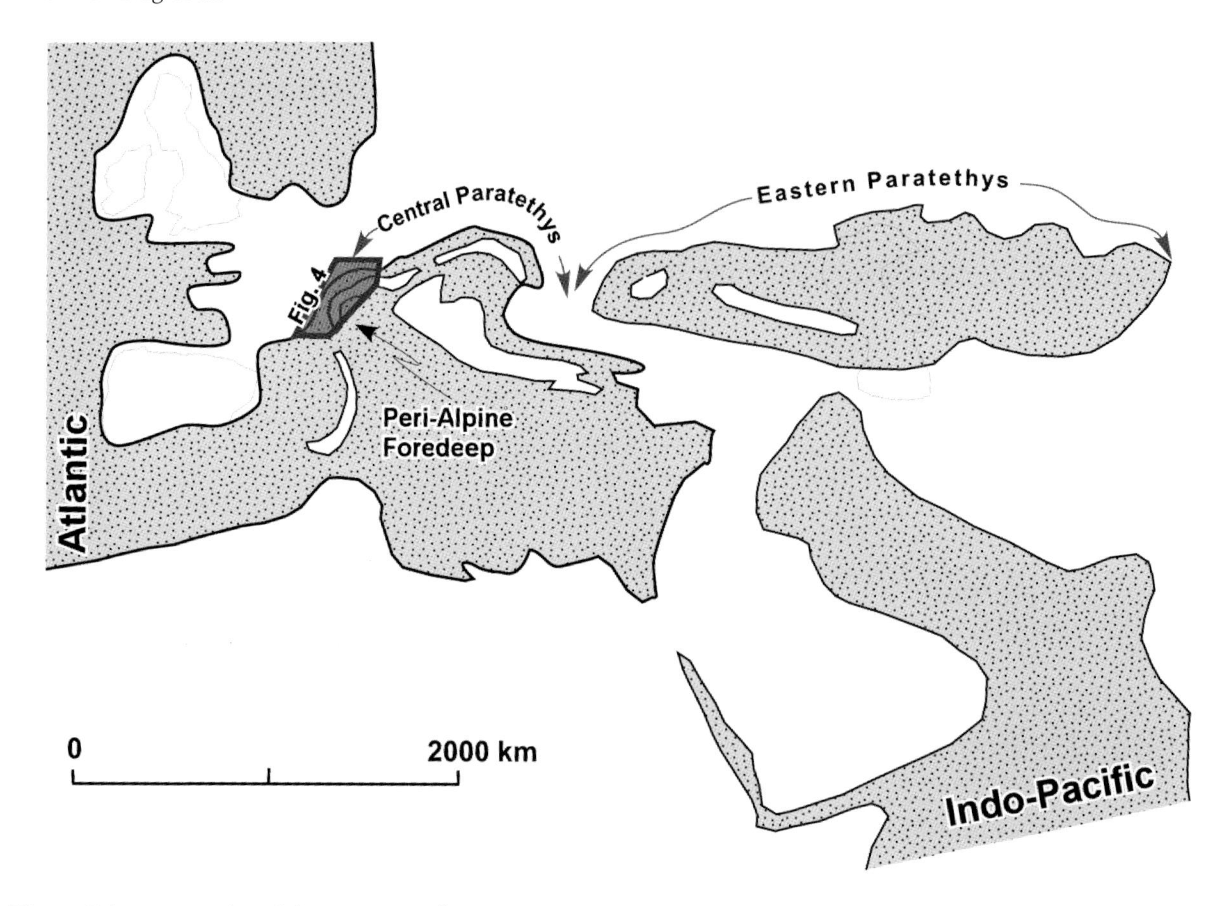

Fig. 1. Palaeogeography of the circum-Mediterranean region during the Early Miocene (Burdigalian, ca. 20 Ma). During this period the shallow marine Burdigalian Seaway (area marked Fig. 4) developed and linked the western Tethys with the Paratethys realm in the northeast (modified from Allen *et al.*, 1985; Rögl, 1998, 1999).

linking the deep marine Western Mediterranean Sea with the Paratethys in the northeast (Rögl & Steininger, 1983). This time is believed to be represented by the northernmost extent of the OMM sea, which cut a prominent cliff into the Swabian and Franconian Alb. We correlate these deposits basinwards with the the Baltringer beds, which can be correlated with similar deposits in Austria and Central Switzerland (Fig. 2).

During the early Ottnangian, tidal waves from the Atlantic Ocean entered the Mediterranean Sea and co-oscillated with the open boundaries of the Burdigalian Seaway (Fig. 1; Bieg *et al.*, 2004). As a consequence, sediments of the Burdigalian Seaway (Figs 1 and 2) were mainly deposited under a strong meso- to macrotidal regime (e.g. Homewood *et al.*, 1985). Tide-influenced deposits such as sand waves, bidirectional cross-stratification and neap-spring tidal-bundle sequences have been recognized within outcrops in France (Tessier & Gigot, 1989; Lesueur *et al.*, 1990; Assemat, 1991; Crumeyrolle *et al.*, 1991; Allen & Bass, 1993; Rögl, 1999, Couëffé *et al.*, 2004), Switzerland (Diem, 1985; Allen, 1984; Homewood & Allen, 1981; Homewood *et al.*, 1986; Allen *et al.*, 1985), southern Germany and Austria (Hülsemann, 1955; Faupl & Roetzel, 1987, 1990; Krenmayr, 1991; Krenmyar & Roetzel, 1996; Salvermoser, 1999). Many of these studies provide information on the local current directions of the ebb and the flood. However, they do not allow the reconstruction of meso-scale current patterns within the Burdigalian Seaway as the stratigraphic resolution and the correlation of OMM sediments from France to Austria is not sufficiently constrained. Thus we concentrate only on the conditions of the early Ottnangian, when the seaway had its maximum extent and connected the western Mediterranean with the Paratethys. Information on current direction and tidal amplitudes is chosen carefully and compared to the model results only where the age control is sufficient. Absolute dating of OMM deposits is of great importance, as fluctuations

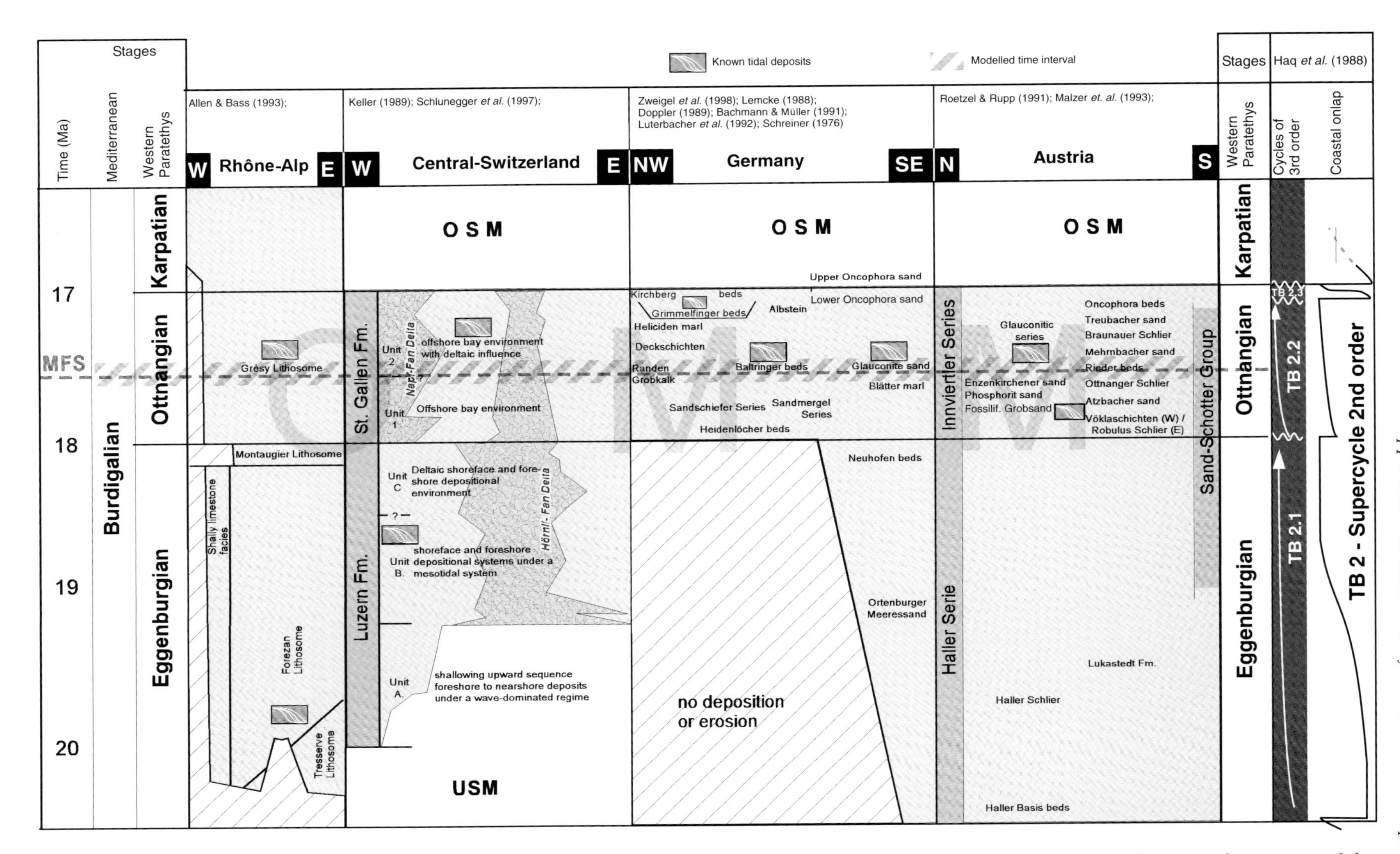

Fig. 2. Simplified stratigraphic overview of the Early Miocene between France (West) and Austria (East). The figure shows the two depositional sequences of the Eggenburgian Ottnangian. The study concentrates on the maximum extent of the Burdigalian Seaway during the Ottnangian (shortly after 18 Ma = time of maximum flooding; MFS). Small boxes indicate known tidal deposits. Stratigraphic timescale after Berggren *et al.* (1995), Rögl (1996) and Steininger *et al.* (1996). OSM, Upper Freshwater Molasse; USM, Lower Freshwater Molasse.

in eustatic, tectonic, and climate conditions of the seaway caused key oceanographic parameters (palaeogeography, palaeobathymetry, stratification, wind stresses and boundary tides) to vary widely throughout the seaway's history. These fluctuations, in turn, must have caused continuous changes in its circulation and sediment-transport regimes.

The large-scale tidal sedimentary structures in the Swiss part of the Burdigalian Seaway stayed unexplained until Martel *et al.* (1994) demonstrated in a numerical model that two propagating tidal waves can generate a meso- to macrotidal system in the Swiss region. As Martel's model is of limited spatial extent (Marseille to Geneva) and provides only information on tidal amplitudes, we expanded the extent of Martel's model, linking the shallow-water Burdigalian Seaway with the deeper marine parts of the Mediterranean Sea in the south and the Paratethys realm in the north (Fig. 3). Sztanó & de Boer (1995) suggested that during the ongoing transgression a series of southward connected shallow marine embayments formed in the foredeep, which allowed resonant amplification of incoming tidal waves and the deposition of tidal deposits from time to time. Later in the paper we evaluate an end member of this model (see Fig. 12), according to which, for a certain period during the flooding, resonant amplification must have developed due to reflection of a tidal wave at the coast of the Bohemian Massif (Sztanó, 1995; Sztanó & de Boer, 1995). In order to define an upper-limit bathymetric boundary condition for the tidal simulation, we reconstructed the potential maximum extent of the marine water body during the OMM (Fig. 2). Using a state-of-the-art finite element circulation model, described in detail by Lynch *et al.* (1996, 1997), we calculated tidal amplitudes and current velocities within the seaway applying different border conditions for the open model boundaries. Vertically averaged residual velocities were derived to predict net sediment transport directions. However, rather then investigating just one stratigraphic scenario it is the aim of our research to use numerical modelling as a tool to find and test general boundary conditions that have influenced the distribution of tidal amplitudes and residual velocities in the seaway. Further the computed residual velocities allow testing suggestions about net sediment transport directions (Büchi & Hofmann, 1960; Schreiner, 1966; Füchtbauer, 1967; Lemcke, 1973; Hofman, 1976; Allen *et al.*, 1985; Freudenberger & Schwerd, 1996).

Stratigraphy and sedimentary facies of the Upper Marine Molasse

The clastic sediments of the Central Molasse Basin traditionally have been subdivided into four lithostratigraphic groups (Matter *et al.*, 1980): Lower Marine Molasse (UMM), Lower Freshwater Molasse (USM), Upper Marine Molasse (OMM), and Upper Freshwater Molasse (OSM). We use the conventional German abbreviations for convenience. In Germany and Switzerland these lithostratigraphic units form two shallowing-upward megasequences (e.g. Schlunegger *et al.*, 1996). The first megasequence comprises the Rupelian UMM and the Chattian and Aquitanian fluvial clastic sediments of the USM. The second megasequence starts with the Burdigalian transgressive sediments, consisting of shallow-marine sand-, silt- and mudstones (OMM), which interfinger with conglomeratic fan-delta deposits adjacent to the thrust front (e.g. Schaad *et al.*, 1992). The top of this megasequence is represented by Mid- to Late Miocene fluvial clastic sediments of the OSM.

The Burdigalian Seaway connects two marine faunal provinces, the Mediterranean and the Paratethys provinces. A correlation of these faunas has been established by Steininger *et al.* (1996) and Rögl (1996). However, a chronostratigraphic correlation of OMM deposits across the entire Burdigalian Seaway is difficult to achieve and still non-existent for all lithosomes and deposits. As a consequence the applied timescale (Fig. 2; Kuhlemann & Kempf, 2002) provides only a 'best fit' between the two differing (bio-) stratigraphic datasets.

The Burdigalian megasequence (Fig. 2) in the western and eastern parts is made up by two depositional sequences, the Eggenburgian (21–18 Ma) and Ottnangian (18–17 Ma) according to the Paratethys stages. The maximum flooding of the Burdigalian Seaway and hence the maximum distribution of marine sediments in the OMM occurred during a basin-wide transgression at the base of the Ottnangian, around 18 Ma (Fig. 2). During the Ottnangian tide-influenced sediments have been deposited widely within the German (e.g. Hülsemann, 1955; Salvermoser, 1999) and Austrian (Faupl & Roetzel, 1987, 1990; Krenmayr, 1991) parts of the seaway. Its correlation is based on the lithostratigraphic framework introduced by Schreiner (1976), Lemcke (1988), Doppler (1989), Keller (1989), Bachmann & Müller (1991), Crumeyrolle *et al.* (1991), Luterbacher *et al.* (1992), Roetzel & Rupp (1991) and Schreiner

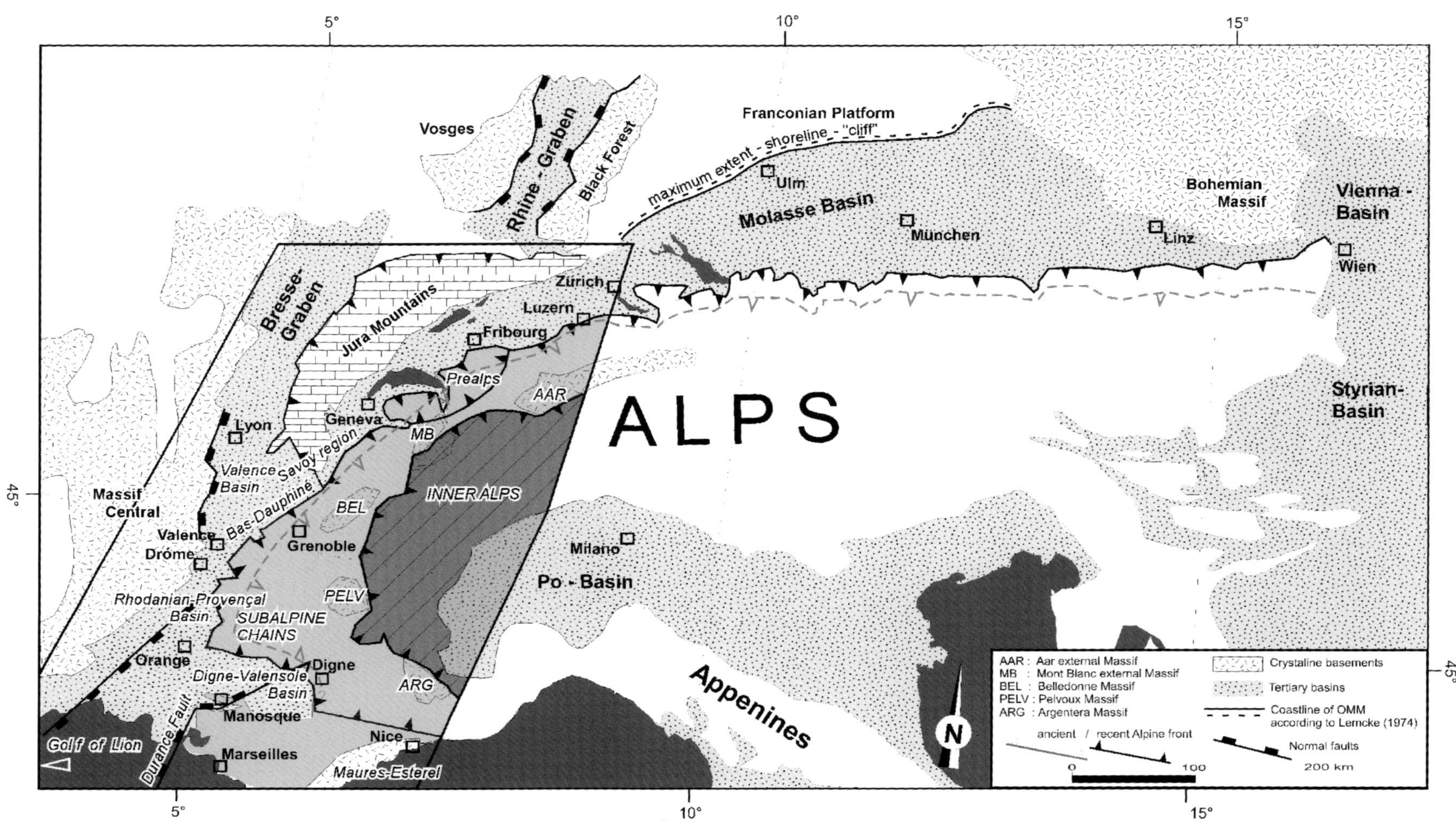

Fig. 3. Overview of the Tertiary basins adjacent to the Alpine orogen. The rectangle highlights a schematic structural map of southeastern France (modified from Phillipe *et al.*, 1998; Keller, 1989; Beck *et al.*, 1998; Ford *et al.*, 1999). The Burdigalian North Alpine thrust front (grey dashed line) was modified from Wagner (1996b, fig. 13), Kuhlemann *et al.* (2002), and Kuhlemann & Kempf (2002) .The ancient and recent Alpine thrust fronts represent the northern limit of the Helvetic/Penninic Units.

& Luterbacher (1999); for further details see also the discussion in Sissingh (1997, 1998, 2001) and Kuhlemann & Kempf (2002). It has to be stated that age constraints especially for the German OMM deposits are weakly constrained and a review of these OMM deposits might come to the conclusion that they are of Eggenburgian age.

Basin shape and palaeobathymetric reconstruction of the Burdigalian Seaway

Basin shape

In southeast France, the Mediterranean Sea occupied an elongated seaway (Demarcq, 1970), with a gulf-like morphology (Demarcq & Perriaux, 1984; Tessier & Gigot, 1989; Crumeyrolle *et al.*, 1991), which was bordered by the Alpine front in the east (Fig. 3). It is suggested that in the course of seafloor spreading in the Mediterranean Sea (Séranne *et al.*, 1995; Ford *et al.*, 1999) an extensional regime was initiated (Figs 3 and 4). This led to enhanced basin subsidence and facilitated the marine transgression towards the north. The Digne-Valensole Basin formed a large embayment connected to the Rhodanian-Provençal Gulf. The Rhodanian-Provençal Gulf was dissected by emergent areas (Demarcq & Perriaux, 1984) and extended northward through the Basse-Dauphiné Graben to the NAFB. In our model we assume that at the beginning of the Ottnangian these emergent areas had been completely flooded allowing the tide to enter the northern areas more easily.

Owing to uplift and erosion, caused by the subsequent Alpine deformation, a full reconstruction of the Digne-Valensole area is difficult to achieve. While Demarcq & Perriaux (1984, fig. 9.34) favoured a narrow, eastward directed embayment, Martel *et al.* (1994) proposed a broad opening towards the Mediterranean Sea. Our reconstruction (Fig. 4) is based on the model of Demarcq & Perriaux (1984), extending their reconstruction of the seaway on the basis of restored sections by Ritz (1991) and Lickorish & Ford (1998).

The seaway extended northward through the peri-Alpine foreland basins (Rhône-Alp, Savoy, Base-Dauphiné) to the NAFB (Debrand-Passard *et al.*, 1984; Martel *et al.*, 1994). The Savoy Basin appears as a strait between the wider and deeper Swiss part of the OMM and the southern French Alps area (Tessier & Gigot, 1989; Allen & Bass, 1993). Based on the data of Allen & Bass (1993) we used a basin width of approximately 30–40 km in the Savoy region for our simulations (Fig. 4).

In Switzerland the seaway was wider and deeper than in the adjacent Savoy region. The time span from 20 to 12 Ma is only poorly constrained in the western Swiss Molasse Basin, since most of the corresponding sediments have been removed by uplift-related erosion after 12 Ma in the course of Jura Mountain deformation. According to palaeogeographical and palinspastic reconstructions for the western Swiss Molasse Basin (Berger, 1996; Kuhlemann & Kempf, 2002) folding and subsequent uplift of the Jura Mountains started during the Middle Miocene. At this time the entire western Molasse Basin became incorporated into the Alpine orogenic wedge and was shortened by up to 30 km and displaced to the northwest (Laubscher, 1965, 1992; Mugnier & Menard, 1986; Guellec *et al.*, 1990). The palinspastic reconstruction of the Burdigalian Seaway hence results in a width of approximately 70 km. This width fits well with results from Allen & Homewood (1984), who reconstructed marine conditions near Fribourg. In western Switzerland the Alpine thrust front was in a fixed position southeast of Lausanne and the 'pinch-out' of Molasse strata to the northwest markedly slowed down between 22 and 12 Ma (Burkhard & Sommaruga, 1998). Towards the east the basin remained at a constant width of about 100 km (Fig. 4).

During its maximum extent, the seaway widened to a maximum of approximately 150 km in the German Molasse Basin. Along the northern margin a 'cliff' (Fig. 2) was cut into the Upper Jurassic limestone of the European foreland (e.g. Lemcke, 1973, 1988). According to Gall (1974, 1975) OMM sediments are also found above the cliff. However, tracing of these deposits is not possible. Hence, we used the cliff as an approximation of the coastline (e.g. Bachmann & Müller, 1991).

In Austria the preserved foreland basin is only a narrow remnant of the original basin width. Facies reconstructions for the Ottnagian have been taken from Wagner (1996b, p. 224, Fig. 9a). The position of the coastlines was taken from Kuhlemann & Kempf (2002, plate 5), indicating a width of 50 km for the Ottnagian. East of the Bohemian Massif the Austrian Molasse basin continued towards the Carpathian foreland.

Reconstructed palaeobathymetry of the Burdigalian

The basin-wide palaeobathymetry was reconstructed using the above described palaeogeography and assumptions, and based on biofacies

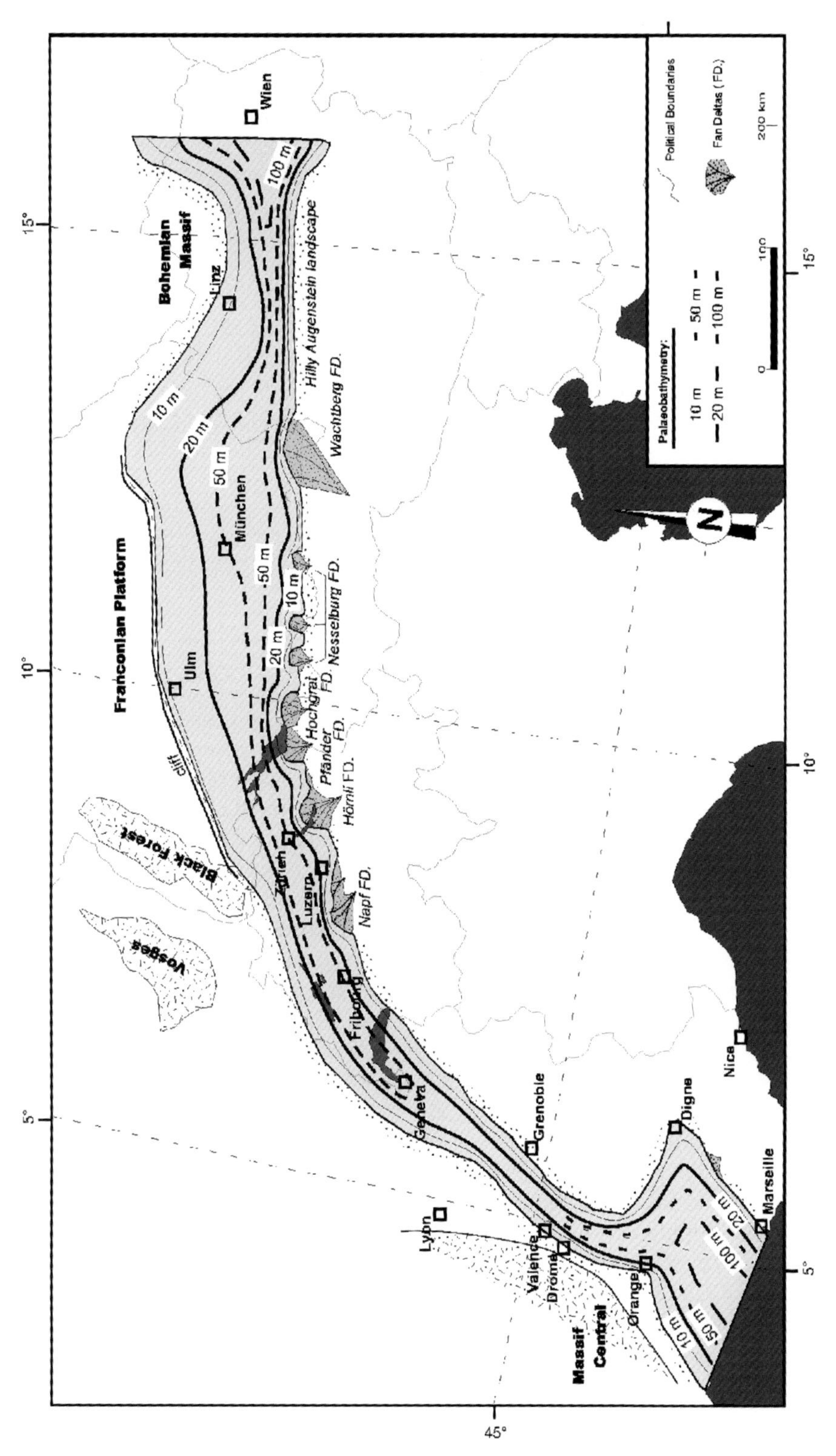

Fig. 4. Palaeogeographical and bathymetric reconstruction of the Upper Marine Molasse during the early Ottnangian shortly after 18 Ma. At this time, a second, more enhanced transgression flooded the entire North Alpine Foreland Basin and generated an extended seaway between the Mediterranean and the Paratethys (from the Rhône area in the southwest to Austria in the northeast). Contour lines indicate the estimated bathymetry. The map is projected in equal area Lambert projection.

Table 1. Palaeobathymetry estimates for the Eggenburgian and Ottnangian deposits of the Molasse Basin. For discussion see text.

Paratethys stages	France		Switzerland		Germany		Austria	
Ottnangian (18–17 Ma)			Keller, 1989; Schaad *et al.*, 1992 Bay environment near Napf dan delta	10–35 m	Wenger, 1987 Grobsand-fazies	30–50 m	Krenmayr, 1991 Rögl *et al.*, 1973 Neritic to deeper shelf	50–150 m
			Allen *et al.*, 1985 Offshore coquina banks	20 m			Faupl & Roetzel, 1987	<50 m
			Allen & Homewood, 1984 Offshore facies	25–60 m				
Eggenburgian (21–18 Ma)	Allen & Bass, 1993 Microfossils Alp-Rhône	30–50 m	Homewood & Allen, 1981 Sand waves near Fribourg	10 m	Erosion		Wagner, 1996a Open marine rim near the Alpine thrust front	500 m

analysis following Martel *et al.* (1994). While during the Ottnangian depth increased in most of the western Burdigalian Seaway, the seaway became shallower in the East (Table 1). For the Austrian part of the Molasse Basin we assume a relatively shallow marine environment of about 20 to 50 m, as discussed in Krenmayr & Roetzel (1996). The association of the *Skolithos* (and *Cruziana*) ichnofacies supports the interpretation of a high-energy environment. Following Wagner *et al.* (1986), Wagner (1998) and Kuhlemann & Kempf (2002), an elongate basin of about 150 m depth close to the Alpine thrust front was reconstructed. The Molasse basin shows an asymmetric, wedge-shaped depth distribution with its deepest part near the Alpine thrust front and continuous shallowing towards its northern coastline. Ottnangian sandstones in Austria and Bavaria were interpreted as subtidal sand waves, deposited under strong tidal current activity (Faupl & Roetzel, 1987, 1990; Krenmayr, 1991, Salvermoser, 1999).

THE PROGNOSTIC TIME-DOMAIN CIRCULATION MODEL

General model set-up and shallow-water physics

We employ a state-of-the-art finite-element circulation model, described in detail by Lynch *et al.* (1996, 1997). The non-linear time-domain model follows the algorithmic approach of Lynch & Werner (1991) with improvements and extensions.

Using a time-stepping finite element method algorithm, the model solves the non-linear, three-dimensional, shallow-water equations with the conventional Boussinesq and hydrostatic assumptions. The model is three-dimensional, with a free surface, partially mixed vertically and fully non-linear. It transports heat, salt and two momentum variables. Both barotropic and baroclinic motions are resolved in tidal time. In this study we use a simple barotropic approach. Vertical mixing is represented by a level 2.5 turbulence closure scheme (Mellor & Yamada, 1982; Galperin *et al.*, 1988; Blumberg *et al.*, 1992). Horizontal mixing is represented by a mesh- and shear-dependent eddy viscosity similar to Smagorinsky (1963). The turbulent Reynolds stress and flux terms in the conservation equations for momentum and energy are parameterized in terms of eddy mixing coefficients. Dirichlet boundary conditions at the free surface for the turbulent quantities are included. Variable horizontal resolution is facilitated by the use of unstructured meshes of triangles (Fig. 5). The comprehensive model provides shelf-scale geographical coverage to resolve local coastal features such as tidal channels, estuaries, headlands and tidal inlets. Because of the representation of

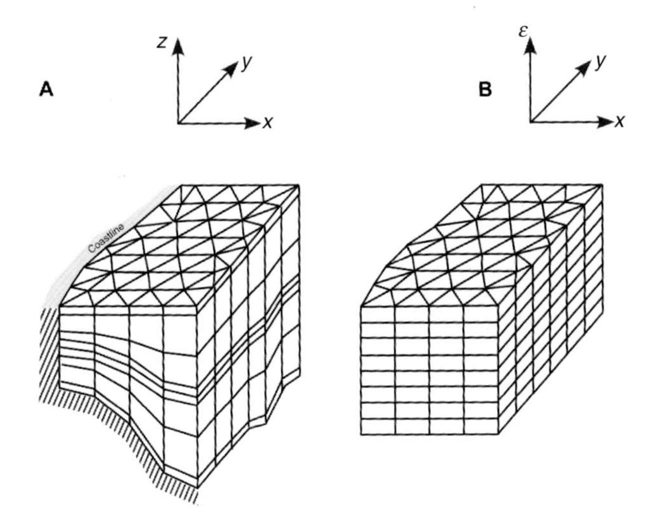

Fig. 5. Main features of the 3D layered mesh: (A) element sides are perfectly vertical, a variable vertical mesh spacing allows adaptation to the boundary and internal layers; (B) mesh spacing is uniform when mapped in the (x, y, ε)-coordinate system (adopted from Lynch & Naimie, 1993).

involved flow features and shallow water physics it is able to capture all important shelf processes in tidal time. Forcing terms include tides, surface wind stress and the barotropic pressure gradients computed from a prognostic evolving density field. Within this study the model is driven only by the M_2 constituents of tides, specified as a Dirichlet elevation boundary condition across the open ocean boundary at both ends of the seaway. The describing elevation time series is a linear sinusoidal function.

The finite element solution is based on the Galerkin weighted residual method, which is discretized by simple linear elements (Fig. 5). The horizontal domain is discretized by triangular elements with linear basis functions in a Cartesian coordinate system. The vertical domain is resolved by a terrain-following coordinate system that consists of a string of nodes connected by 1D linear elements. The nodes are allowed to move vertically to track the time-dependent free surface elevation (Lynch, 1982). The resulting finite element is a six-node prism that can squeeze and stretch vertically in time. The numerical implementation is described in detail by Lynch & Werner (1991).

Computational set-up

Because the calculated current velocities can be roughly described by simple hydrodynamic laws, the resolution of the unstructured finite-element grid is set proportional to both bathymetry and the bathymetric gradient. Generally triangles are smaller in regions with shallow water depth and where the gradients are large (Fig. 6). The variable mesh is especially useful in coastal regions where the water depth varies greatly. The minimum size of the elements is roughly 84 000 m^2 and the maximum size is roughly 23E6 m^2. The model has 20 horizontal layers. The vertical spacing of the layers is not uniform. Layers are closer together near the top and bottom (Fig. 5). Minimum spacing is roughly 1 m in the bottom boundary layer.

Model setup – boundary conditions

The semi-diurnal lunar tide was chosen, as tidal deposits of the OMM show mostly a 14-day neap-spring cycle (e.g. Allen *et al.*, 1985; Tessier & Gigot, 1989). The model is forced exclusively with tidal amplitudes at the open model boundaries (barotropic model). Transport of heat and density (baroclinic), both as forcing due to wind and fresh water point sources (rivers and fan deltas), is so far excluded and a topic of further research.

The simulation is initiated with a timestep of 15 s, four non-linear iterations per timestep and terminated after 12 M_2 cycles, well after dynamical equilibrium has been established. Bottom stress (Table 2) was employed using a non-linear quadratic form with a drag coefficient value C_c. The drag coefficient value C_c has been discussed to range between 0.0025 and 0.0050 (see Sündermann & Lenz, 1983). Most authors have chosen a C_c value of 0.0025. Recently, Werner *et al.* (2003) published drag coefficient values of $3.0 \pm 0.1 \times 10^{-3}$ measured at Georges Bank. Taking a C_d of 0.005, we used a rather high value to account for relatively high bed roughness due to tidal bedforms (see also Martel *et al.*, 1994).

Crumeyrolle *et al.* (1991) proposed a macrotidal regime for the Digne area in southeast France. Initial simulation results showed that this can be reproduced if a tidal amplitude of 2 m is used at the southern open boundary of the model. On the eastern boundary of the model, mesotidal ranges (2–4 m) are assumed based on Faupl & Roetzel (1987) and Krenmayr & Roetzel (1996). During our simulations we tested tidal amplitudes of 2 and 1 m (see Figs 9 and 12). The calculated residual velocities provide the net vector of currents within one tidal cycle and thus give information on the time-averaged and effective palaeotransport directions. Using systematically different phase lags between the two M_2 constituents at either open boundary, we studied how magnitude and direction of the residual velocities develop on a

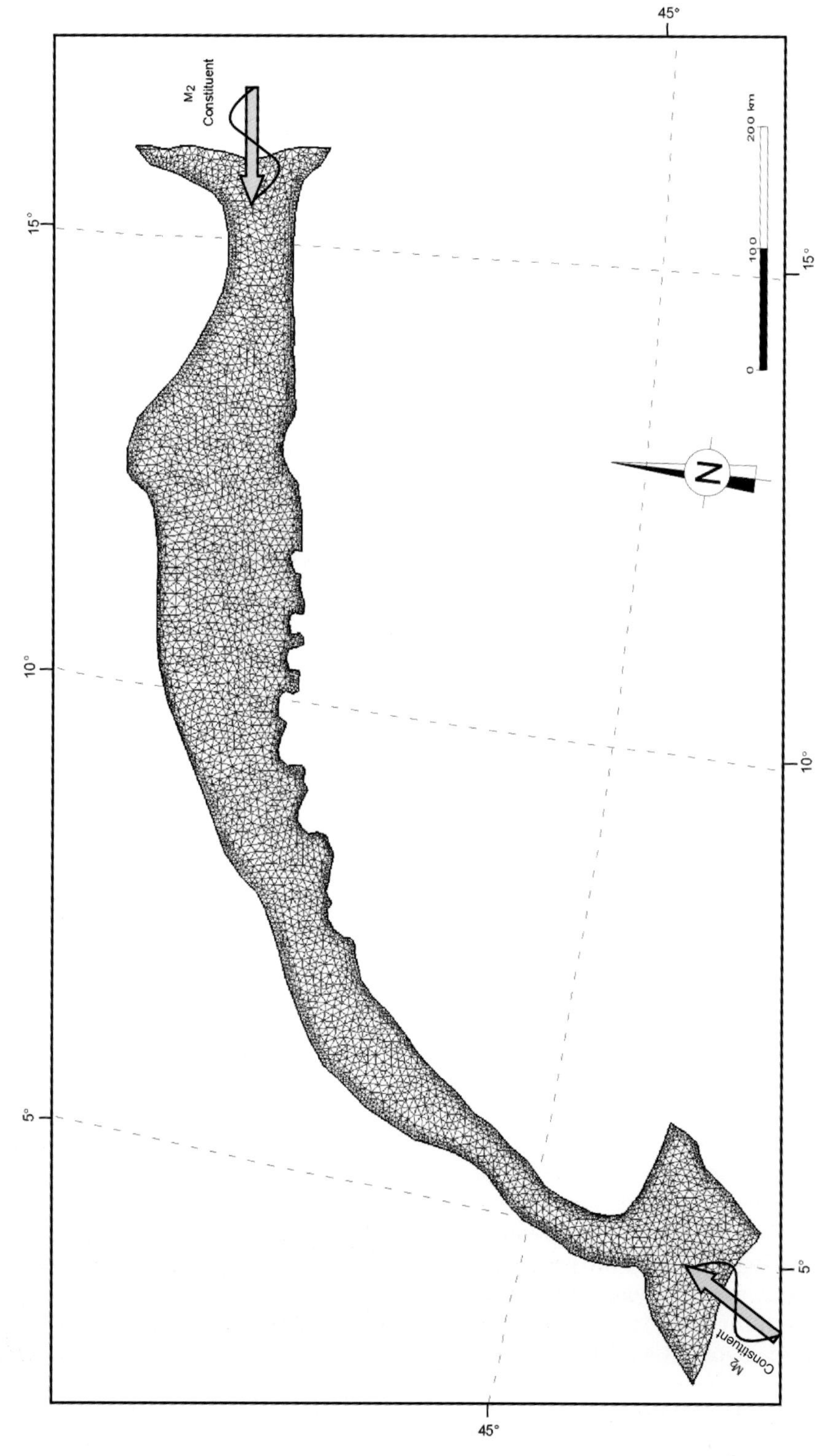

Fig. 6. Horizontal finite element mesh developed for the Burdigalian seaway using linear triangles. *Note:* coastal areas are meshed with a higher resolution than deeper areas, since the triangulation of the unstructured mesh depends on water depth.

Table 2. Listing of a typical parameter set for the tidal model of the Burdigalian Seaway.

Model parameter	Value
Gravity	$9.806\,m^2\,s^{-1}$
Latitude	47°
Coriolis	$0.1067 \times e^{-4}$
	$2.0 \times e^{-4}$
Implicit time weighting factor for vertical diffusion of momentum	0.5
Vertical diffusion of heat and salt	1.0
Vertical diffusion of turbulence	1.0
	0.75
Quadratic bottom stress drag coefficient	$5.0 \times e^{-3}$
Minimum bottom stress coefficient	$1.0 \times e^{-5}$
Smagorinsky coefficient	0.28
Reference temperature	10.0°C
Reference salinity	32 psu
Reference q^2	$1 \times e^{-3}$
Reference $q^2 l$	$1 \times e^{-3}$
Minimum vertical viscosity	$2.0 \times e^{-3}$
Minimum value for $q^2, q^2 l, l$	$1 \times e^{-6}$

basinwide-scale and how far a simple transport pattern may have existed.

Some semi-quantitative considerations about tidal waves

To understand the tidal amplitudes (and tidal ranges) and the complicated pattern of residual current velocities in the Burdigalian Seaway (see Figs 9–13), we start with some basic considerations about tidal waves in a simple set-up. First we consider a one-dimensional case of two propagating two-dimensional, semi-diurnal, lunar M_2 waves, travelling towards each other (Fig. 7). In the presented case both incident waves are initiated in-phase (for an out-of-phase example see Fig. 13). The length of the basin l is a multiple of the quarter wavelength (Sztanó & de Boer, 1995) of the M_2 tidal wave. When basin length is equivalent to 5/4 of the M_2 wavelength, the two propagating waves superimpose on each other, generating a standing wave (Fig. 7A) with two nodes and three antinodes. We obtain a constructive interference between both propagating waves with a maximum amplitude in the middle of the travel line (in our case located near Lake Constance). In this simplified case friction is neglected. Considering friction, the tidal amplitudes and velocity are damped with ongoing travel time. In the 2D case (Fig. 7A) this leads to a significant modification of the tidal system.

In wide gulfs and seaways not located at the Equator, the Earth rotation may produce amphidromic systems; both incoming waves are deflected to the right due to the Coriolis Force (Northern Hemisphere) and the resulting pattern is known as a Kelvin wave. The effect of the Coriolis Force is to convert a standing wave into a rotary wave (Fig. 7B). As the high-tide crest of a wave moves up the channel, the flow of water is deflected to the right, causing a higher tide on the right bank and a lower tide on the left (Fig. 7C). During low tide or at the high-tide maximum of the counter-propagating wave, the flow of water reverses but is again reflected to the right of its path, so that there is now a weaker ebb-phase on the left bank and a stronger ebb-phase on the right (e.g. Pugh, 1987). The net result is a greater range of tidal amplitudes on the right side of the seaway than on the left.

To accommodate a rotary wave and thus to form an amphidromic system, a basin should have a certain width. The so-called internal Rossby radius of deformation and the average water depth determine whether an amphidromic system can develop (Fig. 8). For an averaged basin depth of 50 m, which was used for our study, the seaway should have a width of approximately 200 km to accommodate a complete rotary system. Due to strong frictional effects the amphidromic points are shifted from the centre of the seaway to the border. The theoretical ideas of progressive, standing/Kelvin waves and amphidromic systems are now applied to describe the dynamics of the simulated tides in the Burdigalian Seaway.

Numerical simulation results

Reference model – 2 incident M_2 waves of 2 m amplitude (in-phase)

In accordance with the above geological constraints we used two sinusoidal M_2 waves ($T = 12{,}42$ h) with an amplitude of 2 m being initiated in-phase at the open model boundaries.

Tidal amplitudes. As explained in the previous section the most significant feature of the tidal system under the given geometric boundary conditions is the build-up of several joined half circles, rotating around two amphidromic points (Fig. 9). The simulation shows these two amphidromic points, with no tidal range, one near Lake Neuenburg in Switzerland and the other in

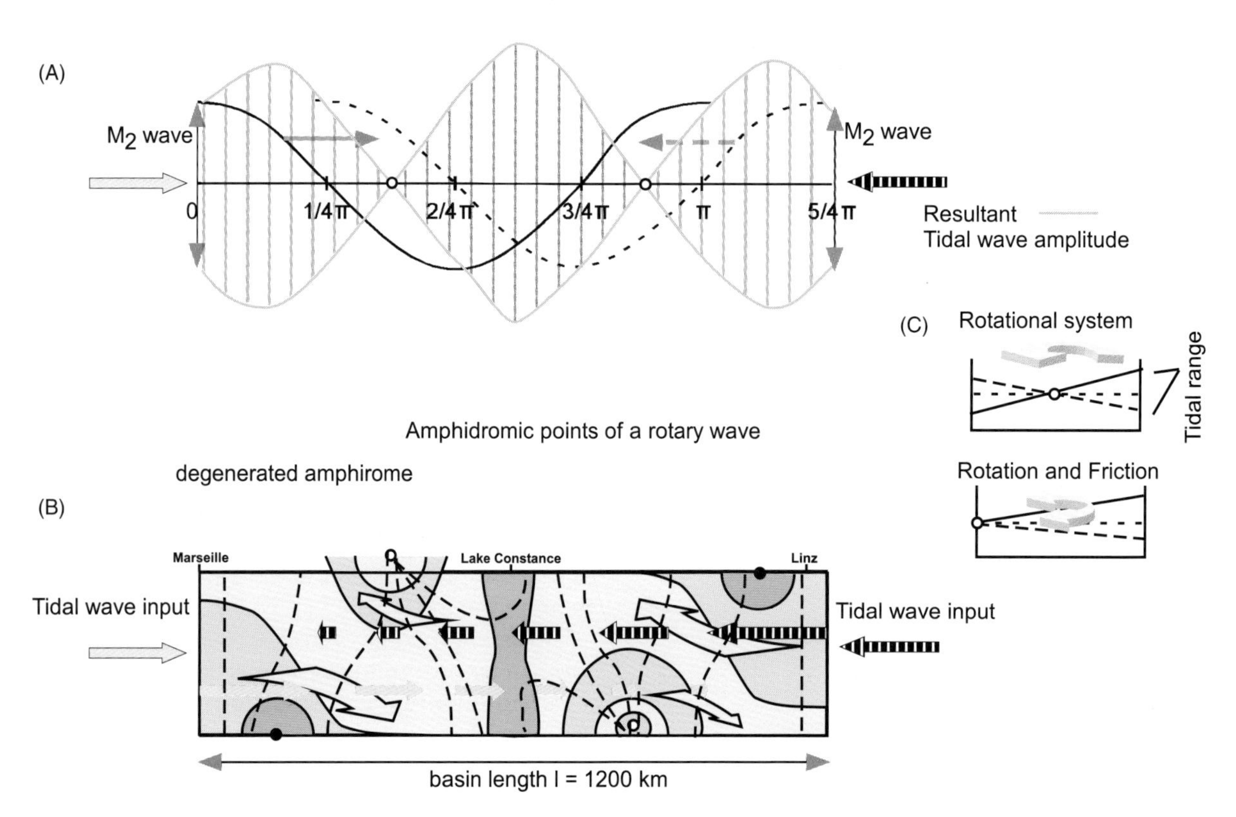

Fig. 7. Simplified models of the Burdigalian Seaway, assuming two propagating M_2 waves from opposite directions. In a one-dimensional case (A) a standing wave (red line) of two superimposing waves forms with two nodal points. In a two-dimensional model (B) a rotary system develops, where the propagating Kelvin waves (yellow arrows) rotate anti-clockwise around two amphidromic points. Friction leads to a shift of the amphidromic point (C). Model (B) shows the combined effects of a standing wave within a rotary system and includes friction. Contours indicate co-range, dashed lines indicate co-tidal lines of constant phase.

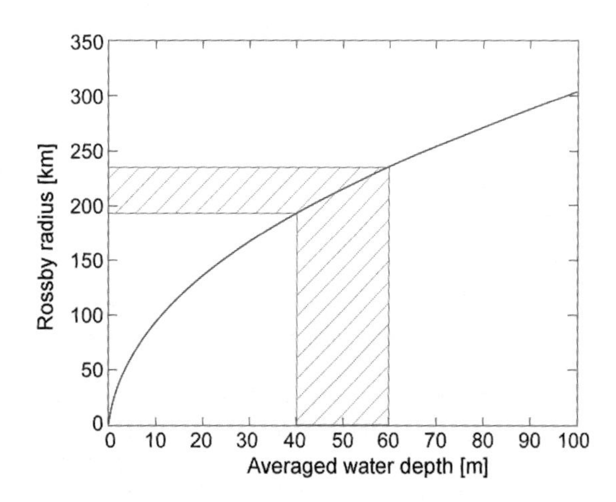

Fig. 8. The Rossby radius of deformation R describes the ideal width and depth of a basin for the development of an amphidromic system; $R = (\sqrt{g} \times d)/f$, where g is acceleration of gravity ($9.81\,\mathrm{m\,s^{-2}}$), d is average water depth, and f is the Coriolis parameter at 45° latitude ($10.3 \times 10^{-5}\,\mathrm{s^{-1}}$).

front of the Nesselburg fan deltas near Bad Tölz in southern Germany. In the Swiss part of the seaway the amphidromic point is located outside the left-hand boundary of the basin and hence the co-tidal lines are focusing on an inland point; this situation is called a degenerated amphidrome. The simulation shows highest amplitudes of up to 3 m in the area of Linz and south of Grenoble. As previously described, the highest tidal amplitudes in these areas result from the deflection of both propagating M_2 waves to the right due to the Coriolis Force. Additionally, a funnelling effect enhances the amplification of tidal amplitudes in these regions due to the northeast directed tidal wave. A rotary amphidromic system, postulated by Salvermoser (1999) for Ottnangian tidal deposits in Bavaria and Austria does not evolve from the model. However, considering residual current directions within this region we can explain his observations. In the model, residual currents are aligned in a nearly

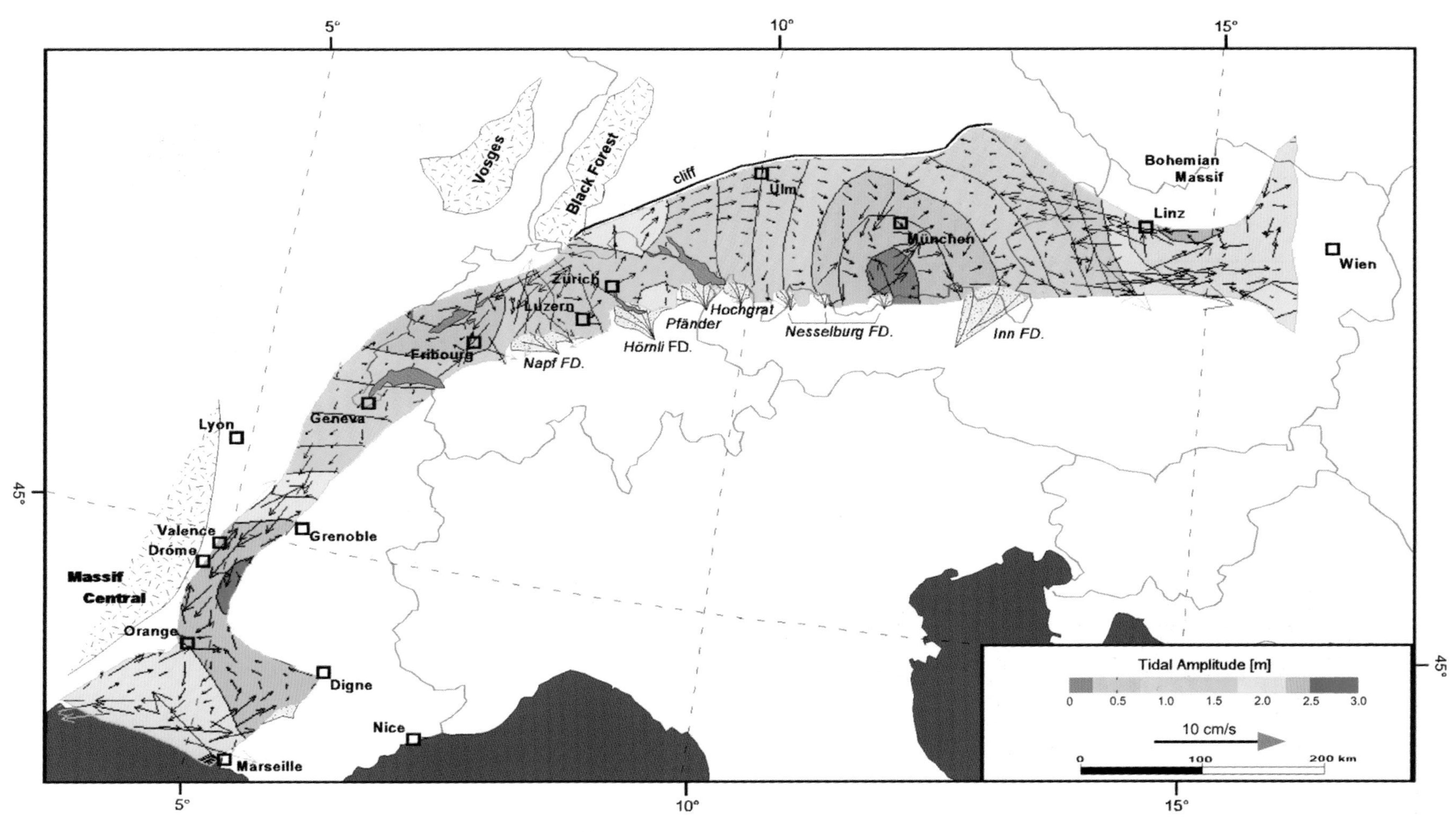

Fig. 9. Celerity of the vertically averaged residual currents (cm s^{-1}), interpreted as the net transport directions of one ebb–flood cycle after the dynamic equilibrium in the reference model has been reached. Contour lines indicate tidal amplitudes (m). The model has been forced with two propagating M_2 waves of 2 m amplitude. Coloured areas indicate tidal amplitudes (m). FD, fan delta.

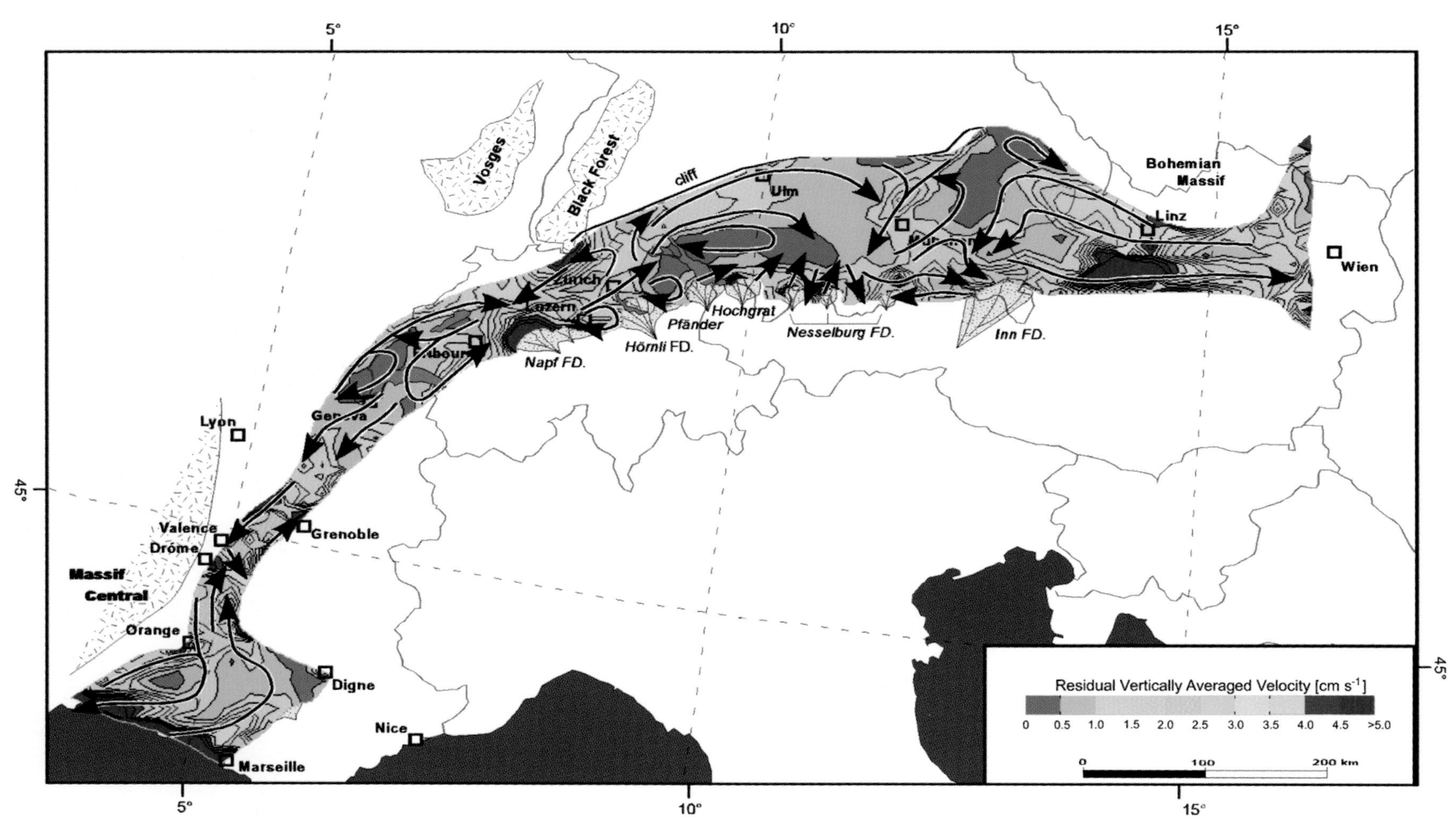

Fig. 10. Generalized mesoscale currents within the Burdigalian Seaway calculated for the reference model. Contour lines indicate the absolute value of vertically averaged residual velocities [cm s^{-1}]. FD, fan delta.

circular way in Bavaria. This results from a transition from in- to outflow currents in the course of the near-open boundary in Austria.

High to mesotidal amplitudes, 1.5 to 2 m, are simulated in the middle of the model domain, near Lake Constance, where the two counter-propagating M_2 waves interfere in a constructive way. Maximum tidal amplitudes within this zone are found in the model between the Napf and Hörnli fan deltas, showing the tidal dominance within this area.

To the west, in the Rhône-Alp region, the banks of the seaway are too close together for a rotary tidal system to develop. Hence co-tidal lines are aligned at a right angle to the coastline.

Residual velocities. After the simulation reached its dynamic equilibrium (after eight tidal cycles), vertically averaged residual current velocities of one ebb–flood cycle were calculated. Residual velocities show the asymmetry between ebb and flood and thus net movement of bedforms. In the given model these residual current velocities (Figs 9 and 10) are interpreted to represent net-transport directions of suspended sediment, and in the case where the residual tidal currents are strong enough, then bedload transport is reached. Residual current velocities of our simulations are within a range of 1–10 cm s^{-1}. Compared with peak tidal/ebb velocities (Fig. 11), residual velocities are in general <10% of the tidal (peak) current velocities.

In the tube-like conditions of the Rhône-Alp region, the model shows strong residual currents of the order of 5–7 cm s^{-1}, with a residual flow towards the south (Figs 9 and 10). In this area Allen & Bass (1993) interpreted palaeocurrents with a complex pattern and without a dominant direction. In the simulation the Rhône area acts like a bottleneck. Ebb and flood tidal currents have to pass this area and no dominant residual current direction can be established. In this area flood and ebb produce strong north- and south-directed currents (Fig. 11) causing strong turbulence and local eddies.

In the vicinity of Geneva, residual velocities are very weak, of the order of 1–2 cm s^{-1}. Residual currents are mainly directed towards the south. Towards the northeast, up to Fribourg, the system is characterized by two counter-propagating current systems and residual currents that are directed to the east. Within the seaway highest residual velocities are found in the vicinity of the Napf fan delta and at the opposite coastline, approximately 5–7 cm s^{-1}. The bays between the Napf, Hörnli and the Pfänder fan delta are characterized by two clockwise rotating cells. North of the Napf fan delta convergent residual currents have been modelled, forming a zone of bedload convergence. Northwest of Lake Constance (Figs 9 and 10), within the area of constructive interference of the two counter-propagating M_2 waves (compare with Fig. 7), a zone of divergent, strong residual currents is found, with bedload parting (BLP). In summary, no general transport direction is modelled for the Swiss part of the seaway, rather there are distinct circulation cells, which might be closed to sediment transport (see also Allen *et al.*, 1985).

In front of the Hochgrat and Nesselburg fan deltas, east of Lake Constance, off- and onshore directed residual currents are observed in the model. The northern part of the German seaway is characterized by northeast-directed residual currents, forming a wide semi-circle with a small-scale vortex in the middle part of the basin (Fig. 10). Residual currents are alongshore towards the northern coastline. In the eastern part of the German Molasse Basin, residual vectors are directed southwestward, forming a strong elongated counterclockwise rotating eddy. Residual currents in Austria are characterized by west- to northwest-directed residual currents (~3 cm s^{-1}) at the northern coastline and east- to northeast-directed currents (~5 cm s^{-1}) at the southern coastline. Overall we observe a net flow towards the Paratethys. The transport from the Swiss area to the southwest can be neglected.

Peak currents. Comparing peak currents of flood and ebb (Fig. 11A and B) we observe that regions of two maximal co-oscillating counter-propagating M_2 waves (Fig. 7) act as a divide for ebb and flood flow directions. Outside this 'ebb–flood divide' current branches continue on both sides in opposite directions. Regions with the strongest asymmetry between ebb and flood have the highest residual velocities. For example, in the vicinity of Fribourg strong ebb currents are of the order of 50–60 cm s^{-1}, clearly stronger than the flood currents of 40–50 cm s^{-1}. Thus, there is a high asymmetry of tides. On the other hand, while highest residual velocities occur next to the southern and northern coastline west of Lake Constance, highest ebb/flood bottom velocities are aligned in the centre of the seaway parallel to its axis. Two exceptions are found, one in the vicinity

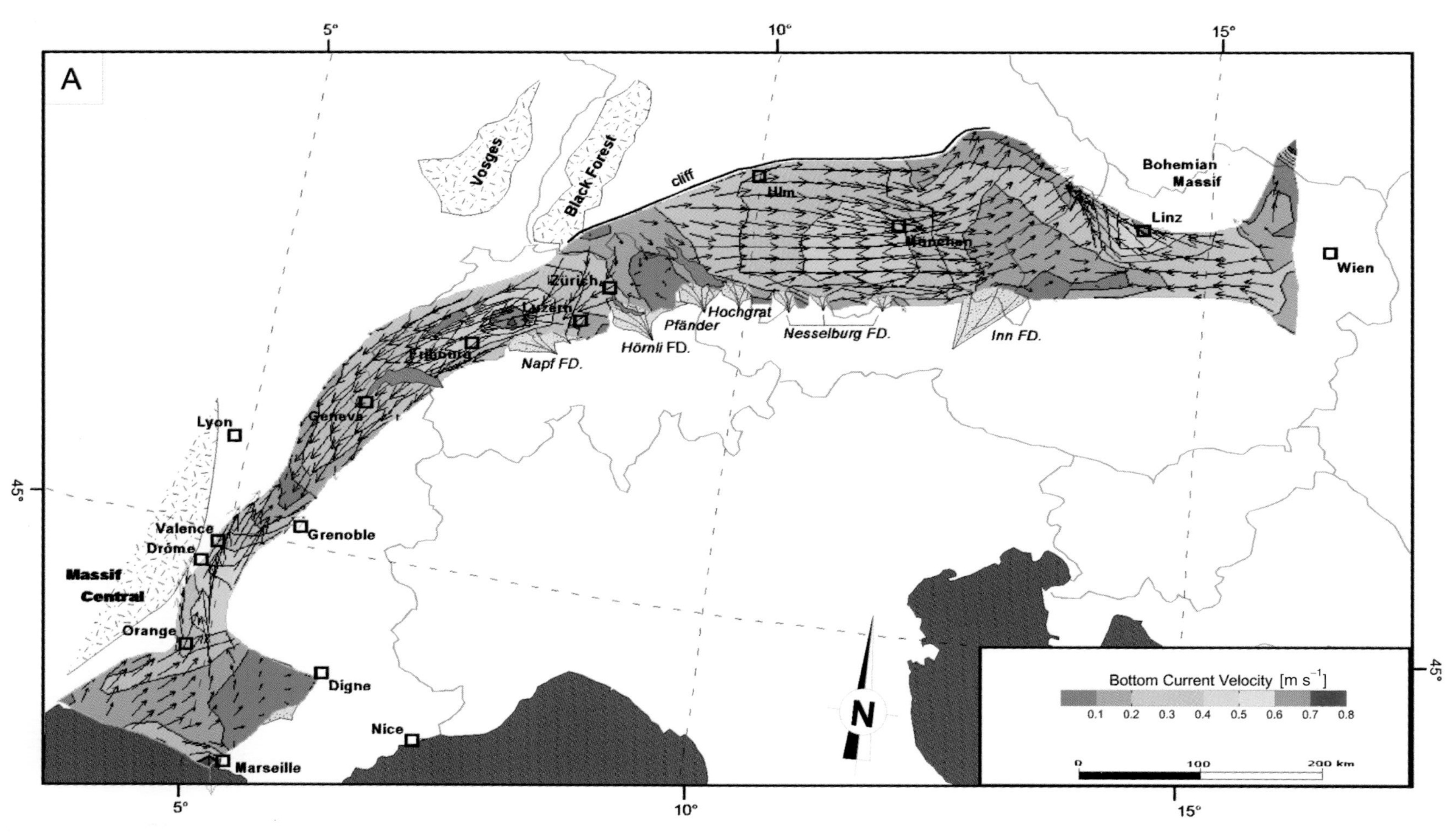

Fig. 11. Simulated bottom velocities during (A) incoming flood and (B) incoming ebb in the reference model. FD, fan delta.

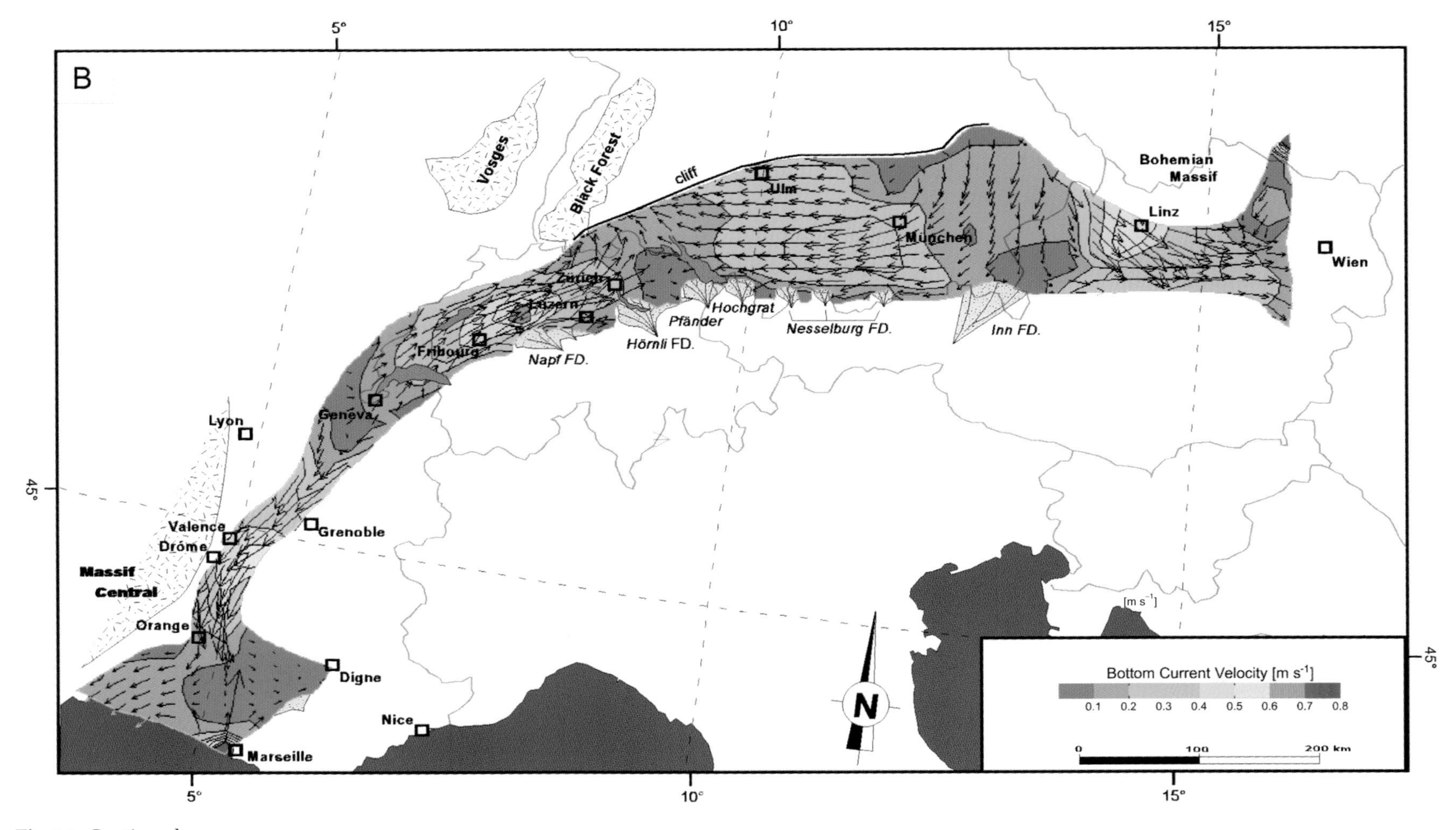

Fig. 11. Continued.

of Linz and the other in the Rhône-Alp region. Here both highest residual velocities and high ebb/flood bottom currents occur nearly at the same place. In the vicinity of Linz, flood currents are stronger than ebb currents, producing an asymmetry towards the flood tide. Proposed flood peak current velocities in Bavaria of 0.5 m s^{-1} (Salvermoser, 1999) and 0.75 m s^{-1} in Austria (Faupl & Roetzel, 1987) are reproduced by the simulation.

Alternative models

In the previous chapter we discussed tidal amplitudes and velocity distributions for in-phase excitation with two co-oscillating M_2 waves having amplitudes of 2 m. The following simulations evaluate alternative out-of-phase excitation and varying M_2 tidal amplitudes for the two M_2 waves, in order to provide insight into the robustness of the model in terms of amplitudes, current strength and flow directions. We also review if it is possible to initiate a southwest directed net transport (Kuhlemann, 2000) by modifying the reference model.

Alternative 1: one incident M_2 wave, initiated at the southwestern gateway. Sztanó (1995) and Sztanó & de Boer (1995) proposed that there was a narrow connection towards the Paratethys, thus the peri-Alpine foredeep acted like a semi-enclosed basin, reflecting the tidal wave coming from the southwest at its eastern boundary (shoreline of the Bohemian Massif). To simplify the numerical problem, we closed the model at its eastern boundary (Fig. 12). As the basin length is a multiple of the quarter M_2 wavelength this geometry should allow a standing wave to develop, as suggested by Sztanó & de Boer (1995). The calculated principal resonant amplification with an antinode at the head of the basin is shown in Fig. 12. However, frictional processes dampen the tidal wave over the 1200 km from southeast France to Austria. Consequently, the generated tidal system within the seaway exhibits only microtidal ranges. As described above, the observed deposits in Austria, Germany and Switzerland (Fig. 2) are interpreted to be deposited under at least mesotidal conditions (e.g. Homewood *et al.*, 1985). Thus we conclude that an incoming tidal wave from the northeast is necessary to generate a mesotidal system within the seaway (see reference model).

Alternative 2: changing the M_2 amplitude. Reducing the M_2 amplitude for the northern open boundary to 1 m causes nearly the same distribution of tidal amplitudes as in the reference model (cf. Figs 11 and 13). A remarkable difference is a general decrease of tidal amplitudes by about 0.5 m within the whole seaway. Residual velocities show nearly the same directions. Noticeable differences to the reference model are diverting residual currents in the Rhône area suggesting a bedload partition zone in this region and a strong eastward-directed residual current along the southern coastline of the Austrian Molasse basin.

Alternative 3: phase lag of 180° between the two M_2 waves. In another test we excited the tidal model with two waves of the same amplitude (2 m), but with a phase difference of 180°. The model (Fig. 14) reacts as the aforementioned, simplified considerations (Fig. 7) would suggest. The two counter-propagating waves counteract destructively in the region of maximum amplification in the reference model, generating an amphidromic point with nearly stationary conditions (Fig. 7A and B). Wave crest and trough are superimposed destructively. The resulting basin-wide tidal ranges are within 1–1.25 m, indicating micro- to low mesotidal conditions. Such a situation contradicts the described sedimentary record in the Burdigalian Seaway (e.g. Homewood *et al.*, 1985). In addition unrealistic ‘no-flow’ conditions in the vicinity of Fribourg are modelled that do not fit the observed data.

CONCLUSIONS

We present a tidal model for the Burdigalian Seaway spanning from southern France to Austria. The model is based on and compared with a palaeogeographical and palaeobathymetric restoration of the seaway during maximum flooding in the Ottnangian. The simulations show that tides in the Burdigalian Seaway depended on the nature of the tides in the adjacent seas with which they co-oscillated. The tidal amplitudes and currents in the basin are a result of the combination of progressive, standing and rotary waves, including modifying effects of friction and reflection. In our reference model the best fit to the sedimentological observations is achieved by two 2 m in-phase M_2 tides, acting at the open model boundaries. The resulting two propagating tidal waves

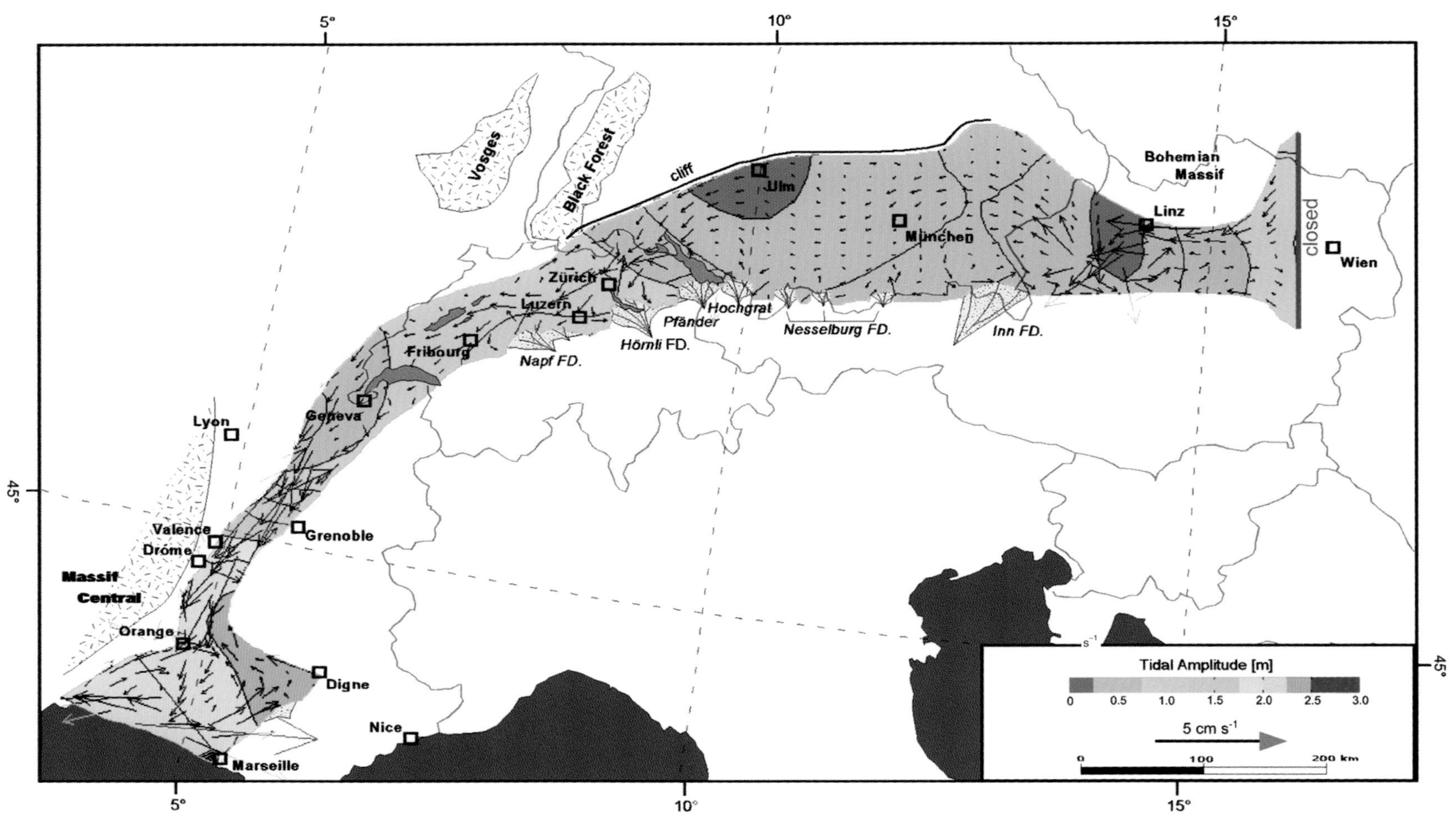

Fig. 12. Residual, vertically averaged velocities and tidal amplitudes of an alternative tidal model being forced by only one incident M_2 wave from the southwest. FD, fan delta.

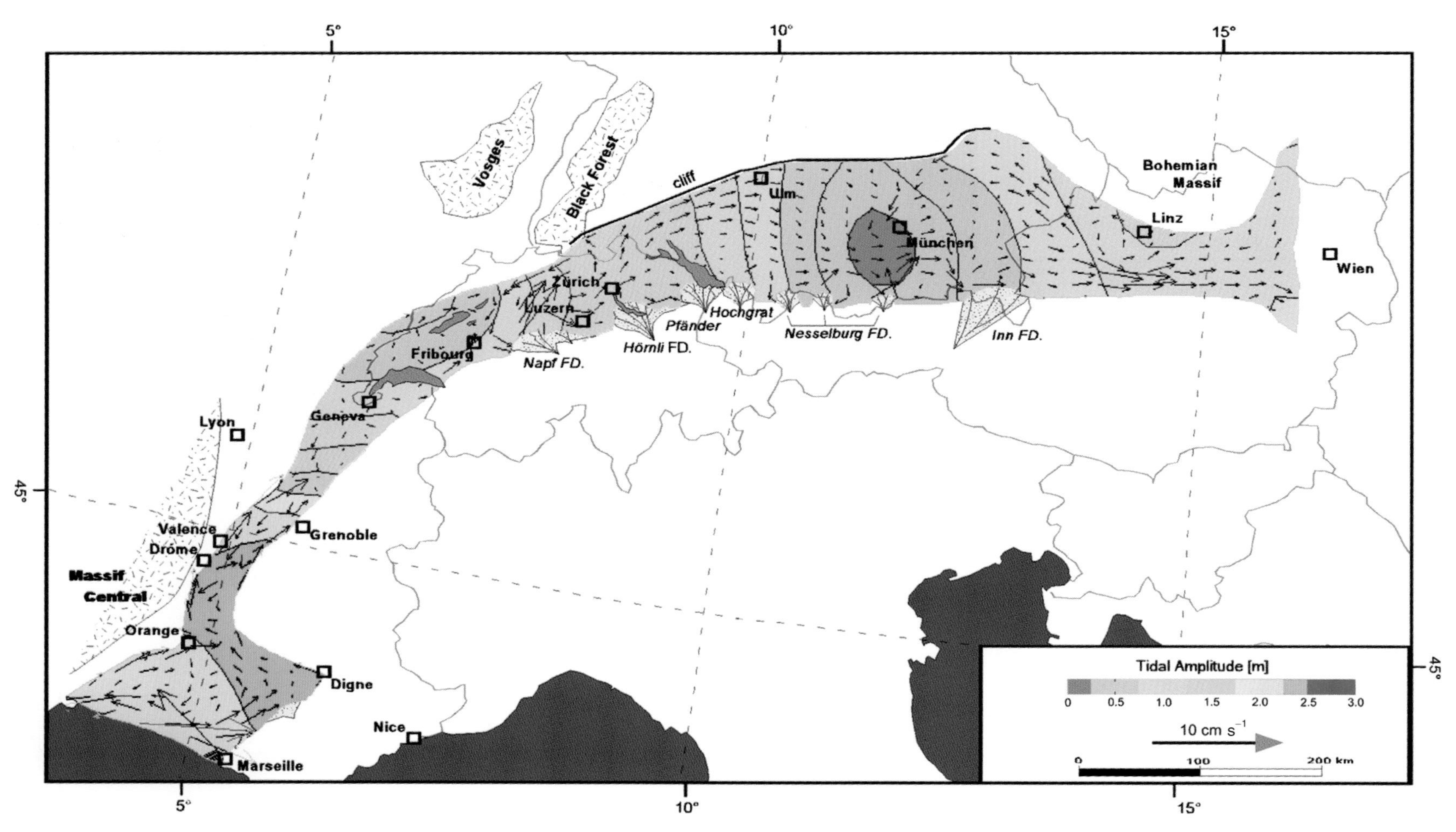

Fig. 13. Residual, vertically averaged velocities and tidal amplitudes of an alternative tidal model being forced by two incident M_2 waves. The M_2 constituent from the south was initiated in phase with an amplitude of 2 m, whereas the northern M_2 wave was initiated with 1 m. FD, fan delta.

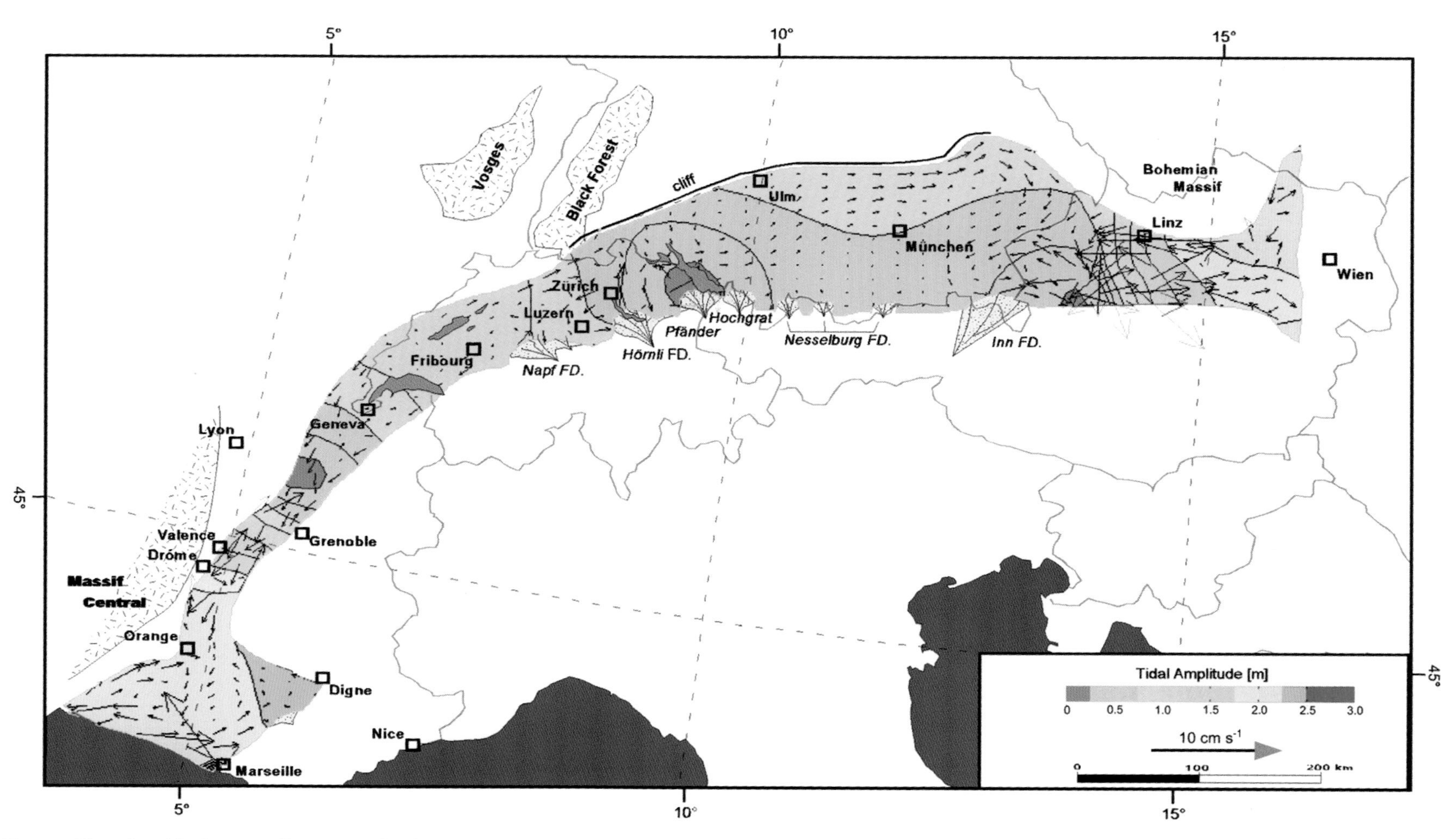

Fig. 14. Plot of residual, vertically averaged velocities and tidal amplitudes of a second alternative tidal model after reaching the equilibrium state. Both incident M_2 tidal waves have been initiated with an amplitude of 2 m, but out-phase with a phase lag of 180°. FD, fan delta.

cause two rotary amphidromic systems in the seaway. Simulated tidal amplitudes indicate high mesotidal ranges for the Swiss area (3–4 m), in agreement with outcrop descriptions and previous simulations by Martel *et al.* (1994), when applying incident tidal waves at the eastern and western boundaries of the seaway. The averaged residual velocities of an ebb–flood cycle are about ten times smaller than ebb/flood peak velocities. Based on the time evolution of velocities it was shown that there are locally symmetric and asymmetric tides depending on the duration and intensity of ebb and flood currents in the basin. Net transport directions follow a complicated pattern of vortices, convergent and divergent currents, similar to what was proposed by Allen *et al.* (1985). The tube-like transfer zone of tidal waves in the Rhône-Alp area is characterized by a diffusive pattern of net flow directions and strengthens with a subordinate flow direction to the south. The simulations so far do not show a dominant southwest transport of sediments as proposed by Kuhlemann (2000). Peak current velocities have been simulated up to 0.8 m s^{-1}. These values compare well with modern tide-influenced seas, where active migration of sand waves occurs (e.g. Dalrymple *et al.*, 1990). Although residual velocities seem to be very small (1–8 cm s^{-1}), recent examples (e.g. Harris, 1991) have shown, that these may still have a significant influence on the net sediment transport.

ACKNOWLEDGEMENTS

Finanacial support was provided by the German Research Foundation (SU 242/3-1). P.L de Boer and in particular O. Sztanó are thanked for their valuable suggestions that substantially improved the manuscript.

REFERENCES

Allen, P.A. (1984) Reconstruction of ancient sea conditions with an example from the Swiss Molasse. *Mar. Geol.*, **60,** 455–473.

Allen, P.A. and **Bass, J.P.** (1993) Sedimentology of the upper marine molasse of the Rhône-Alp region, Eastern France: Implications for basin evolution. *Eclogae Geol. Helv.*, **86,** 121–171.

Allen, P.A. and **Homewood, P.** (1984) Evolution and mechanics of a Miocene tidal sandwave. *Sedimentology*, **31,** 63–81.

Allen, P.A., Mange-Rajetzky, M., Matter, A. and **Homewood, P.** (1985) Dynamic palaeogeography of the open Burdigalian seaway, Swiss Molasse basin. *Eclogae Geol. Helv.*, **78,** 351–381.

Assemat, S. (1991) Récherche et analyse des tidalites dans la molasse marine marine miocène du domaine subalpin, dans la région de Frangy-Rumilly-Aix-les-Bains. *Mém. D.E.A. Univ. de Savoie, Chambéry.*

Bachmann, G.H. and **Müller, M.** (1991) The Molasse Basin, Germany: Evolution of a classic petroliferous foreland basin. In: *Generation, Accumulation, and Production of Europe's Hydrocarbons* (Ed. A.M. Spencer), *Special Publication of the European Association of Petroleum Geoscientists*, **1,** 263–276. EAPG (European Association of Petroleum Geoscientists).

Beck, C., Deville, E., Blanc, E., Phillipe, Y. and **Tardy, M.** (1998) Horizontal shortening control of Middle Miocene marine siliciclastic accumulation (Upper Marine Molasse) in the southern termination of the Savoy Molasse Basin (northwestern Alps/southern Jura). In: *Cenozoic Foreland Basins of Western Europe* (Eds A. Mascle, C. Puigdefàbregas, H. Luterbacher and M. Fernàndez), *Geol. Soc. Sp. Publ.*, **134,** 263–278.

Berger, J.-P. (1983) Biostratigraphie de la transgression de la molasse marine superieure (OMM) en Suisse occidentale. *Eclogae Geol. Helv.*, **76,** 729–732.

Berger, J.-P. (1996) Cartes paléogéographiques-palinspastiques du bassin molassique suisse (Oligocène inférieur – Miocène moyen). *Neues Jb. Geol. Paläontol. Abh.*, **202,** 1–44.

Berggren, W.A., Kent, D.V., Swisher, C.C. and **Aubry, M.-P.** (1995) A revised Cenozoic geochronology and chronostratigraphy. In: *Geochronology, Time Scales and Global Stratigraphic Correlations* (Eds W.A. Berggren, D.V. Kent, M.-P. Aubry and J. Hardenbol), *SEPM, Spec. Publ.*, **54,** 17–28, Tulsa.

Bieg, U., Thomas, M., Cordero, F., Süss, M.P. and **Kuhlemann, J.** (2004) Global M2 simulation for the Early Miocene delivering border conditions for a forthcoming tidal simulation of the circum-mediterranean realm. *Zbl. Geol. Paläont. Teil 1*, **3(4),** 259–276.

Blumberg, A.F., Galperin, B. and **O'Connor, D.J.** (1992) Modeling vertical structure of open-channel flows. *ASCE, J. Hydraul. Eng.*, **118,** 1119–1134.

Burkhard, M. and **Sommaruga, A.** (1998) Evolution of the western Swiss Molasse basin: Structural relations with the Alps and the Jura belt. In: *Cenozoic Foreland Basins of Western Europe* (Eds A. Mascle, C. Puigdefàbregas, H.P. Luterbacher and M. Fernàndez), *Geol. Soc. Spec. Publ.*, **134,** 279–298.

Büchi, U.P. and **Hofmann, F.** (1960) Die Sedimentationsverhältnisse zur Zeit der Muschelsandsteine und Grobkalke im Gebiet des Beckennordrandes der Oberen Meeresmolasse zwischen Aarau und Schaffhausen. *Bulletin der Vereinigung Schweizerischer Petroleum-Geologen und -Ingenieure*, **27,** 11–22.

Couëffé, R., Tessier, B., Gigot, P. and **Beaudoin, B.** (2004) Tidal rhythmites as possible indicators of a very rapid subsidence in a Foreland Basin: An example from the Miocene Marine Molasse Formation of the Digne Foreland Basin, SE France. *J. Sed. Res.*, **74,** 746–759.

Crumeyrolle, P., Rubinio, J.-L. and **Clauzon, G.** (1991) Miocene depositional sequences within a tectonically-controlled transgressive–regressive cycle. In: *Spec. Publ. Int. Assoc. Sedimentol.* (Ed. D.I.M. McDonald), **12**, pp. 373–390.

Dalrymple, R.W., Knight, R.J., Zaitlin, B.A. and **Middleton, R.V.** (1990) Dynamics and facies model of a macro-tidal sand-bar complex, Cobequid Bay-Salmon River Estuary (Bay of Fundy). *Sedimentology*, **37**, 577–612.

Debrand-Passard, S., Courbouleix, S. and **Lienhardt, M.J.** (1984) Synthese geologique du sud-est de la France. Geologic synthesis of southeastern France. *Memoires du B.R.G.M.*, **125–126**, 615.

Demarcq, G. (1970) Etude stratigraphique du Miocène rhodanien. *Mem. BRGM*, **61**, 257.

Demarcq, G. and **Perriaux, J.** (1984) Neogene. In: *Synthèse géologique du Sud-Est de la France: Stratigraphie et paléogeographie* (Eds S. Debrand-Passard and S. Courbouleix), *Bureau de Récherches Géologiques et Minières, Mémoire*, **125**, 469–519.

Diem, B. (1985) Analytical method for estimating palaeowave climate and water depth from wave ripple marks. *Sedimentology*, **32**, 705–720.

Doppler, G. (1989) Zur Stratigraphie der nördlichen Vorlandmolasse in Bayrisch-Schwaben. *Geol. Bavarica*, **94**, 83–133.

Faupl, P. and **Roetzel, R.** (1987) Gezeitenbeeinflußte Ablagerungen der Innviertler Gruppe (Ottnangien) in der oberösterreichischen Molassezone. *Jb. Geol. Bundesanst. Wien*, **130**, 415–447.

Faupl, P. and **Roetzel, R.** (1990) Die Phosphoritsande und Fossilreichen Grobsande; Gezeitenbeeinflusste Ablagerungen der Innviertler Gruppe (Ottnangien) in der oberoesterreichischen Molassezone. *Jb. Geol. Bundesanst. Wien*, **132**, 157–180.

Ford, M., Lickorish, W.H. and **Kusznir, N.J.** (1999) Tertiary foreland sedimentation in the Southern Subalpine Chains, SE France: A geodynamical appraisal. *Basin Res.*, **11**, 315–336.

Freudenberger, W. and **Schwerd, K.** (1996) *Erläuterungen zur Geologischen Karte von Bayern 1:500000.* Bayerisches Geologisches Landesamt, München.

Füchtbauer, H. (1967) Die Sandsteine in der Molasse nördlich der Alpen. *Geol. Rundsch.*, **56**, 266–300.

Gall, H. (1974) Neue Daten zum Verlauf der Klifflinie der oberen Meeresmolasse (Helvet) im südlichen Vorries. *Mitt. Bayer. Stattssamml. Paläont. hist. Geol.*, **14**, 81–101.

Gall, H. (1975) Der III. Zyklus der oberen Meeresmolasse (Helvet) am Südrand der Schwäbisch-Fränkischen Alb. *Mitt. Bayer. Stattssamml. Paläont. hist. Geol.*, **15**, 179–205.

Galperin, B., Kantha, L.H., Hassid, S. and **Rosati, A.** (1988) A quasi-equilibrium turbulent energy model for geophysical flows. *J. Atmos. Sci.*, **45**, 55–62.

Guellec, S., Mugnier, J.L., Tardy, M. and **Roure, F.** (1990) Neogene evolution of the western alpine foreland in the light of ECORS data and balanced cross-section. In: *Deep Structure of the Alps* (Eds F. Roure, P. Heitzmann and R. Polino), *Mémoire du Société Géologique de France*, **156**, 165–184.

Haq, B.U., Hardenbol, J. and **Vail, P.R.** (1988) Mesozoic and Cenozoic chronostratigraphy and cycles of sea-level change. In: *Sea-level Changes: An Integrated Approach* (Eds C.K. Wilgus, B.S. Hastings, C.G.S.C. Kendall, H.W. Posamentier, C.A. Ross and J.C. van Wagoner), *SEPM Spec. Publ.*, **42**, pp. 71–108.

Harris, R.T. (1991) Reversal of subtidal dune asymmetries caused by seasonally reversing wind driven current in Torres Strait, Australia. *Cont. Shelf Res.*, **11(7)**, 979–1003.

Hofmann, F. (1976) Überblick über die geologische Entwicklungsgeschichte der Region Schaffhausen seit dem Ende der Jurazeit. *Bull. Ver. Schweizer Petrol.-Geol. u. -Ing.*, **42**, 1–16.

Homewood, P. and **Allen, P.** (1981) Wave-, tide-, and current-controlled sandbodies of Miocene molasse, Western Switzerland. *AAPG Bull.*, **65**, 2534–2545.

Homewood, P., Allen, P.A. and **Yang, C.S.** (1985) Palaeotidal range estimates from the Miocene Molasse. In: *Abstr. 6th Int. Ass. Sediment. Reg. Mtg.*, pp. 200–201, Lleida.

Homewood, P., Allen, P.A. and **Williams, G.D.** (1986) Dynamics of the Molasse Basin of western Switzerland. In: *Foreland Basins* (Eds P.A. Allen and P. Homewood), *Int. Assoc. Sedimentol. Spec. Publ.*, **8**, pp. 199–217. Blackwell Scientific Publications, Oxford.

Hülsemann, J. (1955) Großrippeln und Schrägschichtungs-Gefüge im Nordsee Watt und in der Molasse. *Senck. Leth.*, **36**, 359–388.

Keller, B. (1989) *Fazies und Stratigraphie der Oberen Meeresmolasse (Unteres Miozän) zwischen Napf und Bodensee.* Unpubl. PhD Thesis, Universität Bern, Bern, 402 pp.

Krenmayr, H.G. (1991) Sedimentologische Untersuchungen der Vöklaschichten (Innviertler Gruppe, Ottnangian) in der oberösterreichischen Molassezone im Gebiet der Vökla und der Ager. *Jb. Geol. Bundesanst. Wien*, **137**, 83–100.

Krenmayr, H.G. and **Roetzel, R.** (Eds) (1996) *Oligozäne und miozäne Becken- und Gezeitensedimente in der Molassezone Oberösterreichs, Exkursionsführer Sediment '96*, Exkursion B2, **33**. Geol. Bundesanst. Wien, 17 Abb., 43 pp.

Kuhlemann, J. (2000) Post-collisional sediment budget of circum-Alpine basins (Central Europe). *Mem. Sci. Geo. Padova*, **52**, 1–91.

Kuhlemann, J. and **Kempf, O.** (2002) Post-Eocene evolution of the North Alpine Foreland Basin and its response Alpine tectonics. *Sed. Geo.*, **152**, 45–78.

Kuhlemann, J., Frisch, W., Székely, B., Dunkl, I. and **Kàzmér, M.** (2002) Post-collisional sediment budget history of the Alps: Tectonic versus climatic control. *Int. J. Earth Sci.*, **91**, 818–837.

Laubscher, H. (1965) Ein kinematisches Modell der Jurafaltung. *Eclogae Geol. Helv.*, **58**, 231–318.

Laubscher, H. (1992) Jura Kinematics and the Molasse Basin. *Eclogae Geol. Helv.*, **85**, 653–675.

Lemcke, K. (1973) Zur nachpermischen Geschichte des nördlichen Alpenvorlandes. *Geol. Bavarica*, **69**, 5–48.

Lemcke, K. (1988) *Geologie von Bayern. – I. Teil: Das bayerische Alpenvorland vor der Eiszeit*, **I**. E. Schweizerbart'sche Verlagsbuchhandlung, Stuttgart, 175 pp.

Lesueur, J.L., Rubino, J.L. and **Girdaudmaillet, M.** (1990) Organisation et structures internes des dépôts tidaux du Miocène Rhodanien. *Société Géologique de France, Bulletin*, **8**, 49–65.

Lickorish, W.H. and **Ford, M.** (1998) Sequential restoration of the external Alpine Digne thrust system, SE France, constrained by kinematic data and synorogenic sediments. In: *Cenozoic Foreland Basins of Western Europe* (Eds A. Mascle, C. Puigdefàbregas, H.P. Luterbacher and M. Fernàndez), *Geol. Soc. Spec. Publ.*, **134**, pp. 299–323.

Luterbacher, H., Köhler, J. and **Winder, H.** (1992) The northern margin of the Molasse Basin in SW Germany. *Eclogae Geol. Helv.*, **85**, 787–788.

Lynch, D.R. (1982) Unified approach to simulation on deforming elements, with application to phase change. *J. Comp. Phys.*, **47**, 387–411.

Lynch, D.R. and **Naimie, C.E.** (1993) The M_2 tide and its residual on the outer banks of the Gulf of Maine. *J. Phys. Oceanogr.*, **23**, 2222–2252.

Lynch, D.R. and **Werner, F.E.** (1991) Three-dimensional hydrodynamics on finite elements. Part II: Non-linear time-stepping model. *Int. J. Num. Meth. Fluids*, **12**, 507–533.

Lynch, D.R., Ip, J.T.C., Naimie, C.E. and **Werner, F.E.** (1996) Comprehensive coastal circulation model with application to the Gulf of Maine. *Cont. Shelf Res.*, **12**, 37–64.

Lynch, D.R., Holbroke, M.J. and **Naimie, C.E.** (1997) The marine coastal current: Spring climatological circulation. *Cont. Shelf Res.*, **17**, 605–634.

Malzer, O., Rögl, F., Seifert, P., Wagner, L., Wessely, G. and **Brix, F.** (1993) Die Molassezone und deren Untergrund. In: *Eröol und Erdgas in Österreich* (Eds B. Friedrich and S. Ortwin) 2nd edn, pp. 281–322. Naturhistorisches Museum Wien, Wien.

Martel, A.T., Allen, P.A. and **Slingerland, R.** (1994) Use of tidal-circulation modeling in paleogeographical studies: An example from the tertiary of the Alpine perimeter. *Geology*, **22**, 925–928.

Matter, A., Homewood, P., Caron, C., Rigassi, D., Stuijvenberg, J.V., Weidmann, M. and **Winkler, W.** (1980) Flysch and Molasse of western and central Switzerland (Excursion No. 5). In: *Geology of Switzerland, a Guide-Book* (Eds R. Trümpy), *Schweizerische Geologische Kommission*, **Part B**, pp. 261–293. Wepf, Basel.

Mellor, G.L. and **Yamada, T.** (1982) Development of a turbulence closure model for geophysical fluid problems. *Rev. Geophys. Space Phys.*, **20**, 851–875.

Mugnier, J.L. and **Menard, G.** (1986) Le developpement du bassin molassique suisse et l'evolution des Alpes externes; un modele cinematique. Development of a Swiss molasse basin in the external Alps; a kinematic model. In: *La subsidence des bassins sedimentaires; Seminaire organise en hommage et a la memoire d'Etienne Winnock. Subsidence of sedimentary basins; a seminar organized in homage to and in memory of Etienne Winnock* (Eds Anonymous. *Bulletin des Centres de Recherches Exploration-Production Elf-Aquitaine*, **10(1)**, 167–180.

Phillipe, Y., Deville, E. and **Mascle, A.** (1998) Thin-skinned inversion tectonics at oblique basin margins: Example of the Vercors and Chartreuse Subalpine massifs (SE France). In: *Cenozoic Foreland Basins of Western Europe* (Eds A. Mascle, C. Puigdefàbregas, H. Luterbacher and M. Fernàndez), *Geol. Soc. Spec. Publ.*, **134**, pp. 239–262.

Pugh, D.T. (1987) *Tides, Surges and Mean Sea-Level – A Handbook for Engineers and Scientists.* Wiley, Chichester, 472 pp.

Ritz, F. (1991) *Evolution du champ de contraintes dans les Alpes du Sud depuis la fin de l'Oligocène. Imlications sismotectoniques.* PhD thesis, Montpellier.

Roetzel, R. and **Rupp, C.** (1991) E/8 - Die westliche Molassezone in Salzburg und Oberösterreich. In: *Exkursionen im Tertiär Österreichs – Molassezone-Waschbergzone-Korneuburger Becken-Wiener Becken-Eisenstädter Becken* (Eds R.N. Roetzel, D.), pp. 114–157. Österreichische Paläontologische Gesellschaft, Wien.

Rögl, F. (1996) Migration pathways between Africa and Eurasia – Oligocene – Miocene Palaeogeography. *EUROPAL*, **10**, 23–26.

Rögl, F. (1998) Palaeogeographic Considerations for Mediterranean and Paratethys Seaways (Oligocene to Miocene). *Ann. Naturhist. Mus. Wien*, **99A**, 279–310.

Rögl, F. (1999) Mediterranean and Paratethys. Facts and hypotheses of an Oligocene to Miocene Paleogeography (short overview). *Geol. Carpath.*, **50**, 339–349.

Rögl, F. and **Steininger, F.F.** (1983) Vom Zerfall der Tethys zu Mediterran und Paratethys. Die neogene Paläogeographie und Palinspastik des zirkum-mediterranen Raumes. *Ann. Naturhist. Mus. Wien*, **85/A**, 135–163.

Rögl, F., Schulz, O. and **Hölzl, O.** (1973) Holostratotypus und Faziostratotypen der Innviertler Schichtengruppe. In: *M2 Ottnangien. Die Innviertler, Salgótarjáner, Bántapusztaer Schichtengruppe und die Rzehakia Formation – Chronostratigraphie und Neostratotypen* (Eds A. Papp, F. Rögl and J. Senes), **3**, pp. 140–196, Bratislava.

Salvermoser, S. (1999) Zur Sedimentologie gezeitenbeeinflusster Sande in der Oberen Meeresmolasse und Süssbrackwassermolasse (Ottnangium) von Niederbayern und Oberösterreich. *Münchner Geol. Hefte (A)*, **26**, 1–179.

Schaad, W., Keller, B. and **Matter, A.** (1992) Die Obere Meeresmolasse (OMM) am Pfänder: Beispiel eines Gilbert-Deltakomplexes. *Eclogae Geol. Helv.*, **85**, 145–168.

Schlunegger, F., Burbank, D.W., Matter, A., Engesser, B. and **Mödden, C.** (1996) Magnetostratigraphic calibration of the Oligocene to Middle Miocene (30–15 Ma) mammal biozones and depositional sequences of the Swiss Molasse Basin. *Eclogae Geol. Helv.*, **89**, 753–788.

Schlunegger, F., Leu, W. and **Matter, A.** (1997) Sedimentary sequences, seismic facies, subsidence analysis, and evolution of the Burdigalian upper marine molasse group (OMM), Central Switzerland. *AAPG Bull.*, **81**, 1185–1207.

Schlunegger, F., Melzer, J. and **Tucker, G.** (2001) Climate, exposed source rock lithologies, crustal uplift and surface erosion: A theoretical analysis calibrated with data from the Alps/North Alpine Foreland Basin system. *Int. J. Earth Sci.*, **90**, 484–499.

Schreiner, A. (1966) Zur Stratigraphie der Oberen Meeresmolasse zwischen der Oberen Donau und dem Überlinger See (Baden-Württemberg). *Jber. u. Mitt. oberrh. geol. Ver.*, **48**, 91–104.

Schreiner, A. (1976) *Hegau und westlicher Bodensee.* Sammlung Geologischer Führer, **62**. Gebr. Bornträger, Berlin - Stuttgart, 337 pp.

Schreiner and **Luterbacher** (1999) Die Molasse zwischen Blumberg und Überlingen (Exkursion J am 9. April 1999). *Jber. Mitt. Oberrh. Geol. Ver.*, **N.F. 81** 171–181.

Séranne, M., Benedicto, A., Labaume, P., Truffert, C. and **Pascal, G.** (1995) Structural style and evolution of the Gulf of Lion Oligo-Miocene rifting. Role of the Pyrenean orogeny. *Mar. Petrol. Geol.*, **12,** 809–820.

Sinclair, H.D., Coakley, B.J., Allen, P.A. and **Watts, A.B.** (1991) Simulation of foreland basin stratigraphy using a diffusion model of mountain belt uplift and erosion: An example from the central Alps, Switzerland. *Tectonics*, **10,** 599–620.

Sissingh, W. (1997) Tectonostratigraphy of the North Alpine Foreland Basin: Correlation of tertiary depositional cycles and orogenic phases. *Tectonophysics*, **282,** 223–256.

Sissingh, W. (1998) Comparative tertiary stratigraphy of the Rhine Graben, Bresse Graben and Molasse Basin: Correlation of Alpine foreland events. *Tectonophysics*, **300,** 249–284.

Sissingh, W. (2001) Tectonostratigraphy of the West Alpine Foreland: Correlation of Tertiary sedimentary sequences, changes in eustatic sea-level and stress regimes. *Tectonophysics*, **333,** 361–400.

Smagorinsky, J. (1963) General circulation experiments with the primitive equations I. The basic experiment. *Mon. Weather Rev.*, **91(3),** 99–164.

Steininger, F.F., Berggren, W.A., Kent, D.V., Bernor, R.L., Sen, S. and **Agusti, J.** (1996) Circum-Mediterranean Neogene (Miocene and Pliocene) marine-continental chronologic correlations of European mammal units. In: *The Evolution of Western Eurasian Neogene Mammal Faunas* (Eds R.L. Bernor, V. Fahlbusch and H.-W. Mittmann), Columbia University Press, New York, pp. 7–46.

Sündermann, J. and **Lenz, W.** (1983) *North Sea Dynamics.* Springer-Verlag, Berlin, 693 pp.

Sztanó, O. (1995) Palaeogeographic significance of tidal deposits; an example from an early Miocene Paratethys embayment, northern Hungary. *Palaeogeogr. Palaeoclimatol. Palaeoecol.*, **113,** 173–187.

Sztanó, O. and **de Boer, P.L.** (1995) Basin dimensions and morphology as controls on amplification of tidal motions (the early Miocene North Hungarian Bay). *Sedimentology*, **42,** 665–682.

Tessier, B. and **Gigot, P.** (1989) A vertical record of different tidal cyclicities: An example from the Miocene Molasse of Digne (Haute Provence, France). *Sedimentology*, **36,** 767–776.

Wagner, L. (1996a) Die tektonisch-stratigrafische Entwicklung der Molasse und deren Untergrundes in Oberösterreich und Salzburg. In: *Ein Querschnitt durch die Geologie Oberösterreichs. Reihe der Exkursionsführer der Oberösterreiches Geologischen Gesellschaften* (Eds H. Egger, T. Hofmann and C. Rupp), **16,** 36–65, Wien.

Wagner, L. (1996b) Stratigraphy and hydrocarbons in the Upper Austrian Molasse Foredeep (active margin). In: *Oil and Gas in the Alpidic Thrustbelts and Basins of the Central and Eastern Europe* (Eds G. Wessely and W. Liebl), *EAGE Special Publication*, **5,** London, pp. 217–235.

Wagner, L. (1998) Tectono-stratigraphy and hydrocarbons in the Molasse Foredeep of Salzburg, Upper and Lower Austria. In: *Cenozoic Foreland Basins* (Eds A. Mascle, C. Puigdefàbregas, H. Luterbacher and M. Fernàndez), *Geol. Soc. Spec. Publ.*, **134**, London.

Wagner, L., Kuckelkorn, K. and **Hiltmann** (1986) Neue Ergebnisse zur alpinen Gebirgsbildung Oberösterreichs aus der Bohrung Oberhofen 1 – Stratigraphie, Fazies, Maturität und Tektonik. *Erdöl-Erdgas-Zeitschrift*, **102(1),** 12–19.

Wenger, W.F. (1987) Die Basis der Oberen Meeresmolasse im westlichen Oberbayern, am Überlinger See, in Vorarlberg und St. Gallen, *Mitt. Bayer. Staatssamml. Paläont. Hist. Geol.*, **27,** 159–174.

Werner, S.R., Beardsley, R.C. and **Williams, A.J.** (2003) Bottom friction and bed forms on the southern flank of Georges Bank. *J. Geophys. Res.*, **108(C11),** 8004, doi:10.1029/2000JC000692.

Zweigel, J., Aigner, T. and **Luterbacher, H.** (1998) Eustatic versus tectonic controls on Alpine foreland basin fill: Sequence stratigraphy and subsidence analysis in the SE German Molasse. In: *Cenozoic Foreland Basins of Western Europe* (Eds A. Mascle, C. Puigdefàbregas, H.P. Luterbacher and M. Fernàndez), *Geol. Soc. Spec. Publ.*, **134,** pp. 299–323.

Spec. Publ. Int. Assoc. Sedimentol. (2008) **40**, 171–189

Predicting discharge and sediment flux of the Po River, Italy since the Last Glacial Maximum

ALBERT J. KETTNER and JAMES P.M. SYVITSKI

Environmental Computation and Imaging Facility, INSTAAR, University of Colorado, Boulder, CO 80309-0450, USA

Correspondence to: Albert J. Kettner, INSTAR, University of Colorado, P.O. Box 450, Boulder, CO 80309, USA, Tel: +1-303 735 5486, Fax: +1-303 492 6388, Email: Kettner@Colorado.edu

ABSTRACT

HydroTrend numerically simulates the flux of water and sediment delivered to the coastal ocean on a daily timescale, based on drainage-basin and climate characteristics. The model predicts how a river may have behaved in the geological past, provided that appropriate assumptions are made regarding past climate and drainage-basin properties.

HydroTrend is applied to simulate a high-resolution discharge and sediment flux record for the Po River in Italy since the Last Glacial Maximum (LGM). A validation experiment of 12 years duration under present conditions shows a high correlation ($r^2 = 0.72$) with 12 years observed daily discharge. Monthly variations in simulated Po River discharge and sediment discharge for this same time period show an even closer agreement ($r^2 = 0.97$) with the observed data.

Community Climate System models (CCSM) indicate that during the LGM the Po drainage-basin climate was much colder and drier than at present. The drainage basin itself was much larger due to a lower sea level. The Bølling period is shown to be an exceptional period in our model study, with significantly higher discharges and sediment loads, as glacier ablation was more dominant.

HydroTrend simulations predict an average suspended sediment flux of 46.6 Mt yr^{-1} and an average bedload of 0.83 Mt yr^{-1} for the Po River during the Late Pleistocene (21–10 cal. kyr BP). This is ~75% greater than in the Holocene (10–0 cal. kyr BP), where simulations indicate a suspended sediment flux of 26.7 Mt yr^{-1} and a bedload of 0.53 Mt yr^{-1}. The Würm Stadial simulations tend to show the highest suspended-sediment concentrations. However, when considering sediment yield where load is normalized to unit area (mass per square kilometre per time), simulations indicate slightly higher yield for the Holocene as compared with the Late Pleistocene, respectively 336 t km^{-2} yr^{-1} versus 283 t km^{-2} yr^{-1}, due to an increase in Holocene precipitation. Despite the overall lower yields during the Pleistocene, the model indicates that glacial influence is a dominant factor on the total sediment flux to the ocean.

Keywords River modelling, climate change, palaeoclimate, *HydroTrend*, drainage basin change, numerical model.

INTRODUCTION

It is important to understand and predict the impact of climate change, sea-level change and human influence on the accumulation of sediment on continental margins (Syvitski & Morehead, 1999). Coastal retreat is directly related to river sediment supply, which might effect large populations since ~35% of the people live within 100 km of a coastline (Syvitski *et al.*, 2005). In the framework of the EUROSTRATAFORM project the Adriatic Sea was selected as a study site where the dynamics of climate and sea-level change and human influence are amplified, because of the semi-closed nature of the basin.

The Po River strongly influences the dynamics of the Adriatic Sea with its great input of fresh water (Cushman-Roisin *et al.*, 2001) and sediment, both presently and during the Last Glacial Maximum (Asioli *et al.*, 2001). Unfortunately, sediment-load measurements are limited. Discharge measurements have been collected during the past 200 years. We cannot extrapolate these records either forward or backward in time, because a number of controlling factors have changed over time. First, analysis of marine cores indicates that climate shifts such as the regional warming since the Last Glacial Maximum (LGM) (21 cal. kyr) have profoundly impacted discharge and sediment flux of the river (Asioli *et al.*, 1999). Second, the large reservoirs originally formed by moraine-dammed lakes of the retreating Alpine glaciers presently control one-third of the total Po River discharge and have regulated the discharge and sediment load differentially over a long time period (Hinderer, 2001). Third, the size of the palaeodrainage basin changed significantly due to sea-level fluctuations in the shallow Adriatic Sea. We advocate that numerical modelling offers a possibility to predict the hydrography and the sediment flux to the coastal ocean by taking into account the shifts of climate and the influence of reservoirs as well as sea-level changes.

The climate-driven hydrological model *HydroTrend* simulates the Po River since the LGM employing three new subroutines for: (1) glacier dynamics, (2) reservoir sediment trapping and (3) fluvial sediment transport. The model output is validated against observations of the modern Po River. Subsequently, we discuss the assumptions used to arrive at the input for climate, sea-level change and drainage-basin characteristics. Finally, we explain the results and investigate which processes are dominant in driving the discharge and sediment flux of the Po River during certain characteristic time periods.

GEOLOGICAL SETTING

Hydrology of the Po River

The present Po River is approximately 650 km long, and has over 141 distributaries. The 74 500 km^2 river drainage basin of the Po is bounded by the Alps with peaks over 4500 m in the north, and by the Apennines mountain chain with peaks generally less than 2000 m in the southwest (Fig. 1). More than a third of the drainage area (30 800 km^2) can be considered mountainous. The Po River has two flood periods, June (freshets caused by snow melting) and November (corresponding to precipitation maxima) and two low-water periods, January and August (Fig. 2) (Marchi *et al.*, 1996; Cattaneo *et al.*, 2003). The average discharge of the Po River is 1.5×10^3 m^3 s^{-1} (Table 1) observed at Pontelagoscuro (near Ferrara) 90 km from the coast and just before the apex of the delta (Fig. 1). Downstream of Pontelagoscuro, the Po forms a delta consisting of six major distributaries: the Levante, Maistra, Pila, Tolle, Gnocca and Goro distributaries. The main channel is the Pila, carrying 60% of the total discharge to the Adriatic Sea.

Although the river-discharge fluctuations are dominated by rainfall, the hydropower management influences the discharge regime considerably. The five largest reservoirs, the Maggiore, Lugano, Como, Iseo and Garda Lakes (Fig. 1) were formed by moraine dams of the retreating Alpine glaciers after the Pleistocene glaciation. A third of the total discharge of the Po is affected by these reservoirs (Camusso *et al.*, 2001). The lakes are currently regulated for hydropower production and irrigation, and are located in the most highly populated and industrialized area of Italy, Insubria, the northern area of the Po catchment (Marchi *et al.*, 1996).

Adriatic Sea

The Adriatic Sea forms the northernmost part of the Mediterranean. It is a relatively shallow, almost rectangular basin bordered to the north by the Alps, to the west by the Apennines and in the east by the Dinaric mountain chain. This temperate warm sea is more than 800 km long in a NW–SE direction and has an average width of about 200 km (Fig. 1). The Adriatic Sea is often divided into three geographical regions, namely the Northern, Middle and Southern Adriatic basins. The Northern Adriatic, defined as the area lying north of the 100 m isobath, has a wide continental shelf, sloping gently south, and is relatively shallow.

The Po River is the main source of fresh water into the northern Adriatic Sea. The river plume hugs the western side of the basin forced by dominant cyclonic circulation (Cattaneo *et al.*, 2003). Along with wave resuspension, the plume is responsible for the formation of a 35 m thick Holocene mud wedge that extends from the

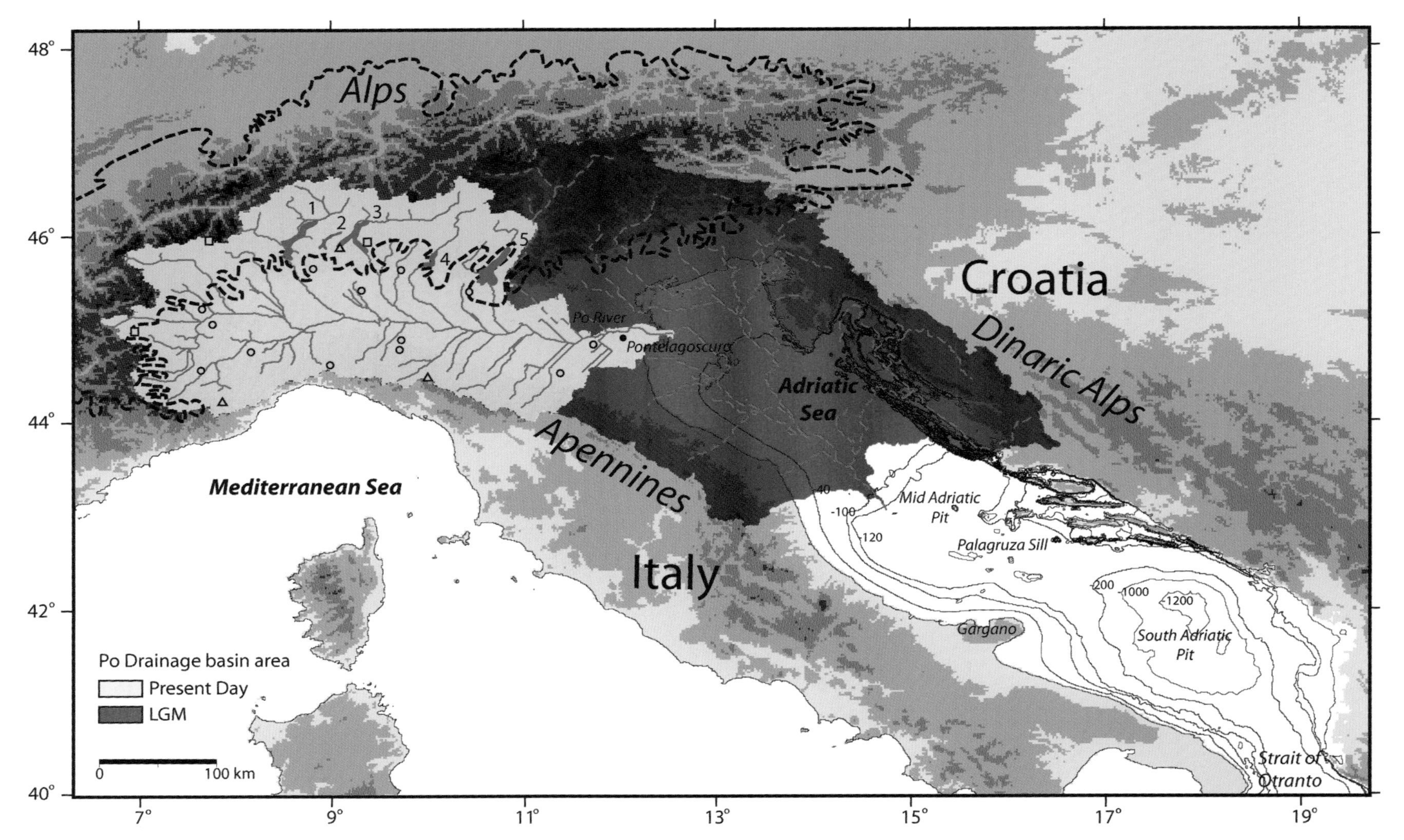

Fig. 1. Po drainage basin during the Last Glacial Maximum (LGM; red and yellow) and today (yellow). The black dotted line represents the maximum extension of the late Würmian ice sheet during the LGM. The blue solid line represents the present day Po River with its mature tributaries; the blue dotted line is a hypothesis of where the Po River might have been during the LGM, based on present-day bathymetry. The green markers are climate stations used to calculate the mean present-day precipitation and temperature. Circles are stations below 1000 m, triangles are stations between 1000 and 2000 m and squares are stations above 2000 m. Numbers 1 to 5 represent the main reservoirs of the Po Basin taken into account for this study: Maggiore Lake, Lugano Lake, Como Lake, Iseo Lake and Garda Lake respectively.

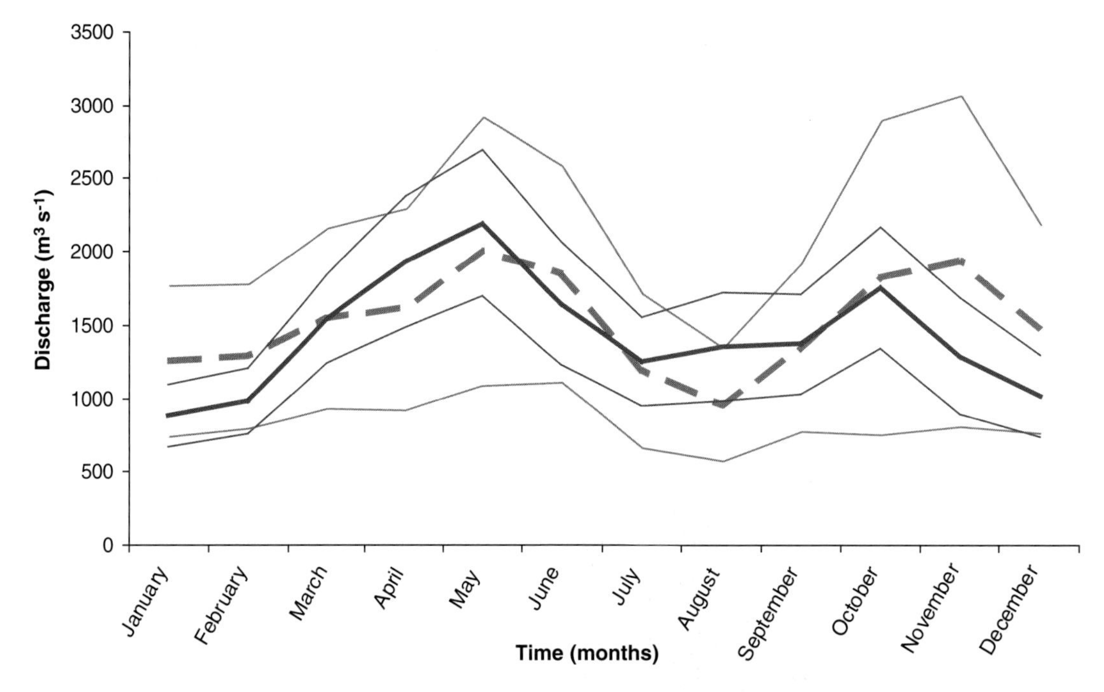

Fig. 2. Monthly average discharge, distribution of 84 years of measured data (dotted orange line) versus modelled (solid blue line). The fine lines represent the variation (plus or minus the standard deviation). Measured field data were obtained from: (a) 1918–1979, http://espejo.unesco.org.uy/part%604/6_europa/italy/6iy'po_at_pontelagoscuro.htm; (b) 1980–1987, ANNALI IDROGRAFICI; (c) 1989–2002, Ufficio Idrografico del Magistrato per il Po (Parma).

Table 1. Characteristics of the Po River, measured values against 12 years daily *HydroTrend* simulations.

	Measured at apex (Pontelagoscuro)	Modelled
Long-term average discharge ($m^3 s^{-1}$)*	1500	1480
Twentieth century floods ($m^3 s^{-1}$)	10 300	10 800
Average suspended sediment load ($Mt\,yr^{-1}$)†	15	17
Peak suspended sediment load ($Mt\,yr^{-1}$)‡	35	39
Average bedload ($Mt\,yr^{-1}$)	–	0.40
Sediment yield ($t\,km^{-2}\,yr^{-1}$)†	201	207

* Nelson (1970).
† Cattaneo *et al.* (2003).
‡ Friend *et al.* (2002).

Po Delta to the Gargano subaqueous delta, 500 km further south (Fig. 1). During the summer the Po River plume generally spreads over the entire northern sub-basin as a thin surface layer, ~5 m thick (Cushman-Roisin *et al.*, 2001).

HydroTrend METHODOLOGY

Model description

HydroTrend numerically simulates discharge and sediment loads at a river mouth on a daily timescale. The model is designed to make discharge predictions based on drainage-basin characteristics and climate, even when field measurements of river flow are not available (Syvitski *et al.*, 1998). Provided that appropriate assumptions are made regarding past climate, the model predicts how any river behaves in prehistoric periods (Syvitski & Morehead, 1999). Syvitski *et al.* (2003) show that sediment-flux predictions are accurate to the same level of accuracy as global field observations.

HydroTrend incorporates drainage-basin properties (river networks, hypsometry, relief, reservoirs) based on high-resolution digital elevation models (e.g. *HYDRO1k* DEM, http://edc.usgs.gov/products/elevation/gtopo30/hydro/index.html). A number of additional biophysical parameters are incorporated to calculate the steady-state hydrological balance (basin-wide temperature, precipitation, potential glacier equilibrium line altitude (ELA), evapotranspiration, canopy, soil depth and

Table 2. Example of key portions of a *HydroTrend* input file of the Po River.

19.1	0.0	0.4			8. Yearly mean temperature: start (C), change/yr (C/a), std. dev.		
0.9	0.0	0.13			9. Yearly precipitation sum: start (m/a), change/yr (m/a/a), std. dev.		
1.0	1.8	7			10. Rain: mass balance coef., distribution exp, distribution range		
365.0					11. Constant annual base flow ($m^3 s^{-1}$)		
Jan	3.25	0.73	45.09	33.77	12–23. monthly climate variables		
Feb	5.02	1.83	40.93	29.15	Column	Variable	Description
Mar	9.83	1.93	69.17	34.86	–	–	–
Apr	13.19	0.73	84.87	56.58	1	moname	Month name (not used)
May	18.76	0.72	98.84	53.60	2	tmpinm	Monthly mean temp. (C)
Jun	22.42	1.12	71.28	24.23	3	tmpstd	Within month std. dev. of temp.
Jul	25.86	1.20	49.35	31.46	4	raininm	Monthly total precip. (mm)
Aug	25.06	1.11	67.11	32.82	5	rainstd	Std. dev. of the monthly precip.
Sep	19.58	1.79	52.75	41.51			
Oct	14.42	0.73	95.27	55.12			
Nov	7.97	1.35	51.66	49.58			
Dec	3.49	0.83	46.73	28.62			
6.540					24. Lapse rate to calculate freezing line (°C km^{-1})*		
3267	0.01				25. Starting glacier ELA (m) and ELA change per year (m a^{-1})		
0.3					26. Dry precip (nival and ice) evaporation fraction		
8.7e−5					27. Delta plain gradient (m m^{-1})		
650					28. River basin length (km)		
24.58	d16335				29. Mean vol., (a)ltitude or (d)rainage area of reservoir (km^3) (m) or (km^2)		
0.5	0.1				30. River mouth vel. coef. (k) and exp. (m); $v = kQ^m$, $w = aQ^b$, $d = cQ^f$		
5.0	0.5				31. River mouth width coef. (a) and exp. (b); $Q = wvd$, $ack = 1$, $b + m + f = 1$		
1.1					32. Average river velocity ($m\,s^{-1}$)		
6.9e10	6.4e10				33. Maximum/minimum groundwater storage (m^3)		
6.35e10					34. Initial groundwater storage (m^3)		
7500	1.0				35. Groundwater (subsurface storm flow) coefficient ($m^3 s^{-1}$) and exp.		
400					36. Saturated hydraulic conductivity (mm d^{-1})		
12.5	44.9				37. Longitude, latitude position of the river mouth (dd)		
0					38. Option 0: $\overline{Q}_s = \alpha_6 \overline{Q}^{\alpha_7} R^{\alpha_8} e^{k_2 \overline{T}}$; Option 1: $\overline{Q}_s = \alpha_3 A^{\alpha_4} R^{\alpha_5} e^{k_1 \overline{T}}$		

* Derived from the NCEP/NCAR Reanalysis of global lapse rates (°C km^{-1}) based on a 2.5 × 2.5 degree grid.

hydraulic conductivity). Table 2 is an example of a *HydroTrend* input file. The most important water discharge and sediment process routines are described below. A more detailed description of *HydroTrend* can be found in Syvitski *et al.* (1998).

The water subroutine

Discharge is generated and controlled by a number of processes in *HydroTrend*. The daily average discharge, Q_T, ($m^3 s^{-1}$) can be written as:

$$Q_{T[i]} = Q_{r[i]} + Q_{ice[i]} + Q_{n[i]} \pm Q_{gr[i]} - Q_{evap[i]} \quad (1)$$

wherein Q_r is the discharge from rainfall as surface runoff, Q_{ice} is ice or glacial melt discharge, Q_n is the snow melt (nival) discharge, Q_{gr} is the groundwater discharge and Q_{evap} is the discharge component lost by evaporation. The daily timestep is defined by i. Variable Q_T (all in $m^3 s^{-1}$) is thus a function of precipitation, basin location, temperature, elevation, and soil properties.

The Po River drainage basin includes Alpine glaciers, which affect the hydrological balance strongly at glacial times and during subsequent warming. The model uses the equilibrium line altitude (ELA), which defines the transition from accumulation and ablation, in combination with the basin hypsometry to determine glacier area, assuming that approximately one-third of a glacier is located below the ELA. Equilibrium line altitude changes over time result in glacier-area changes. *HydroTrend* employs an exponential relationship (equation 2) for glacier area, A_G (km^2) versus glacier volume, V_G (km^3) (Bahr *et al.*, 1997) to simulate glacier ablation or growth, and tracks changes in hydrological balance at the river outlet.

$$V_G = 31.1 \times (A_G)^{1.38} \quad (2)$$

The sediment subroutine

The long-term average suspended sediment load ($\overline{Qs_d}$ in kg s^{-1}) can either be defined as:

$$\overline{Qs_d} = (1 - TE)\alpha_3 A_T^{\alpha_4} R^{\alpha_5} e^{k_1 \overline{T}} \quad (3a)$$

or as:

$$\overline{Qs_d} = (1 - TE)\alpha_6 \overline{Q}^{\alpha_7} R^{\alpha_8} e^{k_2 \overline{T}} \quad (3b)$$

wherein A_T is the drainage basin area (km^2), $\overline{Q}$ is the long-term average discharge (m^3 s^{-1}), R is the maximum elevation in the drainage basin (m), $\overline{T}$ is the basin-average temperature (°C), α_3 to α_8, k_1 and k_2 are dimensionless coefficients which depend on climate zone based on the geographical location of the drainage basin (Syvitski *et al.*, 2003). In the case of the Po River, climate changes from Last Glacial Maximum (LGM) until present do not exceed the threshold values to force changes in the dimensionless coefficients. Variable *TE* is the dimensionless trapping efficiency rate. Po River discharge is strongly affected by five large reservoirs (Fig. 1), and sediment fluxes are likely influenced even more by its sediment trapping. *HydroTrend* has been adjusted to incorporate reservoir effects on sediment load.

The model simulates *TE*, depending on reservoir volume, either by the Brown equation (Verstraeten & Poesen, 2000), for reservoirs smaller or equal to 0.5 km^3, or the modified Brune equation (Vörösmarty *et al.*, 1997), for reservoirs larger than 0.5 km^3 (equations 4 and 5). In this study we only take large reservoirs into account. We assume that the Brune equation developed for artificial reservoirs can also be applied to natural formed reservoir lakes.

$$TE = 1 - \frac{0.05}{\sqrt{\Delta\tau}} \qquad V_i > 0.5\,\text{km}^3 \quad (4)$$

wherein $\Delta\tau$ is the approximated residence time and is estimated by:

$$\Delta\tau = \frac{\sum_1^{n_i} V_i}{Q_j} \quad (5)$$

where V_i is operational volume of the reservoir i (km^3), Q is the discharge at mouth of each regulated sub-basin j (m^3 s^{-1}).

Hallet *et al.* (1996) review a host of field studies on glacial erosion worldwide. Although the data show a wide range of effective rates of glacial erosion, the overall trend is a clear sediment-flux increase when drainage basins contain glaciers. In this study we are particularly interested in the effect of changing glacial cover of the drainage basin due to changing climate conditions (from glacial to interglacial periods) on the suspended sediment flux.

Leonard (1997) found a linear relation between long-term variations in ice extent and sedimentation rates from lake-core records in Alberta, Canada. Hallet *et al.* (1996) summarize studies done by Guymon (1974) and Parks & Madison (1985), which indicate that sediment yields increase pronouncedly with glacial cover. On average, yields for basins extensively covered by glaciers (>30% glacier cover) are about one order of magnitude higher than for glacier-free basins (Lawson, 1993; Guymon, 1974; Hallet *et al.*, 1996). A numerical approach to incorporate sediment derived from glaciers is based on data presented by Guymon (1974):

$$Qs_{g(a)} = A_T \beta_{10}^{(\chi \log((A_G/A_T)100))} - \overline{Qs_d} \quad (6)$$

wherein $Qs_{g(a)}$ is the annual average suspended sediment (kg s^{-1}) *exclusively* derived from glaciers, A_T is the drainage basin area (km^2), β and χ are dimensionless constants (respectively 1.9311e^{-6} and 0.9913) and A_G is the glacier area (km^2). Based on equation (6) the long-term suspended sediment *exclusively* derived by glaciers ($\overline{Qs_g}$ in kg s^{-1}) will be:

$$\overline{Qs_g} = \left(1 - \frac{V_{g(a)}}{P_{g(a)}}\right)\left(\sum_{a=1}^{n} Qs_{g(a)}\right) \Big/ (n) \quad (7)$$

where $V_{g(a)}$ is the water storage (snow and rain transferred into ice) of the glacier (m^3), $P_{g(a)}$ is the total annual precipitation directly on the glacier (m^3) and n is the number of years of a simulated glacier in the drainage basin. The first term in equation (7), dealing with water storage ($V_{g(a)}$) and precipitation ($P_{g(a)}$), reduces sediment flux during glacial advance.

The data used to derive equations (6) and (7) do not include case studies in the Alps, which makes it somewhat uncertain to apply the method straightforwardly to the Po basin. However, Hinderer (2001) corroborates that LGM denudation rates were 5 to 25 times higher than modern rates. We assume that the highest suspended-sediment flux coincides with maximum glacial advance (A_G is maximal). According to Leonard (1997) maximum rates occur one-to-two decades later but this lag time is not taken into account.

The total long-term suspended sediment ($\overline{Qs_T}$ in kg s^{-1}) can be written as:

$$\overline{Qs_T} = \overline{Qs_d} + \overline{Qs_g} \tag{8}$$

A stochastic model (Morehead *et al.*, 2003) is used to calculate the daily suspended sediment load fluxes:

$$\left(\frac{Qs_{[i]}}{\overline{Qs_T}}\right) = \psi_{[i]} m \left(\frac{Q_{[i]}}{\overline{Q}}\right)^{C_{(a)}} \tag{9}$$

where m is a constant of proportionality such that:

$$m = \left[\left(\sum_{i=1}^{n} Qs_{[i]}\right) \Big/ n\right] \Big/ \overline{Qs_T} \tag{10}$$

where n is the number of days being modelled. The following relationships define the terms in equation (9):

$$E(\psi) = 1 \tag{11}$$

$$\sigma(\psi) = 0.763(0.99995)^{\overline{Q}} \tag{12}$$

$$E(C_a) = 1.4 - 0.025\overline{T} + 0.00013R + 0.145\ln(\overline{Qs_T}) \tag{13}$$

$$\sigma(C_a) = 0.17 + 0.0000183\overline{Q} \tag{14}$$

wherein $Qs_{[i]}$ is the daily suspended sediment discharge (kg s^{-1}), $Q_{[i]}$ is the daily discharge (m^3 s^{-1}), $\overline{Qs_T}$ is the long-term average of $Qs_{[i]}$, $\overline{Q}$ is the long-term average of $Q_{[i]}$, ψ is a log-normal random variable and $C_{(a)}$ is a normal random variable.

The daily bedload Q_b (kg s^{-1}) is simulated using a modified Bagnold (1966) equation:

$$Q_b = \left(\frac{\rho_s}{\rho_s - \rho}\right)\frac{\rho g Q_i^{\beta} s e_b}{g \tan f} \quad \text{when} \quad u \geq u_{cr} \tag{15}$$

where ρ_s is sand density (kg m^{-3}), ρ is fluid density (kg m^{-3}), Q_i is daily discharge (m^3 s^{-1}), s is slope of the riverbed (m m^{-1}), e_b is the dimensionless bedload efficiency, β is a dimensionless bedload rating term, g is the acceleration due to gravitation (m s^{-2}), f is the limiting angle of repose of sediment grains lying on the river bed (°), u is stream velocity (m s^{-1}), and u_{cr} is the critical velocity (m s^{-1}) needed to initiate bedload transport.

The *HydroTrend* model with documentation, example files and references, is available on the web: http://instaar.colorado.edu/deltaforce/models/hydrotrend.html.

Model validation

Present-day climate statistics including monthly mean temperature and precipitation and the associated standard deviations for the Po drainage basin are retrieved from daily data records (1977–1991) of 20 climate stations distributed over the drainage basin. The stations have been weighted by elevation to determine characteristic basin-wide values. The yearly mean temperature, precipitation and their standard deviations, similarly weighted by elevation, are obtained from monthly data records (1760–1995) of 13 climate stations (Fig. 1) from the Global Historical Climatology Network of National Oceanic and Atmospheric Administration (NOAA) (Vose *et al.*, 1992) (http://ingrid.ldeo.columbia.edu/SOURCES/.NOAA/.NCDC/.GHCN/).

HydroTrend proved an appropriate model to simulate the fluxes at the Pontelagoscuro gauging station, which is located closest to the river mouth before the main channel splits into distributary channels (Fig. 1). At this delta apex, a high correlation is found between modelled discharges and time series of 12 yr daily discharge data (1990–2001) and monthly discharge (1918–2002) (Figs 2 and 3). Modelled data correlate significantly with the observed data (r^2 of 0.72), as calculated from 100 m^3 s^{-1} discharge intervals.

Figure 2 indicates that the monthly mean discharges are underpredicted during the winter months (November to March) and slightly overpredicted during the summer (April to September). This deviation relates to the well known 'reservoir effect' common in river basins containing major reservoir lakes (e.g. Bobrovitskaya *et al.*, 1996). The river discharge is artificially kept low in summer by storage of water in the reservoirs and extra water is released in winter to generate hydropower, which results in a dampening of the seasonal discharge signal. Over a third of the discharge of the Po River is affected by these hydropower reservoirs (Camusso *et al.*, 2001). The model does not take this policy-driven effect into account.

At the delta apex, *HydroTrend* predicts an average discharge of 1542 m^3 s^{-1} and peak daily discharges of 10 800 m^3 s^{-1} comparable to the

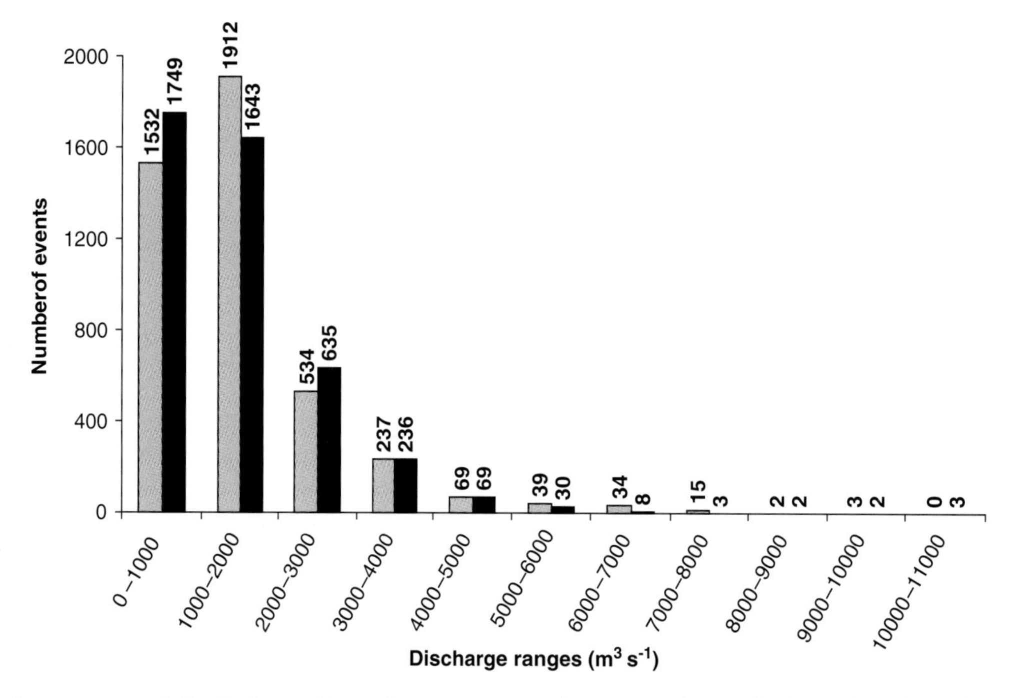

Fig. 3. Twelve-year mean daily discharge. Grey columns represent the measured mean discharge from the Ufficio Idrografico del Magistrato per il Po, 1990–2001, black columns are the modelled mean discharge.

measured floods of 1951 and October 2000 (respectively 10 300 and 9600 $m^3 s^{-1}$, see also Table 1). Average suspended-sediment load of 17 Mt yr^{-1} with peak years of 39 Mt yr^{-1} are predicted, based on the sediment-flux equations explained above. Bedload is predicted to contribute only ~2.5% of the total sediment output of the Po River system (Table 1).

Monthly variations in simulated Po River discharge and sediment discharge are similar to measurements (Fig. 4). The one difference between simulations and observations is that the simulations show less variation in the sediment discharge because it does not include hysteresis effects (e.g. floodplain storage) that provide much of the scatter in the rating plot of the observed data.

Human impact in the form of damming and sediment mining has caused a decrease of 38% in sediment supply during the second half of the twentieth century for the Po (Surian & Rinaldi, 2003). The model captures the trapping of sediment due to the reservoirs (equations 3a, b and 4). The total volume of the mined sediment is estimated to be more than 2.8 Mt yr^{-1} in the 1990s (Marchetti, 2002). Sediment mining is not modelled because this is a 'policy-driven' effect which cannot be included in our physical process model. However, subtracting the mined sediment (2.8 Mt yr^{-1}) from the simulated suspended sediment load (17 Mt yr^{-1})

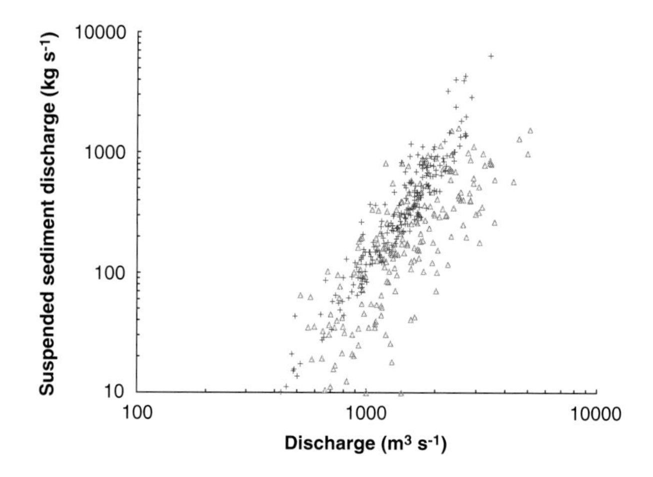

Fig. 4. Seventeen years of observed monthly suspended-sediment load (Δ) and river discharge from ANNALI IDROGRAFICI. Superimposed on the plot (+) are 17 years of simulated suspended sediment discharge and discharge from the *HydroTrend* model.

improves the match with the measured field data (14.2 versus 15 Mt yr^{-1}) (Table 1).

Input and boundary conditions

Both the Mediterranean sea level and the Adriatic sea level were about 120 m below present msl during the LGM (Lambeck & Bard, 2000; Siddall *et al.*, 2003). The evolution of the Po basin area since the

LGM has been reconstructed based on an approximation of the palaeotopography constructed by merging the present-day digital elevation model (http://edcdaac.usgs.gov/gtopo30/gtopo30.asp) and bathymetric data (Fig. 1). Geographical information system (GIS) analysis provides the shifted coastline and subsequently the palaeodrainage basin. During the LGM, the shoreline was about 250 km southeast of the present-day Po River outlet (Amorosi, *et al.*, 1999) because of the relatively shallow gradient of the Northern Adriatic Sea. This implies that part of the Apennine rivers, for example the Metauro, Potenza and the Chienti rivers, as well as present-day northern Italian and Croatian rivers were contributaries to the Po River, which eventually drained into the modern foredeep basin of the Apennine Chain (Fig. 1). We modelled the sea-level fluctuations in 12 steps, each of 10 m. Each step caused a drainage-area decrease (Fig. 5). The total area of the Po drainage basin during the LGM was 2.6 times its present area (Fig. 1).

Monthly climate statistics at the LGM are estimated based on the Community Climate Model1 (CCM1) (Kutzbach *et al.*, 1998). These modelled climate statistics for LGM have been combined with the present climate statistics and interpolated over time using a normalized δ^{18}O GRIP curve as a forcing factor (Dansgaard *et al.*, 1993). According to the CCM1 model predictions the mean basin temperature was much lower (2.4°C vs. 8.8°C) and precipitation was 18% less than at present (Fig. 5). Using the CCM1 climate statistics is justified by the fact that a number of independent proxy studies in northwestern Europe corroborate cool and dry conditions (Zonneveld, 1996; Peyron *et al.*, 1998; Fauquette *et al.*, 1999). Local pollen data also show that the LGM was significantly dryer and colder (Bortolami *et al.*, 1977).

Despite the large latitudinal difference between the location of the GRIP ice core and our study area (respectively 72° vs. 41°N), we are confident of the relevance of the normalized δ^{18}O GRIP curve, because it correlates strongly with local palaeoclimate proxies such as tree rings (Friedrich *et al.*, 2001) stable isotopes in fossil mammal remains (Huertas *et al.*, 1997), marine sediment cores (Asioli *et al.*, 2001; Sbaffi *et al.*, 2001), biostratigraphical pollen data (Asioli *et al.*, 1999) and lake cores (Watts *et al.*, 1996; Ramrath *et al.*, 1999; Allen & Huntley, 2000).

The Alps were almost completely covered by the late-Würmian ice sheet (Florineth & Schlüchter, 1998; Hinderer, 2001) and the glacial drainage divide roughly followed the modern divide (Baroni, 1996). We used the normalized δ^{18}O curve to force the potential glacier equilibrium line altitude (ELA) changes. Potential ELA values are based on global latitude-specific averages both for the LGM and present day (Fig. 5). The ELA dynamically responds when precipitation is insufficient to add enough ice volume to the glacier to keep up with potential ELA changes.

Hinderer (2001) estimates the timing of deglaciation of the peri-alpine lake basins, which form the present-day reservoirs of the Po basin. The deglaciation and consequent activation of the five main reservoirs is unique for each reservoir, due to difference in elevation between the reservoirs (~180 m). The trapping efficiency (*TE*) is a function of basin area (equations (4) and (5)); in such a way that *TE* decreases for the whole basin over time concurrent with the decreasing basin area due to rising sea level (Fig. 5).

Detailed description of four characteristic periods

Four time periods have been studied more closely because we presume that these periods may be distinguished in the stratigraphy of marine cores; the Würm LGM (21–17.5 cal. kyr BP), Bølling (15.6–13.9 cal. kyr BP), Younger Dryas (12.9–11.1 cal. kyr BP) and the late Subatlantic (0.8 cal. kyr BP to present) periods.

The maximum extension of the Late Pleistocene ice-sheet in the Alps was reached during the Würm LGM, when glaciers extended as far as the Alpine foreland (Fig. 1). According to the CCM1 climate simulations, the yearly average temperature for the Po drainage basin was 2.4°C and precipitation was 18% less than at present. The drainage basin was 1.9×10^5 km^2, or 2.6 times larger than today. The average potential ELA was 1920 m and increased slowly at a rate of 0.10 m yr^{-1}.

During the Bølling, major valleys and deeply incised lake basins in the major intra-Alpine valleys and in the Alpine foreland were ice-free. Based on equations (4) and (5), the sediment-load trapping calculated for the river basin as a whole is estimated at ~8%. The potential ELA was rapidly increasing, on average 1.0 m yr^{-1} from 2015 m between 15.6 and 14.4 cal. kyr. Subsequently, the ELA dropped to 0.6 m yr^{-1} until the end of the Bølling. The drainage area did not change significantly compared to the Würm interval, due to the steep gradient of the outermost part of the Northern

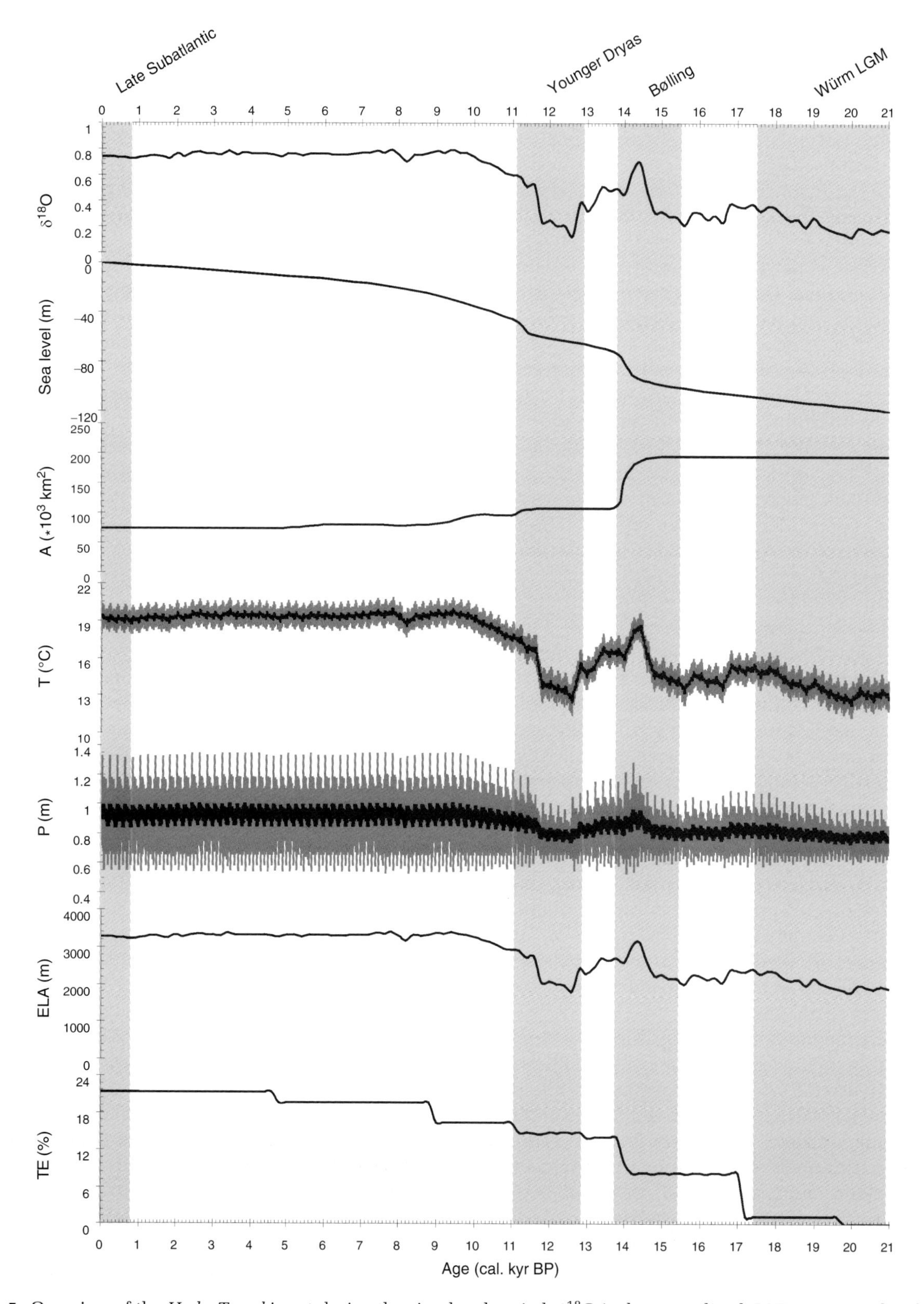

Fig. 5. Overview of the *HydroTrend* input during the simulated period: $\delta^{18}O$ is the normalized GRIP curve used to force T, P and ELA input parameters for the model; A stands for drainage basin area; T is mean annual temperature; P is total annual precipitation; ELA is potential equilibrium line altitude of glaciers in the drainage basin; TE is percentage sediment trapping efficiency of the whole river basin. (Chronozones from Soldati *et al.*, 2003: late Subatlantic, Younger Dryas, Bølling, Würm LGM.)

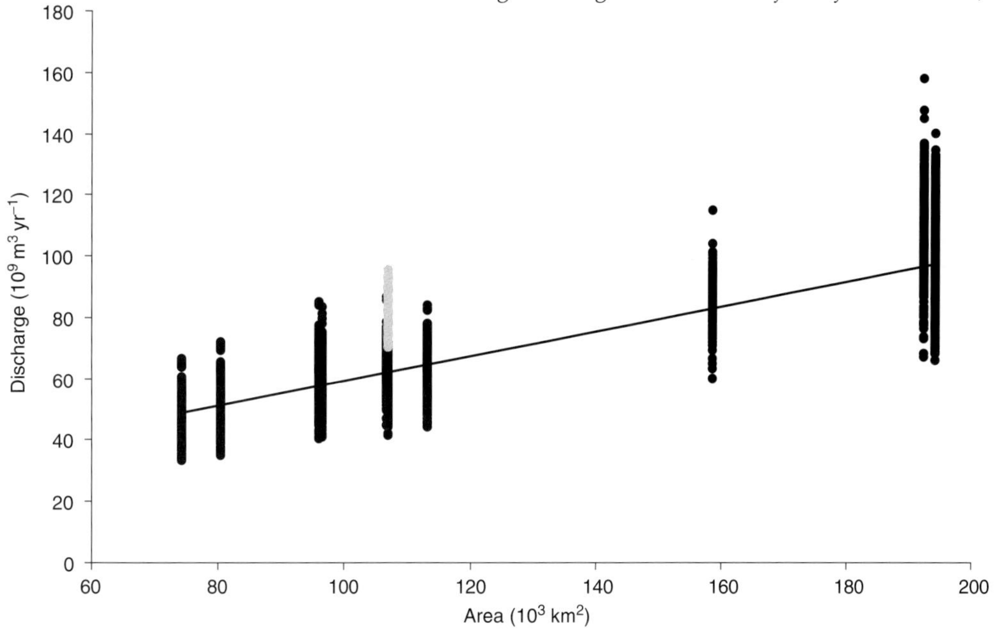

Fig. 6. Relation between simulated discharge and changing drainage area of the Po River due to sea-level fluctuations. The grey dots indicate the melt peak of the Younger Dryas, when glacier ablation had significant impact on the total discharge of the Po River.

Adriatic Sea, which stabilized the coastline position for a long period, despite sea-level rise.

The Younger Dryas (YD) provided the most significant rapid climate-change event during the last deglaciation of the North Atlantic region (e.g. Yu & Wright, 2001). This transition marks the last major climate reorganization during the deglaciation. Potential ELA increased on average by 0.7 m yr^{-1}. Furthermore sea-level rise decreased the drainage basin by almost 50% (1.1×10^5 km^2) compared to the Würm interval. Due to this decrease in drainage area the average *TE* for the drainage basin as a whole increased to 15%, based on reservoir characteristics.

The late Subatlantic (S) period reflects the present climate conditions with less interannual variability than during the Pleistocene. The drainage basin of the Po River is more or less similar as today. The potential glacier ELA is 1300 m higher than during the Würm LGM at 3200 m. Roughly 20% of the sediment was trapped at that time, based on reservoir characteristics.

DISCUSSION: *HydroTrend* LONG-TERM SIMULATION RESULTS

Po River basin area changes due to sea-level fluctuation have a dominant impact on discharge and sediment load. Figure 6 illustrates the relation between total yearly discharge and drainage basin area ($r^2 = 0.80$). A decline in drainage area is reflected in a decrease of the water discharge. The large spread around the mean values is due to the stepwise nature of the drainage basin changes. Other forcing factors (e.g. climate) change at higher time resolution than drainage area and impact the discharge prediction as well, causing the variability over the *y*-axis (Fig. 6). The exception to linear relationship occurs at the end of the Younger Dryas, when the area is $\approx 107 \times 10^3$ km^2. Glacier ablation has a more significant impact on discharge during that time than area change (Fig. 6, grey dots).

Climate studies in the Alpine region indicate that glaciers reached their maximum extent during the LGM and were approximately at their present position at the start of the Holocene (Florineth & Schlüchter, 1998; Hinderer, 2001). *HydroTrend* simulations reflect this trend (Fig. 7). Based on Kotlyakov *et al.* (1997) we mapped the glacier extent into a GIS and found that 15% ($\approx 2.9 \times 10^4$ km^2) of the total Po River drainage basin during the LGM was covered by glaciers. A *HydroTrend* simulation for the LGM, based on the ELA, predicts a similar glacier area to our GIS reconstruction (Fig. 7). Climate changed rapidly during the Bølling and the end of the Younger Dryas. The model simulates several

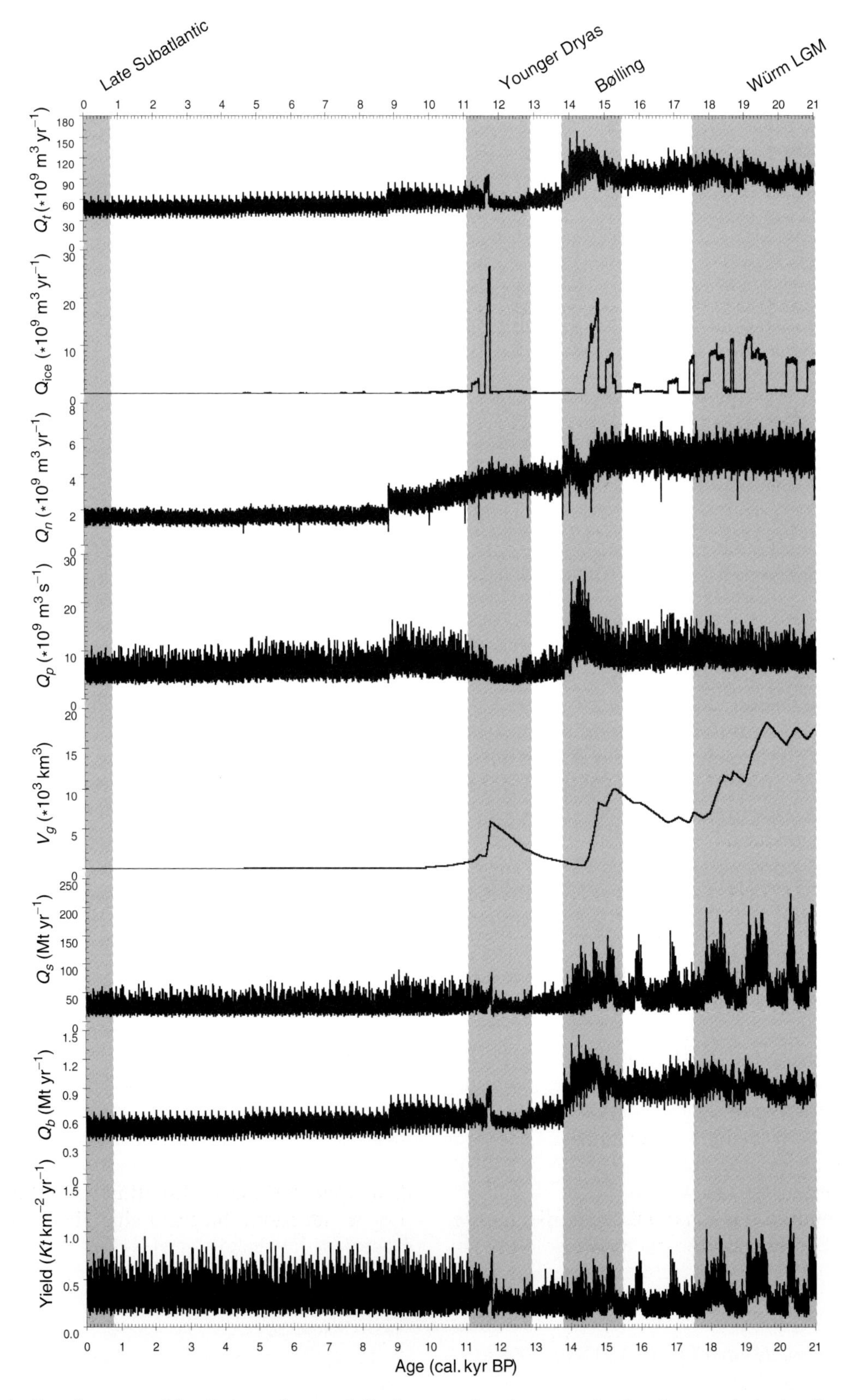

Fig. 7. *HydroTrend* output table: Q_t is total annual discharge at the river mouth; Q_{ice} is annual glacier discharge; Q_n is annual discharge due to snow melt; Q_p is peak discharge; V_g is volume of glaciers in the river basin; Q_s is annual suspended sediment load; Q_b is annual river bedload; *Yield* is the sediment per unit area of the Po drainage basin.

melting phases including during the Bølling and the Younger Dryas that all had significant impact on the total discharge for the Po River. The two highest simulated glacier-meltwater pulses (Fig. 5) coincide with the measured two global major pulses of meltwater entering the world ocean (Bradley, 1999; Clark *et al.*, 2002; Siddall *et al.*, 2003).

Figure 8A shows three years of daily discharge simulations. The first and the third year both represent typical melt events during, respectively, the late Würmian and the Younger Dryas. Meltwater, dominant during the summer months, has significant impact on the total discharge leaving the river. The Younger Dryas melt event had more impact on the total discharge (23% of the total discharge originates from meltwater) than the melt period of the late Würmian (8%) (Fig. 8A). The second year in Fig. 8A represents a hydrograph typical for a growing glacier. In this case, hardly any glacier melt is generated over the summer, and practically all precipitation contributes to the glacier growth.

Figure 8B represents the simulated average daily yield for the same three years. The high sediment yield during the late Würmian period represented in the third year is remarkable. Glaciers have a significant influence on the sediment flux according to our model (see equations 6 and 7). The yield triples during glacier melt as glaciers are covering a significant area in the drainage basin, like during most of the late Würmian period. The influence of glacier coverage is more evident when comparing the plotted year of the Younger Dryas period with the plotted year of the late Würmian period. Although both years represent glacier-melt years, less sediment is produced during the Younger Dryas as the glacial areas in the drainage basin are already much smaller than during the late Würmian period (Fig. 8B, and equation 6). The second year represents a typical glacier-growth year. Although there are still glaciers in the Po drainage basin, little sediment is derived from this growing ice sheet, reflected in the lower sediment-yield values (Fig. 8B).

Figure 9 shows the maximal peak discharges per year plotted against the average discharge for that year. Years with high peak discharges tend to be years with higher than average discharge. One would expect that peak discharges during the Pleistocene occurred most often during glacier-ablation-dominated years. Yet, analysis of the simulation indicates that above-normal snowmelt and rainfall events are the controlling factor for peak discharge during this period. Thus the rapidly fluctuating short-term precipitation signal, based on climate statistics, dominates the peak event generated in the model instead of the medium-term signal of glacier melt. Peak-discharge events are important since they may leave a signal in the stratigraphy, because peak events tend to carry most of the sediment to the ocean (Fan *et al.*, 2004). For example, the event in Fig. 9, highlighted by the green circle, carried more than 64% of the total annual suspended-sediment load to the Adriatic Sea in a 20-day storm event.

For the entire Po River, we estimate an average Pleistocene (21–10 cal. kyr BP) suspended sediment flux of 46.6 Mt yr^{-1} with an average bedload of 0.83 Mt yr^{-1}. This is 1.74 times more than during the Holocene (10–0 cal. kyr BP), when we predict a suspended sediment flux of on average 26.7 Mt yr^{-1} and a bedload of 0.53 Mt yr^{-1}.

Cattaneo *et al.* (2003) summarized the sediment load entering the Adriatic Sea during the late Holocene, for four different catchment areas: eastern Alpine rivers, Po River, eastern Apennine rivers and rivers south of Gargano (Fig. 1). These subsystems have sediment loads of respectively 3 Mt yr^{-1}, 15 Mt yr^{-1}, 32.2 Mt yr^{-1} and 1.5 Mt yr^{-1}. Areas surrounding the Adriatic that are not included (mostly in Croatia) are mainly carbonate-dominated and their sediment supply is negligible. The estimated average suspended-sediment flux draining into the Northern Adriatic Sea by the Po River during the Pleistocene is 34.1 Mt yr^{-1} according to Cattaneo *et al.* (2003). This is less than our estimated 46.6 Mt yr^{-1} sediment flux. However, we estimated that glaciers, covering the Alps during most of the Pleistocene had a large impact on the sediment production, which is not taken into account by the simple extrapolation of Cattaneo *et al.* (2003) estimates (Figs 7 and 10).

Analysis of the four time periods (Fig. 10) confirms that the different conditions (climate and basin area) result in distinctly different water discharge and sediment pulses. We postulate that these time periods are identifiable in the stratigraphy. In addition, the grain-size distributions may carry distinct signals as well. The Würm LGM has the highest suspended sediment concentration of the four time periods (Table 3 and Fig. 10, D2). This is in agreement with the findings of Hinderer (2001), who concluded that the sediment yield of the Alps reached a maximum during deglaciation when large masses of unconsolidated materials were available, vegetation was scarce, and transport capacities were high. During the Bølling water

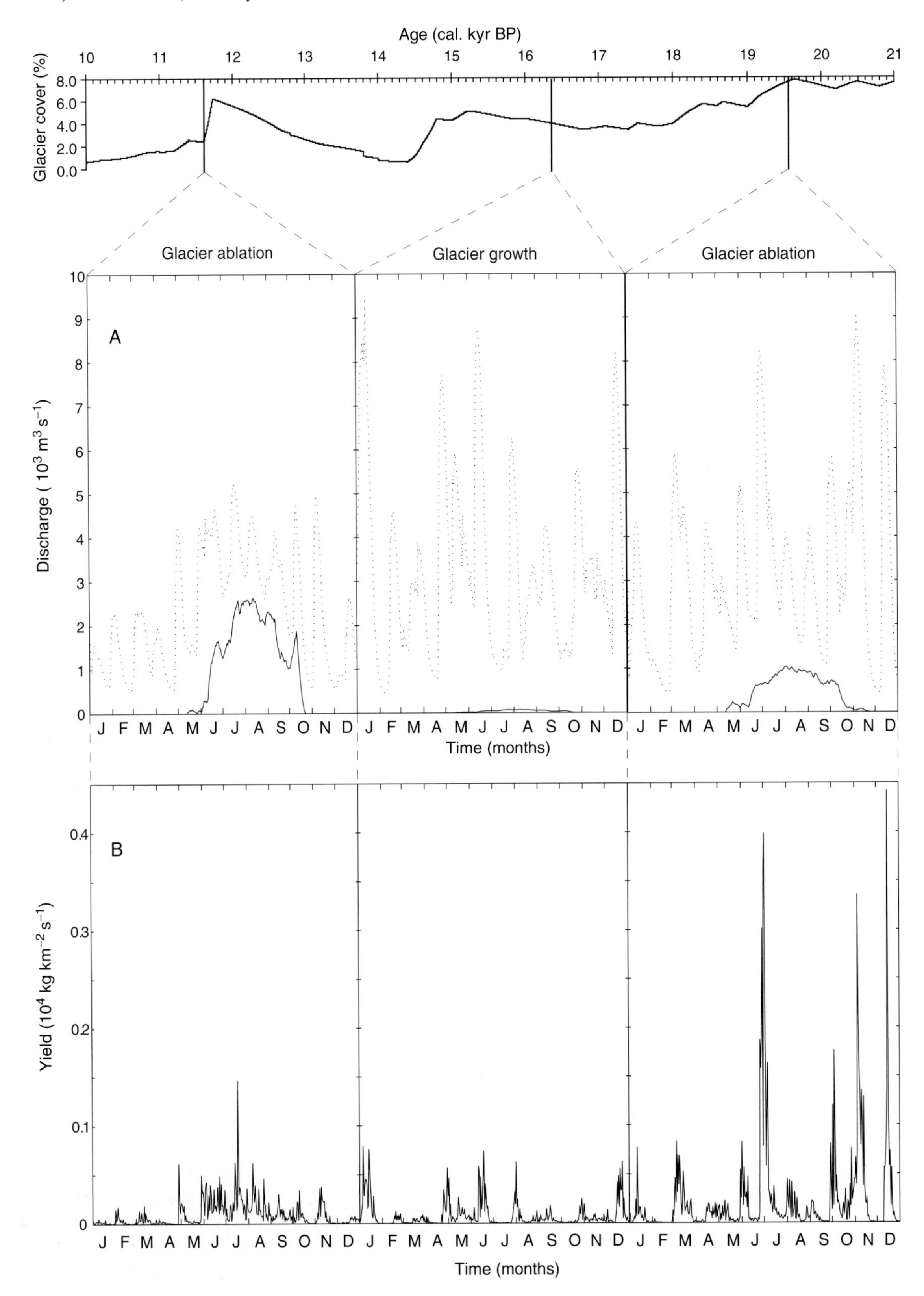

Fig. 8. Influence of glacier ablation on total discharge (A) and suspended sediment load (B) according to the simulation. (A) The dotted line represents the daily discharge ($m^3 s^{-1}$) from all water sources. The solid line represents the daily discharge from glacier ablation. (B) Daily suspended-sediment load increases significantly during glacier ablation.

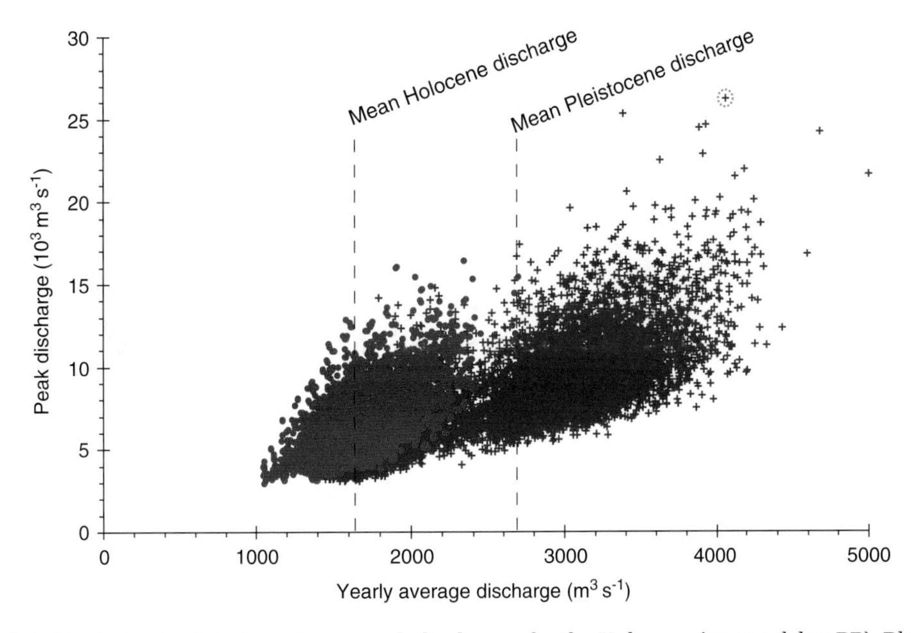

Fig. 9. The red dots illustrate simulated maximum peak discharges for the Holocene (10–0 cal. kyr BP). Blue plus symbols describe simulated maximum peak discharges for the Pleistocene (21–10 cal. kyr BP). The green circle in the upper right is an example of the impact of a flood event. 75.8 Mt suspended sediment load was carried to the Adriatic Sea during that 20-day simulated event, 64.6% of the yearly sediment load.

Table 3. Simulated characteristics of four time periods each based on 50 years daily simulations.

Specific period (cal. yr BP)	Upper Subatlantic 0–50	Younger Dryas 12 340–12 390	Bølling 14 650–14 700	Würm LGM 19 545–19 595
Area (km^2)	74 500	107 300	194 400	194 500
Average discharge (m^3 s^{-1})	1500	1655	3663	3003
Peak discharge (m^3 s^{-1})	10 800	6418	14 686	9887
Average glacier ablation discharge (% of total discharge)	0	1	11.6	8.2
Average suspended sediment load (kg s^{-1})	520	766	1946	3184
Peak suspended sediment load (kg s^{-1})	1235	1102	3944	4978
Average bedload (kg s^{-1})	12.7	17.2	35.0	29.3
Average sediment yield (t km^{-2} yr^{-1})	207	226	317	519

discharges were slightly higher due to glacier ablation, which will have resulted in a high ratio of coarser grain size versus suspended sediment due to a high inflow of bedload (Fig. 10, C2/D2). This increase in discharge during the Bølling was also noticed by Combourieu-Nebout *et al.* (1998).

The sediment supply modelled in this study is only one aspect of the source-to-sink sediment pathway to the marine domain. Sea-level change had a significant impact on the river-mouth location at the shoreline (a 250 km shift over the simulated period) affecting accommodation space. In addition, marine processes will have changed along with climate. This complexity negates direct comparison of our modelled river flux directly with core data. Results of this study are integrated into a stratigraphic modelling study by Kubo *et al.* (2006) and indicate overall agreements with seismic studies of Trincardi *et al.* (1994, 1996) as well as with results from core samples taken from Holocene deposits by Amorosi *et al.* (2003).

Empirical sediment-flux equations used in *HydroTrend* are based on global last-century observations. Obviously, there is uncertainty involved by using this relatively short-term record to hindcast over longer time spans. Another uncertain factor is the long-term input data, which are based on proxy records. Uncertainty analysis for a similar study of the Hudson River has been completed by Overeem *et al.* (2005). One of the findings

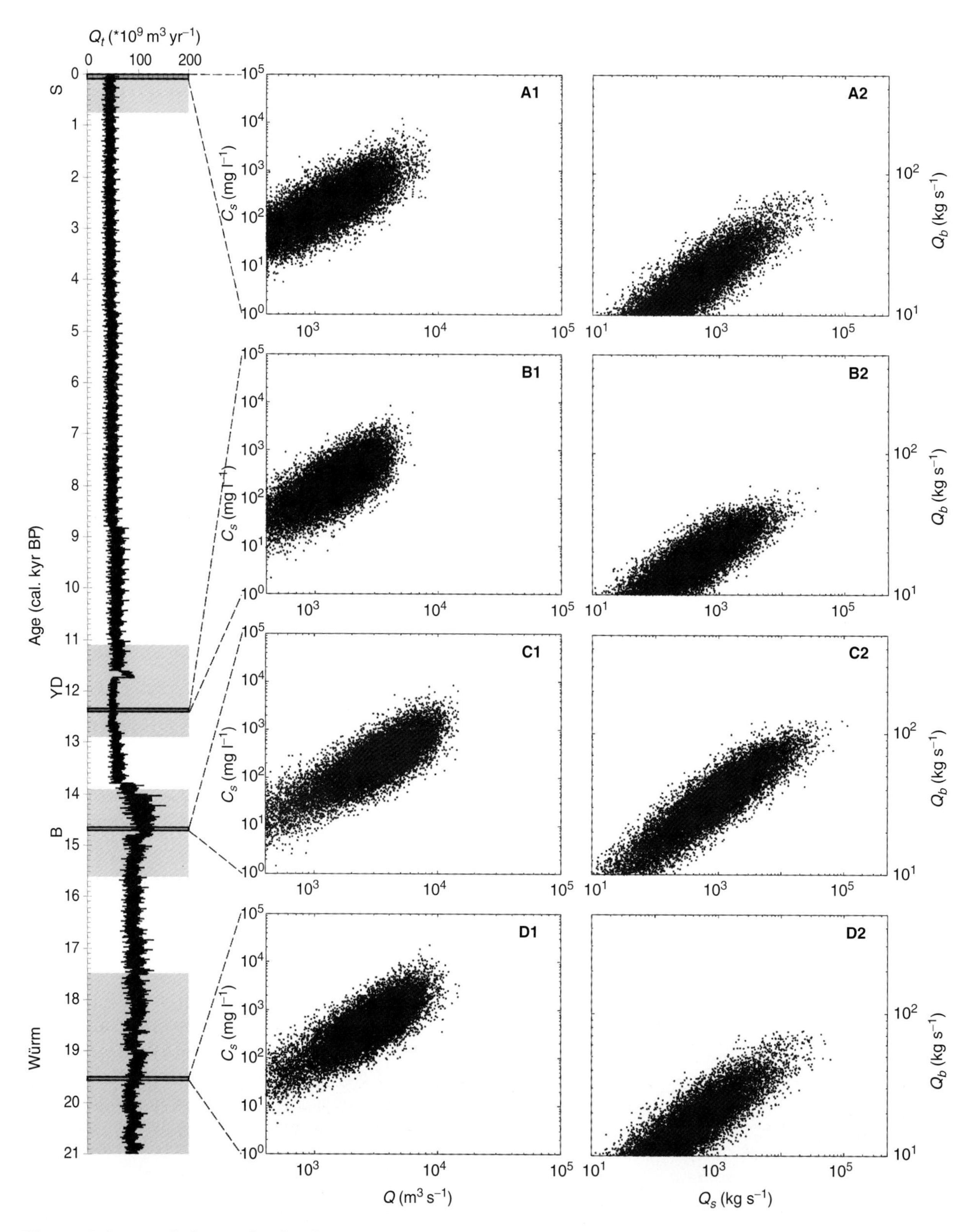

Fig. 10. Fifty-year daily simulated sediment concentration (C_s) versus discharge (Q) and for the same period suspended sediment load (Q_s) versus bedload (Q_b) for four different modelled time periods (A–D): S, late Subatlantic period; YD, Younger Dryas; B, Bølling; Würm, Würm LGM. Q_t is the total annual discharge.

of that study is that the drainage basin area has a dominant impact on the simulations. The change of the extent of the palaeodrainage basin of the Po River as compared to the present is of such magnitude that it overrules the uncertainty in the global sea-level curve and the fact that we use present-day bathymetry.

CONCLUSIONS

HydroTrend is a viable model for predicting the Po River flux of water and sediment under modern conditions. We validated *HydroTrend* predictions on a daily timescale for a 12-year record and found a highly significant correlation with the discharge and sediment-load data at the most downstream gauging station. In addition, a medium timescale validation has been applied by comparing model output to the monthly discharge fluctuations for the record that exists from the period 1760–1995. *HydroTrend* is able to capture the monthly variations over that timescale. Also, the simulated floods are comparable to observed floods over the last 100 years, which is an important ability for the generation of distinct flood deposits in stratigraphic models. *HydroTrend* is designed to be sensitive to hydroclimatological forcing. Human influences such as change in land-use practices are difficult to capture with the present version of the model. Both hydropower management and sediment mining influence the modern Po River. These influences are not relevant for long-term hind casting of river fluxes, the objective of this study.

HydroTrend systematically combines a number of controlling factors: temperature and precipitation changes, basin-area changes due to sea-level rise, ice-cap melt, and evolution of lakes. The model allows us to disentangle the dominance of the individual controls. The long-term simulations show that sea-level fluctuation is a significant control on the fluxes of the Po since the LGM. Sea-level rise controls the 2.6-fold decrease in drainage area of the Po River, which proportionally affects the absolute discharge and sediment flux.

However, the variability in the flux of glacial-derived sediment is even more significant than drainage-area change. The release of meltwater from the ablating Alpine glaciers and snowmelt added substantial water and sediment to the coastal ocean system. In strong glacier ablation years, ~20% of the total discharge can be attributed to glacial melt during summer months. The sediment load more than doubled on average during ablation years in the Würm LGM, with extreme years carrying five times the average sediment load to the river mouth. Fluctuations in suspended load were smaller during the Holocene than in the Würm interval of the LGM. However, the sediment yield increased during the Holocene, due to an increase in precipitation, and thus soil erosion.

ACKNOWLEDGEMENTS

We very much thank Dr Tom Drake for supporting this work through the ONR's EuroSTRATAFORM project. Special thanks to Richard Signell for processing and providing the high-resolution bathymetry data of the Adriatic Sea, and to Annamaria Correggiari for helping obtain Po River observations used in this and other studies. The authors are especially indebted to Irina Overeem for support and her thoughtful review of earlier versions of this manuscript. We thank Mark Dyurgerov and Robert Anderson for discussing the newly developed glacier subroutine in the *HydroTrend* model. The manuscript improved thanks to constructive comments and suggestions by Matthias Hinderer, Bruce Nelson and two anonymous reviews.

REFERENCES

Allen, J.R.M. and **Huntley, B.** (2000) Weichselian palynological records from southern Europe: Correlation and chronology. *Quatern. Int.*, **73–74**, 111–125.

Amorosi, A., Colalongo, M.L., Pasini, G. and **Preti, D.** (1999) Sedimentary response to Late Quaternary level changes in the Romagna coastal plain (northern Italy). *Sedimentology*, **46**, 99–121.

Amorosi, A., Centineo, M.C., Colalongo, M.L., Pasini, G., Sarti, G. and **Vaiani, S.C.** (2003) Facies architecture and latest Pleistocene-Holocene depositional history of the Po Delta (Comacchio Area), Italy. *J. Geol.*, **111**, 39–56.

Asioli, A., Trincardi, F., Lowe, J.J. and **Oldfield, F.** (1999) Short-term climate changes during the Last Glacial-Holocene transition: Comparison between Mediterranean records and the GRIP event stratigraphy. *J. Quatern. Sci.*, **14**, 373–381.

Asioli, A., Trincardi, F., Lowe, J.J., Ariztegui, D., Langone, L. and **Oldfield, F.** (2001) Sub-millennial scale climatic oscillations in the central Adriatic during the Lateglacial: Palaeoceanographic implications. *Quatern. Sci. Rev.*, **20**, 1201–1221.

Bagnold, R.A. (1966) An approach to the sediment transport problem from general physics. *US Geol. Surv. Prof. Pap.*, **422-I**, 37.

Bahr, D.B., Meier, M.F. and **Peckham, S.D.** (1997) The physical basis of glacier volume-area scaling. *J. Geophys. Res.*, **102**, 20,355–20,362.

Baroni, C. (1996) The Alpine 'Iceman' and Holocene climatic change. *Quatern. Res.*, **46**, 78–83.

Bobrovitskaya, N.N., Zubkova, C. and **Meade, R.H.** (1996) Discharges and yields of suspended sediment in the Ob' and Yenisy Rivers of Siberia. In: *Erosion and Sediment Yield: Global and Regional Perspectives* (Eds D.E. Walling and B.W. Webb), *Int. Assoc. Sci. Hydrol. Publ.*, **236**, 115–124.

Bortolami, G.C., Fontes, J. Ch., Markgraf, V. and **Saliege J.F.** (1977) Land, sea and climate in the northern Adriatic region during late Pleistocene and Holocene. *Palaeogeogr. Palaeoclimatol. Palaeoecol.*, **21**, 139–156.

Bradley, R.S. (1999) *Paleoclimatology: Reconstructing Climates of the Quaternary.* Academic Press, San Diego, 610 pp.

Camusso, M., Balestrini, R. and **Binelli, A.** (2001) Use of zebra mussel (*Dreissena polymorpha*) to assess trace metal contamination in the largest Italian sub alpine lakes. *Chemosphere*, **44**, 263–270.

Cattaneo, A., Correggiari, A., Langone, L. and **Trincardi, F.** (2003) The late-Holocene Gargano subaqueous delta, Adriatic shelf: Sediment pathways and supply fluctuations. *Mar. Geol.*, **193**, 61–91.

Clark, P.U., Mitrovica, J.X., Milne, G.A. and **Tamisiea, M.E.** (2002) Sea-level fingerprinting as a direct test for the source of global meltwater pulse 1A. *Science*, **295**, 2438–2441.

Combourieu-Nebout, N., Paterne, M., Turon, J.L. and **Siani, G.** (1998) A High-resolution record of the Last Deglaciation in the Central Mediterranean Sea: Palaeovegetation and palaeohydrological evolution. *Quatern. Sci. Rev.*, **17**, 303–317.

Cushman-Roisin, B., Gačić, M., Poulain, P. and **Artegiani, A.** (2001) *Physical Oceanography of the Adriatic Sea; Past, Present and Future.* Kluwer Academic Publishers, Dordrecht, 304 pp.

Dansgaard, W., Johnsen, S., Clausen, H.B., *et al.* (1993) Evidence for general instability of past climate from a 250-kyr ice core record. *Nature*, **364**, 218–220.

Fan, S., Swift, D.J.P., Traykovski, P., Bentley, S., Borgeld, J.C., Reed, C.W. and **Niedoroda, A.W.** (2004) River flooding, storm resuspension, and event stratigraphy on the northern California shelf: Observations compared with simulations. *Mar. Geol.*, **210**, 17–41.

Fauquette, S., Guiot, J., Menut, J., de Beaulieu, J.-L., Reille, M. and **Guenet, P.** (1999) Vegetation and climate since the last interglacial in the Vienne area (France). *Global Planet. Change*, **20**, 1–17.

Florineth, D. and **Schlüchter, C.** (1998) Reconstructing the Last Glacial Maximum (LGM) ice surface geometry and flowlines in the Central Swiss Alps. *Eclogae Geol. Helv.*, **91**, 391–407.

Friedrich, M., Kromer, B., Kaiser, K.F., Spurk, M., Hughen, K.A. and **Johnsen, S.J.** (2001) High-resolution climate signals in the Bølling-Allerød Interstadial (Greenland Interstadial 1) as reflected in European tree-ring chronologies compared to marine varves and ice-core records. *Quatern. Sci. Rev.*, **20**, 1223–1232.

Friend, P.L., Amos, C.L., Panin, N. and **Trincardi, F.** (2002) Sediment supply and river discharge to the continental shelf – A synthesis of existing data for the Ebro, Rhone, Po and Danube river systems, in Joint European/North American EUROSTRATAFORM meeting (incorporating EURODELTA and PROMESS) Abstract Volume, 8th–13th September 2002, Winchester, UK, 56 pp.

Guymon, G.L. (1974) Regional sediment analysis of Alaska streams. *J. Hydraul. Eng.*, **100**, 41–51.

Hallet, B., Hunter, L. and **Bogen, J.** (1996) Rates of erosion and sediment evacuation by glaciers: A review of field data and their implications. *Global Planet. Change*, **12**, 213–235.

Hinderer, M. (2001) Late Quaternary denudation of the Alps, valley and lake fillings and modern river loads. *Geodin. Acta*, **14**, 231–263.

Huertas, A.D., Iacumin, P. and **Longinelli, A.** (1997) A stable isotope study of fossil mammal remains from the Paglicci cave, southern Italy, 13 to 33 ka BP: Palaeoclimatological considerations. *Chem. Geol.*, **141**, 211–233.

Kotlyakov, V.M., Chernova, L.P., D'yakovo, A.M., Glebova, L.N., Knonvalova, G.I., Osipova, G.B., Rototaeva, O.V., Timofeeva, N.A. and **Varnakova, G.M.** (1997) *World Atlas of Snow and Ice Resources.* Inst. of Geogr., Russ. Acad. of Sci., Moscow, 392 pp.

Kubo, Y., Syvitski, J.P.M., Hutton, E.W.H. and **Kettner, A.J.** (2006) Inverse modeling of post Last Glacial Maximum transgressive sedimentation using 2D-Sedflux: Application to the northern Adriatic Sea. *Marine Geol.*, **234**, 233–243.

Kutzbach, J., Gallimore, R., Harrison, S., Behling, P., Selin, R. and **Laarif, F.** (1998) Climate and biome simulations for the past 21,000 years. *Quatern. Sci. Rev.*, **17**, 473–506.

Lambeck, K. and **Bard, E.** (2000) Sea-level change along the French Mediterranean coast for the past 30,000 years. *Earth Planet. Sci. Lett.*, **175**, 203–222.

Lawson, D.E. (1993) Glaciohydrologic and glaciohydraulic effects on runoff and sediment yield in glacierized basins. *CRREL Report*, **93-2**, 108 pp.

Leonard, E.M. (1997) The relationship between glacial activity and sediment production: Evidence from a 4450-year varve record of neoglacial sedimentation in Hector Lake, Alberta, Canada. *J. Paleolimnology*, **17**, 319–330.

Marchetti, M. (2002) Environmental changes in the central Po Plain (northern Italy) due to fluvial modifications and anthropogenic activities. *Geomorphology*, **44**, 361–373.

Marchi, E., Roth, G. and **Siccardi, F.** (1996) The Po: Centuries of River Training. *Phys. Chem. Earth*, **20**, 475–478.

Morehead, M.D., Syvitski, J.P.M., Hutton, E.W.H. and **Peckham, S.D.** (2003) Modeling the temporal variability in the flux of sediment from ungauged river basins. *Global Planet. Change*, **39**, 95–110.

Nelson, B.W. (1970) Hydrography, sediment dispersal, and recent historical development of the Po river delta, Italy. In: *Deltaic Sedimentation; Modern and Ancient* (Ed. J.P. Morgan) *SEPM Spec. Publ.*, **15**, 152–184.

Overeem, I., Syvitski, J.P.M., Hutton, E.W.H. and **Kettner, A.J.** (2005) Stratigraphic variability due to uncertainty in model boundary conditions: A case-study of the New Jersey Shelf over the last 40,000 years. *Mar. Geol.*, **224**, 23–41.

Parks, B. and **Madison, R.J.** (1985) Estimation of selected flow and water-quality characteristics of Alaskan Streams. *US Geol. Water Resources Invest. Rep.*, **84-4247**, 1–64.

Peyron, A., Guiot, J., Cheddadi, R., Tarasov, P., Reille, M., Beaulieu, J.L. de, Bottema, S. and **Andrieu, V.** (1998) Climatic reconstruction in Europe for 18,000 YR B.P. from Pollen Data. *Quatern. Res.*, **49**, 183–196.

Ramrath, A., Zolitschka, B., Wulf, S. and **Negendank, J.F.W.** (1999) Late Pleistocene climatic variations as recorded in two Italian maar lakes (Lago di Mezzano, Lago Grande di Monticchio). *Quatern. Sci. Rev.*, **18**, 977–992.

Sbaffi, L., Wezel, F.C., Kallel, N., Paterne, M., Cacho, I., Ziveri, P. and **Shackleton, N.** (2001) Response of the pelagic environment to palaeoclimatic changes in the central Mediterranean Sea during the Late Quaternary. *Mar. Geol.*, **178**, 39–62.

Siddall, M., Rohling, E.J., Almogi-Labin, A., Hemleben, Ch., Meischner, D., Schmelzer, I. and **Smeed, D.A.** (2003) Sea-level fluctuations during the last glacial cycle, *Nature*, **423**, 853–858.

Soldati, M., Corsini, A. and **Pasuto, A.** (2003) Landslides and climate change in the Italian Dolomites since the Late glacial. *Catena*, **55**, 141–161.

Surian, N. and **Rinaldi, M.** (2003) Morphological response to river engineering and management in alluvial channels in Italy. *Geomorphology*, **50**, 307–326.

Syvitski, J.P.M., Morehead, M.D. and **Nicholson, M.** (1998) Hydrotrend: A climate-driven hydrologic-transport model for predicting discharge and sediment load to lakes or oceans. *Comput. Geosci.*, **24**, 51–68.

Syvitski, J.P.M. and **Morehead, M.D.** (1999) Estimating river-sediment discharge to the ocean: Application to the Eel margin, northern California. *Mar. Geol.*, **154**, 13–28.

Syvitski, J.P.M., Peckham, S.D., Hilberman, R.D. and **Mulder, T.** (2003) Predicting the terrestrial flux of sediment to the global ocean: A planetary perspective. *Mar. Geol.*, **162**, 5–24.

Syvitski, J.P.M., Vörösmarty, C.J., Kettner, A.J. and **Green, P.A.** (2005) Impact of humans on the flux of terrestrial sediment to the Global Coastal Ocean. *Science*, **308**, 376–380.

Trincardi, F., Correggiari, A. and **Roveri, M.** (1994) Late Quaternary transgressive erosion and deposition in a modern epicontinental shelf: the Adriatic semi-enclosed basin. *Geo-Mar. Lett.*, **14**, 41–51.

Trincardi, F., Cattaneo, A., Asioli, A., Correggiari, A. and **Langone, L.** (1996) Stratigraphy of the late-Quaternary deposits in the central Adriatic basin and the record of short-term climate events. In: *Palaeoenvironmental Analysis of Italian Crater Lake and Adriatic Sediments* (Eds P. Guilizzoni and F. Oldfield), *Mem. Ist. Ital. Idrobiol.*, **55**, 39–70.

Verstraeten G. and **Poesen, J.** (2000) Estimating trap efficiency of small reservoirs and ponds: Methods and implications for the assessment of sediment yield. *Prog. Phys. Geogr.*, **24**, 219–251.

Vose, R.S., Schmoyer, R.L., Steurer, P.M.,Peterson, T.C., Heim, R., Karl, T.R. and **Eischeid, J.** (1992) The Global Historical climatology Network: Long-term monthly temperature, precipitation, sea-level pressure, and station pressure data. Rep. ORNL/CDIAC-53, Carbon Dioxide Inf. Anal. Cent., Oak Ridge Natl. Lab., Oak Ridge, TN, 25 pp.

Vörösmarty, C.J., Meybeck, M., Fekete, B. and **Sharma, K.** (1997) The potential impact of neo-Castorization on sediment transport by the global network of rivers. Human impact on erosion and sedimentation. *Int. Assoc. Sci. Hydrol. Publ.*, **245**, 261–273.

Watts, W.A., Allen, J.R.M. and **Huntley, B.** (1996) Vegetation history and palaeoclimate of the last glacial period at Lago Grande di Monticchio, southern Italy. *Quatern. Sci. Rev.*, **15**, 133–153.

Yu, Z. and **Wright Jr., H.E.** (2001) Response of interior North America to abrupt climate oscillations in the North Atlantic region during the last deglaciation. *Earth-Sci. Rev.*, **52**, 333–369.

Zonneveld. K.A.F. (1996) Palaeoclimatic reconstruction of the last deglaciation (18–8 ka B.P.) in the Adriatic Sea region; a land–sea correlation based on palynological evidence. *Palaeogeogr. Palaeoclimatol. Palaeoecol.*, **122**, 89–106.

Spec. Publ. Int. Assoc. Sedimentol. (2008) **40**, 191–205

Impact of discharge, sediment flux and sea-level change on stratigraphic architecture of river–delta–shelf systems

GEORGE POSTMA and AART PETER VAN DEN BERG VAN SAPAROEA

Utrecht University, Department of Earth Sciences, Faculty of Geosciences, P.O. Box 80.021, 3508TA, Utrecht, The Netherlands (E-mail: gpostma@geo.uu.nl)

ABSTRACT

Variation in sediment transport and temporary sediment storage at large spatial and temporal scales is governed by inherited relief and by the interaction between climate change (water discharge and sediment yield), eustatic sea-level change and tectonics. These processes influence the character of sedimentary successions, and are difficult to reconstruct from stratigraphy alone. Physical modelling can help by allowing observation and quantification of processes that generate model stratigraphy. Sediment transport in the model set-up shows diffusive behaviour. Non-dimensionalizing the relevant physical parameters that underlie model stratigraphy allows a comparison with real-world examples. The effect of (changes in) the forcing parameters (discharge, sediment yield, sea-level change and consequent shoreline position) on a loose bed of sand in a set of experiments is measured by means of high-resolution digital elevation models (DEMs). These DEMs are transformed into geological maps, profiles and Wheeler diagrams that allow detailed reconstruction of sediment flux and routing, storage time, and headward erosion rates. All raw data are stored in the Pangaea database, and are publicly available. A synthesis of all experiment results obtained over the past 10 years has resulted in a first-order prediction of sediment flux that is induced solely by sea-level change. The flux changes with relaxation time (response time) of the system, transport efficiency of the river system (in terms of slope gradient), and the amount that sea level drops below the shelf edge (change in slope gradient); all variables are dimensionless. The results are presented as templates covering systematic trends in the effects of the dimensionless parameters on stratigraphic architecture, which allows for the assessment of the impact of sea-level change on sediment flux relative to the other forcing parameters.

Keywords Analogue modelling, sequence stratigraphy, passive margin, stratigraphic architecture, sea-level change, climate change, sediment supply, accommodation.

INTRODUCTION

Reconstructions of cause-and-effect relationships in basin fills are usually inferred on the basis of deductive reasoning, since there is no direct record of the basin-fill history. Interpretations are qualitative and speculate about the way in which tectonics, eustatic sea-level change and sediment supply mechanisms have influenced the development of accommodation space and sediment supply. This qualitative approach logically implies many uncertainties. The various combinations of variables that can produce sedimentary cycles prohibit detailed reconstructions of the precise forcing mechanisms. A large number of factors, many of which are linked, may contribute to the stratigraphic product, and likely more than anywhere else the notion 'equifinality' is applicable to stratigraphic successions. This has provoked modellers to quantitatively investigate stratigraphic models by constraining input parameters such as sea-level change, tectonics, discharge, sediment flux and grain size (cf. Burgess & Allen, 1996; Kooi & Beaumont, 1996; Burgess & Hovius, 1998; Meijer, 2002; Meijer *et al.*, this volume).

One important and, as yet, underutilized tool is the use of physical experiments on a sedimentary-systems scale. The obvious advantage of a physical experiment on a systems scale is that one can observe from start to end how systems are developed, and how they evolve under various conditions of imposed allocyclic forcing (e.g. Wood *et al.*, 1993; Koss *et al.*, 1994; Milana, 1998; Paola, 2000; Van Heijst *et al.*, 2001; Van Heijst & Postma, 2001). The disadvantage of physical models on a landscape-scale is that there is no commonly accepted scaling strategy, so that the applicability of the results to real-world systems is at best by analogue (e.g. Peackall *et al.*, 1996). Conventional Froude scaling fails for geological-scale sediment transport and alternatives have not yet been investigated to the full extent. Paola (2000) discussed this problem, and Van Heijst *et al.* (2001) proposed a scaling strategy based on the diffusive behaviour of long-term sediment transport (cf. Begin, 1988; Paola *et al.*, 1992).

The first objective of this paper is to address the progress and the outstanding problems in scaling long-term, geological-scale sediment transport. The second objective is to synthesize available data produced over the last ten years by various Utrecht workers and to create a template for flux induced by sea-level change. Sea-level oscillations in the sequence-stratigraphy concept are seen by many as the important independent variable controlling the stratigraphic architecture of sedimentary systems. Changes in relative sea level have a direct effect on accommodation space in the coastal realm and control, together with sediment supply, the large-scale architecture of delta and shelf systems on various timescales (Curray, 1964; Posamentier & Vail, 1988; Van Wagoner *et al.*, 1990). The third objective is to investigate the applicability of the experimental results to real-world systems. We do this using the well studied source–sink example of the Gulf of Mexico, the Colorado River shelf system in Texas.

APPROACH

Basin-fill architecture is controlled by the initial topography and by the ratio of the rate of sediment supply and the rate of change in accommodation space (e.g. Curray, 1964; Posamentier & Vail, 1988). Since the rate of sediment supply at any point is the sum of what is produced and what is cannibalized upstream, it is important to consider the behaviour of the entire source-to-sink system when studying causality in stratigraphy. This will reveal, for instance, the consequences of local and temporal storage of sediment in the river valley and on the shelf. We approached this by modelling an entire source–sink system. The drainage system was represented by a sediment feeder, a fluvial valley system or transfer zone, where sediment can be cannibalized and temporarily stored in a rectangular duct, and the basin by a table covered with sediment that can act as a sink for sediment (Fig. 1). The basic shape and facies belts of the river–delta–shelf system in the flume and its evolution are similar to that of many common real-world systems, where the river gradient is steeper than the coastal plain and shelf gradient. In particular, it represents the well investigated Colorado River shelf system in Texas, which we previously used as a prototype (Van Heijst *et al.*, 2001).

Preparations for each experiment included the levelling of the sand according to a fixed initial topography (cf. Van Heijst & Postma, 2001). Each experiment started at sea-level highstand with a 15-h run to establish a dynamic equilibrium stream profile as the initial condition. The sea-level change was imposed by adjustment of the level of overflow in the main tank at 10-min intervals. Every hour, both discharge and sediment supply were checked, and the stream profile in the fluvial valley was measured by reading the rulers on the valley's transparent wall. At 5-h intervals the topography in the main tank was measured using the automated bed profiler, for which the tank was drained slowly to avoid disturbance of the sediment. The topography was then measured by using a laser following a 2 cm x–y grid. The laser measurements had a vertical (z axis) precision of 300 μm and the background noise in the z direction was less than 1 mm.

A typical example of the evolution of a modelled system forced only by sea-level change is shown in Fig. 2. During the early falling stage, the delta continued to aggrade. The shoreline prograded with the falling sea level. Local gradients and irregularities in the topography controlled the places where headward erosion induced by sea-level fall started incising the emerging shelf. The heads of the incisions (canyons) migrated upslope until one of them connected to the fluvial valley. The relative timing of connection of one of the shelf canyons with the fluvial valley has a strong bearing on the final volume of the slope fan and lowstand delta, because it determines the change from aggradation to degradation of the fluvial valley (Van Heijst & Postma, 2001). Overall, these experiments show

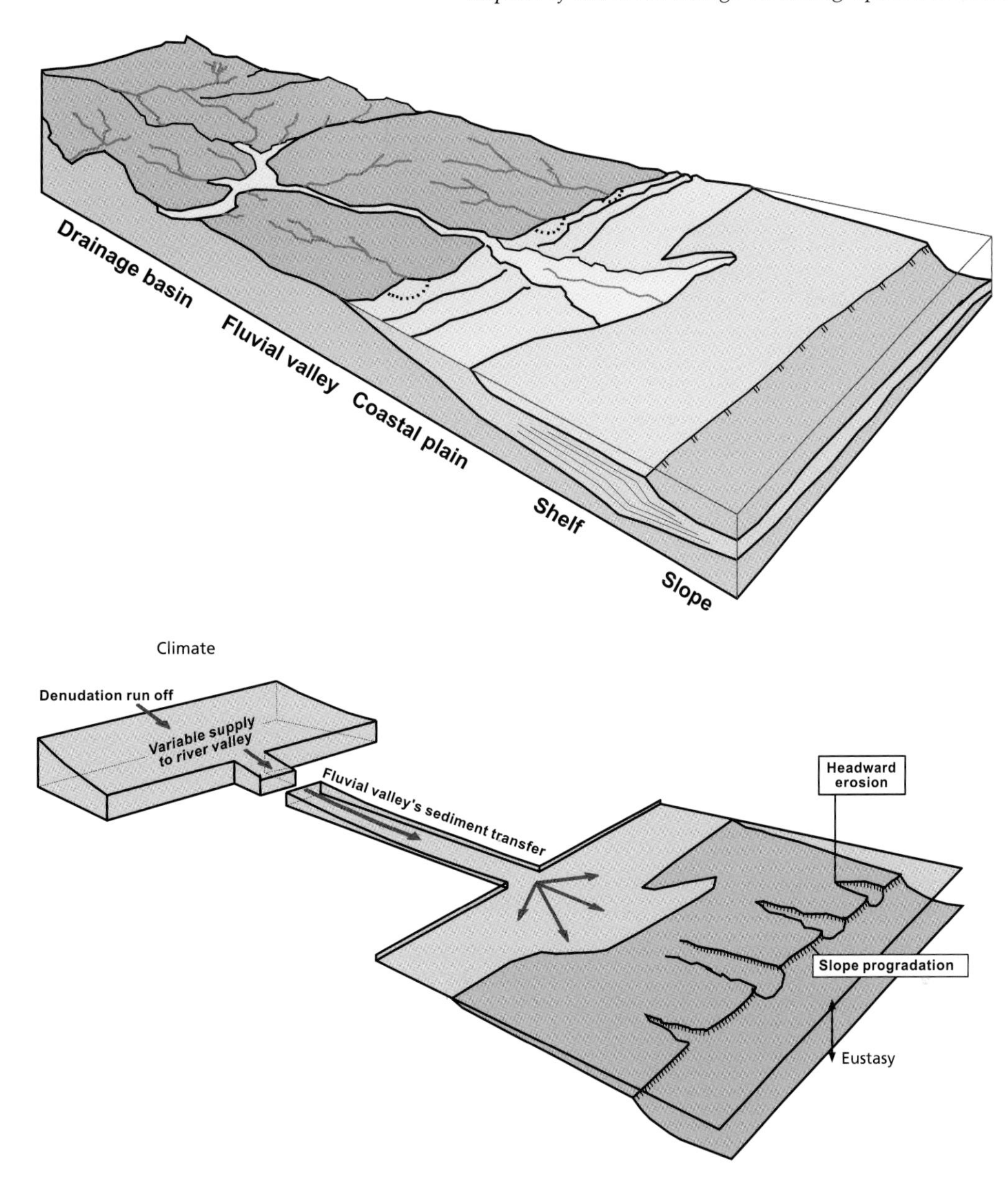

Fig. 1. Real-world source-transfer-basin system simulated in our experiment. The dimensions of the set-up are 3 m × 3.4 m × 1 m for the depositional basin and 4 m × 0.11 m × 0.5 m for the fluvial valley. A water tap with flow meter provides discharge, and a sediment feeder with adjustable conveyor-belt speed controls the sediment supply. Both are located at the upstream end of the fluvial valley and act as a surrogate for the drainage basin. The applied sediment is unimodal, medium, cohesionless quartz sand ($D_{50} = 240\,\mu m$) that is supplied by the feeder and is also used as substrate.

that (1) the dynamic equilibrium gradient of the river profile, (2) the headward erosion rate (defining the amount of cannibalism) and (3) the relaxation time (the time that the system requires to restore equilibrium) all are important variables, which must be considered in a scaling strategy.

Scaling strategy

A critical question and point of ongoing research is the scaling of sediment transport over timescales of landscape evolution. How should we average catastrophic events of various intensities and frequencies combined with normal conditions? Most important is that we do not attempt to accurately model sediment transport, but to model sediment preservation, the net result of sediment transport over geologically relevant timescales. Bedload equations (e.g. Yalin, 1971) do not allow down-scaling of spatial dimensions more than 1:50 (Peackall *et al.*, 1996; Moreton *et al.*, 2002), rendering the Froude scaling for

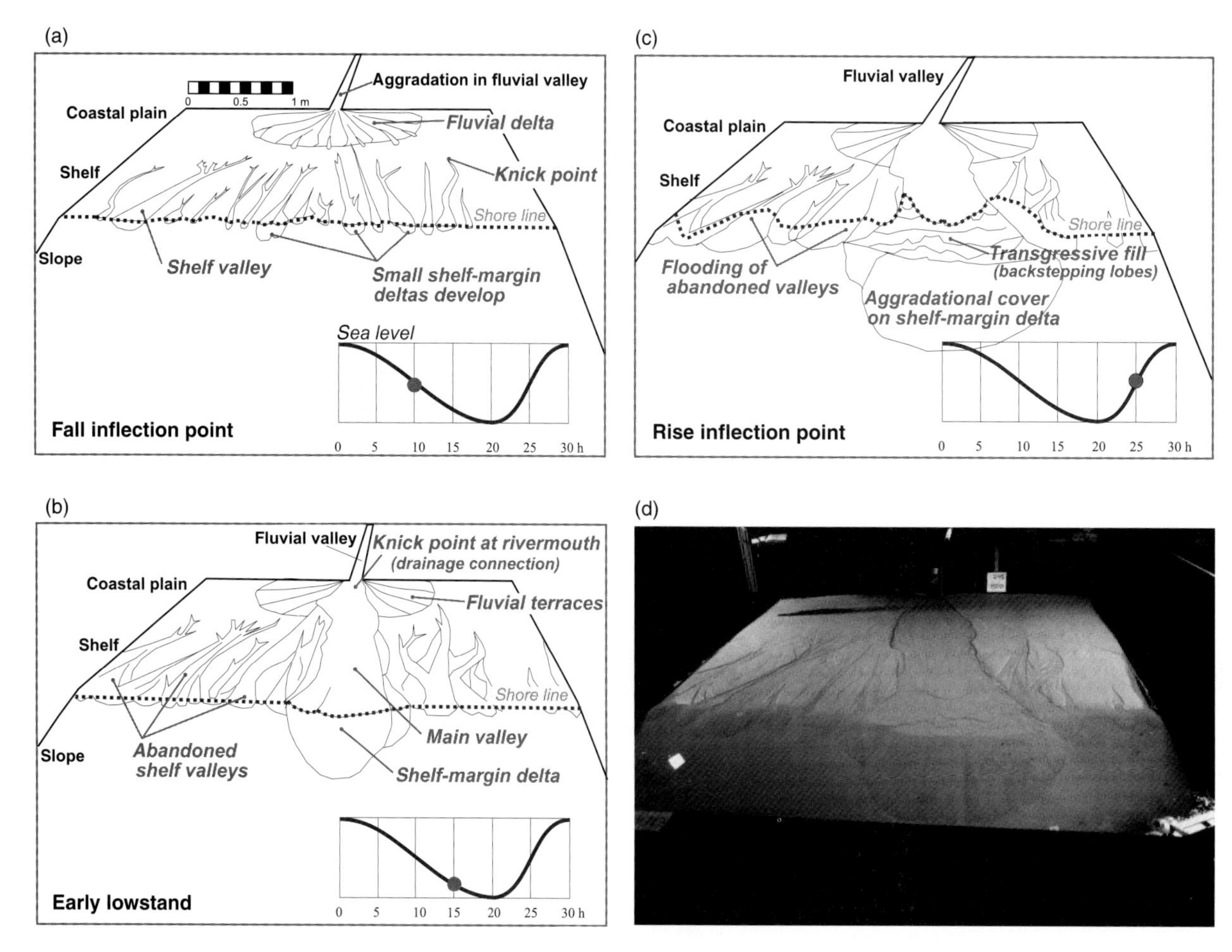

Fig. 2. Photograph and drawings of experiment 240. The drawings show headward erosion of canyons that incise into the shelf. At about 13 h, just after the fall-inflection point, connection of the canyon with the fluvial valley occurred. At this point all the water and sediment discharge is funnelled basinwards directly through the connected valley to the lowstand delta. Headward erosion now starts cannibalizing sediments in the fluvial feeder valley and the drainage basin, increasing the yield at the valley outlet system (modified from Van Heijst and Postma, 2001).

down-scaling entire systems impossible. This is why experiments on river-system scale are often referred to as analogues or analogue experiments. Hooke (1968) used analogue experiments to study evolving landscapes and referred to this scaling strategy as 'similarity of process'. By keeping the ratio between the size of the system and the time-averaged sediment transport rate in experiments and in nature about similar, spatial dimensions (volumes) can be scaled against time (Van Heijst *et al.*, 2001):

$$Q_s = \frac{\Delta V_{(rw)}}{\Delta T_{(rw)}} = \frac{\Delta V_{(exp)} \cdot (\lambda_x \cdot \lambda_y \cdot \lambda_z)}{\Delta T_{(exp)} \cdot (\lambda_t)} \quad \{L^3/T\} \tag{1}$$

where Q_s is the time-averaged volumetric sediment transport rate (including normal and catastrophic events), ΔV is the displaced sediment volume and ΔT is the time period (normally of the order of 500–1000 yr) over which the amount of displaced volume is determined. The scaling factors λ operate on the spatial dimensions (x–y–z) and time (t). Subscripts rw and exp indicate real world and experiment, respectively. In a perfect model, the up-scaled, time-averaged sediment flux observed in the experiment equals the value of the prototype.

Sediment-transport models for long timescales show diffusive behaviour (Begin *et al.*, 1981; Andrews & Buckham, 1987; Begin, 1988; Paola *et al.*, 1992; Kooi & Beaumont, 1996; Martin & Church, 1997; Paola, 2000). Begin *et al.* (1981) and

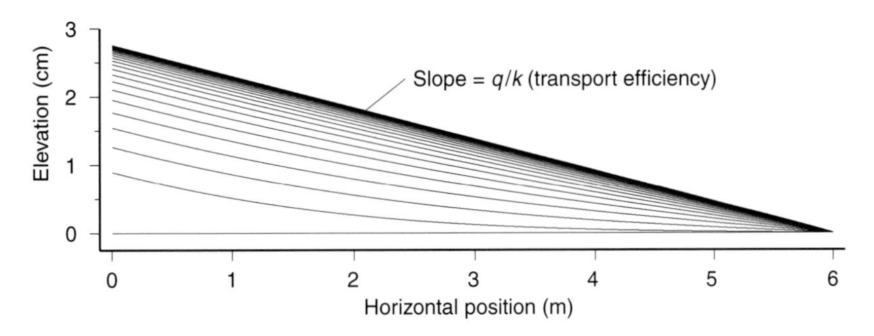

Fig. 3. Numerical solution of the diffusion equation. The gradients in the solution are compared with sediment aggradation patterns (stratigraphy) in the experiment (cf. Begin, 1988; Paola *et al.*, 1992). The equilibrium gradient is a consequence of transport efficiency, which is defined here as the ratio of sediment input (q) over diffusivity (k). The latter coefficient is directly related to discharge.

Begin (1988) showed that channel degradation by base-level lowering closely follows the numerical solution of the simple diffusion equation

$$\frac{\partial\langle h\rangle}{\partial t} = \frac{\partial}{\partial x}\cdot\left(k\cdot\frac{\partial\langle h\rangle}{\partial x}\right) \quad \{\mathrm{m\,s^{-1}}\} \tag{2}$$

where $\langle h\rangle$ is the average streambed height (m) with respect to origin, x (m) is the horizontal position with respect to origin, t is time (s) and k is the diffusivity constant for unit width ($\mathrm{m^2\,s^{-1}}$). The discharge variable appears in the diffusivity coefficient when the diffusion equation is derived from first principles (Paola *et al.*, 1992):

$$k = \frac{-8Q\cdot A\sqrt{c_\mathrm{f}}}{C_\mathrm{o}(s-1)} \tag{3}$$

where Q is discharge per unit width ($\mathrm{m^2\,s^{-1}}$), c_f is drag coefficient [0.025], C_o is sediment concentration of the bed of 0.67, and s is specific density; $\rho_\mathrm{s}/\rho_\mathrm{w} = 2.65$. Changes in diffusivity depend on changes in discharge and riverbank erodibility. Riverbank erodibility constant A has values of 1 for the meandering case and 0.15–0.4 for the braided case.

When the system is in equilibrium, the gradient of the equilibrium surface can be determined from

$$q = k\frac{\partial\langle h\rangle}{\partial x} \tag{4}$$

where q is the sediment flux into the river system (sediment feed). The gradient of the fluvial valley river is dependent on the stream capacity and the available sediment load defined by the ratio q/k. Thus q/k denotes a dynamic equilibrium gradient, which is a measure for the transport efficiency of the system (Fig. 3). Very efficient systems have a low gradient and less efficient systems have a higher gradient (see below).

The time and length scaling is obtained from non-dimensionless analysis of the diffusion equation as shown by Paola *et al.* (1992), who gave a first-order approximation of the relaxation time (T_eq) of the fluvial system by:

$$T_\mathrm{eq} = \frac{L^2}{k} \quad \{\mathrm{T}\} \tag{5}$$

where L is the basin length. Basins with different length L and/or different values for k will have different relaxation times, and thus different response times to imposed changes.

Any allocyclic forcing mechanism can be determined from the stratal record only if its duration is longer than the system's relaxation time. In other words, if the system cannot return to its state of dynamic equilibrium before the next allocyclic change, the allocyclic signal is unlikely to be clearly recognizable is the stratal record. It is thus important to establish the frequency of change relative to the relaxation time of a system. Paola *et al.* (1992) and Heller & Paola (1992) proposed the dimensionless ratio T/T_eq. Van Heijst *et al.* (2001) referred to the ratio as the basin-response factor (Br):

$$Br = \frac{T_\mathrm{(rw)}}{T_\mathrm{eq(rw)}} = \frac{T_\mathrm{(exp)}}{T_\mathrm{eq(exp)}} \quad \{\text{-}\} \tag{6}$$

where T is the duration of one period of allocyclic change (e.g. duration of one sea-level cycle) and the response time of the sedimentary system in the real-world (rw) and the experiment (exp).

For the purpose of geological modelling of sedimentary systems it is important to be able to compare the vertical stacking of depositional environments in the model and the prototype. This is done by considering the time-averaged sedimentation rate per unit area (R_s), which is obtained by dividing the time-averaged sediment deposition ($\Delta Q_s \geq 0$) by the depositional area (A):

$$R_s = \frac{\Delta Q_s}{A} \quad \{L/T\} \tag{7}$$

Applying the concept of Curray (1964), we use a basin fill factor (Bf), a non-dimensional parameter that describes the time-averaged sedimentation rate in relation to the rate of increase in accommodation space:

$$Bf = \frac{R_{s(rw)}}{R_{acc(rw)}} = \frac{R_{s(exp)}}{R_{acc(exp)}} \quad \{-\} \tag{8}$$

where R_{acc} accounts for the rate of change in accommodation space per unit area for both the fluvial and marine realm in real-world (rw) and experiment (exp). Hence, the basin-fill factor is a dimensionless parameter that describes progradation, aggradation and retrogradation of sedimentary systems. Note that R_s and R_{acc} should both be time-averaged over the same time span ΔT.

In studying evolutionary trends in sedimentary systems by experiment we thus focus on the volumetric changes per time slice, that is, time-averaged sediment flux. The sand that we use is merely an isotropic medium, the mass of the grain influencing the value of the diffusivity coefficient, k. It is important to realize that we only value the large-scale topographical features that are built over similar timescales as our time-average sediment transport (equation 1). Hence, we look at the scale of canyon and valley formation over time steps of ~6500 years, which are equivalent to the 5 h between each two measurements, and not at the scale of individual channel incisions. Thus the scale of observation and measurement in the model is well tailored to the scale where allocyclic forcing mechanisms produce regional bounding (erosional) surfaces and depositional units (stratigraphy).

It must be noted, however, that in applying the above outlined non-dimensional scaling ratios to real-world systems there remains much uncertainty in the calculated values due to poor data on precipitation, and valley width and stratigraphy are too poorly known. This is shown below when dealing with the Colorado prototype as an example.

RESULTS

A series of experiments has been conducted to study the behaviour of the system under different frequencies of sea-level change (Fig. 4), that is, different basin-response factors, ranging from 1 to 4, and various transport efficiencies (q/k) (Fig. 5). Observations include time-lapse video recordings, photographs and laser scans taken at 5-h intervals. The laser scans provided digital elevation models (DEMs) of the surface of the model. From the series of DEMs, isopach maps, geological maps and geological profiles were composed to establish the architecture (style) of the deposits in the model (Fig. 5). All DEMs are stored in the publicly accessible Pangaea database (www.pangaea.de) in ASCII files containing x, y and z values. The x, y, z data are compatible with almost any relevant software package. We have used Golden Software Surfer for making topographical maps and calculating volumes in isopach maps, and have developed in-house software for generating the synthetic stratigraphy from the series of DEMs, which is the basis for the geological maps, geological profiles and Wheeler diagrams.

Examining a system's behaviour for various discharges and sediment input, that is, on transport efficiency q/k, we find a strong relationship between preserved stratigraphy in the fluvial and shelf realms and the transport efficiency. For instance, increasing values for q/k results in decreasing thicknesses (and preservation potential) of the highstand and falling stage systems tracts (Fig. 5). Converting the vertical sections to Wheeler diagrams (Fig. 6) reveals that the regional importance (length) of the sequence boundary clearly increases with transport efficiency. Furthermore, inefficient systems (high q/k ratio) are characterized by steep slopes, high preservation potential of all systems tracts in the river valley and poorly developed sequence boundaries that are developed only on the shelf. The opposite is found for highly efficient systems.

Unfortunately, gradients in the experiments are different from those in the real world: to maintain bedload transport with the grain size and densities used in the experimental set up, steeper slopes are required. It is thus not possible to use q/k directly when comparing the experimental results with the real world. However, it can be used in a semi-quantitative way by referring to the preservation of deposits in the valley as shown by the profiles in Fig. 5 and the Wheeler diagrams in

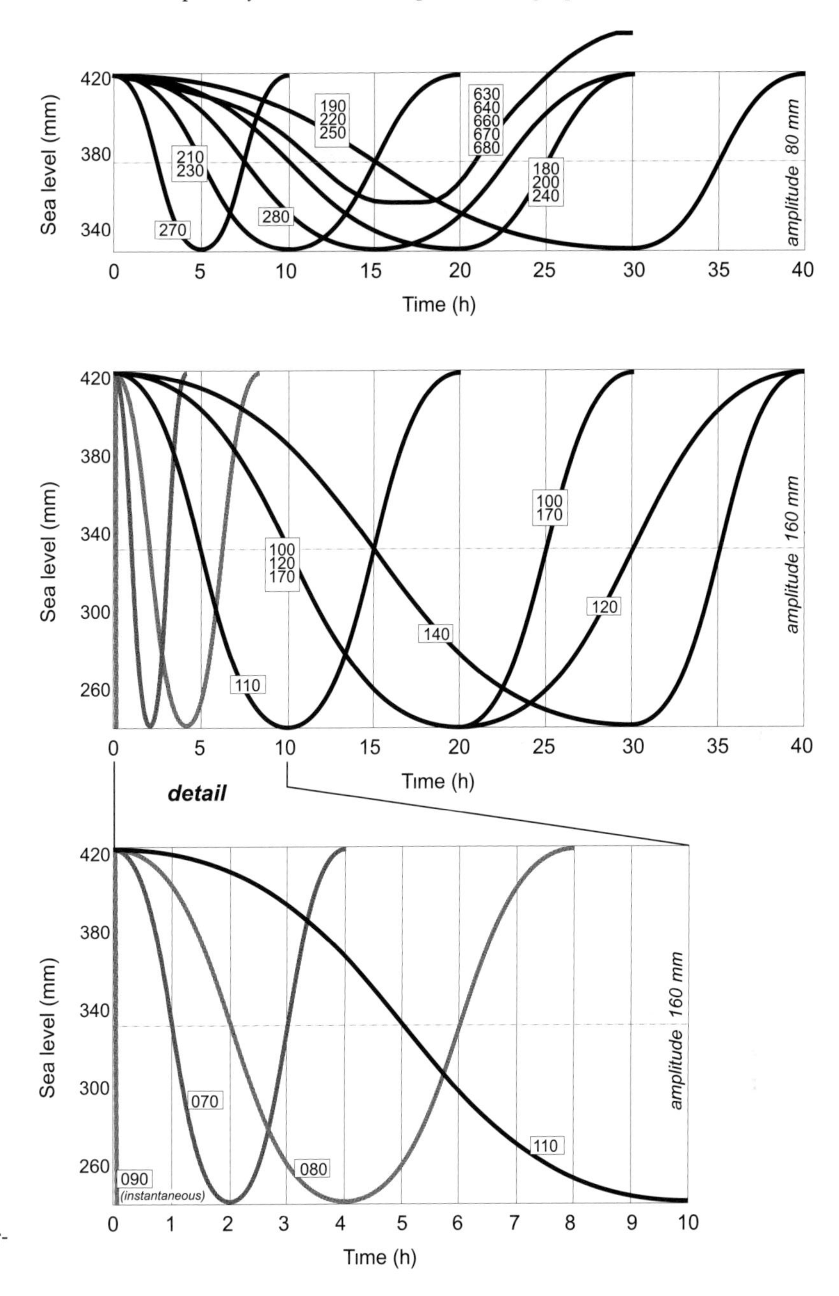

Fig. 4. Sea-level curves used in the experiments plotted in the template in Fig. 7.

Fig. 6. As a first approximation, we have distinguished four categories of transport efficiency in our template. Characteristic preservation patterns (architectural styles) are recognized, and are exemplified by the geological profiles (labelled A–D) in Fig. 5. Type A is characteristic for high-supply systems, as for instance is inferred for the Lorançdel Munt fan-delta system (Steel *et al.*, 2000). Type D is characteristic for a highly efficient system where the entire valley is flushed during lowstand. Types B and C take intermediate positions. Burgess & Allen (1996) came to very similar conclusions through their numerical modelling study. They found that increased gradients (comparable to our case A) in the fluvial domain go hand in hand with reduced vertical and lateral extent of erosion. The sequence boundary resulting from sea-level lowstand in these high-gradient systems is less extensive than in low-gradient systems under identical forcing parameters. It appears difficult to give real-world examples of each, since real-world data on valley systems have not been sufficiently explored.

Checking a system's behaviour in response to sea-level change for various frequencies of sea-level fall using the data of Van Heijst &

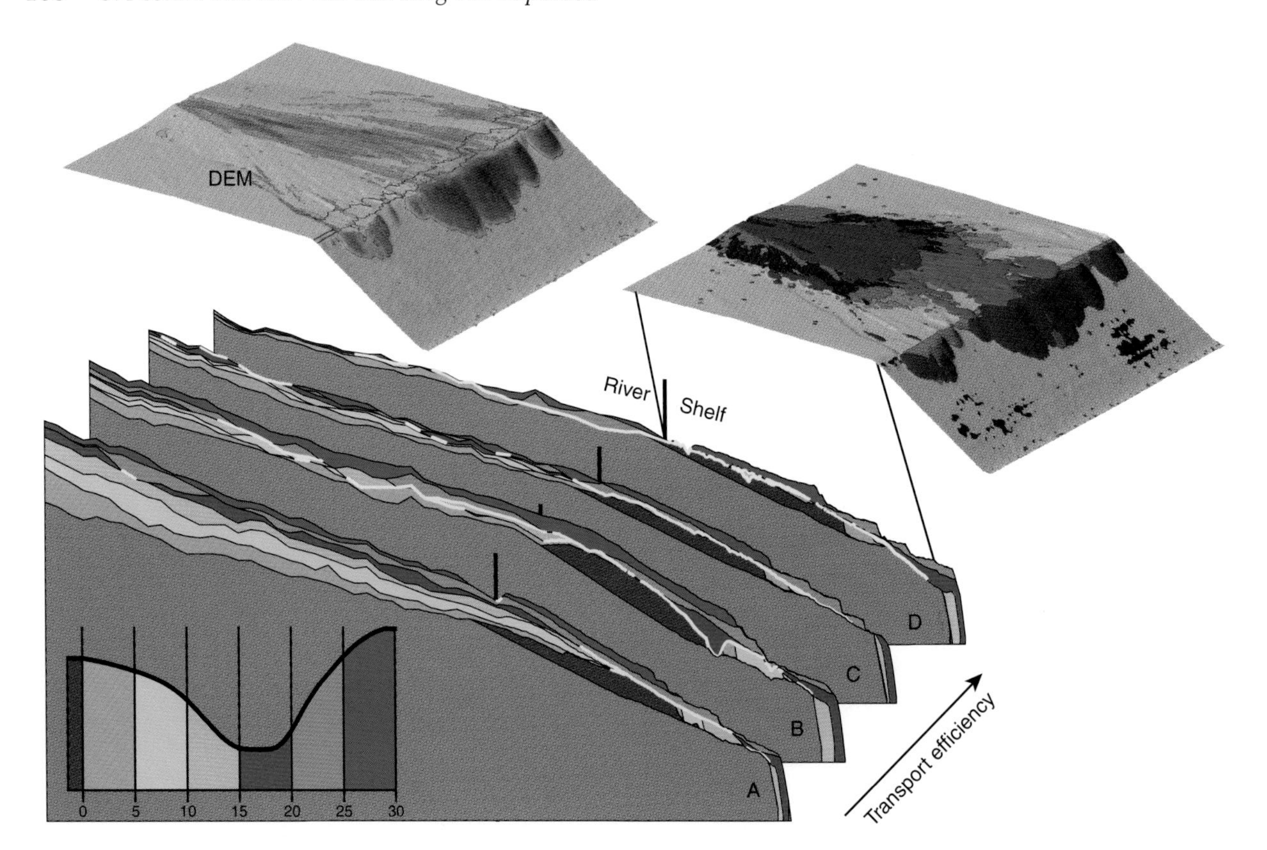

Fig. 5. Longitudinal profiles of the stratigraphy of four experiments with different transport efficiencies. Other parameters were identical for each of these experiments. The sea-level curve given as inset in profile A was used. The colours of the stratigraphic units in the profiles correspond to the colours of the time slices indicated in the sea-level curve.

Postma (2001) (Fig. 4) and additional unpublished data for constant sediment supply, we find that the rate of cannibalization and the related sediment flux towards the shelf edge increases with increasing amplitude of sea-level change, that is, the amount of sea-level fall below the shelf edge, and with the basin-response factor (Fig. 7). Also, with decreasing basin-response factor the connection of incised valleys with the river valley is delayed, resulting in a more diachronous sequence boundary and thicker transgressive systems tracts, which is also enhanced by the increasing amplitude (in particular increasing rate of fall) of sea level.

Figure 7 combines the results into a template for sediment flux induced by sea-level change for various scenarios of transport efficiency, frequency of sea-level change (i.e., basin-response factor) and amplitude of sea-level change. The template indicates what systematic trends in yield relative to the sea-level curve (thus for systems tracts) can be expected in response to the external parameters investigated in this study: amplitude of sea-level change, transport efficiency and basin-response factor (Fig. 7). Sediment fluxes are normalized to the sediment input from the catchment. Under equilibrium conditions, sediment influx from the catchment matches the yield of the river system, since its profile, and thus the volume of sediment in the river valley, does not change in equilibrium. Any deviation from a yield of 100% in Fig. 7 is caused by changes in the external forcing parameters. In natural systems the sediment influx from the catchment must be estimated.

THE COLORADO RIVER-VALLEY–DELTA–SHELF SYSTEM PROTOTYPE

The underlying assumption of the template in Fig. 7 is that external sediment supply remains constant. Thus, variation in sediment flux caused by climate change or tectonics can be assessed on the basis of deviations from the template if other external forcing mechanisms are constant or change in (roughly) the same way as they did in our experiments. We illustrate the application of the

Fig. 6. Wheeler diagrams (deposition and erosion through time) for profiles A (low transport efficiency) and D (high transport efficiency). The solid lines indicate the final product at the end of an experiment (the equivalent of a Wheeler diagram for natural systems). Dashed lines indicate deposits that have been eroded. Sea-level oscillation is indicated on the left. In the low-efficiency system (A) deposition is continuous nearly everywhere. Erosion is limited to the shelf area and occurs only during sea-level fall. In the high-efficiency system (D) deposition is much more localized and intermittent. Most of the shelf and the whole river valley are subject to erosion, which starts immediately with sea-level fall and continues until late in the transgressive phase.

template on the basis of the Colorado river–shelf system of Texas (Gulf of Mexico), which has been studied extensively in the last decades (Anderson *et al.*, 1996, 2004).

The Colorado River is fed by a drainage basin of 110 000 km^2 (Fig. 8). It is a bedload-dominated system that extends over the shelf during periods of sea-level lowstand feeding shelf and shelf-margin deltas (Anderson *et al.*, 1996). The lower Colorado River (downstream of the Balcones escarpment, Table 1) is characterized by high-gradient fluvial terraces (Fig. 8). It was mapped and dated in detail by Blum (1993), Blum & Valastro (1994), Blum *et al.* (1994) and Blum & Price (1998). Deposition during and after the last glacial lowstand (the Eagle Lake Alloformation, 20–14 ka; Blum & Valastro, 1994) has been attributed to a period with high sediment yield that exceeded transport capacity, creating the deposits in which the river could later form fluvial terraces by degradation. If this is true, sediment supply from the drainage basin must have greatly diminished between 14 and 12 kyr BP resulting in abandonment of Eagle Lake Alloformation floodplains and incision of bedrock valleys. Sediment yield from the drainage basin increased again after 5000 yr BP (Blum & Valastro, 1994). Multiple episodes of aggradation, degradation and abandonment of floodplains followed (Columbus Bend Alloformations 1, 2 and 3, respectively). The Colorado River coastal-plain sediments are underlain by a composite basal unconformity that corresponds partly to the lowstand systems tract and in part to the transgressive systems tract (Blum & Valastro, 1994). Thus there is a strong diachroneity along the sequence boundary of the Colorado River. The same coastal prism is truncated by an erosive surface (14–11 ka) that merges with the basal unconformity upstream, which shows that

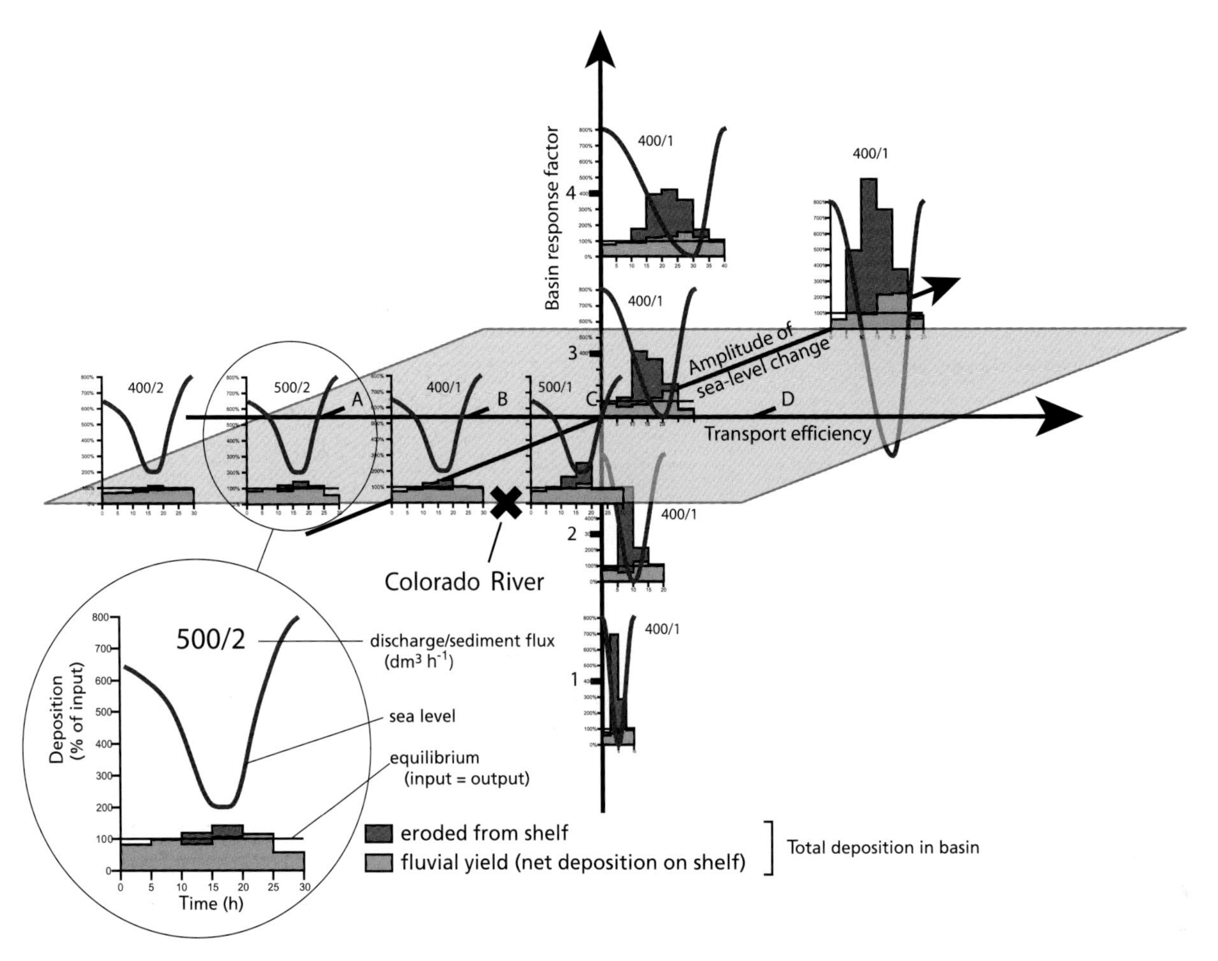

Fig. 7. Template for sea-level-change-induced sediment flux. The flux generated from shelf cannibalization is in dark grey, and the sediment from the fluvial valley is in light grey. Note that the amount of fluvial yield at the valley outlet and the amount of shelf cannibalism follows trends: shelf cannibalism increases with increasing basin-response factor (*Br*) and amplitude of sea-level change (more specifically the degree of sea-level drop below the shelf edge) and transport efficiency. The yield during the transgressive systems tract is enhanced by low *Br* values, low transport efficiency, and increasing amplitude of sea-level change, all delaying the timing of connection of shelf canyon and fluvial valley during the transgressive systems tract relative to the sea-level curve (compare with experiment in Fig. 2). Likewise, very wide shelves also promote connection delay.

the unconformity is strongly time transgressive (Blum & Price, 1998).

The stratigraphy of the Colorado shelf system, created during the last glacial cycle, has also been studied in detail (e.g. Suter & Berryhill, 1985; Berryhill, 1987; Suter *et al.*, 1987; Anderson *et al.*, 1996). Gross sediment volume approximations were derived from seismic studies of the shelf (Anderson *et al.*, 1996) and reassessed by Van Heijst *et al.* (2001) (Fig. 8). The volume enclosed by the stage 3 and 2 isopach maps is 89 km^3. Subtracting the preserved sand volume of the stage 3 highstand delta (11.5 km^3) from the total volume gives an approximation for the volume of stage 2. This yields a total volume of 77.5 km^3 for the stage 2 deposits on the shelf. The Holocene sea-level rise resulted in two phases of transgressive backstepping of delta lobes on the shelf during stage 1. Phase 1 (11.5–9.5 ka) resulted in three fluvially dominated delta lobes with a sand volume of 6.3 km^3. A wave-dominated elongated lobe that contains 4.5 km^3 of sand was formed during phase 2 (9.5–5 ka).

RECONSTRUCTION OF ALLOCYCLIC FORCING IN THE COLORADO RIVER SYSTEM

To compare the Colorado River system to the template of Fig. 7 we need to establish the

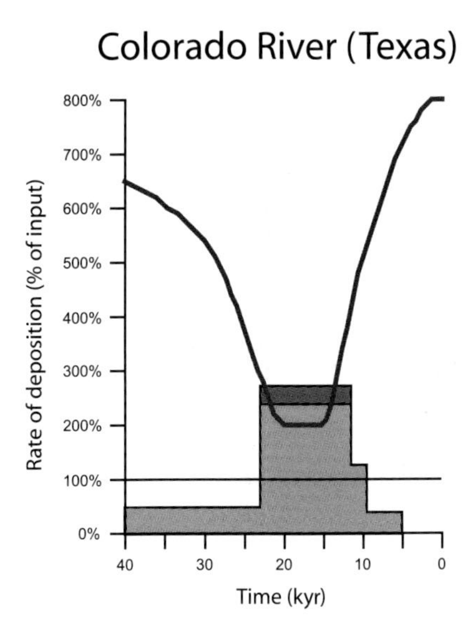

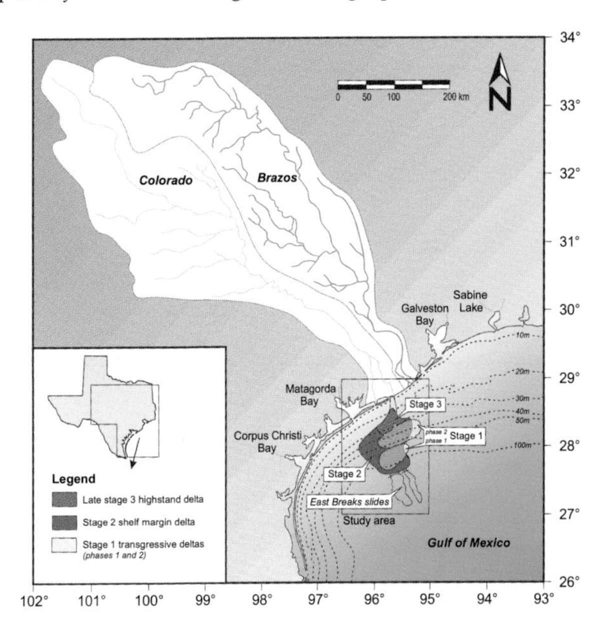

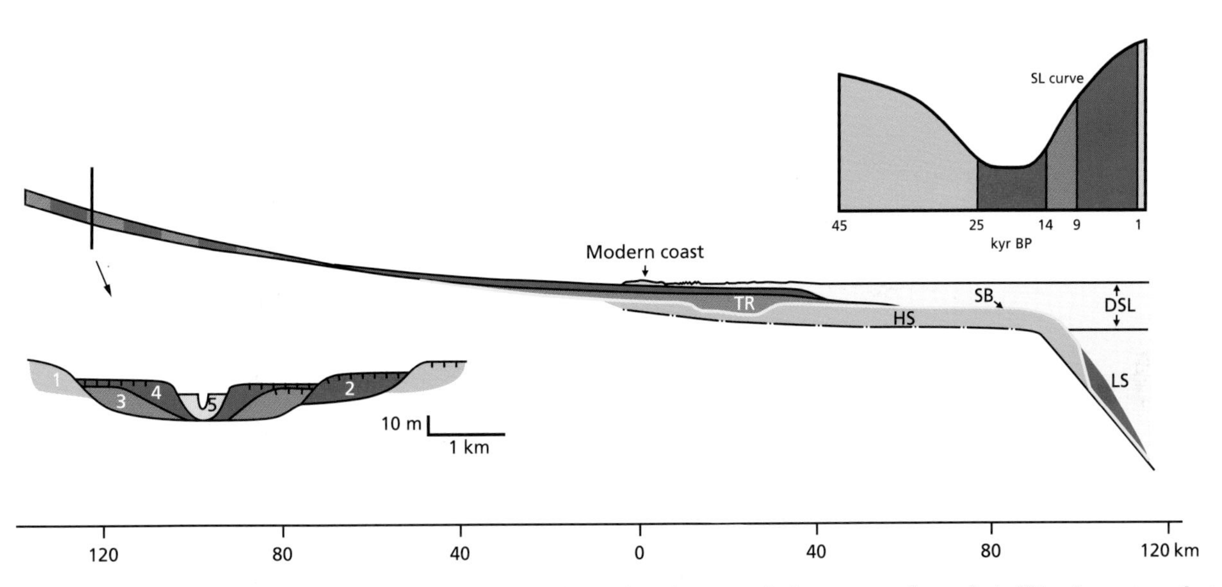

Fig. 8. Map, profile and reconstructed flux diagram of the Colorado River drainage, transfer and shelf basin system that visualize the data summarized in Table 1. In two-dimensional view the river profile compares well with profile D in Fig. 5. Yet, significant preservation of falling stage and lowstand deposits in the form of terraces brings the transport efficiency of the system to a more intermediate position, somewhere in between profiles C and D of Fig. 5. The light grey in the flux diagram is the reconstructed fluvial yield at the valley outlet and the dark grey the amount added by shelf cannibalism.

basin-response factor of the system and the stratigraphic style to which the Colorado River system corresponds. In addition the sediment flux from the drainage basin has to be normalized.

Determination of the basin-response factor (equation 6) for a dynamically changing system is not straightforward. To calculate the equilibrium time of the system (equation 5), both the length of the fluvial part of the depositional system, which changes with sea-level change, and the diffusivity of the system, which changes with discharge, need to be established. The shelf width in the Colorado system is about one-third of the fluvial depositional system during highstand. As sea level drops, the width of the shelf increases, leading to an increased response time. Estimations of discharge based on estimates of drainage-basin size and rainfall deviate on the order of 20–30%. Using

Table 1. Data used to calculate the equilibrium time of the Colorado River and the equivalent experimental model.

	Valley L (km)	Drainage area (km^2)	Active valley width (w) (km)	Rainfall (m yr^{-1})	Braided case $k = 0.1q_w/w$ km^2 yr^{-1} B-Meandering $k = 0.4q_w/w$	$T_{eq} = L^2/k \sim$ kyr	T	Br [-]
Colorado	~350	110 000	4 (Humid)	0.90	B 2.5	49	40	~0.8
			2 (Arid)	0.20	BM 4.4	16.5		~2.5
Flume	0.0062	–	0.000011	–		12 h (observation)	10	~1
							40	~4

Table 2. Estimation of sediment budget of the Colorado River for various time periods based on published stratigraphic data (Blum, 1990; Blum & Valastro, 1994; Blum & Price, 1998; Anderson *et al.*, 1996; Van Heijst *et al.*, 2001).

Age (kyr BP)	River valley km^3 (q_s kyr^{-1})	River–delta plain km^3 (q_s kyr^{-1})	Shelf km^3 (q_s kyr^{-1})	Sea level	Climate
40–23	Erosion	Erosion	21 (1.24)	Falling (~stage 3 FFST)	
23–14	ELA	ELA	77.5 (6.74)	Lowstand (~stage 2 LST)	Humid
	4 × 175 × 0.010	4 × 85 × 0.010			
	7 (0.77)	3.4 (0.38)			
14–11.5	Erosion				Arid
	−2 × 175 × 0.020	2 × 85 × 0.010			
	−7 (−2)	1.7 (0.5)			
11.5–9.5	CBA 1	CBA 1	6.3 (3.15)	Rising (~stage 1 TR 1)	Semi-arid
	2 × 175 × 0.010	2 × 85 × 0.010			
	3.5 (0.54)	1.7 (0.26)			
9.5–5			4.5 (1)	Rising (~stage 1 TR 2)	
5–2.5	Erosion	~2 (0.8)	(0)	Rising (~stage 1 TR 3)	Arid
	1 × 175 × 0.012				
	−2.1 (0.84)				
2.5–1	CBA 2	~2 (1.33)			Humid
	1 × 175 × 0.014				
	2.5 (1.67)				

ELA, Eagle Lake Alloformation.

data from Tables 1 and 2 the best estimate of the basin-response factor of the Colorado River system is between 0.8 and 2.5.

The stratigraphic style of the Colorado system is intermediate between experiments C and D (Figs 5, 7 and 8). The falling stage and lowstand deposits in the fluvial valley have been poorly preserved. The transgressive deposits have been partly preserved. The sequence boundary of the Colorado system can be found up to the catchment area and underlies the Eagle Lake Alloformation (downcutting phase 45–25 kyr BP) with a strongly diachronous development (Fig. 8). Note that in the longitudinal profile of the Colorado River (Fig. 8) the terrace heights are included. These terrace levels converge close to the modern coastline, showing the decrease of the gradient of the river valley during sea-level rise. In our model this reduction of the gradient in the river valley also occurred, but since no terraces have been preserved, no remnants are present in the stratigraphy of our experiments (Fig. 5).

The reconstructed flux of the Colorado River system can now be compared with the template of Fig. 7, where we plotted the system for the maximum basin-response factor of ~2.5 (Table 1) to reflect the dry-climate setting. During the latter periods, the sediment influx from the catchment area is assumed to be twice the yield of the river during highstand and early sea-level fall (Fig. 9),

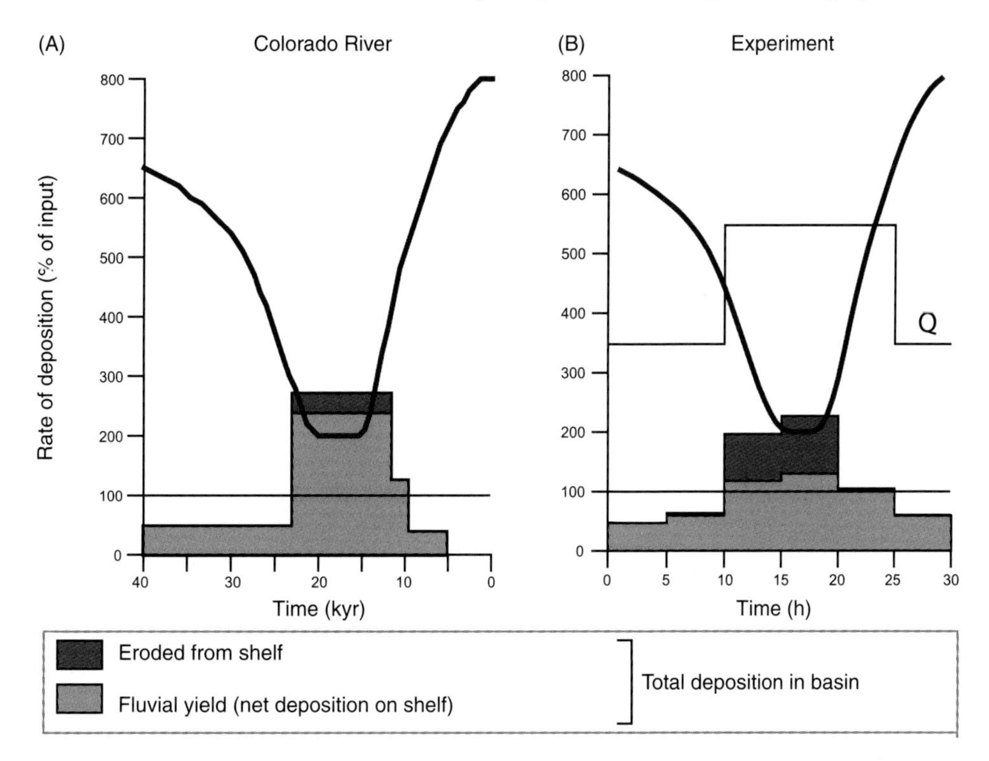

Fig. 9. (A) Flux diagram of session 640, where the discharge was temporarily increased coinciding with the lowstand of the Colorado system. The result is an important flux increase that compares well with the reconstructed flux of the Colorado system. (B) Determination of time-averaged flux from Balcones Escarpment to shelf edge (based on Blum & Valastro, 1994; Blum & Price, 1998; Van Heijst *et al.*, 2001; Anderson *et al.*, 2004; climate data from Toomey *et al.*, 1993). See text for further explanation.

so keeping the dimensionless value of valley flux similar to what we generally find in our models (~50%). To develop a stratigraphic style intermediate between profiles C and D (Fig. 5), the expected flux triggered by the fall and lowstand should be about 50–100% above its highstand and early-fall value as is outlined by the template of Fig. 7. However, the Colorado system shows a much higher increase. The extra sediment influx must have come from another forcing mechanism, probably climate changing to more humid conditions. Tectonics are likely to be unimportant at these relatively short timescales. On the basis of independent evidence such climate forcing was indeed suggested by Toomey *et al.* (1993). An extra experiment was done to simulate the increased discharge during stage 2. This shows a much better match with the real-world data (Fig. 9B).

CONCLUSIONS

Various simple equations for sediment transport over geologically relevant timescales are used by numerical modellers. Many of these depend on slope gradient and on discharge multiplied by a coefficient. None of these equations has been rigorously tested by physical experiment or on the basis of real-world data. Future experimental work in landscape modelling should focus on establishing the best functional relationship for long-term sediment transport in the model, and to test this with real-world examples. Here, a bridge must be built between the classic engineering approach for establishing the physics of sediment transport, valid for short timescales, and the geology approach, where a need exists for time-averaged sediment transport equations that are relevant for stratigraphical and geological problems (focusing on preservation of sediment instead of the transport itself). The diffusion approach is promising, certainly for base-level-induced degradation of landscapes (see Postma *et al.*, 2008). However, it has not yet been established if the same equation can predict the behaviour resulting from high-frequency climate variability within a fluvial–deltaic system.

Any time-averaged transport equation will open up scaling strategies for modelling landscape evolution on geological timescales in flumes, as shown in this paper, by adopting the diffusion equation. An obvious outstanding problem is to verify the value of the equation for real-world prototypes. Values for landscape degradation are for short timescales only (cf. Begin, 1988), while values for sediment transfer over long timescales, caused for instance by climate change, are lacking. Hence, insufficient data are available to verify relaxation (response) times of real-world systems (see further discussion by Castelltort & Van den Driessche, 2003) and to establish reliable basin-response factors.

The gradient of a natural system depends on the discharge per unit width, sediment input and grain size. Both discharge and grain size are included in the diffusion coefficient, the latter in the form of the drag coefficient. Low discharge and large grains will decrease k (see equation 3), which will increase q/k and the rivers equilibrium gradient. The amount of preservation in the river valley depends on the supply rate that causes progradation of the delta system and thus aggradation in the fluvial system as well. Defining stratigraphic styles, which are easily established from both model and prototype and reflect the transport efficiency of the system, eliminates the need for a direct functional relationship between the transport efficiency of the model and that of the prototype.

The amount of sea-level-induced sediment transport can be estimated if the rough dimensions of the entire source–sink system and the amplitude and frequency of sea-level change are known. It means that the contribution of climate change to sediment transport and stratigraphy can be roughly estimated if a well-dated stratigraphy of the whole system is available. Thus, there is a great need for well-integrated source-to-sink databases that include changes in slope gradients of river systems and isopach maps of a resolution of the order of 1000s years. The Colorado system is, in our opinion, one of the best existing examples.

ACKNOWLEDGEMENTS

We are indebted to Dr Max van Heijst for generating a well-organized database during his PhD (1996–2000), which was financed by Shell Research, Rijswijk, The Netherlands. The database set up in Pangaea was complemented with later work by the junior author. Without our technicians Tony van der Gon Netcher, Hans Bliek and Paul Anten experimental data would not have been generated at all. The manuscript benefited from the critical readings of Dr Poppe L. de Boer and reviewers Juan Pablo Milana and Peter M. Burgess.

REFERENCES

Anderson, J.B., **Abdulah, K.C.**, **Sarzaleojo, S.**, **Siringan, F.P.** and **Thomas, M.A.** (1996) Late Quaternary sedimentation and high-resolution sequence stratigraphy of the east Texas shelf. In: *Geology of Siliciclastic Shelf Seas* (Eds M. De Batist and P. Jacobs), *Geol. Soc. London Special Publ.*, **117**, 95–124.

Anderson, J.B., **Rodriquez, A.**, **Abdulah, K.C.**, **Fillon, R.H.**, **Banfield, L.A.**, **McKeown, H.A.** and **Wellner, J.S.** (2004) Late Quaternary stratigraphic evolution of the northern Gulf of Mexico margin: A synthesis. In: *Late Quaternary Stratigraphic Evolution of the Northern Gulf of Mexico Margin* (Eds J.B. Anderson and R.H. Fillon), SEPM Spec Publication, **79**, 1–23.

Andrews, D.J. and **Bucknam, R.C.** (1987) Fitting degradation of shoreline scarps by a non-linear diffusion model. *J. Geophys. Res.*, **92**, 12857–12867.

Begin, Z.B. (1988) Application of a diffusion-erosion model to alluvial channels which degrade due to base-level lowering. *Earth Surface Processes and Landforms*, **13**, 487–500.

Begin, Z.B., **Meyer, D.F.** and **Schumm, S.A.** (1981) Development of longitudinal profiles of alluvial channels in response to base-level lowering. *Earth Surface Processes and Landforms*, **6**, 49–68.

Berryhill, H.L. (1987) The Continental Shelf off South Texas. In: *Late Quaternary Facies and Structure, Northern Gulf of Mexico* (Ed. H.L. Berryhill, J.R. Suter and N.S. Hardin), *AAPG Studies in Geology*, **23**, 11–80.

Blum, M.D. (1993) Genesis and architecture of incised valley fill sequences: A late quaternary example from the Colorado River, Gulf Coastal Plain of Texas. In: *Siliciclastic Sequence Stratigraphy, Recent Developments and Applications* (Eds P. Weimer and H.W. Posamentier), *Am. Assoc. Petrol. Geol. Mem.*, **58**, 259–283. AAPG, Tulsa.

Blum, M.D. and **Price, D.M.** (1998) Quaternary alluvial plain construction in response to glacio-eustatic and climatic controls, Texas Gulf coastal plain. In: *Relative Role of Eustasy, Climate, and Tectonism in Continental Rocks* (Eds K.W. Shanley and P.J. McCabe), *Soc. Economic Paleontol. Mineral. Special Publ.*, **59**, 31–48.

Blum, M.D. and **Valastro, S.** (1994) Late Quaternary sedimentation, lower Colorado River, Gulf Coastal Plain of Texas. *Geol. Soc. Am. Bull.*, **106**, 1002–1016.

Blum, M.D., **Toomey, R.S.** and **Valastro, S.** (1994) Fluvial response to Late Quaternary climatic and environmental change, Edwards Plateau, Texas. *Palaeogeogr. Palaeoclimatol. Palaeoecol.*, **108**, 1–21.

Burgess, P.M. and **Allen, P.A.** (1996) A forward modelling analysis of the controls on sequence stratigraphical geometries. In: *Sequence Stratigraphy in British Geology* (Eds S.P. Hesselbo and D.N. Parkinson), *Geol. Soc. Spec. Publ.*, **103**, 9–24.

Burgess, P.M. and **Hovius, N.** (1998) Rates of delta progradation during highstands: Consequences for timing of deposition in deep-marine systems. *J. Geol. Soc. London*, **155**, 217–222.

Castelltort, S. and **Van den Driessche, J.** (2003) How plausible are high-frequency sediment supply-driven cycles in the stratigraphic record? *Sedi. Geol.*, **157**, 3–13.

Curray, J.R. (1964) Transgressions and regressions. In: *Papers in Marine Geology, Shepard Commemorative Volume* (Ed. R.L. Miller), MacMillan, New York, pp. 175–203.

Heller, P.L. and **Paola, C.** (1992) The large-scale dynamics of grain-size variation in alluvial basins, 2: Application to syntectonic conglomerate. *Basin Res.*, **4**, 91–102.

Hooke, R.L. (1968) Model geology: Prototype and laboratory streams: Discussion. *Geol. Soc. Am. Bull.*, **79**, 391–394.

Kooi, H. and **Beaumont, C.** (1996) Large scale geomorphology: Classical concepts reconciled and integrated with contemporary ideas via a surface process model. *J. Geophys. Res.*, **101**, 3361–3386.

Koss, J.E., **Ethridge, F.G.** and **Schumm, S.A.** (1994) An experimental study of the effects of base-level change on fluvial, coastal plain and shelf systems. *J. Sedi. Res.*, **B64**, 90–98.

Martin, Y. and **Church, M.** (1997) Diffusion in landscape development models: On the nature of basic transport. *Earth Surf. Process. Landforms*, **22**, 273–279.

Meijer, X.D. (2002) Modelling the drainage evolution of a river-shelf system forced by Quaternary glacio-eustasy. *Basin Res.*, **14**, 361–377.

Milana, J.P. (1998) Sequence stratigraphy in alluvial settings: A flume based model with applications to outcrop and seismic data. *Am. Assoc. Petrol. Geol. Bull.*, **82(9)**, 1736–1753.

Moreton, D.J., **Ashworth, P.J.** and **Best J.L.** (2002) The physical scale modelling of braided alluvial architecture and estimation of subsurface permeability. *Basin Res.*, 2002, 265–285.

Paola, C. (2000) Quantitative models of sedimentary basin filling. *Suppl. Sedimentol.*, **47**, 121–178.

Paola, C., **Heller, P.L.** and **Angevine, C.L.** (1992) The large-scale dynamics of grain-size variation in alluvial basins, 1: Theory. *Basin Res.*, **4**, 73–90.

Peakall, J., **Ashworth, P.J.** and **Best, J.L.** (1996) Physical modelling in fluvial geomorphology: Principles, applications and unresolved issues. In: *The Scientific Nature of Geomorphology, Proceedings of the 27th Binghamton Symposium in Geomorphology* (Eds B.L. Rhoads and C.E. Thorn), John Wiley & Sons, Chichester, pp. 221–254.

Posamentier, H.W. and **Vail, P.R.** (1988) Eustatic controls on clastic deposition II – Sequence and systems tract models. In: *Sea-Level Changes: An Integrated Approach* (Eds C.K. Wilgus, B.S. Hastings, C.G. St Kendall, H.W. Posamentier, C.A. Ross and J.C. Van Wagoner), *Soc. Economic Paleontol. Miner. Special Publication*, **42**, 125–154, Tulsa.

Postma, G., **Kleinhans, M.G.**, **Meijer, P.Th.** and **Eggenhuisen, J.T.** (2008) Sediment transport in analogue flume models compared with real world sedimentary systems: A new look at scaling sedimentary systems evolution in a flume. *Sedimentology*, DOI: 10.111/j.1365-3091.2008.00956.

Steel, R.J., **Rasmussen, H.**, **Eide, S.**, **Neuman, B.** and **Siggerud, E.** (2000) Anatomy of high-sediment supply, transgressive systems tracts in the Vilomara composite sequence, Sant Llorenç del Munt, Ebro Basin, NE Spain. In: *High-Resolution Sequence Stratigraphy and Sedimentology of Syntectonic Clastic Wedges* (SE Ebro Basin, NE Spain) (Eds M. Marzo and R.J. Steel), *Sedi. Geol.*, **138**, 125–142.

Suter, J.R. and **Berryhill, H.L.** (1985) Late Quaternary shelf-margin deltas, Northwest Gulf of Mexico. *Am. Assoc. Petrol. Geol. Bull.* **69**(1), 77–91.

Suter, J.R., **Berryhill, H.L.** and **Penland, S.** (1987) Late Quaternary sea-level fluctuations and depositional sequences, Southwest Louisiana continental shelf. In: *Sea-Level Fluctuation and Coastal Evolution* (Eds Nummedal, O.H. Pilkey and J.D. Howard), *Soc. Economic Paleontol. Mineral. Special Publ.*, **41**, 199–219, Tulsa.

Toomey, R.S., **Blum, M.D.** and **Valastro, S.** (1993) Late Quaternary climates and environments of the Edwards Plateau, Texas. *Global and Planetary Change*, **7**, 299–320.

Van Heijst, M.I.W.M. and **Postma, G.** (2001) Fluvial response to sea-level changes: A quantitative analogue, experimental approach. *Basin Res.*, **13**, 269–292.

Van Heijst, M.I.W.M., **Postma, G.**, **Meijer, X.D.**, **Snow J.N.** and **Anderson J.A.** (2001) Analogue flume-model study of the Late Quaternary Colorado/Brazos shelf. *Basin Res.*, **13**, 243–268.

Van Wagoner, J.C., **Mitchum, R.M., Jr.**, **Campion, K.M.** and **Rhamanian, V.D.** (1990) Siliciclastic sequence stratigraphy in Well logs, Cores and Outcrop: Concepts for high resolution correlation of time and facies. *Am. Assoc. Petrol. Geol. Methods in Expl. Ser.*, **7**, 55 pp. Tulsa.

Wood, L.J., **Ethridge, F.G.** and **Schumm, S.A.** (1993) The effects of rate of base-level fluctuation on coastal-plain, shelf, and slope depositional systems: An experimental approach. In: *Sequence Stratigraphy and Facies Associations* (Eds H.W. Posamentier, C.P. Summerhayes, B.U. Haq and G.P. Allen), *Int. Assoc. Sedimentol. Special Publ.*, **18**, 43–54. Blackwell Scientific Publications.

Yalin, M. S. (1971) *Theory of Hydraulic Models*. Macmillan, London, 266 pp.

Spec. Publ. Int. Assoc. Sedimentol. (2008) **40**, 207–222

Grain-size sorting of river–shelf–slope sediments during glacial–interglacial cycles: modelling grain-size distribution and interconnectedness of coarse-grained bodies

XANDER D. MEIJER

Faculty of Geosciences, Utrecht University, P.O. Box 80021, 3508 TA, Utrecht, The Netherlands (E-mail: gpostma@geo.uu.nl)

ABSTRACT

Changing conditions during Quaternary glacial and interglacial stages have had a great influence on the location and amount of sedimentation, and on the grain size of deposits. In this modelling study, the volume and grain-size distribution in passive continental margin strata are investigated. A principal aspect of the basin-scale grain-size sorting process is the formation of a subaerial erosional unconformity. During forced regression, deltaic coarsening-up sequences are deposited on the shelf. The relatively coarse-grained topsets of these successions have a lower preservation potential than the finer prodelta deposits. Therefore, erosion on the exposed shelf results in enrichment of river-transported sediments with coarse material during sea-level lowstand. Thus, shelfal strata are depleted of coarse material, increasing the coarse content of deposits on the upper continental slope. In contrast, in the absence of an erosional unconformity, the composition of sediments on the shelf is relatively coarse. The extent to which sediment within a stratigraphic column has been separated into coarse strata, as opposed to mixed compositions, is expressed as a 'differentiation ratio'. Interconnectedness of coarse-grained sediment bodies in the stratigraphy of the continental shelf and slope is closely related to the palaeogeographical evolution, and is consequently a highly variable property.

Keywords Forward modelling, passive continental margins, glacial–interglacial cycles, stratigraphy, grain-size distribution, interconnectedness.

INTRODUCTION

In the stratigraphy of clastic sedimentary rocks, grain size is without a doubt the most prominent characteristic. However, in stratigraphic modelling it has proved to be a very elusive component.

Sediment grain size is an important aspect in facies classification. Sediments reflect the conditions of their origin and thus are often accorded an environmental significance. A facies model associates lithology and sedimentary structures, amongst others, with transporting agents and consequent depositional and erosional processes. However, since an unknown portion of the supplied sediment is passed on, it is unclear how the grain-size distribution of the deposited material relates to the composition of the original sediment. Dispersal not only depends on process intensity but also on sediment calibre and availability. Furthermore, facies models are mostly qualitative in nature and controlling conditions are poorly constrained. The process activity that leads to sediment sorting is therefore rather intangible for a stratigraphic basin modeller.

In spite of the complexity of the problem, various models have been developed that incorporate dynamic grain transport in one way or another: hydraulic approach (e.g. Bitzer & Harbaugh, 1987; Tetzlaff & Harbaugh, 1989; Syvitski & Hutton, 2001) and rule-based approach (e.g. Angevine *et al.*, 1990; Rivenæs, 1992; Granjeon & Joseph, 1999). Much is to be gained by understanding the development of the petrologic characteristics of sedimentary units. For instance, the location and the architecture of coarse-grained bodies are important constituents in determining the reservoir potential for hydrocarbons and aquifers. Furthermore, such modelling studies will advance the extension of the sequence stratigraphic concept with sediment calibre.

Here, a three-dimensional numerical model is used to examine the grain-size distribution of sediments on a passive continental margin under Quaternary glacio-eustatic conditions. The occurrence of coarse-grained units on continental shelf and slope and their interconnectedness are investigated for several scenarios of boundary conditions.

METHOD

The following is a brief description of the conceptual model. The equations used are outlined in Appendix A. For a more comprehensive explanation of the model scheme, see Meijer (2002a).

The model used in this study is a cellular representation of a landscape and subsurface in which the relations between adjacent grid cells are laid down in a number of rules of sediment transport, distribution and grain-size sorting. The set of rules constitutes a three-dimensional process–response system prompted by boundary conditions (e.g. sea level, subsidence), which is capable of complex behaviour (Meijer, 2002a). Elevation changes in grid cells are evaluated in order of their altitude, starting with the highest. The elevation change due to sediment transport is derived by balancing deposition (or erosion) and sediment outflux on the basis of the 'resulting' gradient (Fig. 1a). The amount of sediment (if available) that is deposited is such that a gradient will arise that is exactly competent to transport any remaining (i.e., bypassing) sediment downslope. Similarly, erosion produces a gradient that is exactly capable of transporting the amount of influxed plus denudated sediment downslope. Sediment outflux to all (up to eight) downward grid cells is determined concurrently. However, these gradients may subsequently be altered in response to elevation changes that can occur in downslope grid cells. They may be further modified by tectonic movements (and other processes not included in the model used here, such as sediment compaction and isostasy). Short-range, diffusive transport is related to gradient, whereas long-range, stream transport is governed by the degree of undercapacity or overcapacity of the river related to water discharge as well as slope. Transport parameters depend on the depositional environment related to bathymetry called 'terrestrial', 'coastal', 'marine' and 'deep marine' (Table 1). River sediment and water discharge enter the model landscape at an inlet grid cell, mimicking the outflow of a drainage basin (Fig. 2a). Rainwater is distributed evenly over the entire landscape. To minimize edge effects, the grid boundaries perpendicular to the basin axis are joined. Certain processes, such as waves, tides and turbidity currents, are not included in the model but are briefly touched upon in the discussion.

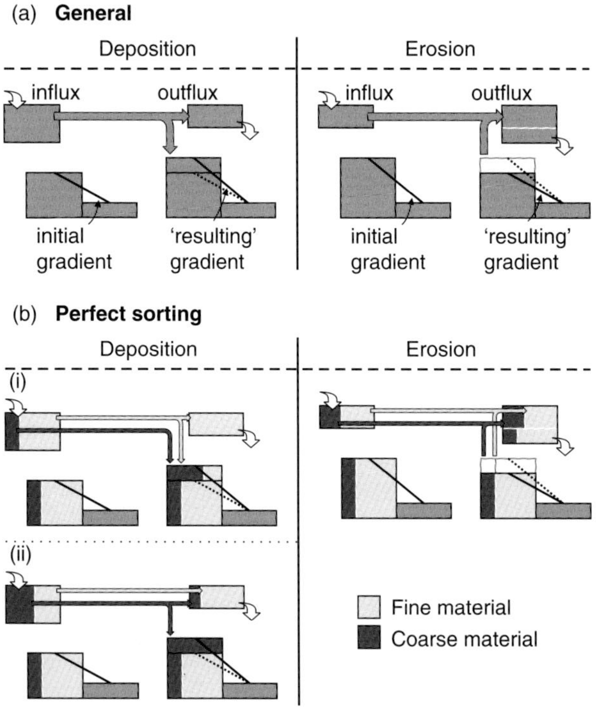

Fig. 1. A schematic diagram of a portion of the model grid explaining the fundamentals of the model. (a) General sediment transport algorithm. (b) The perfect sorting sediment transport algorithm. Two situations are possible: (i) the influx of coarse material is less than the amount of deposition – the deposited layer contains material of both grain-size classes; (ii) the influx of coarse material equals or exceeds the amount of deposition – the layer contains only material of the coarse grain-size class.

Modelling grain-size sorting

Whereas the general model treats sediment size as constant, real-world processes sort fine and coarse materials, leading to deposits that may be dominantly coarse textured (gravel, sand) or fine (clay). Because the distribution of coarse and fine stratigraphic units is important for predicting storage of water or hydrocarbons, it is interesting to include grain-size sorting in the model.

Most modelling approaches proceed by representing a grain-size distribution by a finite number of classes rather than by a continuous population.

Table 1. Values of parameters applied in the model runs. See Appendix A and Meijer (2002a) for an explanation of the parameters and the accompanying equations.

Parameter	Symbol	Environment	Value: General	Value: Perfect sorting	Value: Imperfect sorting – Fine fraction	Value: Imperfect sorting – Coarse fraction	Unit
Diffusive transport power	U		1.0				–
Diffusive transport coefficient	K_D	Terrestrial		0.30	0.24	0.06	$km^3\,kyr^{-1}$
		Coastal		0.40	0.32	0.08	$km^3\,kyr^{-1}$
		Marine		0.40	0.32	0.08	$km^3\,kyr^{-1}$
		Deep marine		0.32	0.26	0.06	$km^3\,kyr^{-1}$
Threshold slope	S^*_{thr}	Terrestrial		0.0005	0.0005	0.0005	–
		Coastal		0.0040	0.0030	0.0050	–
		Marine		0.0200	0.0200	0.0250	–
		Deep marine		0.0200	0.0200	0.0250	–
Transition depth with respect to sea level		Terrestrial – coastal	0				m
		Coastal – marine	25				m
		Marine – deep marine	100				m
Discharge distribution threshold	k	Terrestrial	0.75				–
		Non-terrestrial	0.00				–
Stream transport power	V		1.0				–
Stream transport power	W		1.0				–
Discharge threshold	Q_{thr}		2.0×10^8				$m^3\,yr^{-1}$
Stream transport coefficient	K_S			0.40	0.36	0.04	–
Stream attenuation constant	L_S	At overcapacity		10	20	5	–
		At undercapacity		50	50	50	–
Incision criterion power	υ		1.00				–
Incision criterion power	ω		0.25				–
Incision threshold	P_{thr}		0.224				$(m^3\,yr^{-1})^{0.25}$
Grid cell size			2000				m
Timestep			10				yr
White-noise range			0.05				m

Such subdivision enables conservation of mass of each class. The divisions may correspond to classes of the grain-size scale (gravel, granule, sand, silt and clay) or are left unresolved.

There are two prevalent process-based sorting algorithms (Den Bezemer *et al.*, 2000): 'perfect' and 'imperfect' sorting. Perfect sorting is a more elementary representation of sorting processes than imperfect sorting. However, the perfect sorting algorithm is considerably less time-consuming and the resulting large-scale stratigraphies generated with both approaches have comparable distribution patterns (Meijer, 2002b). Since the requirement of the model is to be able to recognize diagnostic first-order trends, perfect sorting was considered to be adequate for our present purposes and has been used for the model experiments in this paper.

The perfect sorting algorithm (Angevine *et al.*, 1990; Paola *et al.*, 1992) breaks sediment down into a coarse and a fine class, although it can be extended to more than two classes. The relative abundance of the two classes is expressed as the 'grain-size fraction' (with zero signifying solely fine material and unity only coarse). The underlying assumption of perfect sorting is that the coarser class takes precedence over the finer one at sedimentation (Fig. 1b). When the situation requires sediment entering the grid cell to be deposited, the perfect sorting algorithm will first draw upon the supplied coarse-grained portion. Fine sediment will not be deployed until all available coarse material is exhausted. No such grain-size preference exists when material is eroded.

Model procedure

For this paper, the standard model setting is a passive continental margin with a major river and a wide shelf (a 'river–shelf system'). The

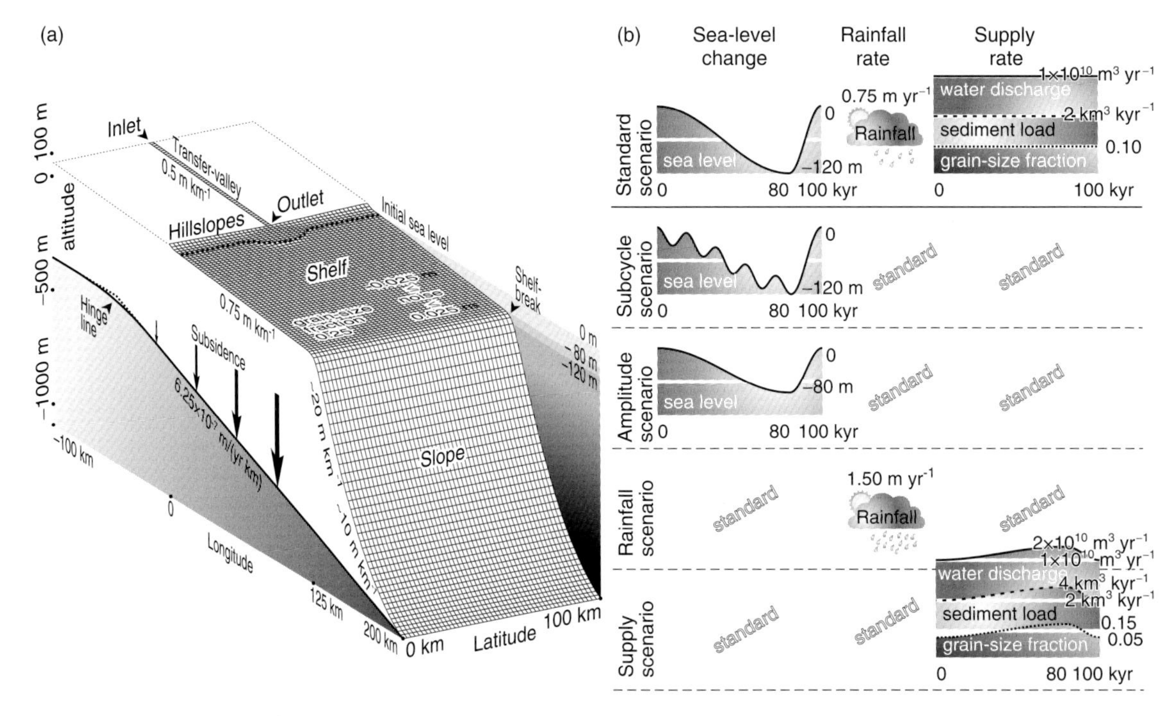

Fig. 2. Overview of the model. (a) Model grid set-up. (b) Model variables of the various scenarios. Scenarios 2–5 are variants of the standard scenario 1. One glacial–interglacial cycle is depicted. Model runs typically cover five cycles.

applied forcing variables, such as high-amplitude eustatic sea-level change and sediment-supply signals, are characteristic of late Quaternary glacial–interglacial cycles, each spanning 100 kyr. Every model run completes five of such cycles. Various scenarios of allogenic forcing have been examined (Fig. 2b). For the benefit of statistical analysis, each model scenario was run 50 times, each time with a different white-noise pattern superimposed on the initial topographic surface. This enables results of different scenarios to be compared (Meijer, 2002a).

Values of model parameters are listed in Table 1. In the interest of data reduction, new strata are compiled by joining a number of layers (each spanning one model time step) in a column and taking the weighted average of their properties. These compiled strata are used for calculations on stratigraphy and are depicted in the stratigraphic sections. For the benefit of explanation, the terms inner and outer shelf are used as geographical designations related to the interglacial (highstand) situation, rather than physiographical definitions. Furthermore, in this paper, depositional environments correspond to bathymetric zones (Table 1).

The objective of this study is not to replicate any specific geological history, but instead, the experiments explore model solution space and the system's sensitivity in order to draw general conclusions for a set of geologically realistic conditions.

RESULTS AND DISCUSSION

In order to evaluate the effects of grain-size sorting on the development of passive continental margins, the model results are considered in terms of sediment distribution, grain-size distribution, and stratigraphic architecture for a standard scenario. In addition, the effects of different boundary conditions are examined using various scenarios with sea-level change, increased precipitation, and increased sediment supply during glacial periods.

Sediment distribution

The standard model scenario was applied to the passive margin system in order to examine patterns emerging in the stratigraphy. The diagrams in Fig. 3 outline the chronostratigraphy created after five glacial–interglacial cycles. The total sediment volume in the system between consecutive 1 kyr spaced isochrons (i.e., strata between particular age limits) is summarized in Fig. 4a. A volume of sediment of 2 km^3 enters the system every 1 kyr. Whenever in the stratigraphy there is less than this amount preserved, net erosion

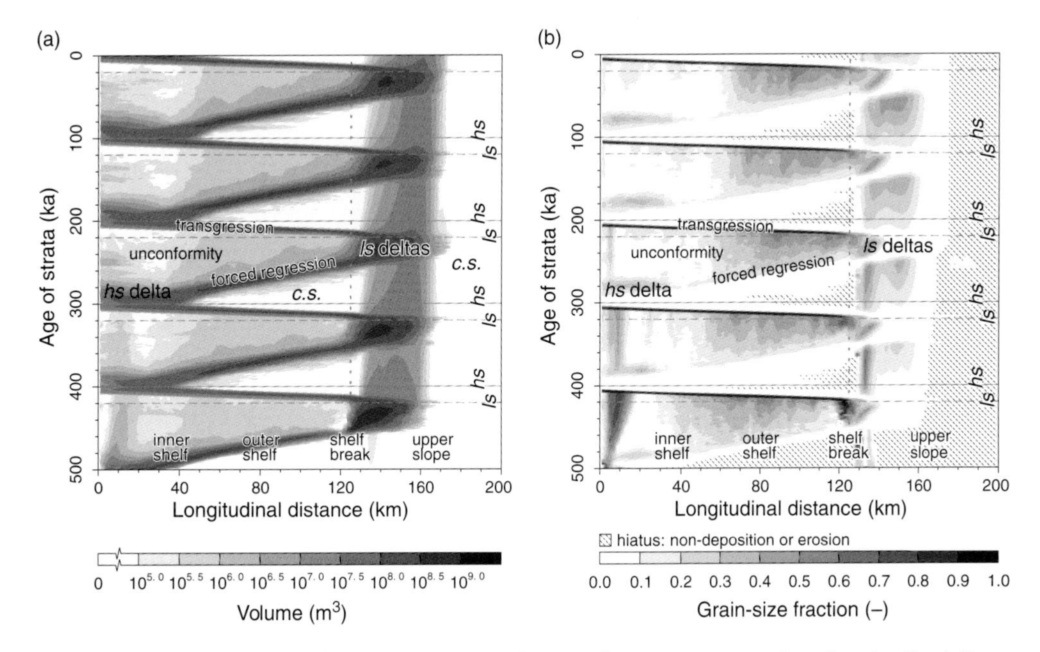

Fig. 3. Chronostratigraphy displaying sediment properties of strata of a given age, at a given longitudinal distance. Average of 50 similar model runs implementing the standard scenario: *hs*, highstand; *ls*, lowstand; *c.s.*, condensed section. (a) Total volume of sediment. (b) Mean grain-size fraction (relative abundance of the coarse grain-size class) of the sediment.

has occurred. Any volume exceeding this value represents remobilized material from earlier strata or from the initial substratum. It is evident that deposits from roughly the first half of the cycle have a lower preservation potential than those from the second half (Fig. 4a).

A detailed description of the river–shelf system development during a Quaternary glacio-eustatic cycle emphasizing the landscape and drainage evolution is presented in Meijer (2002a). Here, the stratigraphic characteristics are emphasized and the stages of deposition are sketched only briefly, starting at sea-level highstand. In Fig. 4b, highstand-delta aggradational and progradational units reveal themselves in the considerable volume of terrestrial sediments from the earliest phase (~20 kyr) of every new cycle. Gradually, normal regression turned into forced regression (Posamentier *et al.*, 1992). Terrestrial deposits from this time dwindle because aggradation diminished under the influence of falling base-level and sediment was removed during subsequent subaerial erosion. Instability along the shelf edge induced by basin subsidence and sea-level fall resulted in marine redeposition (Posamentier & Allen, 1993). Later, about halfway through the cycle, sediments were no longer laid down on the gentle shelf that supports a wide prodelta. On the relatively steep continental slope there was less room for shallow deposits, and they were more readily eroded. This led to a decline in coastal deposition (Fig. 4b). River sediment was subsequently supplied directly to deeper water, bypassing the shelf.

At that time, formation of an unconformity had already begun. The few terrestrial deposits that date back from falling sea-level stage originated primarily from the degradation of topography on the exposed shelf (i.e., delta lobes). Note that the volume of these sediments (Fig. 3a) corresponds to a mean thickness of less than 5×10^{-3} m or a mean deposition rate of 5×10^{-6} m yr^{-1}. At such low deposition rates, processes such as aeolean transport, bioturbation, soil formation and vegetation growth are likely to play an important role locally (Galloway, 1989). Erosion concentrated mainly around incising rivers of the main, hinterland-fed and the secondary drainage systems. The inner shelf in particular was subject to erosion since it was exposed longer and because the higher gradient of the shelf with respect to the transfer valley promoted incision (e.g. Miall, 1991; Posamentier *et al.*, 1992; Schumm, 1993; Wescott, 1993).

As the incising river system eventually neared equilibrium, supply decreased. Shortly before eustatic lowstand, relative sea level started to rise owing to subsidence, resulting in transgression and a depocentre shift from deep, upper continental slope waters to the outer shelf. The low preservation potential of (shallow) shelf-edge

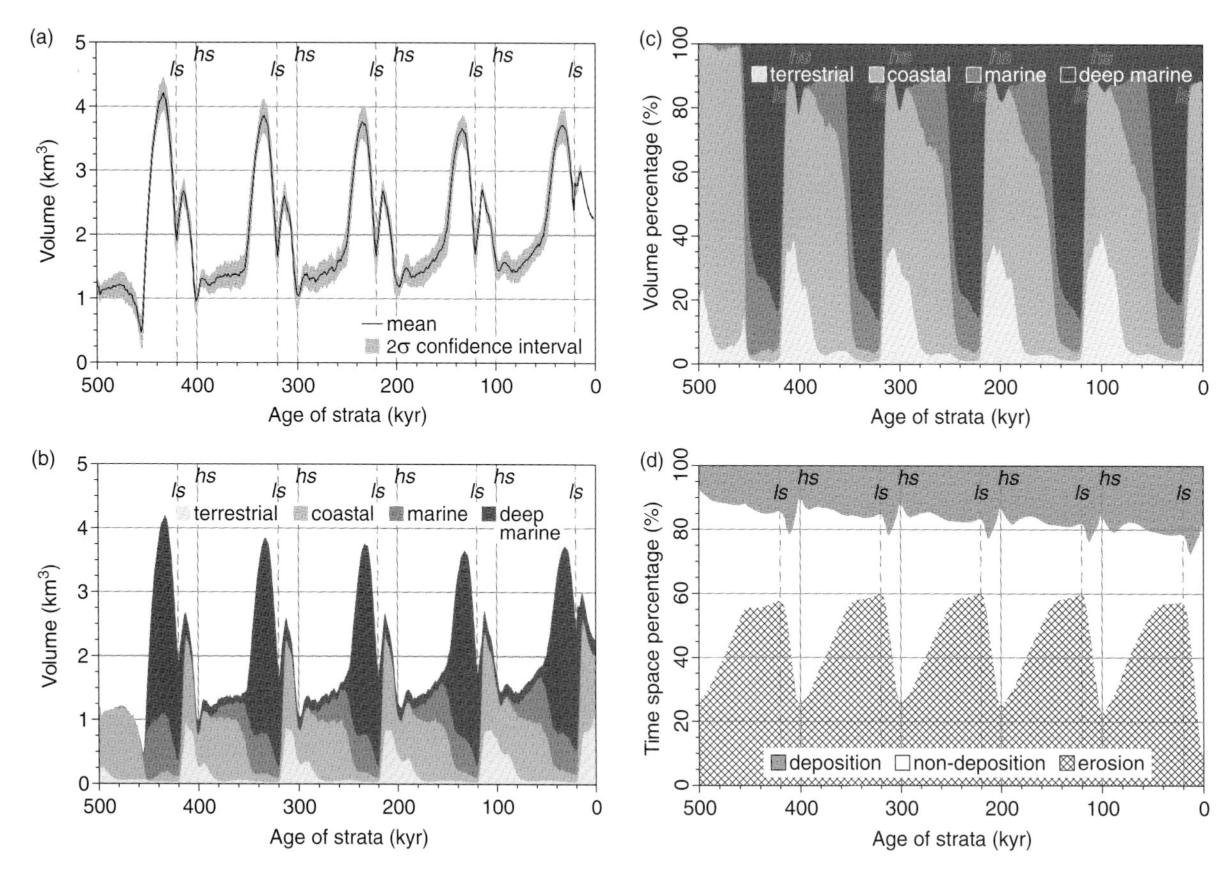

Fig. 4. Sediment properties as a function of stratal age. Average of 50 similar model runs implementing the standard scenario: *hs*, highstand; *ls*, lowstand. Strata represent intervals of 1 kyr. (a) Volume of sediment preserved in the stratigraphy. (b) Volume of sediment subdivided by depositional environment. (c) Volume percentage subdivided by depositional environment. (d) Percentage of stratigraphy taken up by deposition or hiatus owing to non-deposition or erosion.

deposits manifests itself in a depression in stratigraphic volume around lowstand (Fig. 4a and b). Back-filling of fluvial valleys accompanied by coastal onlap explains the large amount of terrestrial and coastal sediments dating from the period of sea-level rise. At the same time, the submerged shelf-edge deltas acted as a source for deposits on the upper continental slope. Degradation of the shelf edge continued throughout the cycle in the model (Fig. 4a) (Vanney & Stanley, 1983). However, the volume of these deep marine deposits, consisting entirely of remobilized sediment, is relatively low (cf. Fig. 4b and c). As sea-level rise decelerated, a coastal prism (Talling, 1998) was built up on the inner shelf and in the drainage basin outlet. Fluvial onlap proceeded into the transfer valley where deposits have a high chance of being removed in subsequent cycles.

The distribution of the total stratigraphic volume over the depositional environments (Fig. 4c) reflects the dichotomy typical of Quaternary passive margins: highstand depositional systems on the inner shelf where terrestrial/coastal deposits dominate, alternate both temporally and laterally with lowstand upper-slope depocentres with mainly marine sediments. If we envisage the stratigraphic record as a three-dimensional Wheeler-diagram (with units ‘time–space’), it can be broken down into the categories: preserved deposition, hiatus owing to non-deposition and hiatus owing to erosion (Fig. 4d). Preserved strata do not occupy more than a fifth of the chronostratigraphic record; the decrease with age indicates the role of remobilization of sediment during subsequent cycles. The bulk is taken up by non-deposition, corresponding to the formation of condensed sections in the real world, and hiatuses owing to unconformity formation and degradation of the shelf edge.

Grain-size distribution

The grain-size evolution through the stratigraphy fluctuates around the composition of 10% coarse sediment (i.e., a grain-size fraction of 0.10) that enters the system (cf. Burgess & Hovius, 1998) (Fig. 5a). During the first half of each cycle, deposits are finer than average owing to erosion of coarse topsets of highstand and forced-regressive units. These intervals are followed by periods with strata that are coarser than average owing to the supply of sediment enriched with coarse material to the ocean basin where preservation potential is higher. The trend is similar to that of volume (cf. Fig. 4a) because the stratigraphic properties grain size and preservation potential are correlated. For instance, since deltaic sequences tend to coarsen upwards, coarse sediments are more likely to be removed upon erosion. Therefore, differential preservation during large sea-level oscillations in a situation with practically decoupled depositional systems on shelf and slope is the main mechanism for basin-scale grain-size sorting.

In Fig. 5b, the grain-size fraction is depicted per depositional environment. At the start of every glacial cycle the mean grain size of preserved terrestrial sediments is relatively fine. This value is dictated by the few hillslope sediments (Fig. 3a and b), because concurrent highstand (and preceding transgressive) coarse material in the transfer valley was largely flushed out by river incisions during subsequent falling stage and lowstand. In reality, terrace formation could preserve some of the latter deposits (Blum & Price, 1998); that process, however, is below the resolution of the model. Aggradational and progradational paralic successions on the shelf were responsible for high terrestrial grain-size values at times of falling sea-level. Coarse deposits found in deep water environments dating back from the same periods are due to instabilities on the upper continental slope, that is, the initial substratum and the front of drowned shelf-edge deltas.

From halfway through to the cycle, the river system delivered directly to the shelf edge. Fine river-plume sediment preceded deposits of the advancing coastline. The river load then became enriched with coarse material originating from shelf erosion (Posamentier & Allen, 1993). Most of the coarse fraction in the supplied sediment has been incorporated in coastal and marine deposits; mainly fines went through to deeper water. Although coarse sediment was delivered to the basin edge during lowstand-delta construction, it is striking that in the model, transport of relatively coarse material further into the basin occurred during periods of falling sea-level (Fig. 3b).

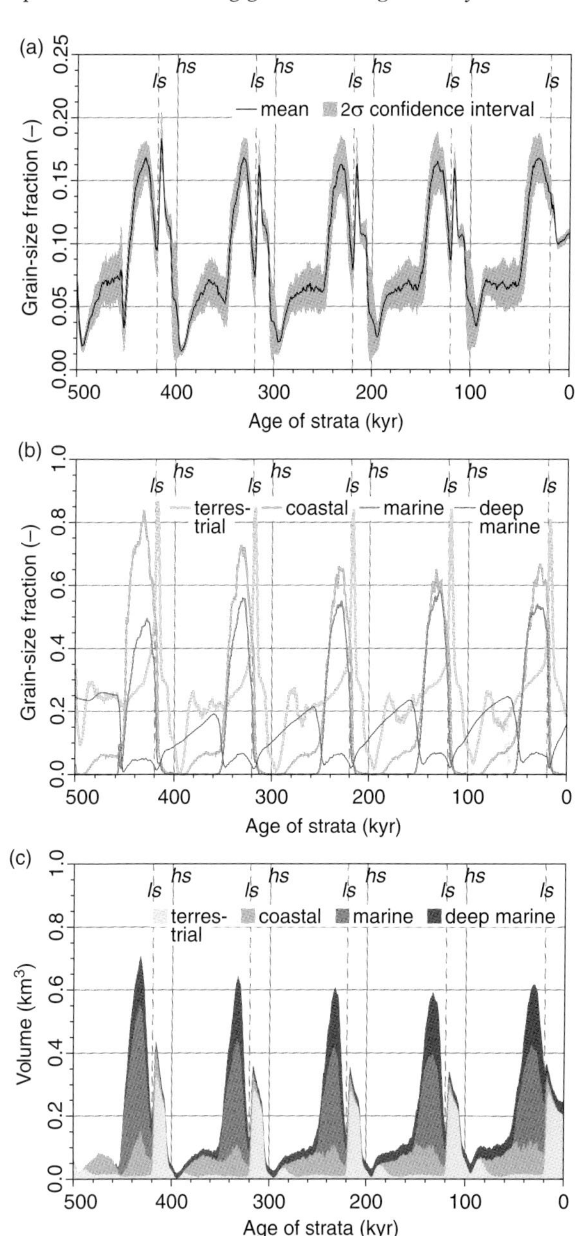

Fig. 5. Sediment properties as a function of stratal age. Average of 50 similar model runs implementing the standard scenario: *hs*, highstand; *ls*, lowstand. Strata represent intervals of 1 kyr. (a) Grain-size fraction (relative abundance of the coarse grain-size class) of sediment in the stratigraphy. (b) Grain-size fraction of sediment of each depositional environment. (c) Volume of coarse sediment fraction subdivided by depositional environment (volume (Fig. 4b) multiplied by grain-size fraction (Fig. 5b)).

As relative sea level began to rise, the proximal parts of the shelf-edge deltas were covered with relatively fine-grained transgressive strata. Transgressive fill of incised river valleys trapped almost all available coarse material from the hinterland (Posamentier & Allen, 1993) (Fig. 5c). 'Estuarine' deposits are fine, overlaying the coarse fills at the base of the valley. The transition from coarse to fine fluvial deposition was upstream from the delta plain during early highstand, leaving the delta starved of coarse material until late highstand. This would be similar to many Holocene coasts and estuaries that were withheld by coarse sediment from a fluvial source.

Figure 7a recapitulates the stratigraphy by plotting mean grain size of the five cycles against distance along the river–shelf system. The high values on the most landward part of the profile reflect highstand-delta topsets, mostly originating from the late sea-level highstand. Directly basinward of this, fine sediments of the highstand prodelta occur, which were rapidly crossed and bypassed after the transition from normal to forced regression. On the shelf, substrate cannibalization has resulted in a mean grain size below the hinterland supply value, increasing basinward (Posamentier & Morris, 2000). Deltaic sorting-processes have created coarsening-up sequences during sea-level fall, and subsequent unconformity formation has removed their top. Shelf-edge deltas are responsible for the coarse composition beyond the shelf break. There is a second, smaller peak (Fig. 6a) where slope-dependent flows, as opposed to more proximal river-plume transported sediments, are deposited. In reality, these deposits can also be transported to greater depths, for instance by turbidite systems depending on the ratio of coarse to fine material (Posamentier & Allen, 1993).

Although the increase of sediment flux and sand-to-mud ratio during forced regression has been recognized (Posamentier & Morris, 2000), some researchers (e.g. Galloway, 1989; Burgess & Hovius, 1998; Blum & Törnqvist, 2000) argue that delivery of sediment to the shelf edge at times of sea-level lowstand is not altered significantly relative to hinterland yield in terms of volume and grain size. However, a river can adjust to changing base level in ways other than incision by adapting its channel characteristics (Schumm, 1993; Wescott, 1993; Leeder & Stewart, 1996). Widespread incision into the exposed shelf does not occur during a large part of sea-level fall (Van Heijst & Postma, 2001; Meijer, 2002a). Therefore, the generated additional supply will be a surge during a limited period of the interglacial–glacial cycle, rather than a sustained, gradual one, and the consequences of valley formation for the grain-size distribution and stratigraphic architecture can still be substantial.

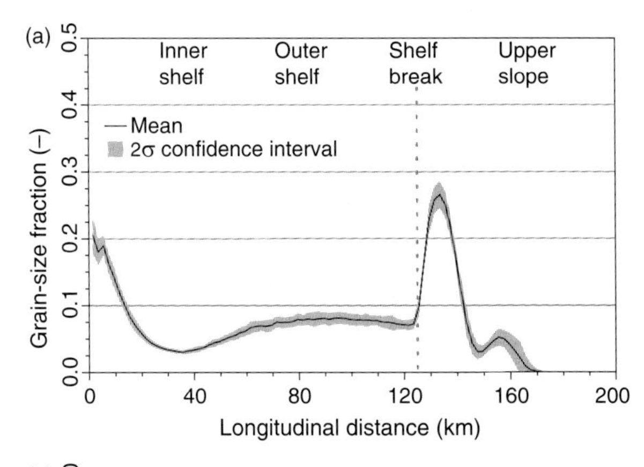

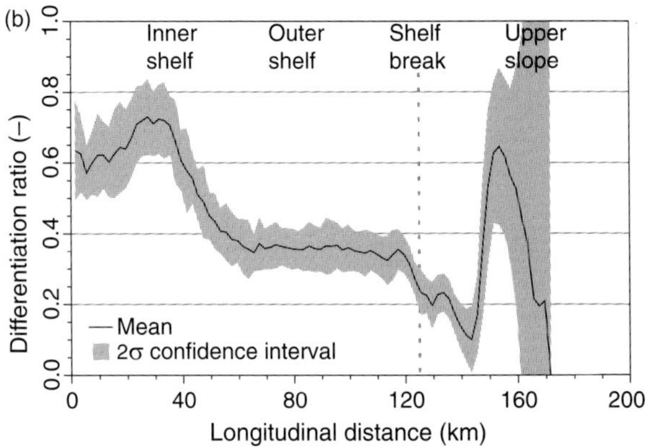

Fig. 6. Sediment properties as a function of longitudinal distance. Average of 50 similar model runs implementing the standard scenario. (a) Grain-size fraction (relative abundance of the coarse grain-size class) of sediment in the stratigraphy of five cycles. (b) Differentiation of coarse material into coarse bodies within the stratigraphy.

Stratigraphic architecture

Meijer (2002a) demonstrated that the landscape evolution of river–shelf systems is sensitive to initial conditions. Seemingly insignificant random noise on the initial morphology leads to a great variation in development. Similarly, the spatial distribution of stratigraphic characteristics can vary considerably under identical forcing conditions owing to non-linear behaviour. For instance, both stratigraphies depicted in Fig. 7 have developed under exactly the same standard scenario, but with different distributions of white noise superimposed on the initial topography. Even though the details of the stratigraphic architecture in the individual model runs deviate, the stratigraphies exhibit comparable coarse-calibre features. The most prominent are the voluminous delta-front deposits located on the upper continental slope. Secondly, the shelf is criss-crossed with elongate bodies at the base of incised valleys

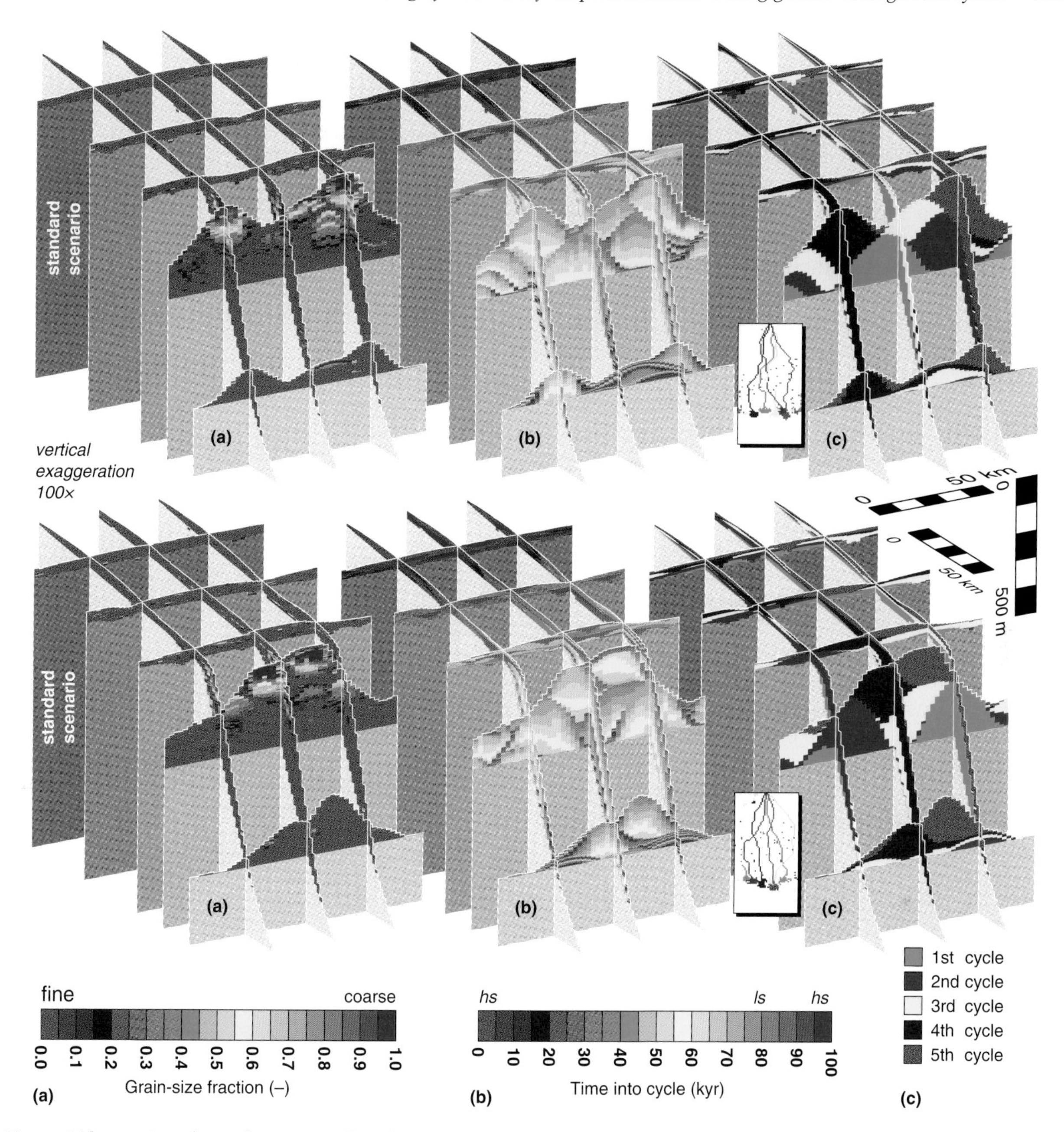

Fig. 7. Lithostratigraphies of two examples of similar model experiments run implementing the standard scenario. (a) Grain-size fraction (relative abundance of the coarse grain-size class) of sediment. (b) Time of deposition within a 100 kyr cycle. (c) Glacial–interglacial cycle in which deposition occurred. Inset: map view of coarse bodies thicker than 1 m.

deposited during transgression and covered by fine material, which makes them potential hydrocarbon reservoirs (Posamentier & James, 1993). Finally, coarse sediments derived from the degradation of shelf-edge deltas occur, enveloped in fine material, at relatively great depths.

An aggregate of connected sequences in a model stratigraphy with grain-size fraction ≥ 0.95 (i.e., coarse-grained blocks in adjacent grid columns that share part of a side or an edge) is defined here as a 'coarse body'. The criterion threshold is set rather high, since the perfect sorting algorithm is predisposed to unmixing, and thus to the formation of coarse bodies. These coarse bodies can be determined in two different ways. The first method is to define them solely based on lithostratigraphy, whereas the second is more restrictive as it also takes chronostratigraphy into account: coarse-sediment successions that are spatially connected but are separated by a hiatus, either erosional or non-depositional, are considered to be separate units. The presence of coarse bodies is an indication of potential resource reservoirs.

The ratio of the thickness of coarse bodies and the total sediment thickness divided by the mean grain size is a measure of separation of the two grain-size classes. This 'differentiation ratio' is plotted in Fig. 6b as a function of longitudinal distance. On the inner shelf, for example, roughly 60% of the total amount of coarse grains is part of a coarse body. In contrast, outer shelf successions are less differentiated owing to the relative scarcity of aggradational deposits, both at the time of deposition owing to the rapidly retreating coastline, and after preservation. Prolonged dumping of river sediment on the continental slope during lowstand led to oversteepening of the delta front (Coleman *et al.*, 1983), and thus to a greater influence of mass transport relative to river-plume transport. Consequently, deposits exhibit a greater range of blends of coarse and fine grains (Fig. 7), resulting in a low differentiation ratio despite a high mean grain-size value. More distal sediments become more differentiated owing to continuous sorting that took place under less turbulent circumstances when the depocentre was removed from the shelf edge. The variation, however, is great (Fig. 6b).

The probability of finding coarse bodies in the stratigraphy of a given volume and average thickness based on 50 model runs is plotted in Fig. 8. Figure 8a considers aggregates of strictly genetically related strata. Most of these bodies are small. Exceptions are the bases of major incised-valley fills, large clusters of coarse material on the continental slope, and sheet-like highstand deposits. However, the coarse valley fills and the shelf-edge bodies can connect with each other in various combinations, resulting in a wide band of possible connected coarse bodies (Fig. 8b). The fills of cross-shelf river belts are usually joined at or near the transfer-valley outlet to form large bodies dipping basinward. Although their winding paths often intersect in map-view on the shelf (Fig. 7), incisions during successive glacial cycles do not reach the same depth owing to subsidence and sediment accumulation, and valley fills rarely connect there (Talling, 1998). In reality, shelfal deposits, including transgressive valley fills, may be removed partially or entirely, or reworked into coarse-grained sheets, depending on the effectiveness of wave or tidal ravinement during sea-level rise (Zaitlin *et al.*, 1994). On the other hand, the coarse bodies in the model occur at

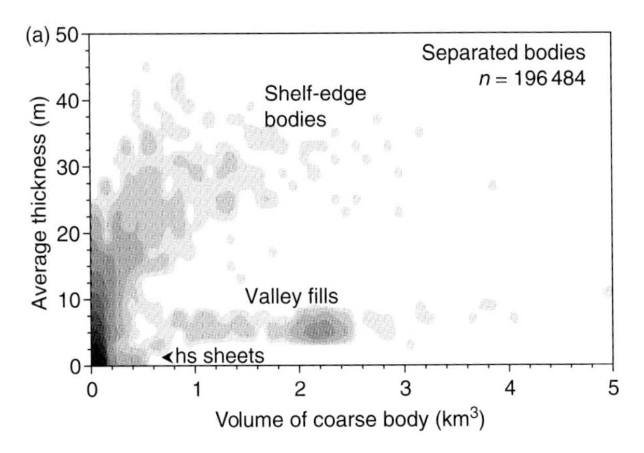

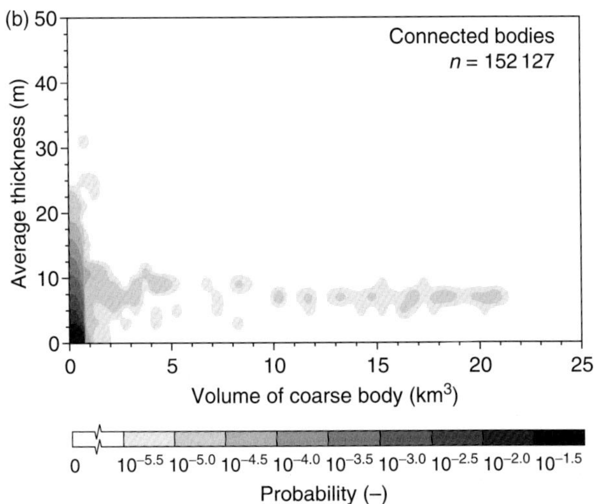

Fig. 8. Probability of the occurrence in the stratigraphy of coarse bodies of a given volume and average thickness. Total of 50 similar model runs implementing the standard scenario. Strata represent intervals of 1 kyr. (a) Coarse bodies that are separated by non-depositional or erosional hiatuses. (b) Connected coarse bodies.

the base of incised valleys, which have a relatively high preservation potential. Similarly, when subaerial accommodation is reduced, for instance during highstand, fluvial deposits are prone to floodplain reworking (Wright & Marriott, 1993). These effects may significantly alter the interconnectedness on the shelf. The large shelf-edge bodies are blanketed with early transgressive sediment (with a low preservation potential) and, more importantly, they may have been covered by prodelta fines of subsequent cycles (see also Fig. 3b). Whether these bodies are connected with each other or to transgressive bodies is highly variable. That makes the interconnectedness of individual passive margin coarse bodies extremely unpredictable, but statistically fair prediction may be possible.

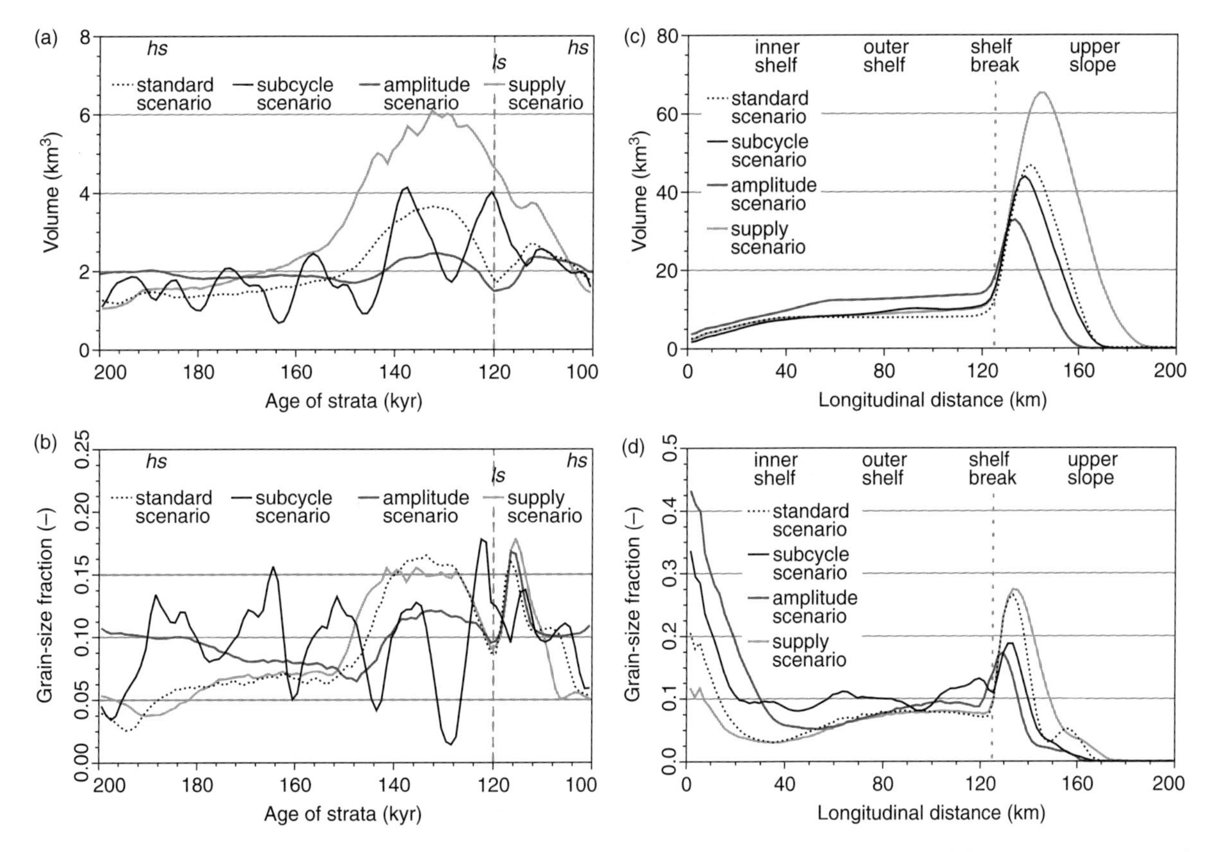

Fig. 9. Sediment properties as a function of stratal age or longitudinal distance for various model scenarios. Average of 50 similar model runs: *hs*, highstand; *ls*, lowstand. Strata represent intervals of 1 kyr. (a) Volume of sediment preserved in the stratigraphy versus stratal age. For clarity, only one of the five cycles is shown. (b) Grain-size fraction (relative abundance of the coarse grain-size class) versus stratal age. Only a single cycle is shown. (c) Volume versus longitudinal distance. (d) Grain-size fraction versus longitudinal distance.

Other scenarios

Scenario with sea-level change including subcycles

The first variant of the standard scenario (Fig. 2b) is one with five additional sea-level 'subcycles' to each glacial–interglacial cycle. These higher order oscillations are clearly expressed in the record of preserved volume and grain-size fraction (Fig. 9a and b). Similar to the case of a single cycle, sub-lowstand sediments have a greater preservation potential than sub-highstand strata, and transgressive valley fills that are particularly vulnerable during the subsequent sea-level falls. However, sub-lowstand deposits on the shelf, in contrast to the upper continental slope, are in reality likely to be reworked by waves during transgression, promoting the formation of coarse bodies. Note that, because of this differential preservation of deposits from periods within the subcycles, there are more 'subsequences' in the volume record than there are sea-level forced subcycles. In the standard scenario, relative sea-level change was stagnant before lowstand, allowing the river system to achieve a near-equilibrium profile, and resulting in early transgressive strata with low preservation potential on the shelf edge. However, with the subcycles superimposed, major river incision, that had been temporarily interrupted during the preceding sub-highstand, was still going on at ultimate lowstand under the influence of the much higher rate of sea-level fall. Moreover, knickpoint migration continued upstream during the rapid relative sea-level rise that immediately followed. Consequently, deposition on the continental slope occurred later than in the standard scenario. Since the coastline resided on the shelf for a longer period of time, more sediment as well as more coarse material remained on the shelf (Fig. 9c and d). Coarse, sheet-like bodies deposited during subcycle transgression and highstand are found there (Fig. 10). The succession of relative sea-level rises and falls resulted in

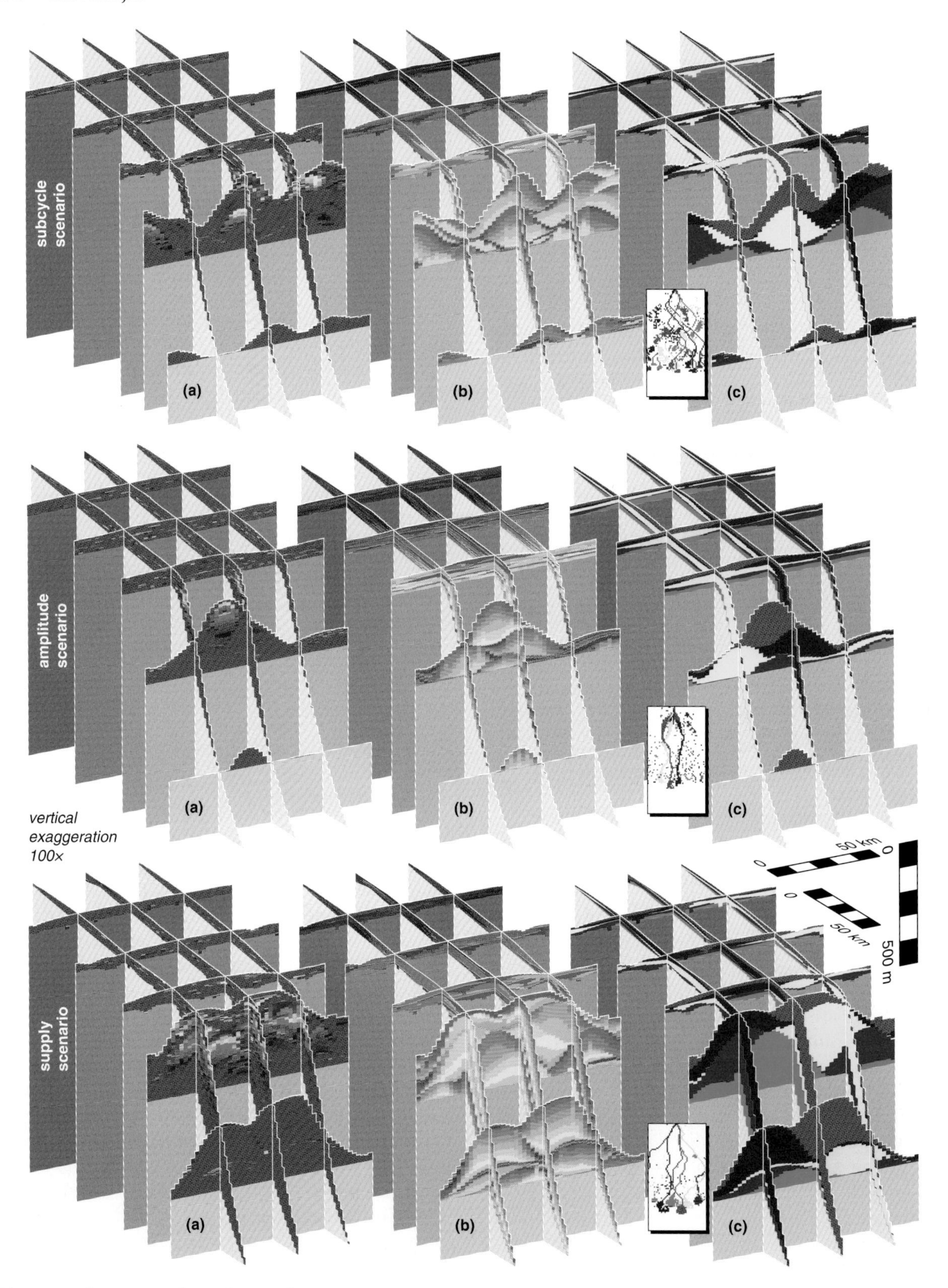

Fig. 10. Lithostratigraphies of examples of model experiments run implementing various scenarios. Strata represent intervals of 1 kyr. (a) Grain-size fraction (relative abundance of the coarse grain-size class) of sediment. (b) Time of deposition within a 100-kyr cycle. (c) Glacial–interglacial cycle in which deposition occurred. Inset: map view of coarse bodies thicker than 1 m. Colour scales are as in Fig. 7.

different levels of incision on the shelf (cf. Thomas & Anderson, 1994) and in a stepped topography on the exposed shelf (cf. Posamentier & Morris, 2000), although large-scale delta lobe switching leads to very similar configurations (Meijer, 2002a).

Scenario with lower amplitude sea-level change

The second variation on the standard scenario has a lower amplitude of eustatic sea-level change: 40 m instead of 60 m (Fig. 2b). Initially, sea level remained above the level of the shelf break throughout the cycle. Many river belts distributed sediment over the shelf. Avulsions were frequent, but their number decreased as sea-level fall progressed. Regressive delta build-up resulted in numerous thin sheets of coarse-grained bodies (Fig. 10). Incisions away from the shelf edge only occurred very locally in convex topographic features and were never deeper than roughly 2 m. Several hinterland-fed river sources supplied sediment to the shelf edge, approximating a line source. During transgression, there was no fluvial onlap, except near the river mouth. Marine onlap occurred by widespread coastal deposits, rather than in the form of valley fills as in the standard scenario. However, by filling much of the accommodation space on the shelf, the sedimentary system evolved toward a situation where the level of the shelf break matched the sea-level lowpoint. Major erosion was initially restricted to the rim of the shelf. In the last two to three cycles, incision increasingly took place over the entire shelf and eventually as far back as halfway into the transfer valley. Excavation limited itself to recent falling stage deposits, playing a relatively small role in the formation of an unconformity. Nevertheless, the pattern of preserved sediment in the stratigraphy becomes more similar to the standard scenario in the latest cycles (Fig. 9a). In the present scenario there is $\sim$ 30% less hiatuses owing to erosion in the final stratigraphy after five glacial cycles than in the standard scenario, where sea level fell well below the shelf break. Furthermore, as a consequence of the reduced role of erosion, there is less separation of the grain-size classes (Fig. 9b). This emphasizes the significance of differential preservation as a mechanism for basin-scale grain-size sorting in the development of passive margin stratigraphy.

Scenario with increased precipitation

The third alternative scenario differs from the standard one by having a wetter climate on the continental margin. The exposed shelf has a considerable surface area that also acts as a drainage basin. However, because of the relatively permeable shelf subsurface and low gradients, in the real world groundwater-flow plays an important role in the water cycle, as well as vegetation. Therefore, the model precipitation value must be considered the effective amount of surface water. Moreover, in the model, the contribution of rainwater to the main hinterland-fed river system is limited, as these rivers tend to build a (weakly) convex delta-topography around them during regression, thereby directing local rainwater flow largely away (Schumm, 1993). Unless fault structures (Coleman *et al.*, 1983) or exposed submarine canyons create large-scale topographic depressions, or basin physiography forces rivers to join (e.g. southern North Sea, Adriatic; Mulder & Syvitski, 1996), accumulated rainwater is in reality likely to form a separate, secondary drainage system or is added to sediment-poor streams from small, proximal basins (Schumm, 1993). Channel belts that were abandoned when a competing river branch captured the water discharge at the transfer-valley outlet (i.e., when drainage connection was established; Meijer, 2002a) remained active as 'overfitted valleys' for the little runoff they received. Because of lower discharge, incisions of the rain-fed drainage system occurred primarily on the outer shelf and shallowed landward. As a source of sediment, they were relatively unimportant. However, they greatly influenced the formation of unconformities.

Scenario with increased supply during glacials

The effect of climate change in the hinterland on water discharge and sediment yield from the drainage basin during transitions between glacial and interglacial periods is still unclear (Leeder, 1997; Leeder *et al.*, 1998; Goodbred & Kuehl, 2000). Even when a shift from one climate regime to another is identified, chaotic behaviour precludes inference of the supply development. Modified by local vegetation, a cooler and/or wetter climate can lead to both an increase and a decrease of supply (e.g. Walling & Webb, 1983; Schumm, 1993; Leeder *et al.*, 1998). Whether the supply signal changes gradually, abruptly or in pulses, needs to be determined from the sedimentary record. Therefore, sediment budgeting of continental margin stratigraphy is essential for resolving the control of climate on sediment transport. Pending such empirical datasets, water discharge, sediment load and grain-size fraction in the fourth model variant scenario are proportional to sea level and in phase (Fig. 2b). Despite the different supply patterns, the volume of preserved sediment in the stratigraphy with respect to the influx value is similar to that in the standard scenario (Fig. 9a). However, since in both scenarios the river profile was controlled by sea level, the amount of cannibalized material was reduced in relation to the river load. Therefore, it had less influence on river dynamics and basin-scale grain-size sorting. Moreover, the coarse sediment influx was less during interglacial times, resulting in inner shelf sequences with a lower coarse sediment content (Fig. 9d). Hence, there is relatively less enrichment of coarse sediment on the continental slope at the expense of falling-stage deposits (Fig. 9b). Incised river valleys are both deeper and steeper (Fig. 10). Furthermore, coarse bodies at the base of valley fills have a decreasing thickness landward, reducing the chances of connection.

CONCLUDING REMARKS

The model presented here is a conceptual one, and the applied scenarios are just a selection from the vast array of natural forcing variables and spatial configurations. To interpret the model results in a particular setting, local factors have to be taken into account. Future basin-scale field studies of passive margins can act as model-prototypes. This may eventually lead to the development of a sequence stratigraphic concept of clastic continental margin systems with grain calibre incorporated as a parameter, similar to the approach of Orton and Reading (1993).

A bipartite sedimentary architecture with a proximal highstand system and a distal lowstand system coupled by a long-distance regressive shelfal system is a distinctive feature of many Quaternary passive continental margins. This dual character is clearly reflected in the modelled development of volume and grain-size distribution in the stratigraphy. Concentrations of coarse-grained material are found on the innermost shelf and beyond the shelf break, with sediments finer than average in between.

In the model, small-scale grain-size sorting is controlled by deltaic processes, through the build-up of coarsening upward sequences during normal and forced progradation, and by fluvial/estuarine processes, resulting in onlapping upward-fining valley fills during sea-level rise and transgression. However, basin-scale, long-term grain-size sorting is closely related to the formation of an unconformity on the exposed shelf and in the fluvial transfer valley. Coarse paralic strata and fine-grained prodelta sediments have a different preservation potential; erosion during lowstand preferentially affects coarse falling-stage deposits and subsequently enriches shelf-edge successions with coarse material. This is the main mechanism for basin-scale sorting.

In the modelled stratigraphy, large coarse bodies are found on the upper continental slope and across the shelf at the base of incised river valleys. The individual coarse river-channel belts are connected within the complex of highstand deltas, but rarely make contact on the rest of the shelf. Shelf-edge bodies are nearly completely encapsulated by fine material delivered during falling sea-level stage and during early rise, and therefore poorly or not connected to each other. However, they are usually connected to the transgressive valley fills of the same interglacial–glacial cycle.

When relating the model results to real-world situations, caution must be taken with regard to the activity of processes not incorporated into the model that have the ability to modify the local interconnectedness of coarse-grained deposits, such as turbidite currents and wave/tidal ravinement. However, from the model it is evident that interconnectedness of major coarse bodies is very unpredictable owing to the large variability of landscape, and thus stratigraphic development.

APPENDIX A

Here, the outlines of the model are described. A comprehensive description is given in Meijer (2002a). The basis of the model is a grid of cells that constitute a landscape over which volumes of sediment can be transported under the influence of flowing water and gradients. Diffusive transport is a *short-range* mechanism for interfluvial, delta-plain, coastal and marine transport processes. This kind of transport takes place only on grid cells where a constant 'threshold discharge' value Q_{thr} is not exceeded. The outflux in a particular grid direction (dir) is proportional to a 'resulting slope' S^*_{dir} to a given power U, provided a certain threshold slope S^*_{thr} is exceeded (equation 1). For each of the depositional environments (env) in the system, ranging from terrestrial to deep sea, a threshold slope and a 'transport-efficiency constant' K_D is defined.

$$\Psi^{OUT}_{D\,dir} = K_D(\text{env}) \cdot [(S^*_{dir} - S^*_{thr}(\text{env})) \times H(S^*_{dir} - S^*_{thr}(\text{env}))]^U \quad (1)$$

where H is the Heaviside function. The resulting slope equals the 'initial slope' S_{dir} plus the balance of influx and total outflux, that is, erosion or deposition. The total diffusive outflux is:

$$\Psi^{OUT}_{D} = K_D(\text{env}) \cdot \sum_{dir} \left[\left(\frac{\Psi^{IN} - \Psi^{OUT}_D}{\textit{grid cell area}} \cdot \frac{\textit{time step}}{\textit{distance}_{dir}} + S_{dir} - S^*_{thr}(\text{env}) \right) \cdot H(S^*_{dir} - S^*_{thr}(\text{env})) \right]^U \quad (2)$$

Stream transport represents *long-range* transport by rivers and river-plumes. It applies only to terrestrial grid cells where the threshold discharge is exceeded. In the model, sediment and discharge are distributed relative to the initial slopes in the respective grid directions (equation 3). Thus, it allows for channelization with an opportunity to avulse and bifurcate the divergent stream networks. In view of the size of the grid cells, slopes should be considered valley gradients rather than river gradients.

$$\textit{portion}_{dir} = \frac{S_{dir} \cdot H(S_{dir} - k \cdot S_{max})}{\sum_{dir} \{S_{dir} \cdot H(S_{dir} - k \cdot S_{max})\}} \quad (3)$$

The stream transport algorithm requires an amount of discharge to be present in the grid cell in order to be effective. The transport equation is based on the common formulation of stream power. The stream in a given grid direction has a 'carrying capacity' $K_S \cdot [Q \cdot \textit{portion}_{dir} \cdot H(Q \cdot \textit{portion}_{dir} - Q_{thr})]^W \cdot S^{*\,V}_{dir}$ where Q is the river water 'discharge'. The carrying capacity entails the sediment load that the stream is able to carry for the present conditions. The volume that is eroded or deposited by the river further depends on the current sediment load and the ability to change its load L_S.

$$\Psi^{OUT}_{S\,dir} = [\Psi^{IN} \cdot \textit{portion}_{dir}] + \{K_S \cdot [Q \cdot \textit{portion}_{dir} \times H(Q \cdot \textit{portion}_{dir} - Q_{thr})]^W \times S^{*\,V}_{dir} - [\Psi^{IN} \cdot \textit{portion}_{dir}]\}\{L_S(\textit{capacity})\}^{-1} \quad (4)$$

The equation of the total stream outflux is:

$$\Psi^{OUT}_{S} = \Psi^{IN} - \frac{\Psi^{IN}}{L_S(\textit{capacity})} + \frac{K_S}{L_S(\textit{capacity})} \times \sum_{dir} \left\{ [Q \cdot \textit{portion}_{dir} \cdot H(Q \times \textit{portion}_{dir} - Q_{thr})]^W \times \left[\frac{\Psi^{IN} - \Psi^{OUT}_S}{\textit{grid cell area}} \cdot \frac{\textit{time step}}{\textit{distance}_{dir}} + S_{dir} \right]^V \right\} \quad (5a)$$

In the model an expression is used to prescribe a geomorphic threshold to initiate erosion. This formulation implies that a river below incision-threshold does not incise (equation 5b). On the other hand, aggradation is allowed in streams above as well as below threshold.

$$\Psi^{OUT}_{S} = \Psi^{IN} \text{ if capacity} = \text{undercapacity and} \times \sum_{dir} \{[Q \cdot \textit{portion}_{dir} \cdot H(Q \cdot \textit{portion}_{dir} - Q_{thr})]^{\omega} \cdot S^{\upsilon}_{dir}\} < P_{thr} \quad (5b)$$

The algorithm for river-plume transport is an adaptation of that of stream transport, which takes underwater relief (e.g. estuaries or submarine canyons) into account when distributing the sediment. Within the plume the carrying capacity equals zero (i.e., exponential decrease of transport) and diffusive transport can occur as well.

ACKNOWLEDGEMENTS

Financial support was provided by The Netherlands Organization for Scientific Research (NWO). I thank P.L. de Boer and G. Postma for their constructive comments on earlier versions of the manuscript. D. Granjeon and A. Kettner are thanked for their reviews of this paper.

REFERENCES

Angevine, C.L., **Heller, P.L.** and **Paola, C.** (1990) Quantitative sedimentary basin modeling. *AAPG Continuing Educ. Course Note Ser.*, **32,** 140.

Bitzer, K. and **Harbaugh, J.W.** (1987) DEPOSIM: A Macintosh computer model for two-dimensional simulation of transport, deposition, erosion, and compaction of clastic sediments. *Comp. Geosci.*, **13,** 611–637.

Blum, M.D. and **Price, D.M.** (1998) Quaternary alluvial plain construction in response to glacio-eustatic and climatic controls, Texas Gulf Coastal Plain. In: *Relative Role of Eustasy, Climate, and Tectonism in Continental*

Rocks (Eds K.W. Shanley and P.J. McCabe), *SEPM Spec. Publ.*, **59**, 31–48.

Blum, M.D. and **Törnqvist, T.E.** (2000) Fluvial responses to climate and sea-level change: A review and look forward. *Sedimentology*, **47**, 2–48.

Burgess, P.M. and **Hovius, N.** (1998) Rates of delta progradation during highstands: Consequences for timing of deposition in deep-marine systems. *J. Geol. Soc. London*, **155**, 217–222.

Coleman, J.M., Prior, D.B. and **Lindsay, J.F.** (1983) Deltaic influences on shelfedge instability processes. In: *The Shelfbreak: Critical Interface on Continental Margins* (Eds D.J. Stanley and G.T. Moore), *SEPM Spec. Publ.*, **33**, 121–137.

Den Bezemer, T., Kooi, H., Kranenborg, J. and **Cloetingh, S.** (2000) Modelling grain-size distributions. A comparison of two models and their numerical solution. *Tectonophysics*, **320**, 347–373.

Galloway, W.E. (1989) Genetic stratigraphic sequences in basin analysis I: Architecture and genesis of flooding-surface bounded depositional units. *AAPG Bull.*, **73**, 125–142.

Goodbred, S.L. and **Kuehl, S.A.** (2000) Enormous Ganges-Brahmaputra sediment discharge during strengthened early Holocene monsoon. *Geology*, **28**, 1083–1086.

Granjeon, D. and **Joseph, P.** (1999) Concepts and applications of a 3-D multiple lithology, diffusive model in stratigraphic modeling. In: *Numerical Experiments in Stratigraphy: Recent Advances in Stratigraphic and Sedimentologic Computer Simulations* (Eds J.W. Harbaugh, W.L. Watney, E.C. Rankey, R. Slingerland, R.H. Goldstein and E.K. Franseen), *SEPM Spec. Publ.*, **62**, 197–210.

Leeder, M.R. (1997) Sedimentary basins: Tectonic recorders of sediment discharge from drainage catchments. *Earth Surf. Proc. Land.*, **22**, 229–237.

Leeder, M.R. and **Stewart, M.D.** (1996) Fluvial incision and sequence stratigraphy: Alluvial responses to relative sea-level fall and their detection in the geological record. In: *Sequence Stratigraphy in British Geology* (Eds S.P. Hesselbo and D.N. Parkinson), *Geol. Soc. London Spec. Publ.*, **103**, 25–39.

Leeder, M.R., Harris, T. and **Kirkby, M.J.** (1998) Sediment supply and climate change: Implications for basin stratigraphy. *Basin Res.*, **10**, 7–18.

Meijer, X.D. (2002a) Modelling the drainage evolution of a river–shelf system forced by Quaternary glacio-eustasy. *Basin Res.*, **14**, 361–377.

Meijer, X.D. (2002b) Quantitative three-dimensional modelling of Quaternary passive continental margins. *Geologica Ultrajectina*, **221**, 88 pp.

Miall, A.D. (1991) Stratigraphic sequences and their chronostatigraphic correlation. *J. Sed. Petrol.*, **61**, 497–505.

Mulder, T. and **Syvitski, J.P.M.** (1996) Climatic and morphologic relationships of rivers: Implications of sea-level fluctuations on river loads. *J. Geol.*, **104**, 509–523.

Orton, G.J. and **Reading, H.G.** (1993) Variability of deltaic processes in terms of sediment supply, with particular emphasis on grain size. *Sedimentology*, **40**, 475–512.

Paola, C., Heller, P.L. and **Angevine, C.L.** (1992) The large-scale dynamics of grain-size variation in alluvial basins, 1: Theory. *Basin Res.*, **4**, 73–90.

Posamentier, H.W. and **Allen, G.P.** (1993) Variability of the sequence stratigraphic model: Effects of local basin factors. *Sed. Geol.*, **86**, 91–109.

Posamentier, H.W. and **James, D.P.** (1993) An overview of sequence-stratigraphic concepts: Uses and abuses. In: *Sequence Stratigraphy and Facies Associations* (Eds H.W. Posamentier, C.P. Summerhayes, B.U. Haq and G.P. Allen), *Spec. Publ. Int. Assoc. Sedim.*, **18**, 3–18.

Posamentier, H.W. and **Morris, W.R.** (2000) Aspects of the stratal architecture of forced regressive deposits. In: *Sedimentary Responses to Forced Regressions* (Eds D. Hunt and R.L. Gawthorpe), *Geol. Soc. London Spec. Publ.*, **172**, 19–46.

Posamentier, H.W., Allen, G.P., James, D.P. and **Tesson, M.** (1992) Forced regressions in a sequence stratigraphic framework: Concepts, examples, and exploration significance. *AAPG Bull.*, **76**, 1687–1709.

Rivenæs, J.C. (1992) Application of a dual-lithology, depth-dependent diffusion equation in stratigraphic simulation. *Basin Res.*, **4**, 133–146.

Schumm, S.A. (1993) River response to baselevel change: Implications for sequence stratigraphy. *J. Geol.*, **101**, 279–294.

Syvitski, J.P.M. and **Hutton, E.W.H.** (2001) 2D SEDFLUX 1.0C: An advanced process-response numerical model for the fill of marine sedimentary basins. *Comput. Geosci.*, **27**, 731–753.

Talling, P.J. (1998) How and where do incised valleys form if sea level remains above the shelf edge? *Geology*, **26**, 87–90.

Tetzlaff, D.M. and **Harbaugh, J.W.** (1989) *Simulating Clastic Sedimentation.* Van Nostrand Reinhold, New York, 202 pp.

Thomas, M.A. and **Anderson, J.B.** (1994) Sea-level controls on the facies architecture of the Trinity/Sabine incised-valley system, Texas continental shelf. In: *Incised-Valley Systems: Origin and Sedimentary Sequences* (Eds R.W. Dalrymple, R. Boyd and B.A. Zaitlin), *Spec. Publ. Soc. Econ. Paleont. Miner.*, **51**, 63–82.

Van Heijst, M.W.I.M. and **Postma, G.** (2001) Fluvial response to sea-level changes: A quantitative analogue, experimental approach. *Basin Res.*, **13**, 269–292.

Vanney, J.R. and **Stanley, D.J.** (1983) Shelfbreak physiography: An overview. In: *The Shelfbreak: Critical Interface on Continental Margins* (Eds D.J. Stanley and G.T. Moore), *SEPM Spec. Publ.*, **33**, 1–24.

Walling, D.E. and **Webb, B.W.** (1983) Patterns of sediment yield. In: *Background to Palaeohydrology* (Ed. K.J. Gregory), John Wiley & Sons, Chichester, pp. 69–100.

Wescott, W.A. (1993) Geomorphic thresholds and complex response of fluvial systems – Some implications for sequence stratigraphy. *AAPG Bull.*, **77**, 1208–1218.

Wright, V.P. and **Marriott, S.B.** (1993) The sequence stratigraphy of fluvial depositional systems: The role of floodplain sediment storage. *Sed. Geol.*, **86**, 203–210.

Zaitlin, B.A., Dalrymple, R.W. and **Boyd, R.** (1994) The stratigraphic organization of incised-valley systems associated with relative sea-level change. In: *Incised-Valley Systems: Origin and Sedimentary Sequences* (Eds R.W. Dalrymple, R. Boyd and B.A. Zaitlin), *SEPM Spec. Publ.*, **51**, 45–60.

Spec. Publ. Int. Assoc. Sedimentol. (2008) **40**, 223–238

Modelling the preservation of sedimentary deposits on passive continental margins during glacial–interglacial cycles

XANDER D. MEIJER, GEORGE POSTMA, PETER A. BURROUGH and POPPE L. DE BOER
Faculty of Geosciences, Utrecht University, P.O. Box 80021, 3508 TA, Utrecht, The Netherlands (E-mail: gpostma@geo.uu.nl)

ABSTRACT

The preservation potential of sedimentary deposits is a reflection of the probability that a stratigraphic level will escape reworking. In this paper, the preservation of strata on passive continental margins during glacial–interglacial cycles is investigated by means of a three-dimensional, dynamic numerical model. The reworking of sediment is primarily driven by glacio-eustatic base-level changes, and varies across the shelf. Local topography and the existing drainage pattern influence the distribution of erosion. Statistically the coastal wedge (highstand delta), the adjacent inner shelf and the shelf edge are most vulnerable. The differential preservation, especially of shelf and upper continental slope deposits, is also manifested in the chronostratigraphy, leading to a considerable deviation in the volume distribution from the sediment supply signal. The formation of a subaerial erosional unconformity commences during falling sea level, continues until early rise, and greatly affects paralic deposits. Various scenarios of sea-level change, different gradient contrasts between coastal prism and shelf, and sediment supply and discharge have been investigated.

Keywords Forward modelling, passive continental margins, glacial–interglacial cycles, stratigraphy, preservation potential.

INTRODUCTION

Much of the geological time represented by sedimentary successions is taken up by hiatuses. Reworking and remobilization are basic aspects of sedimentary processes, and are especially associated with recurrent events over various timescales. During the Quaternary, glacial–interglacial sea-level and climate changes have been important cyclic driving forces on a relatively long timescale of 40–100 kyr. Basin margins have been particularly affected by the large, glacio-eustatically driven base-level changes which led to periodic emergence and submergence of continental shelves. As a result of subaerial erosion, fluvial incision and wave ravinement, some deposits are better preserved than others. A measure of the likelihood of a stratigraphic level to escape reworking is termed its 'preservation potential'.

The actual preservation of ancient sediments after a certain time interval following initial deposition gives a quantitative assessment of the preservation potential of similar, modern deposits. For example, remnant highstand deposits from a previous glacial cycle indicate how much of the Holocene coastal-plain deposits may remain after the next glacial cycle. However, it is difficult to estimate the volume percentage of sedimentary deposits that survived from previous glacial cycles, since the volume of the original strata usually cannot be reconstructed. Dynamic sediment transport models are able to record the history of depositional and erosional stages, and thus can assist in evaluating the preservation potential of different sedimentary facies deposited during glacial–interglacial cycles.

Here, a numerical stratigraphic model is used to investigate the preservation of deposits on a general passive continental margin during Quaternary glacial–interglacial cycles. In anticipation of three-dimensional, volumetric data sets from passive margins, the model is applied here as a conceptual tool to study the dynamics of the sedimentary system. The effects of various boundary conditions are examined, including eustatic sea-level change, contrast in gradients of the coastal prism and the continental shelf, and river discharge and sediment load.

METHOD

Appendix A in Meijer (this volume, p. 220) provides an outline of the equations that are used in the model. See Meijer (2002) for a detailed description of the model and its calibration. The model is cellular: the fluxes of water and sediment over a topographic surface from one grid cell to neighbouring cells are defined by a number of straightforward transport rules. The rules describe time-averaged sediment transport in several depositional environments, defined by bathymetric zones, as a function of local conditions such as slope and discharge. There are two transport algorithms: one for short-range, slope-dependent transport and another for long-range, slope and discharge-dependent river transport. Wave action is not modelled explicitly, apart from increased transport capability in the shallow subaqueous environment. Deposition, erosion, and tectonic subsidence change the surface elevation. Feedback through the relationships between topography and transport results in an intricate process–response system, driven by the boundary conditions set by sea-level change, tectonics, precipitation, the supply of sediment, and discharge. The model produces a morphological (palaeogeographical) evolution and builds a three-dimensional lithostratigraphy and chronostratigraphy by recording the amount and properties (for instance, age, provenance and grain size) of deposited sediment, and the presence of hiatuses.

Model procedure

The set-up of the model is outlined in Fig. 1, and parameter values are given in Table 1. Combinations of variables acting as boundary conditions that are applied to the model are referred to as 'scenarios' (Fig. 2). The scenarios used in this paper are listed in Table 2. For the benefit of statistical analysis, each model scenario was run 50 times, each time with a different white-noise pattern superimposed to roughen the initial topographic surface. This allows the results of different scenarios to be compared statistically.

From the model stratigraphy, the volume of sediment between two 'stratigraphic levels' (i.e., the total volume of deposits between certain age

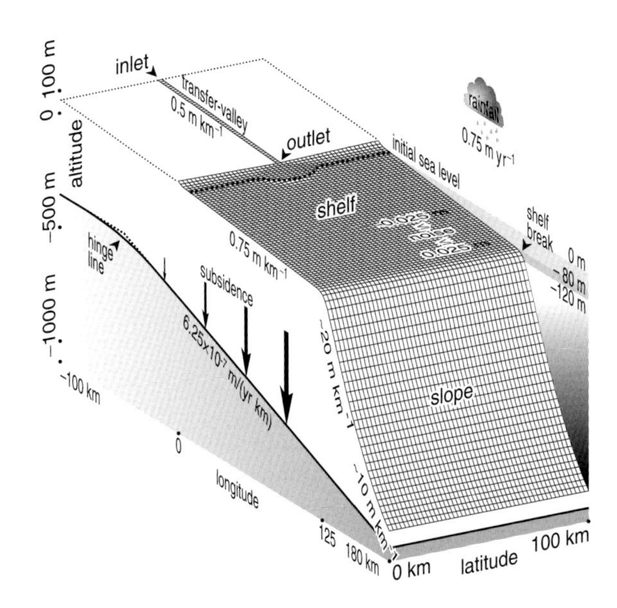

Fig. 1. Overview of the grid set-up of the model. The initial topography outlines a transfer valley emanating from the drainage basin, an initial highstand delta, a continental shelf and slope. The scale of the passive continental margin system is 280 × 100 km.

limits) is summed for the entire area of interest, for a specific site, or for a depositional environment, and recorded. Subsequently, in order to determine the preservation of a particular stratigraphic interval, the decrease of this initial amount through time owing to erosion is registered. A model run covers five glacial–interglacial cycles, each with a duration of 100 kyr (with the exception of the gradient scenarios). The first cycle is a preparatory one; the second cycle is the main object of study. Preservation of deposits from the second cycle is monitored both during and after this cycle.

RESULTS AND DISCUSSION

Geomorphology and preservation

Before analysing the preservation of deposits laid down on a passive margin during a glacial–interglacial cycle, the general character of geomorphological evolution and the associated sediment distribution are briefly described on the basis of an arbitrary model run. In Fig. 3, the topographical development of this example for the first two cycles is depicted at 10 kyr intervals, as well as the net sedimentation that has occurred during the preceding 10 kyr.

The model evolution starts during an interglacial with a submerged shelf and a shelf delta. During highstand, river sediment is deposited

Table 1. Values of parameters applied in the model runs. See Appendix A and Meijer (2002, 2008 – this Volume) for an explanation of the parameters and the accompanying equations.

Parameter	Symbol	Environment	Value	Unit
Diffusive transport power	U		1.0	–
Diffusive transport coefficient	K_D	Terrestrial	0.30	$km^3\,kyr^{-1}$
		Coastal	0.40	$km^3\,kyr^{-1}$
		Marine	0.40	$km^3\,kyr^{-1}$
		Deep marine	0.32	$km^3\,kyr^{-1}$
Threshold slope	S^*_{thr}	Terrestrial	0.0005	–
		Coastal	0.0040	–
		Marine	0.0200	–
		Deep marine	0.0200	–
Transition depth with respect to sea level		Terrestrial – coastal	0	m
		Coastal – marine	25	m
		Marine – deep marine	100	m
Discharge distribution threshold	k	Terrestrial	0.75	–
		Non-terrestrial	0.00	–
Stream transport power	V		1.0	–
Stream transport power	W		1.0	–
Discharge threshold	Q_{thr}		2.0×10^8	$m^3\,yr^{-1}$
Stream transport coefficient	K_S		0.40	–
Stream attenuation constant	L_S	At overcapacity	10	–
		At undercapacity	50	–
Incision criterion power	υ		1.00	–
Incision criterion power	ω		0.25	–
Incision threshold	P_{thr}		0.224	$(m^3\,yr^{-1})^{0.25}$
Grid cell size			2000	m
Timestep			10	yr
White-noise range			0.05	m

subaerially and on the inner shelf (Fig. 3a and b, at ~10 and ~110 kyr). Aggradation on the highstand delta decreases as accommodation decreases under the influence of falling sea-level. Avulsions on the delta plain are initially frequent, but decrease in number. Forced regression (Posamentier *et al.*, 1992) leads to rapid progradation of the coastline over the continental shelf (at ~20–50 and ~120–150 kyr). Relatively shallow incisions occur in convex delta morphologies, in particular in the front of the coastal prism (Posamentier *et al.*, 1992; Talling, 1998) that has emerged. Thus, river belts become locally embedded in shallow valleys in some reaches. Although upstream avulsion events occur sporadically, the main locus of avulsions shifts basinward along with the active deltas. Delta-lobe switching produces a complex of stacked delta bodies on the shelf, forming a forced regressive wedge. Deposition is largely limited to the coastal region. As the shoreline approaches the shelf break, accommodation space increases dramatically and the progradation rate diminishes (at ~50 and ~150 kyr). Incipient shelf-edge deltas are formed on the upper continental slope. A continuously falling base level results in the formation of upstream migrating knickpoints in the river belts, and the exposed shelf is incised. Headward erosion produces a diachronous subaerial erosional unconformity across the shelf. As the knickpoints migrate upstream, one of the competing river belts connects to the outlet of the hinterland, capturing all discharge from the drainage basin, and focusing deposition in a major delta along the shelf edge. During the glacial period, this shelf-edge delta remains the main depocentre on the basin margin. Avulsions are restricted to this delta system. At lowstand, as relative sea-level begins to rise, deposition occurs near river mouths in relatively shallow water (at ~80 and ~180 kyr). This submarine aggradation gradually passes into retrogradational coastal and fluvial onlap (at ~90 and ~190 kyr). Sedimentation is initially confined in incised valleys with limited space for avulsions. As sea-level rise decelerates, a new coastal prism is built with frequent avulsions on the highstand delta (at ~100 and ~200 kyr).

Figure 3c shows the topographic evolution during the third glacial–interglacial cycle. In addition, the preservation of the deposits of the 10 kyr time slices of the second cycle (cf. Fig. 3b) are depicted

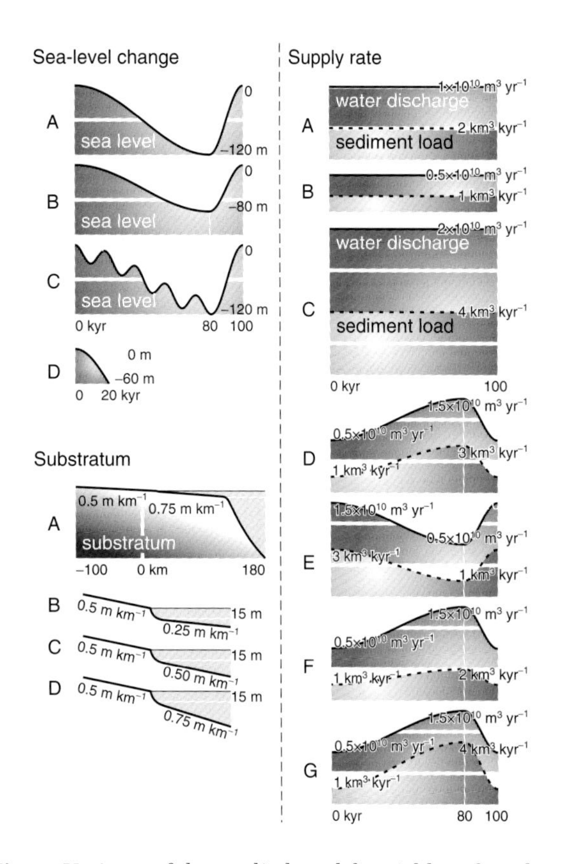

Fig. 2. Variants of the applied model variables of sea-level change (A–D), initial topography (A–D), and discharge and sediment supply rate (A–G). Table 2 lists the combination in which they are used in the different scenarios. For the sea level (except D) and supply variables, only single glacial–interglacial cycles of 100 kyr are depicted. Model runs cover five identical cycles, where the preservation of the second cycle is the object of study.

after 100 kyr. The preservation at a particular time is defined as the ratio of the stratal thickness at that time to the original thickness.

At 100 kyr after its formation, the highstand delta has been bisected by the incised valley connected to the drainage-basin outlet that formed before the sea-level lowstand of the second cycle (Fig. 3c, at ~210 kyr). Similarly, the forced regressive wedge on the shelf has been partitioned as a result of several incisions (at ~220–250 kyr). River incision during late falling stage and lowstand of the third cycle (at ~270–280 kyr) again erodes 'recent' forced-regressive deposits, but does not greatly affect the preservation of the 100 kyr old deposits. This is because the small amount of sediment from this period of the second cycle that was deposited on the shelf has long since been reworked or buried. Deposits that are affected by third-cycle river incision are the highstand delta and early regressive wedge (the result of which can be seen in Fig. 3e, at ~310 and ~320 kyr), and the transgressive strata (Fig. 3c, at ~280 and ~290 kyr). The preservation of the transgressive strata does not only decrease owing to river incision before lowstand, but also by erosion of aggradational strata on top of the shelf-edge deltas. Shelf-edge degradation is slow when the shelf is submerged, but is intensified during emergence. Transgressive shelf strata are probably less preserved in the real world than shown here, because of wave ravinement during sea-level rise. Likewise, reworking during falling sea-level is likely to further diminish the preservation of shelf sediments (Plint, 1988; Posamentier *et al.*, 1992; Plint

Table 2. Model scenarios used in this paper. The sea-level curves, initial topographies and supply curves referred to in the table are depicted in Fig. 2.

Scenario	Variables		
	Sea-level curve	Initial topography	Supply curve
Standard scenario	A	A	A
Sea-level scenarios			
'Amplitude scenario'	B	A	A
'Subcycle scenario'	C	A	A
Gradient scenarios			
All permutations	D	B,C,D	A,B,C
Supply scenarios			
'Low supply scenario'	A	A	B
'High supply scenario'	A	A	C
'High glacial-supply scenario'	A	A	D
'Low glacial-supply scenario'	A	A	E
'Low glacial-load scenario'	A	A	F
'High glacial-load scenario'	A	A	G

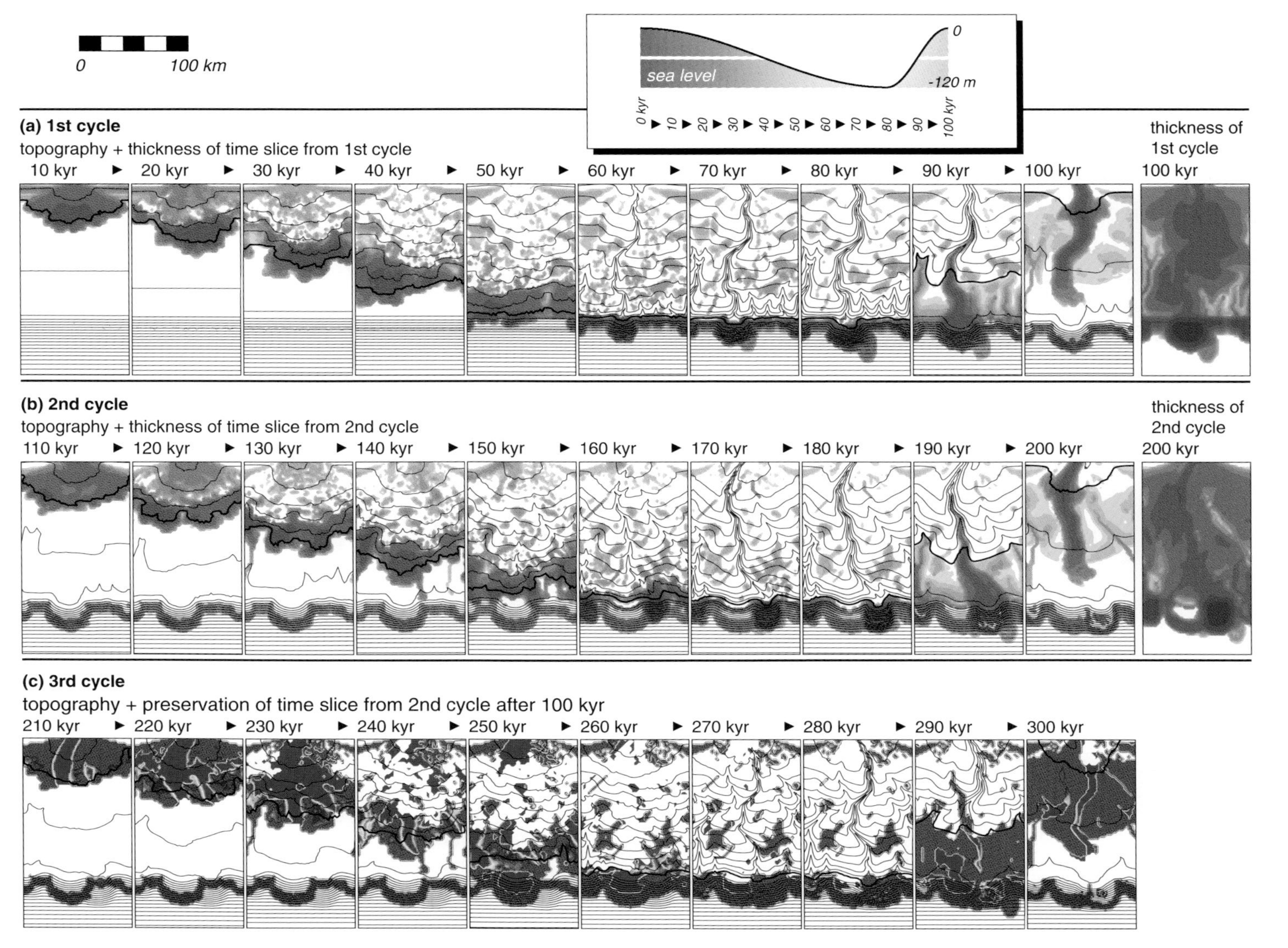

Fig. 3. Maps of the palaeogeographical evolution, sediment thickness and preservation during four glacial–interglacial cycles. Topographic contour interval above sea level is 10 m; below sea level 50 m. The shoreline is indicated by a thick line. Preservation is defined as the sediment thickness relative to the initial thickness. The inset provides an overview of the applied sea-level curve. (a) The first cycle (0–100 kyr) depicted in intervals of 10 kyr. In each frame, the surface topography at the end of a 10 kyr interval and the final thickness of deposits accumulated during the preceding 10 kyr 'time slice' are shown. On the far right, the thickness of deposits of the entire first cycle measured at the end of the first cycle is given. (b) The second cycle (100–200 kyr), the preservation of which is monitored in subsequent cycles. Shown are topographies and thicknesses of time slices similar to those in (a). (c) The third cycle (200–300 kyr), within each frame, the surface topography at the end of a 10 kyr interval is shown as well as the preservation of sediments of the time slice from the previous cycle.

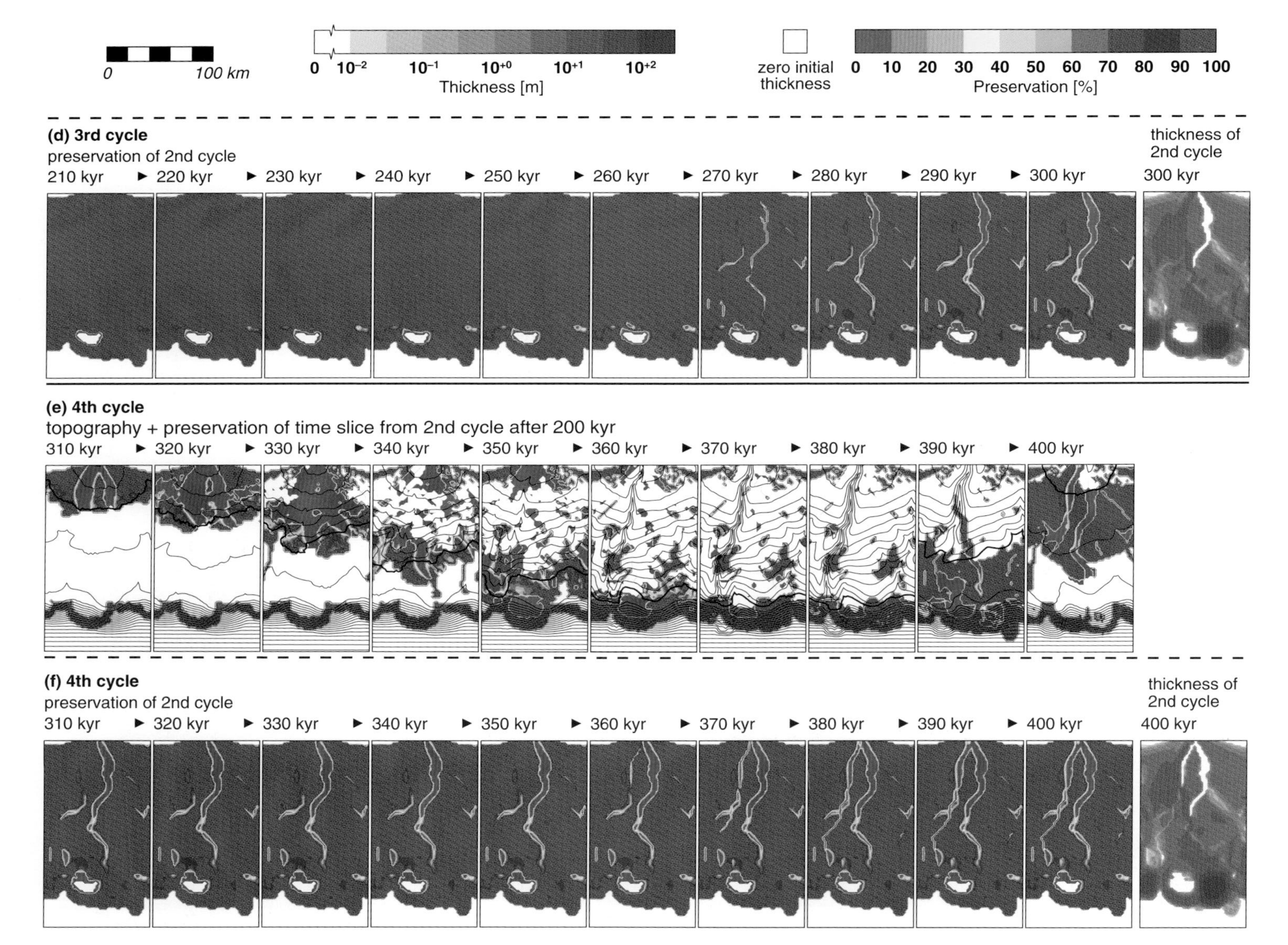

Fig. 3. (d) The third cycle, with the preservation of sediments of the entire second cycle since the end of the second cycle. On the far right, the preserved thickness of deposits of the entire second cycle. (e) The fourth cycle (300–400 kyr), similar to (c) but with the preservation of sediments of time slices from two previous cycles. (f) The fourth cycle, similar to (d).

& Nummedal, 2000; Posamentier & Morris, 2000). The model results therefore indicate an upper limit to preservation and are especially applicable for natural situations with a mild wave regime.

Instead of looking at time slices of a cycle, we can also determine what deposits are present after a full cycle (Fig. 3b, right), and see how they survive the next glacial–interglacial cycles. The preservation of the deposits of the entire second cycle can be followed with respect to the stratigraphic thickness at the end of the second cycle, shown in Fig. 3d and f. Fluvial incision and shelf-edge degradation are the main causes of reworking; subaerial erosion is limited to valley walls and the shelf edge. The effects of incision are especially apparent on the inner shelf, where rivers cut into the former highstand coastal prism. With the burial by younger sediments under the influence of basin subsidence, the intensity of reworking decreases with every new cycle, just as erosion is less vigorous than it was during the second cycle itself.

Erosion and remobilization of sediment can be quantified in the model by measuring the time that elapses between the entrance of sediment from the drainage basin into the system and its eventual deposition (i.e., the actual age of the deposit in the stratigraphy). In Fig. 4b the stratigraphy after five glacial–interglacial cycles (shown in Fig. 4a) is depicted with this difference in 'supply age' and stratal age. The higher the value of the differential age, the older the sediment that is incorporated into the deposit relative to the age of the stratigraphic level. This property can be related to the maturity of the grains. Figure 4b shows that material on the shelf was deposited relatively shortly after it left the drainage basin, whereas markedly more cannibalized sediment is found on the upper continental slope. Moreover, the differential age is low at the base of a shelf-edge delta sequence and increases upward as progressively more cannibalized sediment was supplied from the shelf during the formation of a subaerial erosional unconformity. Reworked sediment also ends up in fills of incised valleys, since cannibalization continues upstream during fluvial and coastal onlap.

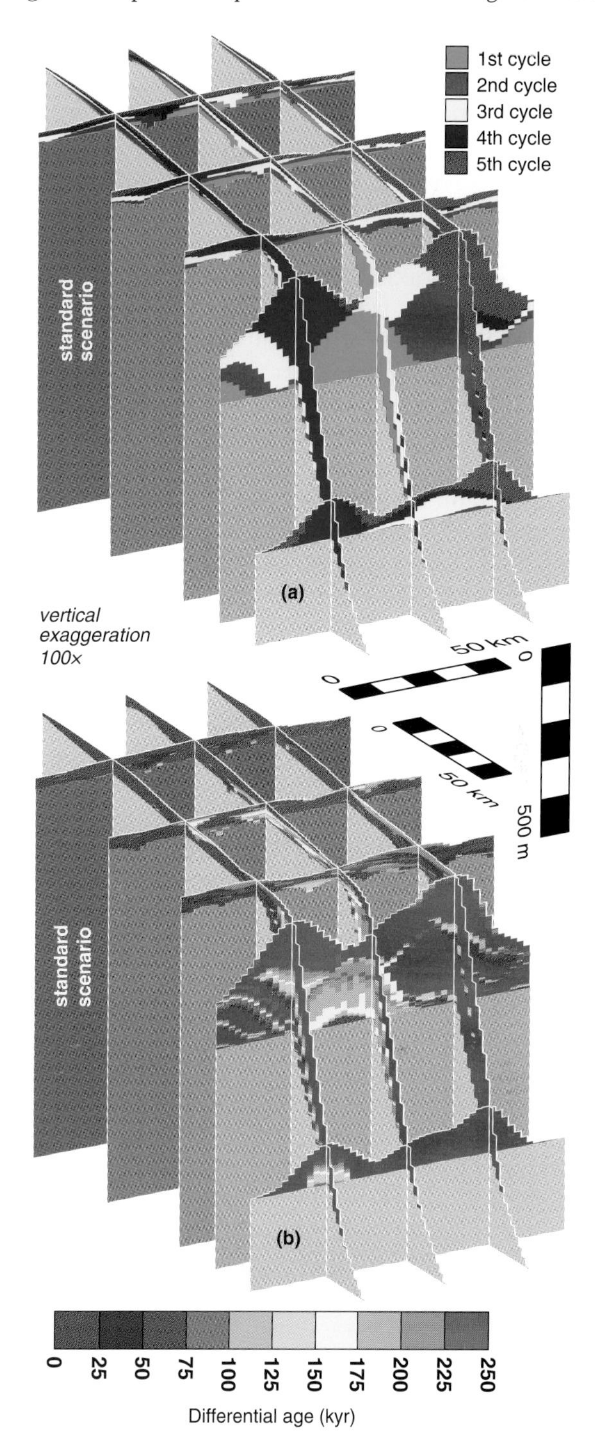

Fig. 4. Lithostratigraphy of an example of a model experiment run implementing the standard scenario. (a) The glacial–interglacial cycle in which deposition occurred. (b) The 'differential age' of deposits, that is, the difference between the time the sediment entered the system from the hinterland and the time it was incorporated into the deposit in the final stratigraphy. This property indicates cannibalization and remobilization of sediment and is a rough measure of the maturity of the sediment. Note the difference in differential age between shelf and shelf-edge strata.

Stratigraphy and preservation

In addition to the spatial distribution of erosion discussed in the previous section, we can also consider preservation from the perspective of the chronostratigraphy of the entire system. The mean of 50 model runs implementing the same standard scenario, but with different random noise on the initial topography, is determined to examine the trends in preservation.

Figure 5b and c shows how the cumulative volume of the whole study area and the volume relative to the initial amount (indicated on the vertical axes) of deposits of a certain age (indicated on the horizontal axis) decreases with time (the succession of bands from dark to light grey). On the basis of their time of deposition, strata are allocated to stratigraphic intervals with a duration of 1 kyr. Highstand deposits from the beginning of the second cycle initially endure, as erosion is usually limited to minor incision into the front of the highstand delta during the early part of forced regression. The immediate reworking of freshly deposited, forced regressive material increases as the rate of sea-level fall increases. However, after the shelf break emerges above sea level, erosion in the system accelerates (cf. Schlager, 1993; Posamentier & Morris, 2000). Headward incising rivers start to affect the inner shelf from around 160 kyr, and highstand deposits are rapidly cannibalized. Close to 40% of the sediments dating from the first half of the second cycle are eroded before the end of the cycle (cf. Figueiredo & Nittrouer, 1995). Deposition after ~150 kyr occurs on the upper continental slope. The depocentres are situated in deep water and rapid burial increases the preservation potential considerably (Fig. 5). Deposits laid down around lowstand are mostly shallow-water, aggradational shelf-edge strata, which are relatively poorly preserved. In the absence of ravinement, transgressive strata are not eroded until late in the next cycle (~260–280 kyr), after having been covered by regressive sediments.

Figure 6 shows the preservation of the stratigraphic volume that is left at the end of the second glacial–interglacial cycle. As a consequence of previous geomorphic work and burial, reworking is generally not as intense as during the second cycle itself. This is particularly the case for forced-regressive (shelf) deposits. They are only affected during the second halves of following cycles, related to major shelf incision. Strata laid down around lowstand (shelf-edge deposits) and during rising sea-level (valley fills and highstand delta), on the other hand, are relatively severely reworked in later glacial–interglacial cycles. The coastal prism that is built up mainly during rise is vulnerable to incision. Because of its confinement, the transfer valley is greatly affected by fluvial erosion. In the real world additional traces of previous cycles will remain in terraces, which are

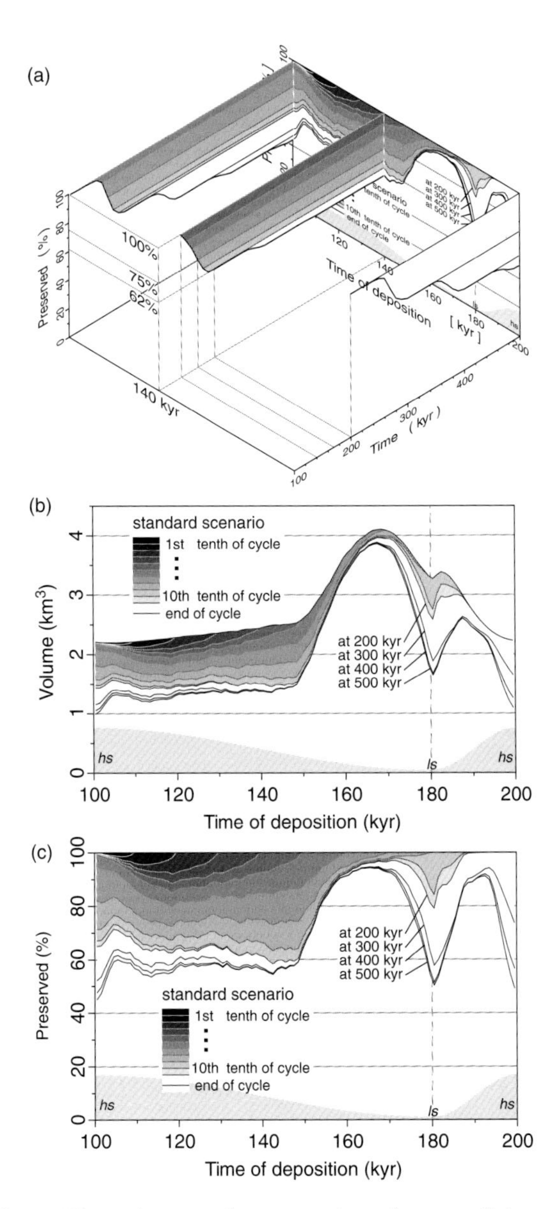

Fig. 5. The volume and preservation of strata of the second glacial–interglacial cycle deposited in intervals of 1 kyr through time. (a) Explanation of the method of construction of the diagrams in (b) and (c), and similar figures. At 1 kyr intervals during the model run, the volume of all deposits of 1 kyr and younger is determined and recorded. For example at 140 kyr, deposits laid down between 139 and 140 kyr were found with a total volume of 2.43 km^3. Through time, the volumes of sediments in the aged 1 kyr stratigraphic intervals are remeasured, yielding lower values when erosion has occurred and corresponding decreased values of preservation. For the above example, at 170 kyr the volume of the 139–140 kyr stratigraphic interval is 1.83 km^3, or 75% of the initial volume; at 200 kyr it is 1.5 km^3, or 62%. The volume and preservation evolutions of all 1 kyr intervals are combined to create diagrams like Fig. 5a and b, respectively. (b and c) The graphs are means of 50 runs implementing the standard scenario. Also indicated in light grey is the shape of the sea-level curve: *hs*, highstand; *ls*, lowstand. Note that the upper boundaries of the volume and preservation graphs are diachronous.

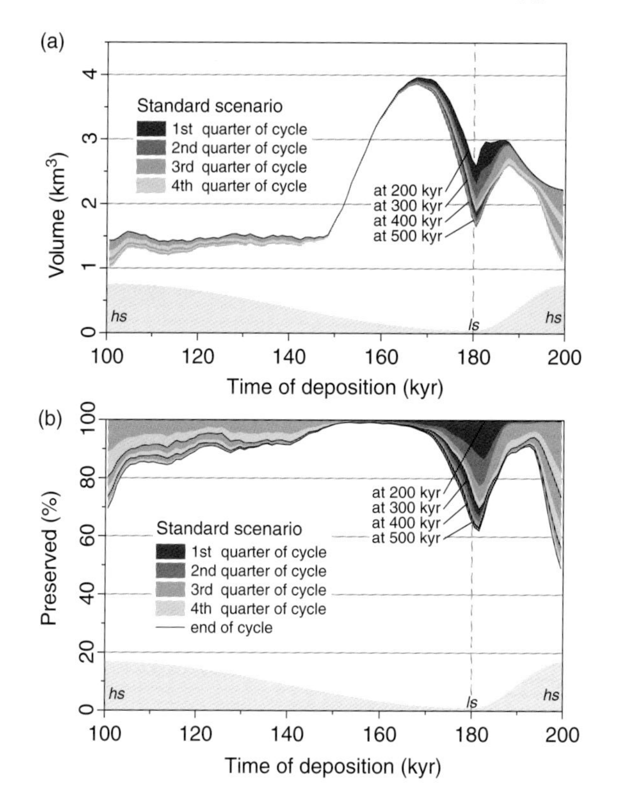

Fig. 6. The volume and preservation of strata of the second glacial–interglacial cycle deposited in intervals of 1 kyr through time. Preservation at a given time is defined as the volume at that time relative to the initial volume. The graphs are similar to those in Fig. 5, except that the 'initial' reference volumes are determined at the end of the second cycle, instead of directly after the creation of the stratigraphic intervals. The graphs are means of 50 runs implementing the standard scenario. Also indicated in light grey is the shape of the sea-level curve.

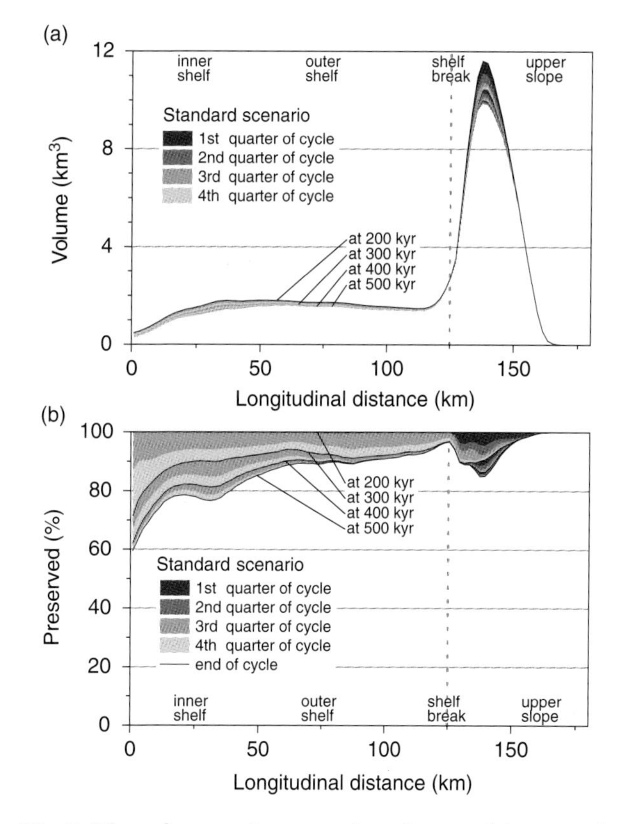

Fig. 7. The volume and preservation of strata of the second glacial–interglacial cycle deposited at a longitudinal distance through time. Preservation at a given time is defined as the volume at that time relative to the initial volume. The graphs are similar to those in Fig. 6, except that the independent variable is longitudinal distance rather than the stratigraphic interval. The graphs are means of 50 runs implementing the standard scenario.

below the resolution of the model. Lowstand shelf-edge deposits are eroded both subaerially and submarine during the entire cycle (Figs 6 and 7). The poor preservation of these deposits leads to a marked depression in the chronostratigraphic volume around the time of lowstand, which should not be interpreted as a decrease in hinterland supply around that time.

The preserved volume of sediment of the second glacial–interglacial cycle decreases variably with time (Fig. 8). Note the relatively slow and steady decrease of preservation during the first halves of the cycles, followed by high rates of decrease during the second halves when the continental shelf was subaerially exposed. There is also a considerable contrast in preservation of sediments of the different depositional environments. Especially terrestrial deposits, and coastal ones to a lesser extent, suffer great losses in volume during shelf emergence. In contrast, as

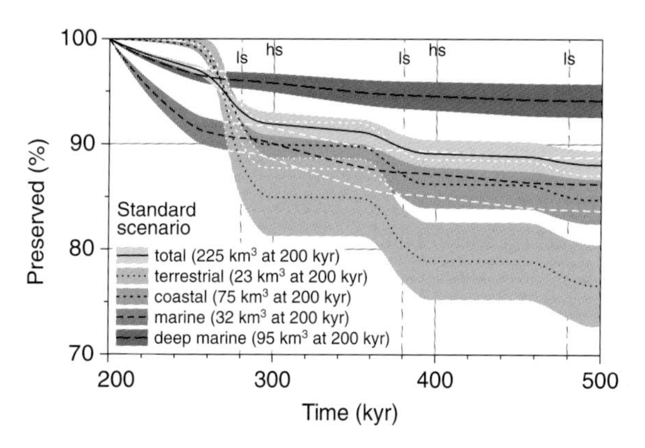

Fig. 8. The evolution of the preservation of the volume of second-cycle sediments present at the end of the second cycle, in total as well as subdivided by depositional environment. Shown are the means and 2σ-confidence intervals of 50 runs implementing the standard scenario.

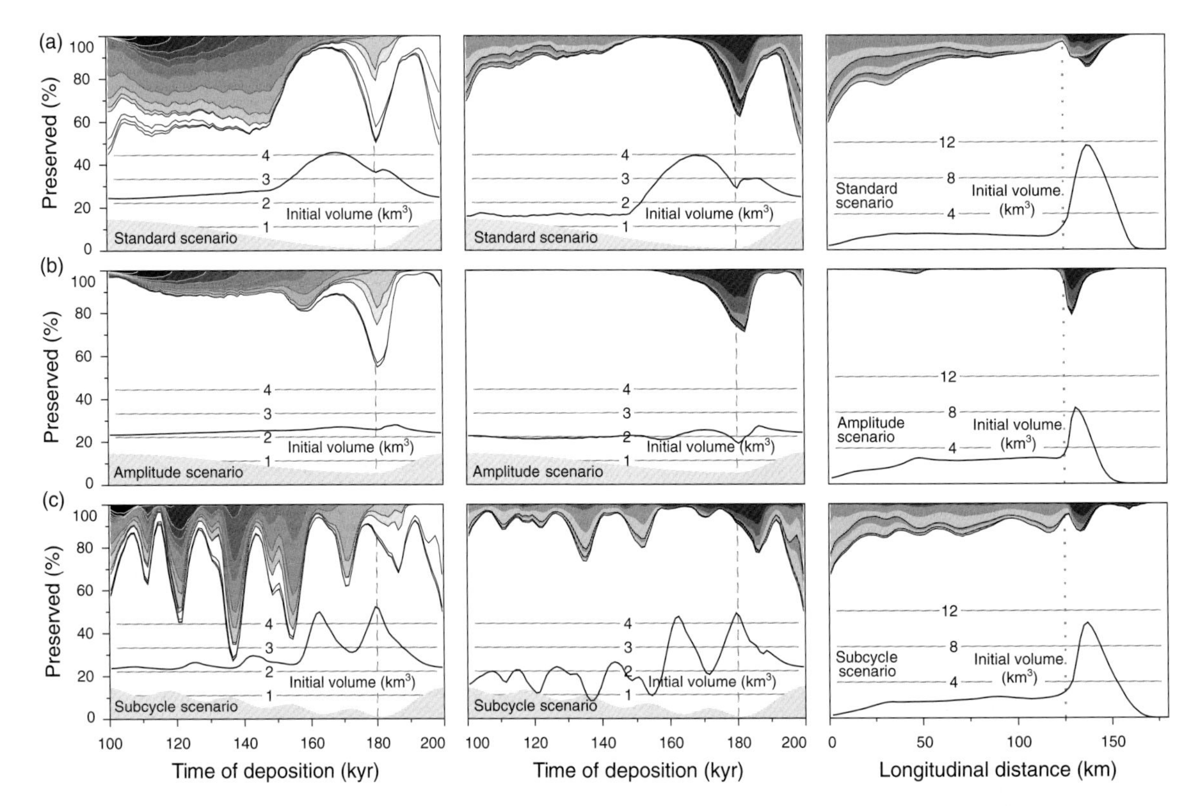

Fig. 9. The volume and preservation of strata of the second glacial–interglacial cycle through time for the standard and sea-level scenarios (Table 2). The graphs on the left are similar to Fig. 5, those in the middle to Fig. 6, and the ones on the right to Fig. 7. Also indicated in light grey are the shapes of the respective sea-level curves. (a) The standard scenario. (b) The 'amplitude scenario'. (c) The 'subcycle scenario'.

expected, deep marine sediments on the continental slope seem unaffected directly by the sea-level fluctuations.

Other scenarios

Sea-level scenarios

Glacio-eustasy has a great influence on the preservation of deposits on passive continental margins as it drastically modifies base level and may result in subaerial exposure of the shelf. There is, however, a marked difference between systems where sea level drops below the shelf break and systems where it does not. The 'amplitude scenario' (Table 2), with a lowstand sea-level of only 80 m, is an example of the latter situation. The lower rate of sea-level fall results in sustained fluvial and deltaic aggradation on the inner shelf for ~40 kyr from the start of the cycle. Because there is more accommodation space on the shelf than in the standard scenario, progradation of the forced regressive wedge is slower, and shelf-edge deltas are not formed until about 60 kyr into the cycle. Little subaerial erosion occurs since not much relief was built during sea-level fall (cf. Fig. 9a and b), and the subaerial unconformity represents a smaller hiatus than in the standard scenario. Incision occurs mostly on the outermost shelf (resulting in a decreased preservation of strata from around 160 kyr shown in Fig. 9b), but is rarely deep enough to cut through deposits underlying the forced regressive wedge. In the absence of deep river valleys, sedimentation is not confined during sea-level rise, which results in widespread sheet-like transgressive strata. In subsequent cycles, sediment of the second cycle is almost fully preserved owing to burial, with the exception of the shelf edge that is degrading with decreasing intensity (Figs 9b (right) and 10b).

In contrast, in the 'subcycle scenario' (Table 2), higher-order sea-level fluctuations lead to a pronounced, stepped topography on the shelf (cf. Posamentier & Morris, 2000). The combination of the formation of topography and base-level change results in marked differential preservation of successive deposits (Fig. 9c). For instance, deposits laid down around 107 kyr are relatively

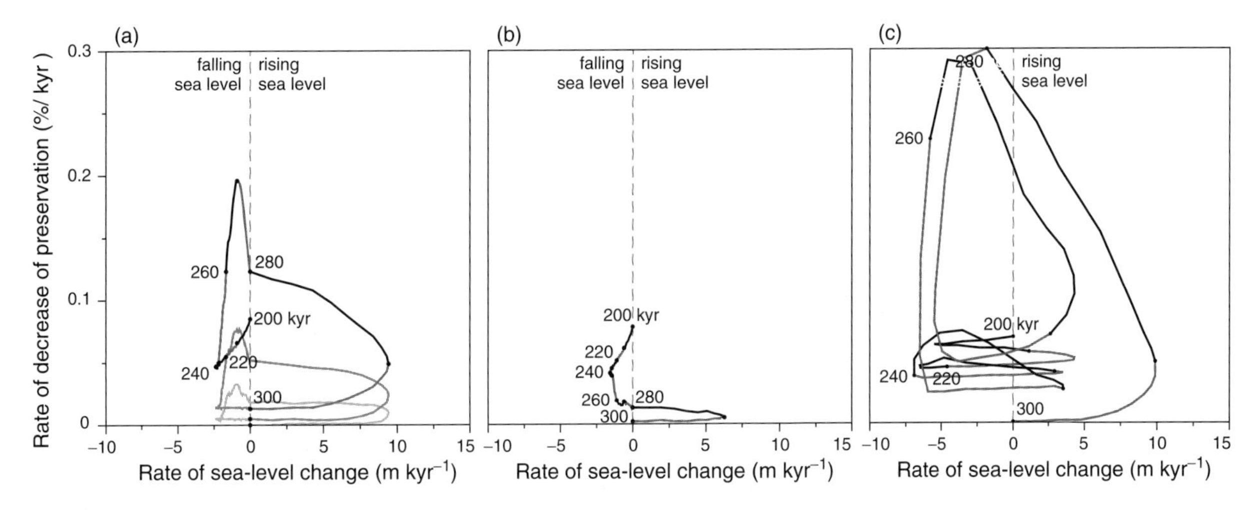

Fig. 10. The relation between the rate of sea-level change and the rate of change of preservation of sediment from the second glacial–interglacial cycle for the standard and the sea-level scenarios (Table 2). The graphs are means of 50 runs. (a) The standard scenario. (b) The 'amplitude scenario'. (c) The 'subcycle scenario'.

well preserved because they were buried by a sub-highstand delta from ~118 kyr, in contrast to ~111 kyr sub-lowstand strata that are exposed (Fig. 9c (left)). Later in the cycle, erosion particularly affects sub-transgressive and sub-highstand strata whenever sea level falls. Large-scale shelf erosion after emergence of the shelf break is interrupted by a short transgression (~164–172 kyr) with onlap in the incised valleys. During the subsequent sea-level fall, the river belts still occupy the same valleys and completely flush out the valley fills, resulting in low overall preservation. In contrast to the standard scenario, the rate of decrease of preservation is high at ultimate lowstand (Fig. 10c). The chronostratigraphic distribution of volume is greatly modified after the second cycle (Fig. 9c (middle)).

Gradient scenarios

The difference in gradient between the river profile and the submerged shelf profile has often been put forward as a governing factor in the preservation of the inner shelf when sea level falls (Miall, 1991; Posamentier *et al.*, 1992; Schumm, 1993; Wescott, 1993). However, in that portrayal, the coastal prism has been oversimplified by omitting the presence of a steeper shoreface (Törnqvist *et al.*, 2000) and the morphological development takes no account of submarine deposition in the coastal area. In fact, the trajectory of the coastline and the ensuing longitudinal river profile is much more complex, especially in three dimensions. Figure 11 exemplifies this by means of a number of model runs with varying gradient contrast (vertical) and varying sediment supply (horizontal). Indeed, erosion is dependent on the gradient contrast between coastal prism and shelf, but the amount of sediment supply appears to be a more important control on erosion (Table 3). With increasing shelf gradient, fluvial aggradation on the highstand delta becomes less prominent and incision into the highstand delta rim and the forced regressive deposits right in front of it becomes more common for a given sediment supply. For increasing sediment supply, subaerial exposure of the highstand delta plain and front decreases because they are buried more extensively (Fig. 11). Increasing supply improves preservation of initial topography, despite the accompanying increase in erosive power of rivers because deposition inhibits the development of steep slopes. Therefore, incision is more likely with increasing shelf gradient relative to the gradient of the coastal prism and with decreasing sediment supply.

When in the course of sea-level fall, a second convexity in the margin profile basinward of the coastal prism is exposed, be it a shelf break or a curved shelf, a second phase of incision may occur that is more widespread, which will decrease preservation further. Only in situations where the shelf break is relatively deep or the shelf exceptionally wide, do major cross-shelf incisions fail to occur and the formation of large hiatuses is restricted to the inner shelf and coastal prism (Talling, 1998; Wallinga, 2001, Chapter 10). An additional influence on the preservation of the

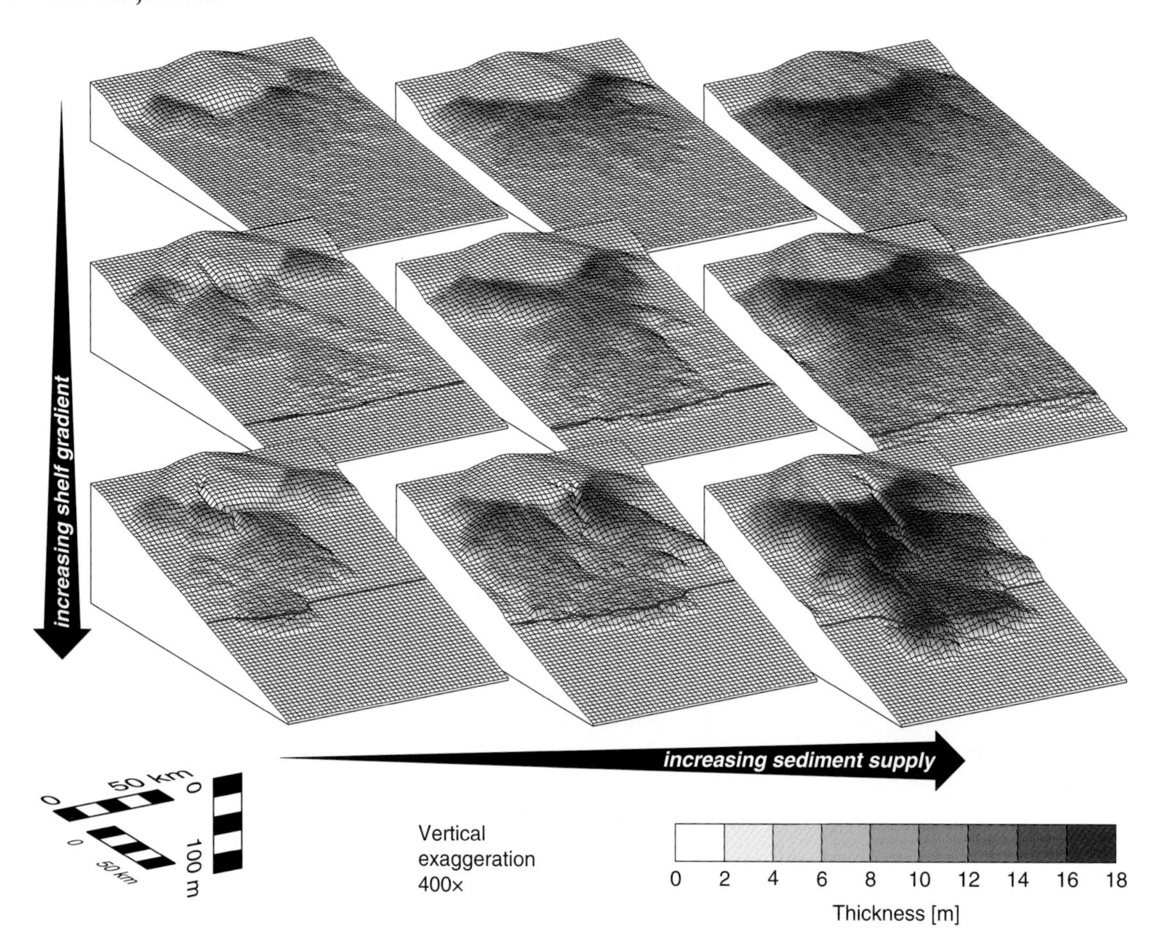

Fig. 11. Surface plots of the coastal prism and shelf produced with the gradient scenarios (Table 2), illustrating the relation between preservation of the coastal prism and the shelf gradient following sea-level fall, and how this relation is modified by varying sediment supply. The shelf gradient is less steep, equally steep or steeper than the coastal prism, respectively from top to bottom. The coastline is indicated with a thick line. The depth of incision decreases landward and basinward of the highstand delta rim (cf. Posamentier *et al.*, 1992; Talling, 1998; Törnqvist *et al.*, 2000). The maximum depth of incision increases with increasing shelf gradient and with decreasing sediment supply.

Table 3. The amount of erosion of the initial topography expressed as a percentage of the reference case, which is the underlined mean value, for varying sediment supplies and differences between the gradient of the coastal prism and the shelf.

	Sediment supply		
Gradient contrast	Low $1\,km^3\,kyr^{-1}$	Moderate $2\,km^3\,kyr^{-1}$	High $4\,km^3\,kyr^{-1}$
Shallower shelf $0.25\,m\,km^{-1}$	174 ± 14	99 ± 9	56 ± 4
Equally steep $0.50\,m\,km^{-1}$	180 ± 26	$\underline{100} \pm 9$	55 ± 4
Steeper shelf $0.75\,m\,km^{-1}$	199 ± 39	115 ± 29	78 ± 35

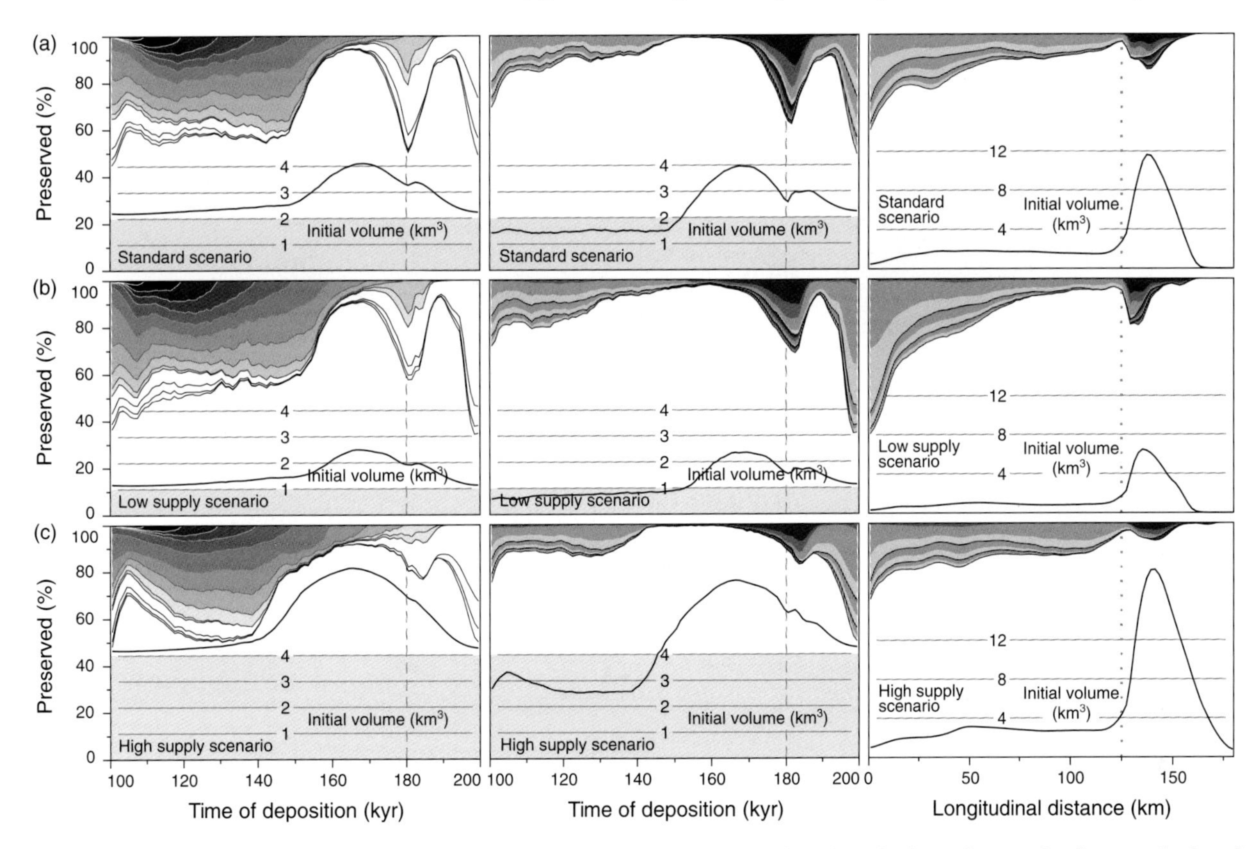

Fig. 12. The volume and preservation of strata of the second glacial–interglacial cycle through time for the standard and two supply scenarios (Table 2). The graphs on the left are similar to Fig. 5, those in the middle to Fig. 6, and the ones on the right to Fig. 7. Also indicated in light grey are the shapes of the respective supply curves. (a) The standard scenario. (b) The 'low supply scenario'. (c) The 'high supply scenario'.

coastal prism which is likely to occur is river erosion owing to changes in the palaeohydrology as a result of climate change in the drainage basin (Blum & Price, 1998; Blum & Törnqvist, 2000).

Supply scenarios

Sea-level change may be the most important driving force in coastal areas (Posamentier & James, 1993; Blum & Törnqvist, 2000), while the amount of sediment supply can modify its effect on the formation and preservation of stratigraphy (Schlager, 1993; Posamentier & Allen, 1993). In this section, the modelled results of a number of scenarios of discharge and sediment load are discussed in terms of morphological development and processes that control preservation, which are distinct from the standard scenario.

In the 'low supply scenario' (Table 2), sediment load and discharge are halved relative to the standard scenario. As a consequence of low supply, there is less progradation and aggradation at initial highstand. Furthermore, because of slow delta deposition, relatively steep gradients arise on the inner shelf as sea-level fall forces the coastline to migrate basinward, which promotes (small-scale) incision despite the low discharge. The thin forced regressive wedge, as well as underlying deposits from earlier glacial–interglacial cycles, is greatly affected by river incision (Fig. 12b) (cf. Posamentier & Morris, 2000). The local, rain-fed drainage system on the shelf plays a more important role relative to the discharge from the hinterland. The system is not able to reverse retrogradation and start progradation before the end of the cycle, and there is a funnel-shaped river mouth at final highstand (Talling, 1998). In addition, fluvial onlap does not reach as far up the river as in the standard scenario, and downward profile adjustment still continues during the early part of the following cycle. Because the incised valleys are not completely filled during transgression, there is a greater influence of antecedent topography on the development of forced regressive deposits during the subsequent sea-level fall. When the supply is double the standard value,

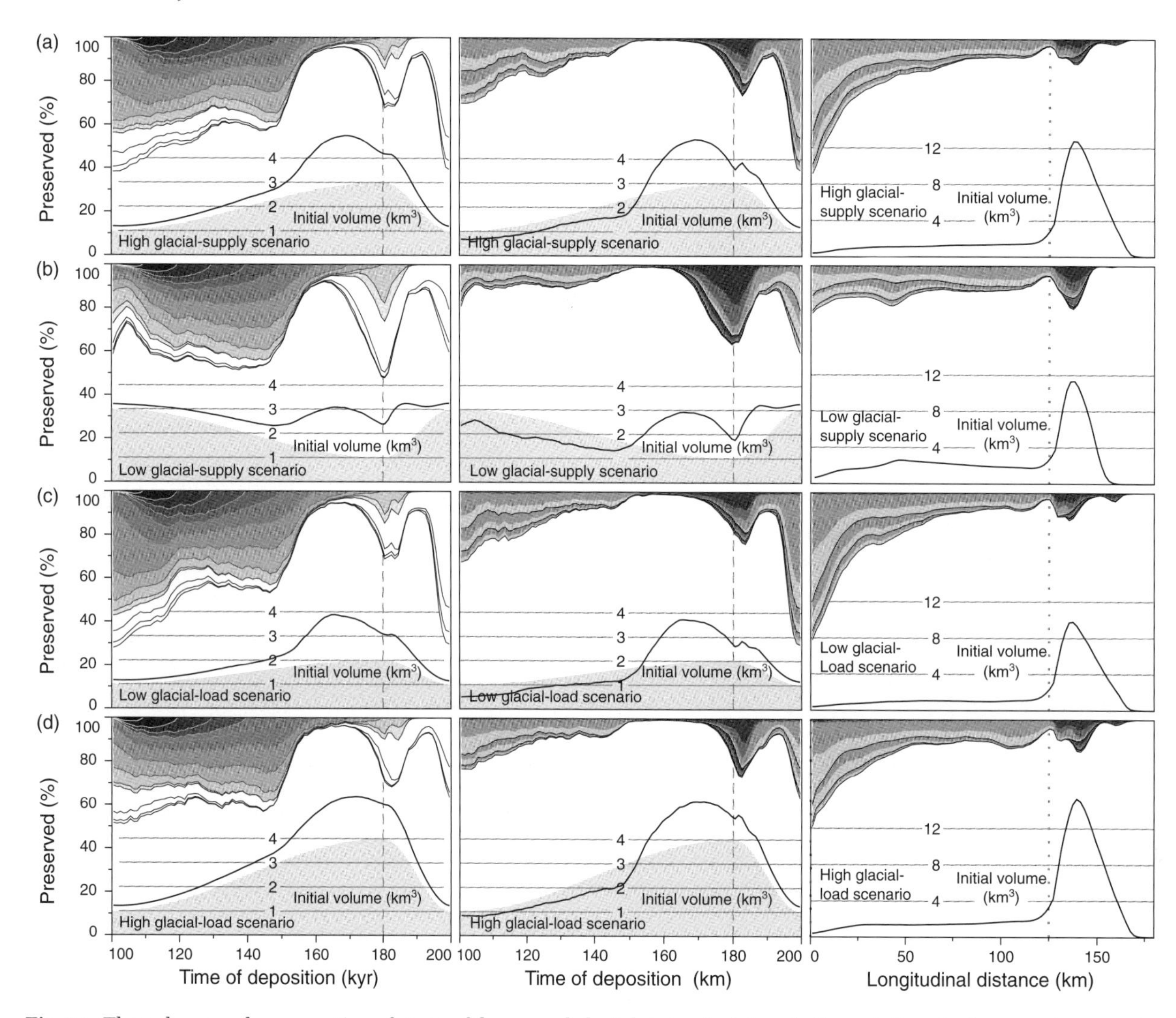

Fig. 13. The volume and preservation of strata of the second glacial–interglacial cycle through time for four supply scenarios (Table 2). The graphs on the left are similar to Fig. 5, those in the middle to Fig. 6, and the ones on the right to Fig. 7. Also indicated in light grey are the shapes of the respective supply curves. (a) The 'high glacial-supply scenario'. (b) The 'low glacial-supply scenario'. (c) The 'low glacial-load scenario'. (d) The 'high glacial-load scenario'.

as in the 'high supply scenario' (Table 2), rapid, forced progradation occurs that is not as strongly dictated by sea-level fall. Protruding delta lobes reach the shelf edge as soon as ~40 kyr into the cycle, resulting in increased preservation of subsequent upper continental slope deposits (Fig. 12c, left). The high erosive capacity of rivers causes relatively prominent incision before the shelf break emerges. The thickness of the forced regressive wedge is considerably more than in the 'low supply scenario', which leads to deep river erosion. Fast fluvial onlap and highstand progradation produce a voluminous coastal prism (Talling, 1998).

The 'high glacial-supply scenario' (Table 2) has a sinusoidally increasing sediment load and discharge from the hinterland. The model results combine features of the 'low supply scenario' during interglacial periods and the 'high supply scenario' during glacials. The opposite is the case for the 'low glacial-supply scenario' (Table 2), which has a decreased supply during glacial times. The relatively low rate of sedimentation during initial highstand in the former scenario results in low preservation of inner shelf strata (Fig. 13a). In both scenarios the supplied discharge and sediment load were changed proportionally, keeping carrying capacity and river load balanced, except for changes en route to the river mouth. However, in the 'low glacial-load scenario' (Table 2), the discharge increases more than the drainage basin yield, resulting in a tendency for rivers to pick up sediment and to lower their gradient. Moderate incision occurs steadily during sea-level fall in response to the increasing undercapacity, and subsequent major incision after emergence of the shelf break results in a relatively low preservation of inner shelf deposits (Fig. 13c). In contrast, in the 'high glacial-load scenario' (Table 2), the increased yield in proportion to the water discharge leads to overcapacity of the rivers. Rivers are less inclined to incise, and even show local aggradation during forced regression, making mid-shelf river avulsions, as opposed to avulsions on the downstream active delta, more likely to occur. Knickpoint migration from the shelf edge landward is slow and since the achieved equilibrium gradient is steeper than in the standard scenario, knickpoints do not reach as far upstream. Because of this slow connection, deposits in the transfer valley are relatively well preserved and more shelf-edge deltas remain active during lowstand. In this scenario, there is an overall greater preservation of deposits on the shelf (Fig. 13d).

CONCLUDING REMARKS

The nature of Quaternary glacio-eustasy favours the formation of depositional units with a relatively low preservation potential, despite the fact that the sediment accumulates on a constantly subsiding basin margin. Because positions of coastlines are repeatedly reoccupied during a succession of glacial–interglacial cycles, strata must regularly endure cannibalization. The model results show that preservation of deposits from a glacial–interglacial cycle varies both spatially and in time, dependent on the palaeohydrology (discharge and sediment supply), sea-level change and its relation to the surface morphology. There is a marked difference in preservation between deposits on the shelf and the continental slope, which has consequences for basin-scale grain-size sorting (see Meijer, this volume).

The model results presented here are largely conceptual. In the future, data sets will become available that allow further elaboration of the model and that will lend more empirical support. Given the importance of lateral and temporal variability illustrated in this paper, these data sets will need to be three-dimensional (i.e., they should include volumetric data), covering the coastal plain, continental shelf and continental slope, and have good age constraints.

Unravelling the history of deposition on a basin margin is not a straightforward task, especially for Quaternary deposits generated under the influence of drastic base-level fluctuations and shoreline shifts. However, eventually this research may help interpret the stratigraphy of passive continental margins in terms of sediment yield from the hinterland, which then can be related to climate changes.

ACKNOWLEDGEMENTS

Financial support was provided by the Netherlands Organization for Scientific Research (NWO). Y. Bartov and G. Plint are thanked for their reviews of this paper.

REFERENCES

Blum, M.D. and **Price, D.M.** (1998) Quaternary alluvial plain construction in response to glacio-eustatic and climatic controls, Texas Gulf Coastal Plain. In: *Relative Role of Eustasy, Climate, and Tectonism in Continental Rocks* (Eds K.W. Shanley and P.J. McCabe), *SEPM Spec. Publ.*, **59**, 31–48.

Blum, M.D. and **Törnqvist, T.E.** (2000) Fluvial responses to climate and sea-level change: A review and look forward. *Sedimentology*, **47**, 2–48.

Figueiredo, A.G., Jr. and **Nittrouer, C.A.** (1995) New insights to high-resolution stratigraphy on the Amazon continental shelf. *Marine Geol.*, **125**, 393–399.

Meijer, X.D. (2002) Modelling the drainage evolution of a river-shelf system forced by Quaternary glacio-eustasy. *Basin Res.*, **14**, 361–377.

Miall, A.D. (1991) Stratigraphic sequences and their chronostatigraphic correlation. *J. Sed. Petrol.*, **61**, 497–505.

Plint, A.G. (1988) Sharp-based shoreface sequences and 'offshore bars' in the Cardium Formation of Alberta: Their relationship to relative changes in sea level. In: *Sea-Level Changes: An Integrated Approach* (Eds C.K. Wilgus, B.S. Hastings, C.G.S.C. St Kendall, H.W. Posamentier, C.A. Ross and J.C. Van Wagoner), *Soc. Economic Paleontol. Mineral. Special Publ.*, **42**, 357–370.

Plint, A.G. and **Nummedal, D.** (2000) The falling stage systems tract: Recognition and importance in sequence stratigraphic analysis. In: *Sedimentary Responses to Forced Regressions* (Eds D. Hunt and R.L. Gawthorpe), *Geol. Soc., London Spec. Publ.*, **172**, 1–17.

Posamentier, H.W. and **James, D.P.** (1993) An overview of sequence-stratigraphic concepts: Uses and abuses. In: *Sequence Stratigraphy and Facies Associations* (Eds H.W. Posamentier, C.P. Summerhayes, B.U. Haq and G.P. Allen), *Spec. Publ. Int. Assoc. Sedim.*, **18**, 3–18.

Posamentier, H.W. and Morris, W.R. (2000) Aspects of the stratal architecture of forced regressive deposits. In: *Sedimentary Responses to Forced Regressions* (Eds D. Hunt and R.L. Gawthorpe), *Geol. Soc. London Spec. Publ.*, **172**, 19–46.

Posamentier, H.W., Allen, G.P., James, D.P. and **Tesson, M.** (1992) Forced regressions in a sequence stratigraphic framework: Concepts, examples, and exploration significance. *AAPG Bull.*, **76**, 1687–1709.

Posamentier, H.W. and **Allen, G.P.** (1993) Variability of the sequence stratigraphic model: Effects of local basin factors. *Sed. Geol.*, **86**, 91–109.

Schlager, W. (1993) Accomodation and supply – A dual control on stratigraphic sequences. *Sed. Geol.*, **86**, 111–136.

Schumm, S.A. (1993) River response to baselevel change: Implications for sequence stratigraphy. *J. Geol.*, **101**, 279–294.

Talling, P.J. (1998) How and where do incised valleys form if sea level remains above the shelf edge? *Geology*, **26**, 87–90.

Törnqvist, T.E., Wallinga, J., Murray, A.S., De Wolf, H., Cleveringa, P. and De Gans, W. (2000) Response of the Rhine-Meuse system (west-central Netherlands) to the last Quaternary glacio-eustatic cycles: A first assessment. *Global Planet. Change*, **27**, 89–111.

Wallinga, J. (2001) *The Rhine-Meuse system in a new light: Optically stimulated luminescence dating and its application to fluvial deposits.* Unpublished PhD Thesis, Utrecht University.

Wescott, W.A. (1993) Geomorphic thresholds and complex response of fluvial systems – some implications for sequence stratigraphy. *AAPG Bull.*, **77**, 1208–1218.

Spec. Publ. Int. Assoc. Sedimentol. (2008) **40**, 239–274

Modelling source-rock distribution and quality variations: the organic facies modelling approach

UTE MANN and JANINE ZWEIGEL[1]

Basin Modelling Department, SINTEF Petroleum Research, N-7465 Trondheim, Norway (E-mail: ute.mann@iku.sintef.no)

ABSTRACT

This paper demonstrates the concept of process-based organic facies modelling (OF-Mod) software, and discusses general conceptual ideas and their translation and numerical description in the different modules of the program.The OF-Mod concept is based on the idea that (1) the source type, (2) the preservation conditions of organic matter during deposition and burial, and (3) sequence-stratigraphic aspects are the most important elements defining the organic-facies–source-rock distribution. Each of these elements and their interactions are reviewed and their translation in the program is outlined. Modelling results from a case study of the Late Jurassic Spekk Formation of the mid-Norwegian Continental Shelf are presented in this study. The challenge here was to mimic realistically the special environmental conditions during deposition of the lower 'rich' Spekk Formation, and to reproduce the subsurface data adequately, thereby testing the validity of OF-Mod's process description.

Keywords Source rock modelling, organic sedimentation, Late Jurassic, Spekk Formation.

INTRODUCTION

In exploration, the ability to predict hydrocarbon occurrence and quality variations within a prospect, prior to drilling, is of great economic importance. Particularly, recently developed 3D modelling techniques are gaining significance with respect to volumetric hydrocarbon predictions. Nevertheless, in basin modelling studies, source rocks are often the least constrained input parameters, even though the source rock is the first prerequisite for a hydrocarbon accumulation. Often a conceptual approach or simple models applying average geochemical values describing source rock properties are used. This often is insufficient, particularly in areas with heterogeneous geological conditions and/or variable depositional environments. Hence, the original idea to develop a simulation program for organic facies modelling (OF-Mod) evolved from the needs in the petroleum industry to obtain better estimates of source-rock characteristics and variability in basin kitchen areas.[1]

Several attempts for predicting source-rock occurrences have been made (e.g. Passey *et al.*, 1990; Carpentier *et al.*, 1991; Schwarzkopf, 1993; Mallick & Raju, 1995). Calculation of organic carbon (OC) content based on well-log data (resistivity, density, sonic or gamma-ray logs; Carbolog software by Carpentier *et al.*, 1991) is widely utilized by explorationists. This method is fast and easy to apply, and therefore commonly used to obtain a first estimate of source-rock distribution. However, it gives only an estimate of source-rock distribution at well positions (1D). To obtain a spatial impression, good subsurface (well) control for extrapolation is required. Further, it does not provide any information on source-rock quality. Schwarzkopf (1993) used a process-based approach to predict marine siliciclastic source rocks. Based on calculations of marine productivity, carbon flux and burial efficiency he estimated OC content, and combined this with Monte Carlo simulation to account for the uncertainties in the input parameters (water depth, sedimentation

[1]Present address: Statoilhydro ASA, Strandveien 4, N-7501 Størdal, Norway.

rate, marine productivity). Also this approach gives little indication of quality variations of the organic matter, and the distribution and variability of the source rock in space are not considered.

For a better appraisal of spatial source-rock distribution and also of spatial quality variations, source-rock occurrences have been discussed in a sequence-stratigraphic framework (e.g. Wignall, 1991; Katz & Pratt, 1993; Pasley *et al.*, 1993; Bessereau *et al.*, 1995; Tyson, 1996; Mann & Stein, 1997). This integrated approach describes the variability of a source-rock system as a function of the depositional setting and it depicts the relationship between the accumulation of organic rich strata and sea level. Based on a realistic understanding of the distribution of various facies types, predictability away from the actual sample site is provided.

Few attempts have been made to incorporate these ideas into the modelling of source-rock distribution, even though several 2D and 3D tools for stratigraphic modelling have been published (e.g. SedSim, Tetzlaff, 1989; DEMOSTRAT, Rivenæs, 1993; SEQUENCE, Steckler *et al.*, 1996; Fuzzim, Norlund, 1996; DIONISOS, Granjeon *et al.*, 1999; PHIL, Petrodynamics Inc.; SedFlux, Syvitzki & Hutton, 2001; Griffith & Dyt, 2001; Repro, Huessner *et al.*, 2001; Carbonate 3D, Warrlich *et al.*, 2001; among many others). Carpentier *et al.* (1993) published a model combining evaporite with organic sedimentation and applied it to the cyclic evaporite sequences of the Mulhouse Basin. Their model, SIMSALT, was the first to consider the interaction of basin-fill stratigraphy (here evaporite sequences) and organic sedimentation. Some of the stratigraphic models started to incorporate marine productivity, mainly applying the approach that Schwarzkopf (1993) used in his model. However, little has been published and it seems that this approach is not flexible enough to account for the various processes influencing source-rock distribution and quality. Terrestrial organic matter is not included and variations in preservation conditions during OC flux and burial are not considered.

In 1998 the OF-Mod project was started at SINTEF Petroleum Research. The goal was to describe the spatial distribution of a source rock in terms of thickness, and spatial quality/facies variations in terms of OC content and generation potential (hydrogen index – HI) for input in basin-modelling studies. Source-rock modelling should account for two main aspects: (1) The program should provide the initial properties and characteristics of the modelled source rock to be used as direct input into subsequent basin modelling because often only maturity-altered data exist; and (2) there is a particular need to obtain information on source-rock characteristics in deeper graben areas, that is, in areas where little or no well data are available. Based on these requirements a statistical modelling approach was excluded as it always requires very good well control. A process-based approach was favoured.

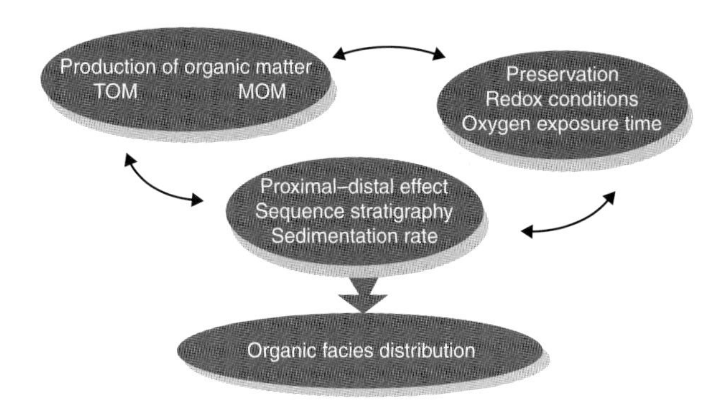

Fig. 1. Main aspects affecting organic facies distribution: TOM, terrestrial organic matter; MOM, marine organic matter.

THE CONCEPT

The OF-Mod concept is based on the idea that (1) the organic-matter source type, (2) the organic-matter preservation conditions during deposition and burial, and (3) basin fill/sequence-stratigraphical aspects are the most important features influencing the organic facies/source-rock distribution in space and time in marine environments. Therefore these three main aspects had to be considered in the modelling approach (Fig. 1).

(1) The two principal *organic-matter source types*, marine (autochthonous) and terrestrial (allochthonous) organic matter, are important for source-rock assessment even though the marine/aquatic organic matter is the more relevant part in terms of oil-proneness. However, more than 90% of OC burial presently occurs in continental margin sediments (Hartnett & Devol, 2003) and up to 50% of the organic matter deposited at modern shelves can be of terrestrial origin (Schluenz & Schneider, 2000) and even more at huge river mouths such as at the Amazon Shelf (Showers & Angle, 1986). Therefore, the terrestrial organic matter fraction should also be included in the modelling as it can constitute a significant part of the

source rock and may contribute considerably to the source-rock quality in terms of composition.

(2) The source-rock quality is further affected by the *preservation conditions* during deposition and burial. Because different types of OC are produced and recycled, a differentiation between the source fractions (marine and terrestrial) has to be made, also when discussing organic matter preservation. Generally, the terrestrial organic matter (TOM) has already experienced previous degradation and microbial breakdown in soils and during transport before entering the marine system. It is therefore considered to be more resistant to further extensive degradation at sea (Hedges & Keil, 1995; Hedges & Oades, 1997). Once entered into the marine environment, transport behaviour, mechanical grain-size reduction and sorting of the terrestrial/allochthonous organic particles is similar to those of mineral grains, even though different densities and grain shapes have to be considered (Littke *et al.*, 1997). With increasing distance from the shoreline, grain size and also quality of the allochthonous organic particles decrease. Littke *et al.* (1991) determined the grain sizes of vitrinite and inertinite particles in marine sediments in the central Indian Ocean and found that there are virtually no particles larger than 20 μm. Generally, with increasing transport distance the intertinite/vitrinite ratio increases indicating a greater degree of chemical degradation (Littke *et al.*, 1997). An exception may be turbidity currents which are able to transfer larger terrestrial particles into distal offshore regions (Degens *et al.*, 1986). Also Bergamaschi *et al.* (1997) and Keil *et al.* (1998) examined intensively the relationship of source, composition and diagenesis of organic matter with grain size and mineralogy. They concluded that particle size and mineralogy coincide with variations in composition and distribution of the organic matter. Fresh woody debris as well as lignin-phenol enriched organic matter is concentrated in the sand-sized fraction, pollen grains in the silt-sized fraction whereas organic matter in the finer fractions is progressively more degraded. Tyson and Follows (2000) investigated particularly the relation of microscopic plant debris and sequence stratigraphy in an Upper Cretaceous section of the Pyrenees. They found that sedimentological sorting based on particle size and density better predicts the composition of phytoclasts along an onshore–offshore transect than do palaeoecological gradients. Summing up, we consider the distribution and preservation of TOM in the marine environment mainly

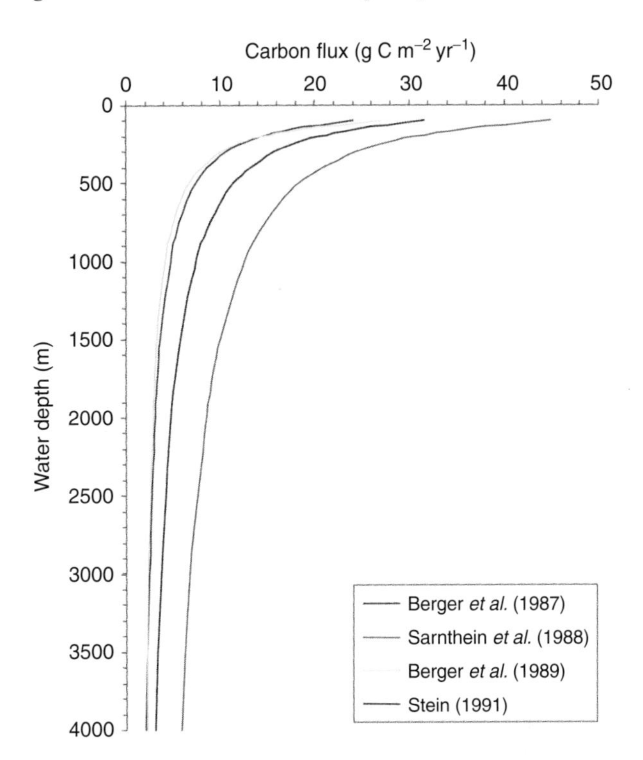

Fig. 2. Calculations of carbon flux with a constant primary productivity (PP) value of 150 applying equations by Berger *et al.* (1987, 1989), Sarnthein *et al.* (1988) and Stein (1991).

a function of transport distance, mechanical grain-size reduction and sorting similar to those of inorganic sediment particles.

In contrast, autochthonous produced marine organic matter (MOM) may be severely degraded on its way through the water column and in the uppermost sediment layers depending on water depth, sinking velocity and redox conditions. Measured carbon-flux data from sediment traps show that most of the primary produced organic matter is degraded in the uppermost part of the water column. Based on such data, several authors have estimated the carbon flux by correlating the primary produced organic matter (PP in $g\,C\,m^{-2}\,yr^{-1}$) to water depth (Fig. 2). However, when incorporated in faecal pellets or aggregates or due to increased vertical particle flux (high organic-matter rain rate), organic particles can be transported much faster to the sea floor and accumulate in the sediment (Ittekkot *et al.*, 1992; Knies & Stein, 1998). Also sorption of organic particles on mineral surfaces (organic coatings) might be an important process enhancing preservation of the organic matter (Keil *et al.*, 1994; Mayer, 1994; Hedges & Keil, 1995). Further, increased oxygen-exposure time and burial efficiency at the sediment

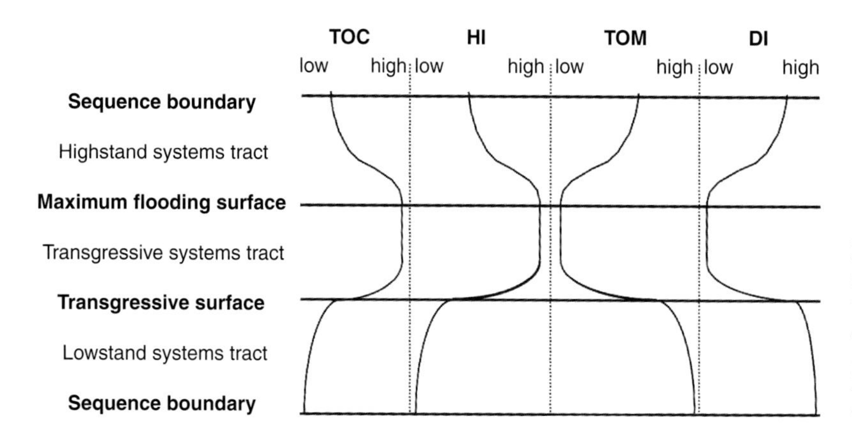

Fig. 3. Relationship between organic matter characteristics and position within a depositional sequence (Mann & Stein, 1997; modified after Pasley *et al.*, 1993): TOC, total organic carbon; HI, hydrogen index; TOM, terrestrial organic matter; DI, degradation index.

surface and in the uppermost centimetres of the sediment can play a major role in preservation of the labile marine organic matter (e.g. Henrichs & Reeburgh, 1987; Betts & Holland, 1991; Stein, 1991; Hartnett *et al.*, 1998). However, whether high marine productivity and high OC flux rates or oxygen deficiency and preservation conditions play a more important role for the accumulation of organic-matter-rich sediments is an ongoing debate (e.g. Pedersen & Calvert, 1990; Parrish, 1995 (and references therein); Littke *et al.*, 1997).

(3) Sequence stratigraphy, or *basin-fill stratigraphy*, is the third aspect which we regard a main element when discussing spatial organic/source-rock facies distribution. Recent papers have examined the close relation between sequence stratigraphy and organic-matter-rich strata (aforementioned references). The two organic-matter-source fractions, marine/aquatic and terrestrial, are genetically related to parasequence stacking (Fig. 3). Each systems tract in a depositional sequence is affected by the position of the transgressive or regressive shoreline and consequently by the rate of terrestrial input (both inorganic and organic matter) to the shelf (Pasley *et al.*, 1993). During seaward movement of the shoreline, sediments are generally deposited in delta, strand-plain and pro-delta environments, and are subsequently redistributed and transported to the shelf as storm deposits and by shelf plumes. Therefore the amount of terrestrial organic matter (TOM) delivered to the shelf during regressions is also relatively high, but both OC and hydrogen index (HI) values are often relatively low due to dilution effects and poor preservation. During a transgression, relatively small amounts of terrigenous sediment are delivered to the shelf because most of the sediments are trapped in proximal environments. In contrast, autochthonous organic matter is proportionally more common in transgressive systems tracts, because less TOM is delivered to the shelf. In addition, fine-grained facies in the transgressive systems tract contain more hydrogen-rich organic matter and may be more prone to anoxia (Pasley *et al.*, 1993) and may therefore enhance preservation of the autochthonous fraction. Primary production of marine OC, however, is related to shoreline position. In coastal and near-shore areas, where nutrient supply from the continent is much higher than in distal open ocean regions, primary productivity (PP) is elevated as a consequence of increased phytoplankton growth.

Furthermore, preservation of marine OC in the sediment may increase with increasing sedimentation rate due to short oxygen exposure time and rapid passage through the near-surface zone of intense organic degradation (e.g. Müller & Suess, 1979; Stein, 1991; Hartnett *et al.*, 1998). Poor preservation occurs especially in deep-sea distal offshore regions with extremely low sedimentation rates. In areas with a very high clastic input (high sedimentation rates), for example in delta regions or during turbidite deposition, sediment dilution effects may occur and lower the OC fraction in the sediment as well. To conclude, the above relations provide the link between the nature of organic-matter deposition and depositional systems tracts because the type and preservation of organic matter deposited on the shelf is directly associated with parasequence stacking. Therefore, organic-sedimentation or source-rock modelling has to consider basin-fill stratigraphy if a comprehensive and complete model of the organic facies distribution in time and space is intended.

There are other, more indirect very important aspects such as palaeocenographic (water-mass

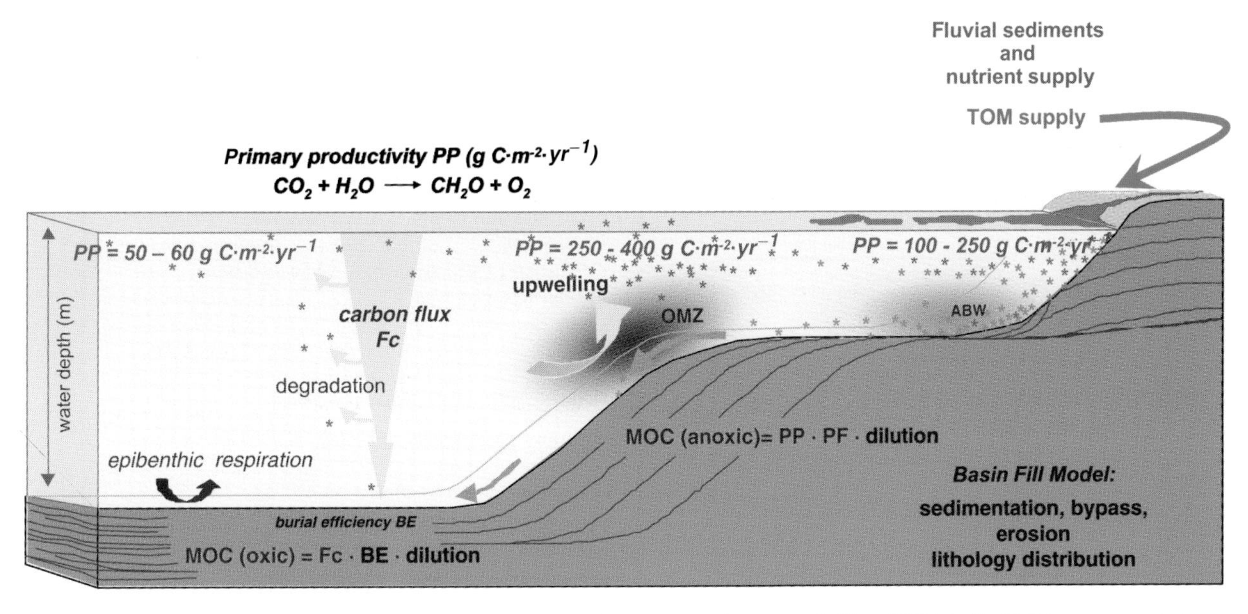

Fig. 4. Sketch showing parameters and processes included in organic facies modelling (OF-Mod): OMZ, oxygen minimum zone; ABW, anoxic bottom water; MOC, marine organic carbon; TOM, terrigenous organic matter.

circulation, ocean currents, etc.) and palaeoclimate conditions (wind systems, climate belts, humidity, aridity, etc.) affecting OC accumulation in sediments which are not directly considered in our model but influence the aforementioned aspects. For instance, a shift in climate conditions from a humid to a more arid climate may be expressed in a distinct decrease of TOM input, that is, a decrease of higher land plant debris simultaneously with an increased input of clastic material (e.g. Stein, 1991; Wagner, 2000). Further, climate shifts can be simultaneous with or can be the cause of sea-level changes and are therefore expressed in the basin-fill stratigraphy. We do not consider such palaeoclimate and palaeoceanographic parameters directly because they are still very difficult to constrain, particularly for ancient sediments. That type of modelling concerns a much larger scale (ocean scale, ≫basin scale) than organic facies/source-rock modelling (basin, kitchen-scale). The development of palaeoceanography and palaeoclimate models (e.g. Barron *et al.*, 1995; Parrish, 1995; Sellwood *et al.*, 2000; Handoh *et al.*, 2003) has made enormous progress during recent years but they still contain large uncertainties regarding the boundary conditions. However, the present OF-Mod concept does not ignore the application of palaeoceanographic modelling, even though a direct modelling of these parameters in OF-Mod is excluded. A more appropriate approach would be to use results of these models and construct together with other local parameters a probability distribution, for example upwelling occurrences.

IMPLEMENTATION

In this article we will outline how we considered, translated and interpreted the aforementioned aspects into the OF-Mod program.

Figure 4 outlines the processes and parameters accounted for in OF-Mod. Only processes in the marine realm are included in OF-Mod. Aeolian transport of organic particles, coal deposition and peat growth, evaporites and other chemical sediments are not included.

As mentioned above the two OC source fractions were treated separately; we model distribution of marine primary productivity (PP) in the surface layer of the water column and input/supply of terrestrial organic matter (TOM) into the marine/aquatic system via rivers and run off. Also preservation of the two organic matter types is treated independently. The primary produced marine organic matter (MOM) is modelled under aerobic conditions to accumulate as a function of carbon flux and burial, while under oxygen depleted conditions two optional preservation scenarios can be chosen. An oxygen-minimum-zone scenario (OMZ) develops as the result of intensive phytoplankton growth, for example induced by upwelling conditions. Alternatively, one can chose an anoxic bottom water scenario (ABW),

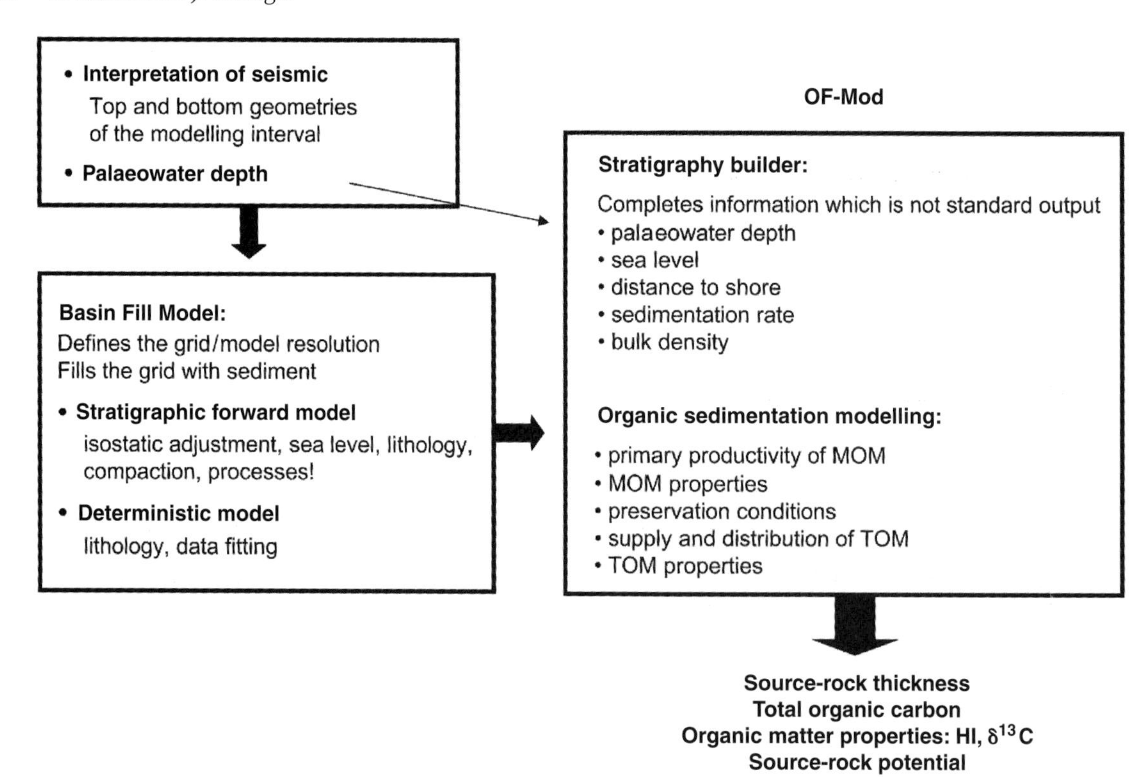

Fig. 5. Work flow of an organic facies modelling (OF-Mod) study: MOM, marine organic matter; TOM, terrestrial organic matter; HI, hydrogen index.

which is independent of PP values and depends on reduced oxygen content in the bottom water due to sluggish circulation or stagnation of water masses.

In the modelling of the distribution, accumulation and preservation of terrestrial OC, the mineral grain distribution is taken as a measure. Coarse grain-size fractions (sand) are more common in near-shore areas as is discrete/particulate terrestrial plant debris. The heavily degraded and/or residual allochthonous organic matter is associated with the fine grain-size fractions (shale). The grain size or lithology distribution is provided by the basin-fill stratigraphy.

The basin-fill stratigraphy can be provided by stratigraphic forward modelling tools or as deterministic models created by reservoir modelling tools and is then imported into OF-Mod. Alternatively, simple deterministic models can be directly created inside OF-Mod with the stratigraphy builder module. These latter are comparable to those from other reservoir modelling tools. Having imported/created the stratigraphy/basin-fill model, it has to be completed with the additional parameters, needed for calculation of the organic sedimentation (sedimentation rate, distance to the shore, etc.; see discussion below). Subsequently, OF-Mod uses each time layer of the basin-fill model as a basis for the subsequent organic sedimentation modelling.

Figure 5 shows the work flow of an OF-Mod study. To run an OF-Mod study the interpreted geometries from seismic information and the initial palaeobathymetry/basin topography are required. Based on these, one can choose between a deterministic and a forward modelling tool to fill the lithology between the bounding surfaces. The definition of the grid resolution and all further steps can be handled in OF-Mod.

Basin-fill stratigraphy

As mentioned above, the (inorganic) basin-fill stratigraphy is handled separately from the organic modelling. We chose this approach because there already exist several powerful and sophisticated forward and deterministic stratigraphic modelling tools (cf. references in the introduction). Furthermore, the option to handle the stratigraphy modelling independent from the organic modelling allows choosing between forward or deterministic models from case to case. The choice of a forward stratigraphic model or a deterministic model is certainly dependant on data, time, and software availability. An advantage of the forward modelling is that these models include process understanding, and processes such as isostatic adjustment, sea-level changes, compaction processes, and so on, are considered. They only need to be 'imported'. A deterministic model as provided by reservoir modelling tools needs much better well control to achieve the same detail, since variations on a smaller scale might not be covered if well control is not available. However, these models are generally much more flexible with respect to the grid building and are also easier to handle with respect to the fitting/calibration of data to existing well data.

Another advantage of decoupling the stratigraphic and the organic modelling is that it makes the source-rock model definitely faster, allowing quick 'organic' parameter testing and multiple-scenario runs, because the sedimentary model does not need to be re-run. This is especially applicable when the basin-fill model is produced by a forward modelling tool as these kinds of models often require considerably more computing time.

A disadvantage of modelling the basin-fill stratigraphy and organic sedimentation separately is that erosion and redeposition of organic material cannot be accounted for. Further, volumetric changes of a stratigraphic layer because of an extreme richness in organic matter cannot be considered. As this applies only at very high OC values ($\gg$20% TOC), or in coal deposits and because such high OC values are quite exceptional for marine source rocks, this problem can be neglected for most applications.

Deterministic reservoir models and most of the present-day stratigraphic forward modelling tools have been designed for topics related to reservoir rock. Several parameters, essential for source-rock modelling, are not standard output. Therefore, additional information (e.g. sea-level history, well data/log datings, and palaeobathymetry maps) can be imported directly into OF-Mod and the necessary parameters are then generated. Besides lithology distribution, other parameters needed by OF-Mod are age, sedimentation rate, dry bulk density, palaeowater depth and distance-to-shore.

In this context the shoreline position or distance-to-shore parameter is especially important, as several input parameters are defined as a function of shoreline distance (e.g. marine PP, particulate terrestrial OC supply, etc., see discussion below). We therefore included the option to account also for information from outside the study area by either extrapolation of the slopes or by importing the shoreline position outside the study area (Fig. 6).

Organic matter source types

Marine organic matter fraction

The MOM fraction is defined as marine/aquatic primary-produced organic matter. Primary productivity (PP) in modern oceans shows a great contrast between coastal areas and the open sea as well as an asymmetry between the Atlantic and the Pacific Ocean. Generally, PP is higher in coastal and inner shelf areas than in outer shelf or distal open ocean regions (Fig. 7). In OF-Mod, the PP of MOM is therefore defined as a function of shoreline distance. For example, with increasing distance from the shoreline, PP decreases until a constant value is reached. This general trend is in accordance with the present-day PP distribution, though absolute values may vary significantly with latitude, and the location within an ocean basin (west or east side).

Areas with particularly high PP may occur locally due to increased nutrient supply caused by upwelling conditions or river input (Fig. 7). However, location, duration and intensity of these features cannot be predicted directly as this would require ocean and atmosphere circulation models. Therefore, in OF-Mod a flexible approach is used, allowing the user to place a high PP area of a certain size and intensity at a distinct position. The program automatically adapts the high PP lenses to the shoreline position (Fig. 8).

Preservation and sedimentation of marine OC is calculated using the three main processes controlling the accumulation under 'normal' aerobic conditions: (I) PP of MOM and its flux through the water column; (II) dilution by inorganic sediment at the sea bottom; (III) decomposition/preservation during burial (burial efficiency).

$$\text{MOC} = \text{carbon flux (I)} \times \text{dilution by organic sediment (II)} \times \text{burial efficiency (III)}$$

where MOC is the marine organic carbon, PP is the primary productivity and carbon flux is PP $\times$ degradation during sinking.

Numerically this term can be expressed as follows (Knies & Mann, 2002; Mann & Zweigel,

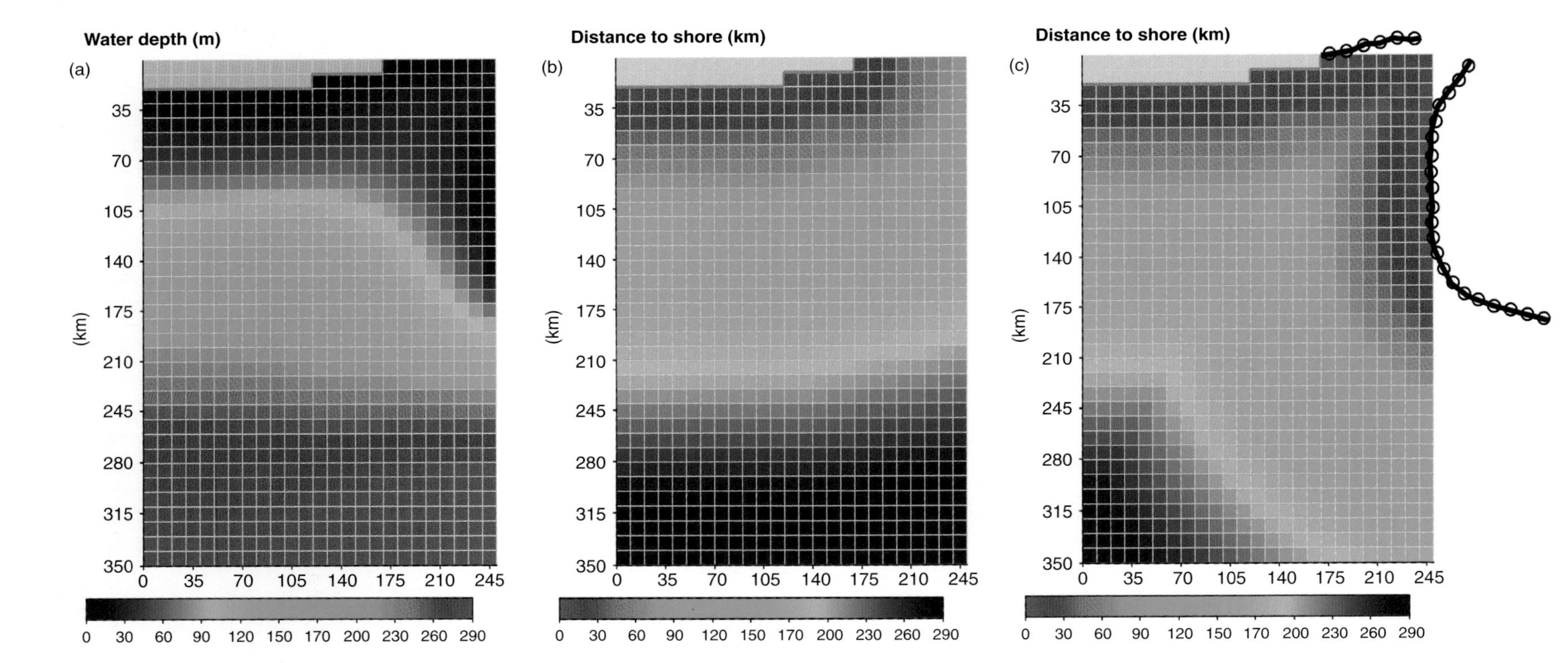

Fig. 6. Influence of the shoreline position on the distance-to-shore parameter. (a) Water-depth distribution and shoreline position of the modelling area are shown in a surface plot. (b) Distance-to-shore and shoreline position are calculated simply based on water depth (0 m water depth = shoreline). (c) The shoreline distance is extrapolated and edited based on additional information from the outside of the modelling area. The shaded area in the upper part of the figures indicates on-shore area within the modelling area, the grey line limiting the shaded area indicates the shoreline, the thick black line with circles in (c) indicates the edited shoreline outside the modelling area.

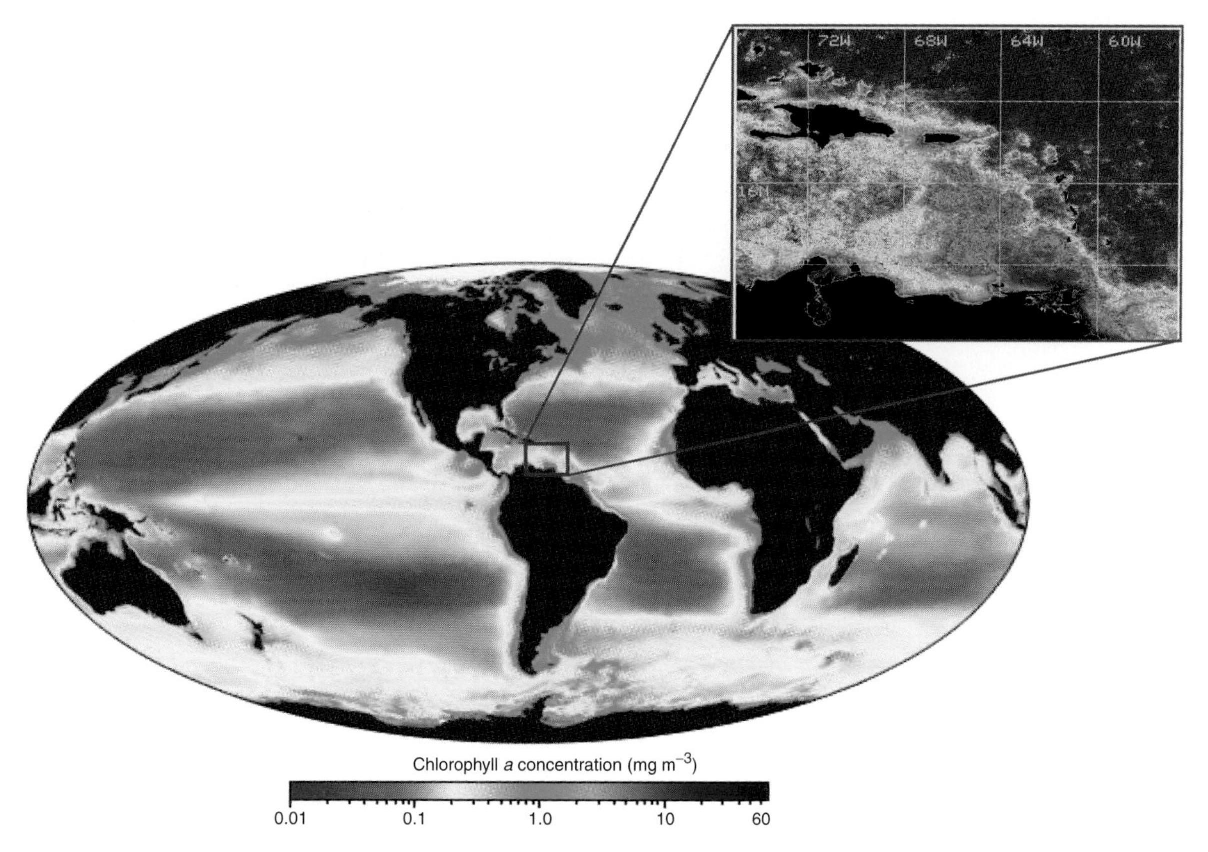

Fig. 7. Present-day global chlorophyll distribution in the sea-surface layer of the world ocean as a measure for phytoplankton growth, i.e., marine primary productivity. Note the increased chlorophyll intensity in coastal and in upwelling areas. Enlarged area shows the Orinoco River plume. The river outflow, enriched in nutrients and because of its low density, remains at the surface, and stimulates phytoplankton growth, so that a plume of higher productivity waters is observed. (Source – http://earthobservatory.nasa.gov)

2002a; Knies *et al.*, 2003):

$$\mathrm{MOC} = \overbrace{\left(\frac{0.409PP^{1.41} \times z^{-0.63}}{10}\right)}^{\text{Carbon flux}} \times \overbrace{\left(\frac{100}{\mathrm{DBD} \times \mathrm{LSR}}\right)}^{\text{Dilution}} \times \overbrace{\left(0.54 - 0.54 \times \left(\frac{1}{0.037 \times \mathrm{LSR}^{1.5} + 1}\right)\right)}^{\text{Burial efficiency}} \quad (1)$$

where PP is primary productivity, *z* is water depth (m), DBD is dry bulk density, and LSR is linear sedimentation rate, of which the latter three can be derived from the basin fill model.

Equation (1) is based on studies by Müller & Suess (1979), Johnson-Ibach (1982), Betzer, *et al.* (1984), Stein (1991) and Betts & Holland (1991). The carbon-flux term is based on data from sediment-trap studies, whereas the burial-efficiency term is based on a huge collection of data sets from mainly Cenozoic deep-water – and from DSDP/ODP sediment cores. However, since all these terms are empirical, their application is limited, and with decreasing water depth (<200 m) uncertainty increases significantly.

As pointed out, increased preservation of MOM may occur under anoxic conditions, when the oxygen consumption by degradation exceeds the oxygen supply. Aerobic degradation processes will be reduced/stopped and less efficient anoxic processes take over. This applies to the degradation in the water column as well as in the sediment. Therefore, the above carbon-flux and burial-efficiency approach cannot be applied in these environments. Instead we apply an OC preservation factor (PF) that describes the relation between the primary produced OC (PP) and the accumulation of OC in the sediment. This factor is independent of water depth (Brumsack, 1980; Bralower & Thierstein, 1984):

$$\mathrm{PF} = ((\mathrm{MOC}/100) \times \mathrm{LSR} \times \mathrm{DBD})/\mathrm{PP} \quad (2)$$

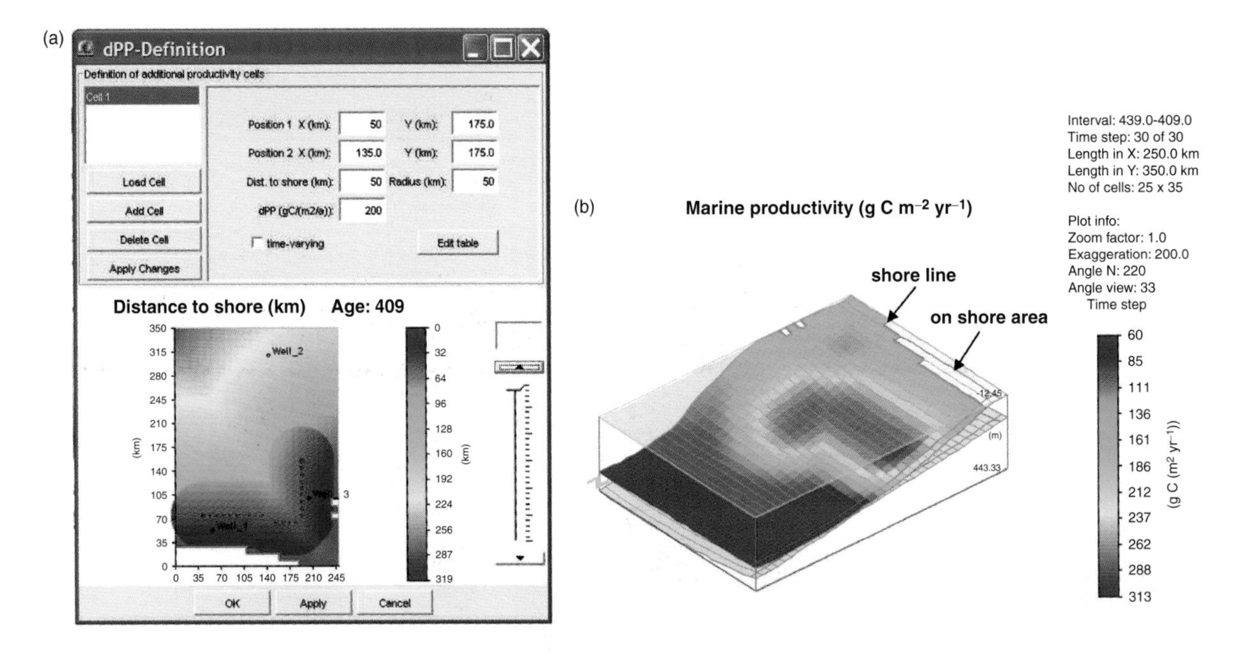

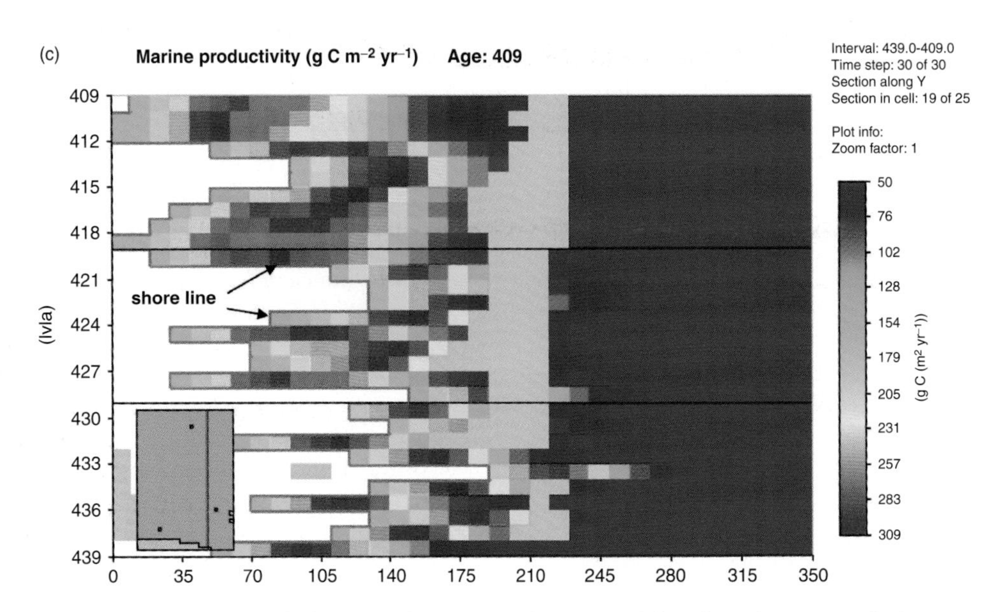

Fig. 8. Synthetic scenario showing modelled PP distribution as a function of shoreline distance including an additional high-productivity area caused by coastal upwelling. (a) How the user can define an area of higher productivity in terms of position, extent and intensity; the small map shows where the defined high-productivity area is located. (b) The primary productivity (PP) distribution including the high-productivity area is shown on a surface plot. (c) A chronostratigraphic plot of PP. It illustrates how the zone of high productivity automatically fluctuates with the shoreline movements through time. The small inset map shows the position of the transect in the study area: light grey, the undefined onshore area; light blue, areas of non-deposition or later erosion.

where PF is the preservation factor (%), and the expression ((MOC/100) × LSR × DBD) is the organic carbon accumulation rate (in $\mathrm{g\,cm^{-2}\,kyr^{-1}}$), and PP is primary productivity.

For the forward modelling we rearranged equation (2) for the calculation of the MOC under anoxic conditions into:

$$\mathrm{MOC} = (\mathrm{PP} \times \mathrm{PF})/(\mathrm{DBD} \times \mathrm{LSR}) \qquad (3)$$

Thus, under anoxic conditions the modelled MOC content is the result of the primary productivity PP, the preservation factor PF, and the dilution. The PF describes the simplified summary effect of carbon flux and burial efficiency under anoxic conditions.

OF-Mod uses values for DBD and LSR extracted from the basin-fill model, these values relate only to the inorganic part and not to the total sediment. Therefore, in cases of very high PP values and low inorganic sedimentation rates, equation (3) leads to unrealistically high MOC values; when inorganic LSR approaches zero the calculated MOC will approach infinity. If one assumes that the porosity of the organic matter is the same as the porosity of the inorganic sediment (for reasons of simplification), this problem can be avoided by calculating the total as follows:

$$LSR_{tot} = LSR_{inorg} + LSR_{org} \quad (4)$$

The LSR_{org} can be derived from PP

$$LSR_{org} = \frac{PP \times PF}{DBD_{org}} \quad (5)$$

and

$$DBD_{org} = \rho_{org} \times (1 - \Phi) \quad (6)$$

where ρ_{org} is the density of the organic matter and Φ is the porosity.

The density of organic matter can be easily derived from literature data. In Tyson (1995) a comprehensive summary of density of particulate organic matter, kerogen, and associated materials is given. Densities for MOM are about 1.1 g cm^{-3}.

The DBD can be expressed as:

$$DBD_{tot} = \frac{DBD_{org} \times LSR_{org} + DBD_{inorg} \times LSR_{inorg}}{LSR_{org} + LSR_{inorg}} \quad (7)$$

By combining this with equation (3) for calculation of MOC deposited under anoxic conditions the following equation results:

$$MOC = \frac{PP \times PF}{(PP \times PF) + 10 \times (DBD \times LSR)} \quad (8)$$

In order to test equation (8) for a range of PP values (5–150 g C m^{-2} yr^{-1}) and LSR_{inorg} (0.1–100 cm kyr^{-1}), MOC values were calculated and compared with the initially used formula (3). Applying a PF of 0.02% (open ocean, pelagic environment), a porosity of 60% and the density of MOM of 1.1 g cm^{-3}, almost no differences are observed except for very extreme conditions. For low sedimentation rates (0.1 cm kyr^{-1}) and PP values of 50 g C m^{-2} yr^{-1}), the effect reaches a difference of 0.81% MOC. But, using a PF of 4% (as an example for Black Sea conditions; Bralower & Thierstein, 1984) even for sedimentation rates of 1 cm kyr^{-1} and moderate PP values (150 g C m^{-2} yr^{-1}), the different equations produce a difference in MOC of 20%. Using equation (3) at sedimentation rates lower than 1 cm kyr^{-1} calculated MOC values were higher than 100% at almost all tested PP rates.

The main challenge in the modelling process is to define where anoxic conditions occur. As mentioned above we differentiate between two scenarios with different controlling mechanisms (Fig. 4). In oxygen minimum zones (OMZ) the amount of the primary-produced organic matter controls the intensity and extent of anoxia. The effect of anoxia is recognizable where the OMZ impinges on the sea floor (anoxic response). The coupling between productivity, water depth and oxygen depletion can be directly defined in the program. The lateral extension is controlled by the PP distribution in the model. Anoxia is only modelled where PP in the surface waters, and thus oxygen consumption below, exceeds a threshold value, and where the resulting OMZ impinges on the sea floor (Fig. 9a–c).

Anoxic bottom water conditions (ABW) may develop due to reduced water circulation in (semi-) enclosed basins and/or through stratification of the water column, and are independent of PP values (Fig. 10). Reasons for a stratified water column can be variable. Without an oceanographic model, these conditions cannot be simulated directly. Again a flexible approach is used. The user has to define, based on well data and/or geological knowledge, where an ABW could be present. The program will then automatically centre the ABW-lens in the sub-basin or at a river mouth at the coast and calculate the degree of anoxia through the interaction of the lens' lateral and vertical extent and the topography (i.e., anoxia will be modelled only when and where the basin reaches a sufficient depth; Fig. 11).

Terrestrial organic matter fraction

In contrast to the productivity of marine/autochthonous organic matter, the production of terrestrial organic matter is not simulated because

OF-Mod only considers OC deposition in the marine realm. Modelling TOM production would indeed require information about climate, atmosphere and drainage area for realistic simulation of vegetation cover and type. Here, the 'terrestrial organic fraction' represents all allochthonous organic matter which is not produced *in situ* in the marine environment and enters the marine system via rivers and run-off. It is assumed to have already experienced some degradation and microbial breakdown in soils and during transport before entering the marine system and is

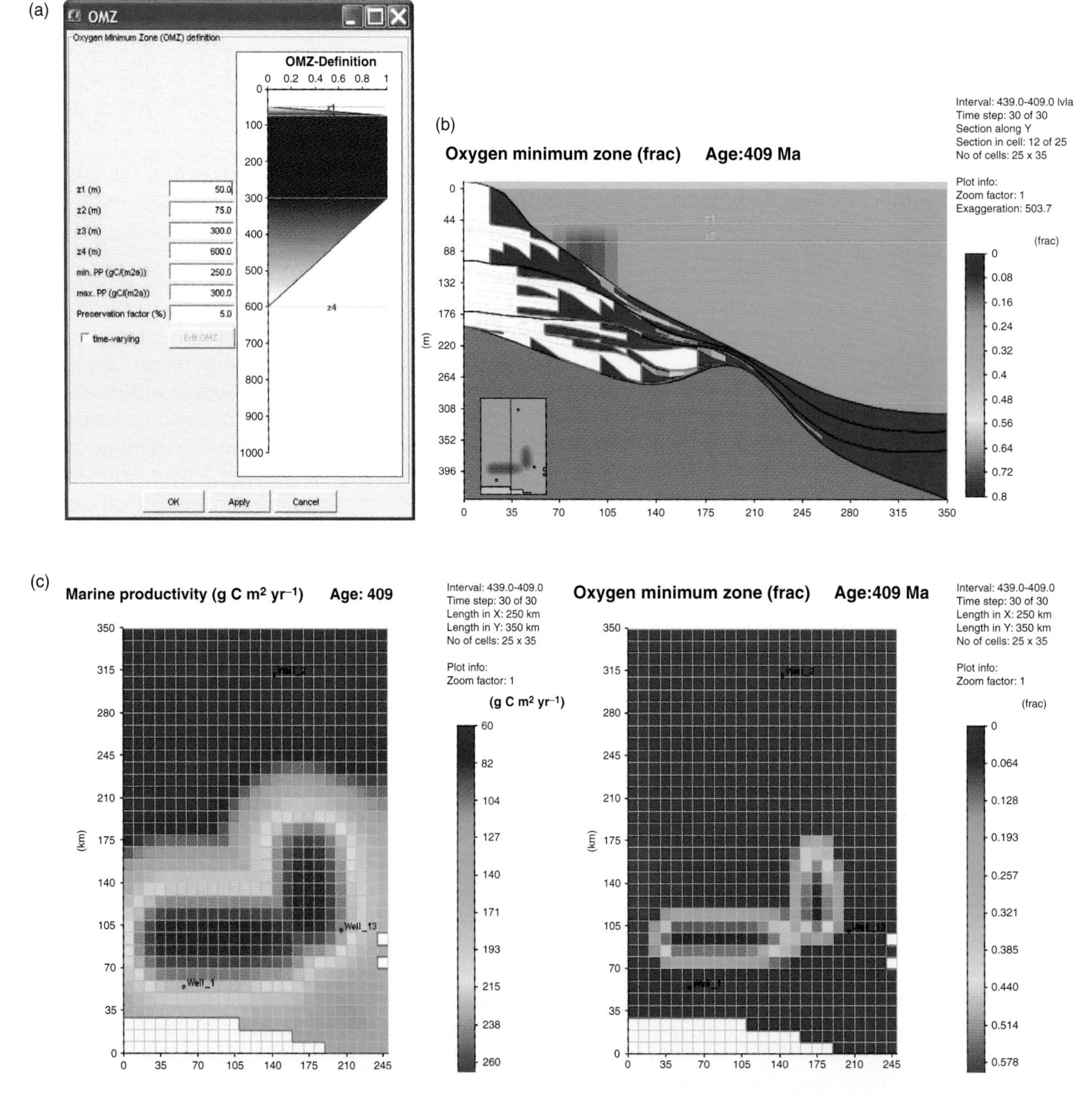

Fig. 9. Development of an oxygen minimum zone (OMZ) as a result of intense organic matter productivity. (a) How the user can define the OMZ in terms of vertical extension, the productivity-threshold values and the preservation factor. (b) The defined vertical extent of the OMZ is shown in a cross-section including the anoxic response at the sea floor (i.e., where the oxygen depletion actually affects the carbon deposition at the sediment surface). Full anoxic conditions (anoxic response = 1) only apply where primary productivity exceeds 300 g C m^{-2} yr^{-1}, and are therefore not reached in the modelling area (maximum anoxic response 0.58, cf. scale). (c) The surface plots of the primary productivity and the respective development of the oxygen minimum zone for one time step is shown.

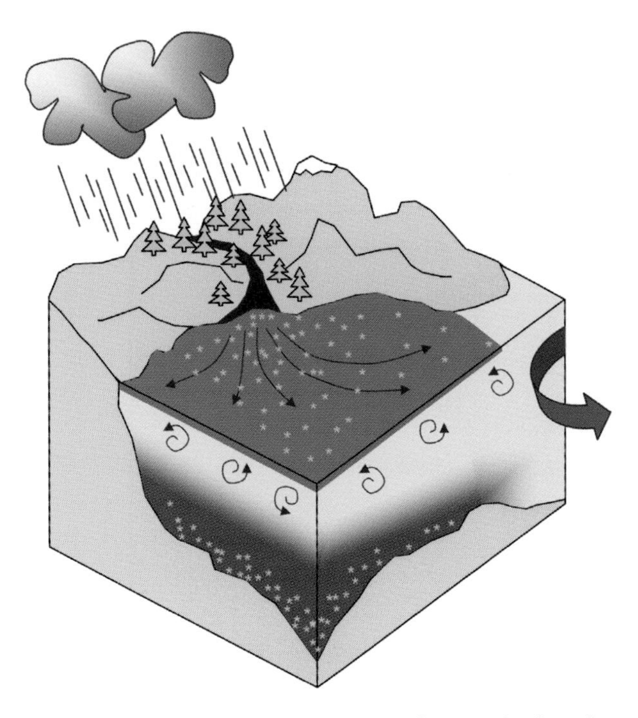

Fig. 10. Sketch illustrating processes that can lead to the development of anoxic bottom water conditions. Reduction of water-mass mixing can be the result of (semi-) restricted basin topography, salinity differences because of high freshwater input from rivers or high precipitation, high terrestrial organic carbon input consuming oxygen and leading to oxygen depleted conditions, or a combination of several of these processes.

therefore regarded resistant to further extensive degradation. We differentiate two subfractions based on grain size, transport behaviour and sorting: a discrete/particulate fraction derived from higher land-plant material and a soil or residual OC fraction. Depending on their particle size they are modelled to be associated with the distribution of the coarse mineral grains (sand fraction) and with the fine grain size (shale fraction) of the inorganic basin fill, respectively. Based on end-member values for the respective terrestrial sub-fraction, the discrete and/or residual terrestrial OC content is calculated for each cell filled with a distinct sand–shale mixture, that is the higher the shale content of a grid cell, the more the modelled residual OC content will approach the user-defined end-member/maximum value. However, the maximum end-member user-defined particulate terrestrial OC content is not necessarily associated with pure sand, but with lower values, since pure sands often do not contain any kind of organic matter as a result of winnowing of the lower density organic particles in highly hydrodynamic regimes (Fig. 12a and b).

The input of TOM may vary laterally in a basin as does water and sediment discharge. Part of this lateral variation is already accounted for by the basin-fill model. Higher sand discharge leads to a higher deposition rate, therefore OF-Mod will automatically model higher amounts of the coarse or particulate land-plant-derived organic matter in these areas. However, input of TOM type may be significantly higher close to river mouths. These scenarios can be modelled by defining areas (lenses) of higher terrestrial input, which can vary in position and intensity over time, and which are automatically located at the shoreline (Fig. 12b). The modelled terrestrial content is then the result of the extent of the lens(es), a defined maximum particulate terrestrial fraction to be applied within that area, and the sand content of the grid cells within this area (Fig. 12c).

Source-rock quality

As mentioned previously, source-rock quality is dependent on organic matter source types and also on preservation. Quality was briefly discussed earlier in connection with the organic-matter source fractions, marine and terrestrial. The source-rock quality in terms of hydrocarbon richness (hydrogen index, HI), is simulated in OF-Mod applying an end-member mixing model. For each organic-matter type (marine primary produced, particulate terrestrial, and residual organic matter) a HI end-member value has to be defined. Taking into account the modelled amount of each fraction in every grid cell, the HI parameter is calculated. Various end-member mixing models are proposed to show mixing processes of two source fractions (marine–terrestrial) or two preservation fractions (well preserved–poorly preserved) to explain the actual component distribution in cores and also to estimate complete organic carbon budgets of certain areas (Huc *et al.*, 1992; Westerhausen *et al.*, 1993; Prahl *et al.*, 1994; Villanueva *et al.*, 1997; Dean & Gardner, 1998; Ransom *et al.*, 1998; Knies, 1999; Rachold & Hubberten, 1999). In this context, the HI values of the pure particulate terrestrial fraction and the marine primary-produced organic matter represent the marine and terrestrial end-members while the residual organic-matter fraction represents the degraded end member of the two source fractions. However, applying a pure end-member mixing model in a 2D or 3D model implies that the influence of preservation on the marine-organic-matter fraction is considered to be similar at all locations

(a)

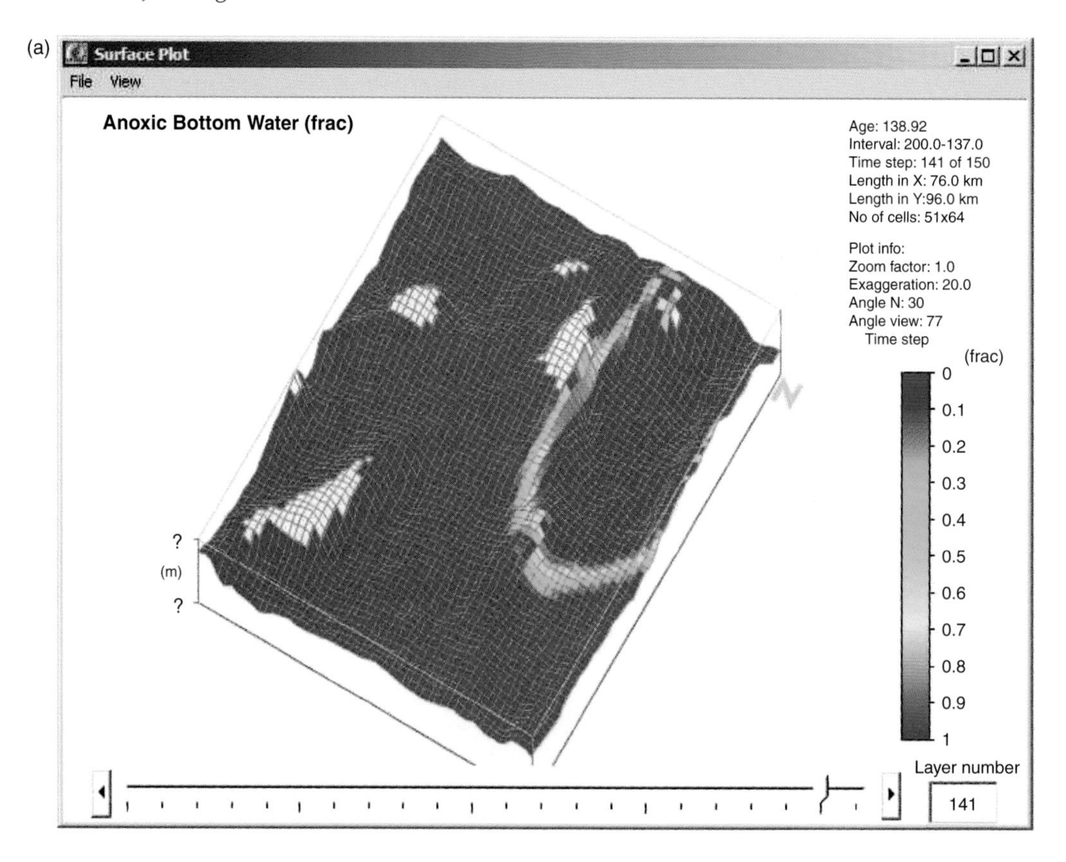

(b)

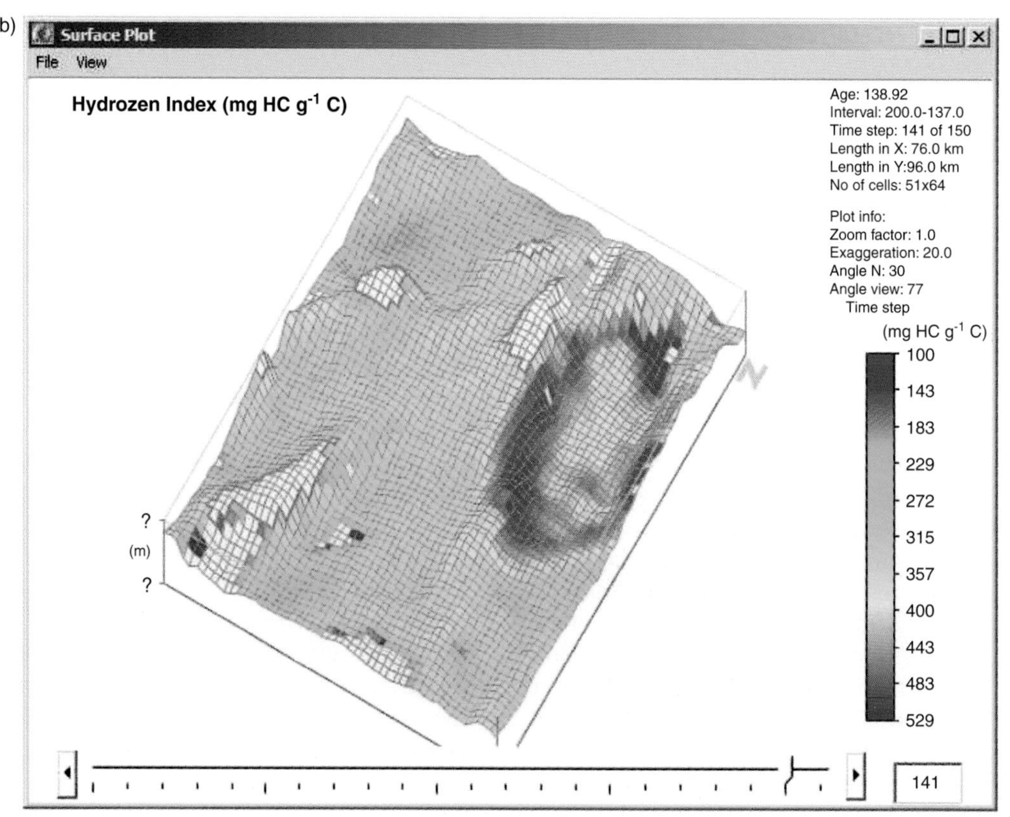

Fig. 11. Surface plot illustrating anoxic bottom water distribution in one sub-basin of the modelling area. (a) Extent of the anoxic conditions restricted to the extent of the sub-basin and dependent on depth. (b) The effect of anoxic conditions on quality distribution (i.e., hydrogen index, within and outside the area of oxygen depletion/enhanced preservation).

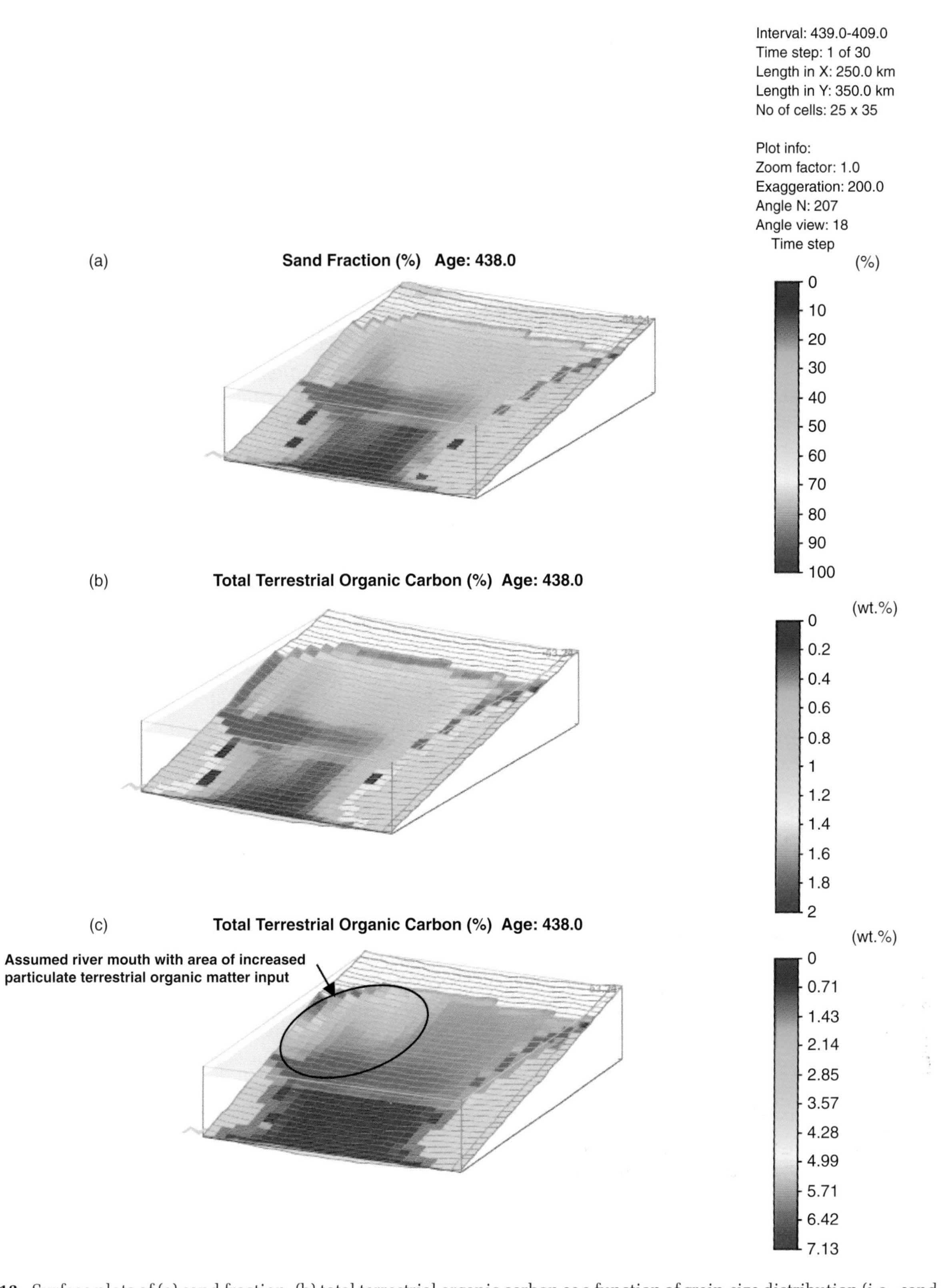

Fig. 12. Surface plots of (a) sand fraction, (b) total terrestrial organic carbon as a function of grain-size distribution (i.e., sand and shale fraction) and (c) total terrestrial organic carbon distribution including an area with increased particulate terrestrial organic carbon input. Note different scales.

in the modelling area (basin). This is not necessarily the case: the HI value for the marine fraction may vary significantly between the maximum value (end-member HI marine fraction) and the minimum value depending on the degree of preservation. OF-Mod therefore offers, in addition to the HI end-member mixing, an empirical calculation of the HI-value of the marine fraction coupled to the preservation of this fraction. The degree of preservation or the preservation factor, describes how much of the produced MOM actually gets entombed in the sediment. We therefore use this ratio also to describe how much of the initial potential of the MOM is finally preserved as MOC in the sediment.

The HI-value of the marine fraction is calculated by the following equation for each location:

$$HI_{\mathrm{MOC}} = HI_{\mathrm{SOM}} + (HI_{\mathrm{MOM}} - HI_{\mathrm{SOM}}) \times P^{0.1} \quad (9)$$

where HI_{MOC} is the *HI* value for the marine OC fraction, HI_{MOM} and HI_{SOM} the HI end-member values given by the user for each of the source fractions, and *P* the preservation factor (fraction) of the MOM.

The relationship between preservation, the given end-members of HI_{SOM} and HI_{MOM}, and the resulting HI value of the marine OC fraction is shown in Fig. 13.

In addition to the hydrogen index, other parameters could be added as well to improve the predictive nature of the simulations in terms of variations in source-rock-facies trends and/or fingerprinting

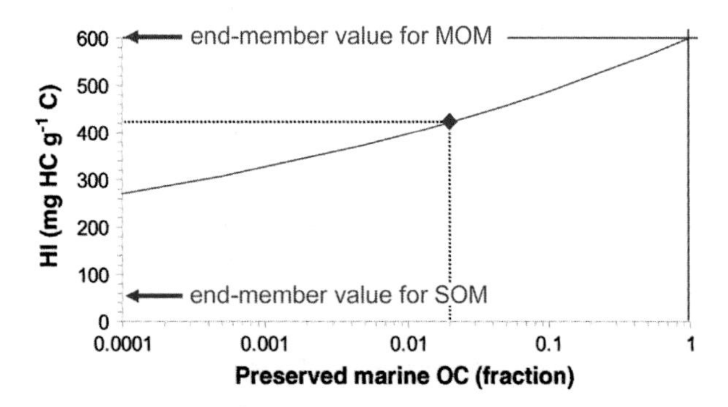

Fig. 13. Plot illustrating the relationship between the preservation (P) and the resulting hydrogen index (HI) of the marine organic carbon (MOC) fraction, based on the user-defined end-member values for the marine and the soil organic matter. Note the logarithmic scale. The red diamond shows that a HI of 422 mg HC g^{-1} C is modelled for a preservation of 2% using end-members of HI MOM (600 mg HC g^{-1} C) and HI SOM 50 (mg HC g^{-1} C), respectively. SOM = soil or residual organic matter.

of distinct components resulting in a specific hydrocarbon phase distribution.

MODELLING THE LATE JURASSIC SPEKK FORMATION AT THE MID-NORWEGIAN CONTINENTAL SHELF: THE FIRST OF-MOD APPLICATION TO A COMPREHENSIVE EMPIRICAL DATA SET

The Late Jurassic Spekk Formation is the major source rock at the mid-Norwegian Shelf. During a shallow-drilling campaign by SINTEF Petroleum Research (former IKU) detailed seismic exploration and drilling was carried out in the Froan Basin, a small semi-restricted sub-basin of Late Jurassic age (Skarbø *et al.*, 1988). The cored Late Jurassic to earliest Cretaceous succession (Spekk and Rogn Formations) provides one of the best documented reference sections for this time interval in the region.

The comprehensive data set from seismic line IKU125-87 and the cores of the Froan Basin (6307/07-U-02 and -03A) were used as a base for this OF-Mod 2D study (Fig. 14). The cored sections are shale-dominated, thermally immature, and have a good biostratigraphic control (Krokstad & Monteil, 1995). A comprehensive organic-geochemical data set (Skarbø *et al.* 1988; Krokstad & Monteil, 1995: Monteil, 1997; Tyson, 1997) served as a base for source-rock modelling in order to test if the OF-Mod software is capable of realistically simulating the deposition of the Spekk Formation and if the processes in OF-Mod adequately reproduce the well data. A challenge during this study was to simulate the special depositional conditions of the locally occurring lower 'rich Spekk Formation', which is characterized by high amounts of *Botryococcus*-like algal organic matter and exceptional preservation conditions.

Based on the IKU wells 6307/07-U-02 and -03A, the Spekk Formation can be locally divided into an upper and lower section separated by the sandy Rogn Formation (Skarbø *et al.*, 1988) (Fig. 15). The part of the Spekk Formation below the Rogn Sand (lower Spekk Formation) reaches a thickness of 61.5 m in IKU-well 6307/07-U-03. However, the base of the Spekk Formation was not reached. The source-rock quality in this lower part of the Spekk Formation is variable, with a very rich middle part showing single TOC values up to 18% and

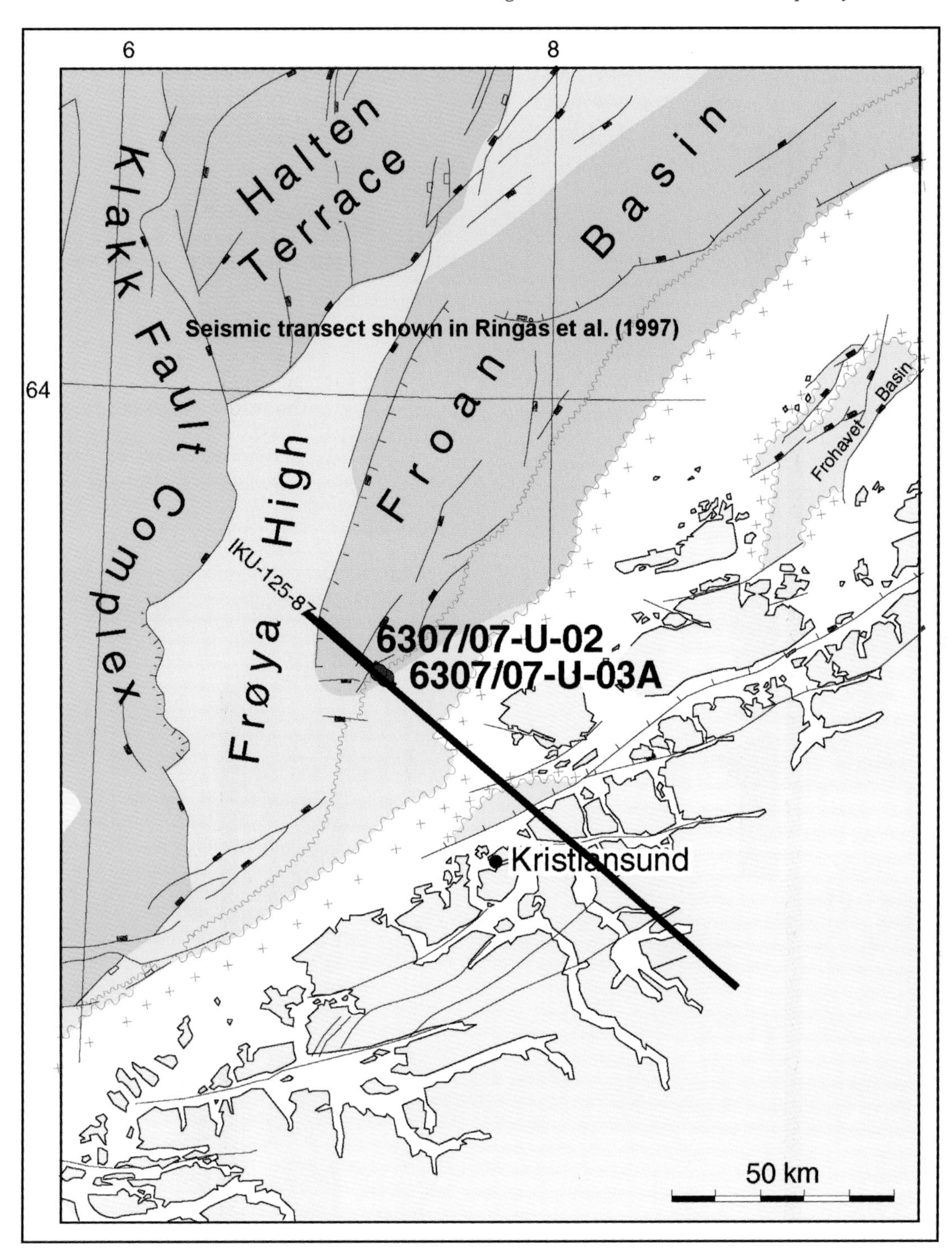

Fig. 14. Map showing the study area and the position of the studied wells and modelled transect (from Ringås *et al.*, 1997).

HI values up to 900 mg HC g^{-1} C (lower 'rich Spekk Formation').

The sandy Rogn Formation was drilled in well 6307/07-U-03A and partly in well 6307/07-U-02. It reaches a thickness of 79 m in well 6307/07-U-03A. The deposits consist of fine-grained, well sorted, subangular sandstones, which are interpreted as mass flow deposits with short transport distances but an effective sorting mechanism (Skarbø *et al.*, 1988). Some shale intercalations occur, in particular in the upper part of the Rogn Formation.

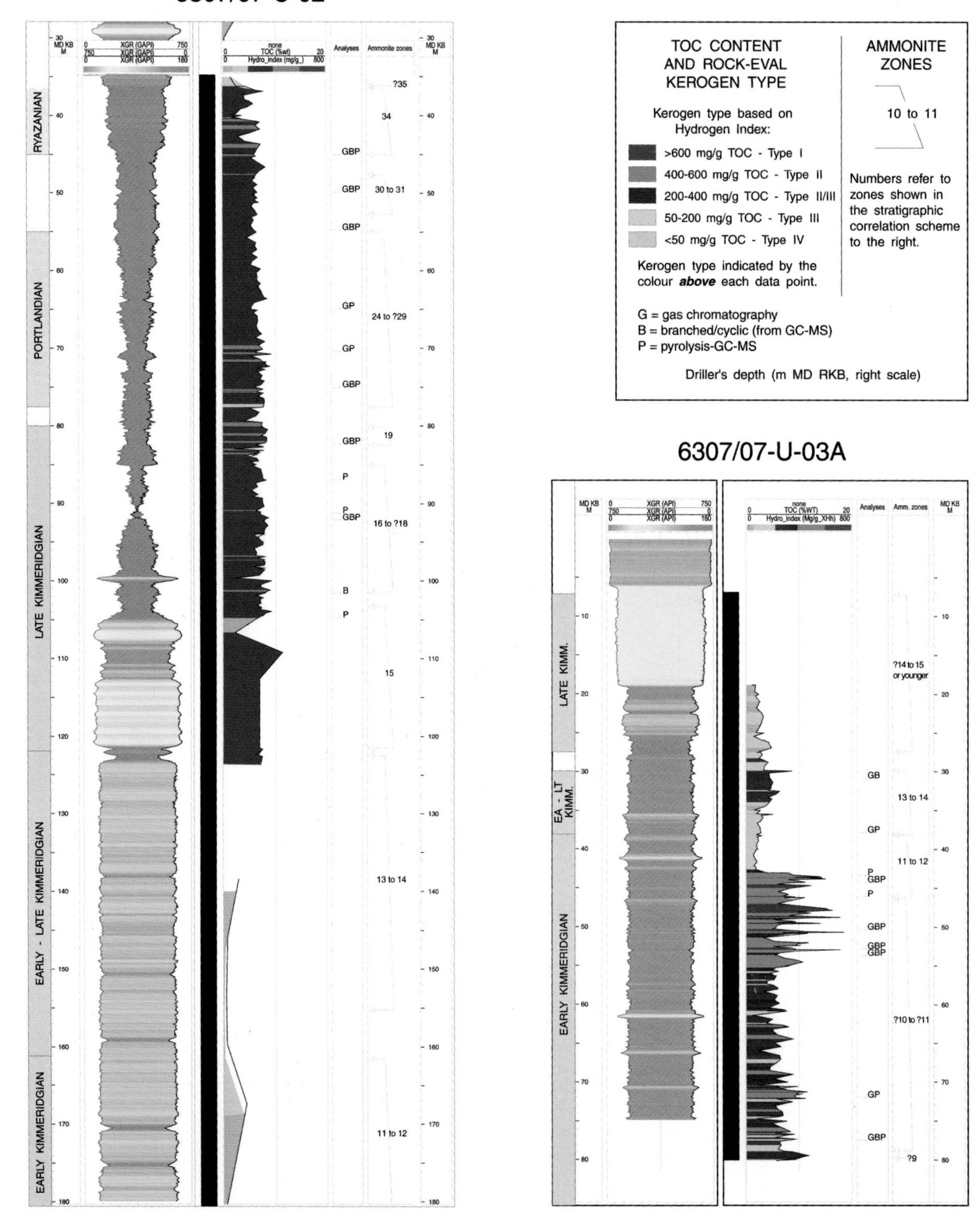

Fig. 15. Summary logs and correlation of the two IKU-wells, which serve as control wells in the present study. Well 6307/07-U-02 includes the sandy Rogn Formation in the lower part and the Upper Spekk Formation in the upper part. Well 6307/07-U-03 comprises the lower Spekk Formation and the lower part of the sandy Rogn Formation in the upper part. The lithology is indicated to the left, interpreted from the gamma-ray logs. To the right total organic carbon contents (TOC) and interpreted kerogen types are shown.

The upper part of the Spekk Formation recovered in IKU-well 6307/07-U-02 reaches a thickness of 68.5 m. It is of higher source rock quality with TOC-values above 6% and HI values around 350 mg HC g^{-1} C. Two condensed sections in this succession are indicated by palaeontological data. This had, however, no clear impact on source-rock quality.

The basin was rather shallow with water depths less than 200 m throughout the Upper Jurassic, as indicated by micropalaeontological data (Kjennerud & Gillmore, 2001).

Basin-fill stratigraphy

First, a structural restoration of the initial basin geometry was carried out based on data from seismic section IKU-125-87 at the southern tip of the Froan Basin (Kjennerud & Gillmore, 2001). This was used as a basis for the successive stratigraphic modelling. However, since part of the former basin fill has been eroded, a reconstruction of the complete initial basin shape was difficult. The available seismic data (seismic line IKU-125-87) cover a stretch of ca. 21 km. Micropalaeontological proxies of palaeowater depth and basin geometry constrain the position of the centre of the Late Jurassic basin to west of the drilling sites. This implies that deposition of the Rogn Sand occurred on the eastern basin flank with the probable sediment source east of the reconstructed section (i.e., the mainland; Fig. 16).

The basin fill was modelled applying the process-based forward modelling tool

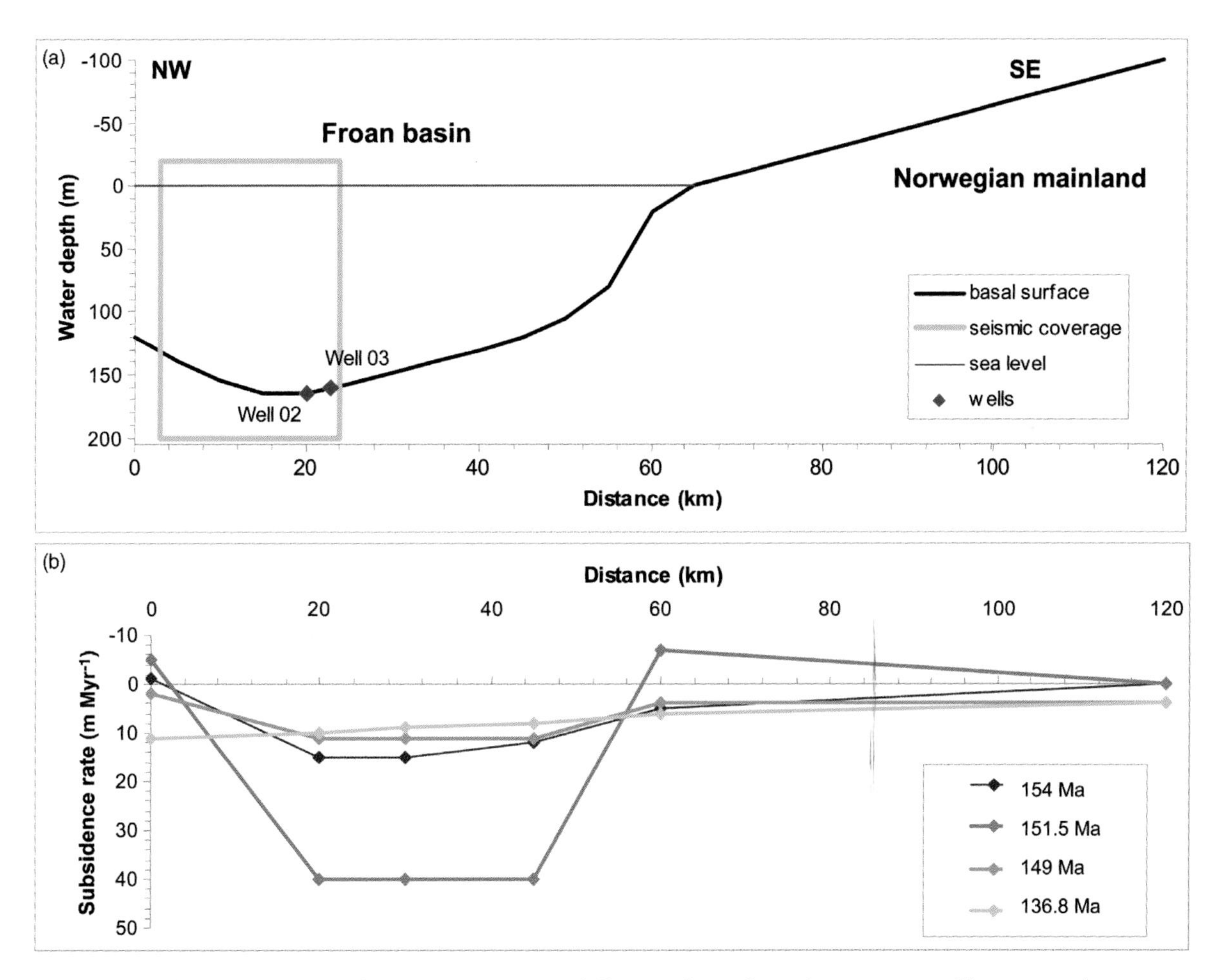

Fig. 16. (a) Initial basin shape for the DEMOSTRAT modelling. Indicated are the area covered by seismic line IKU-125-87 and positions of wells 6307/07-U-02 and -03. The reconstructed base was slightly modified, smoothed and extended towards the mainland to include the palaeoshoreline throughout the simulation. Note that the topography is exaggerated. (b) Subsidence values used along the section. The different curves show the subsidence at given times; linear interpolation between these estimates were made for other age estimates. Note that the main period of differential subsidence was during the Rogn Sand deposition.

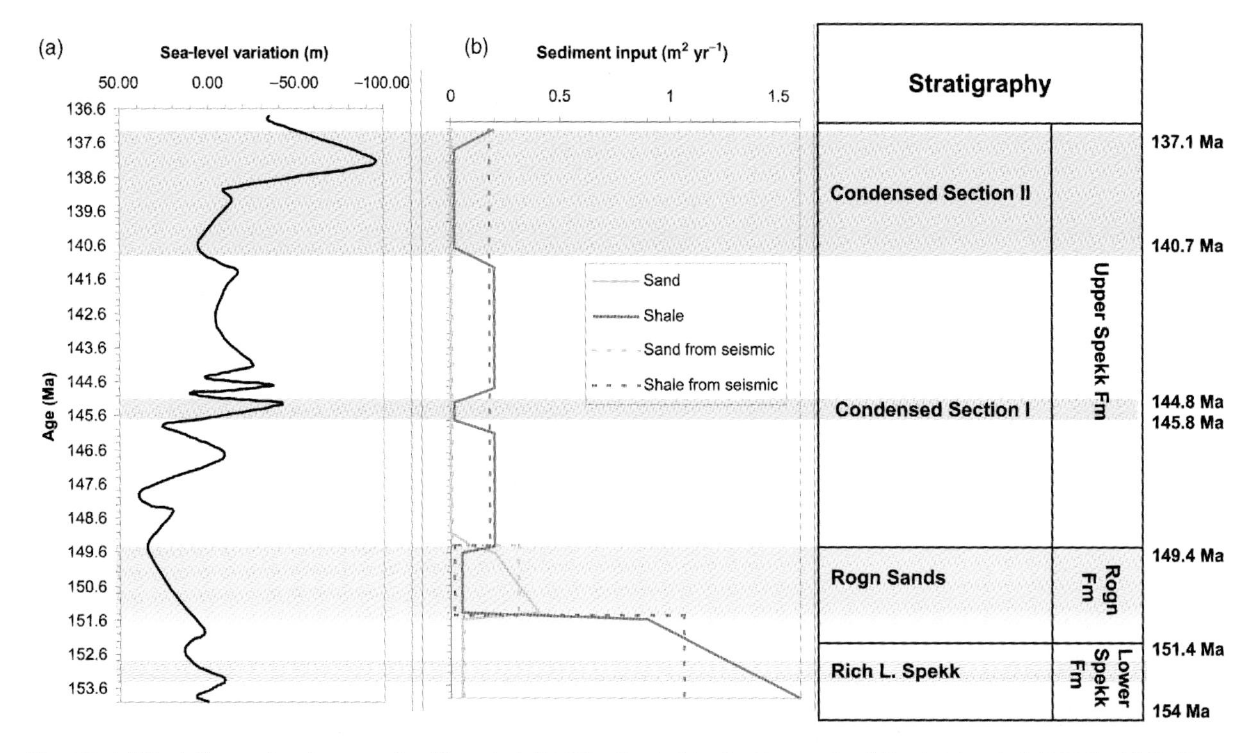

Fig. 17. (a) Eustatic sea-level curve after Haq *et al.* (1987) adjusted to the timescale of Gradstein *et al.* (1999). (b) Sediment input curve and stratigraphy. Note that the deposition of the Rogn Sands occurred during a period of eustatic sea-level rise and that the condensed sections correspond to phases of eustatic sea-level falls.

DEMOSTRAT (Rivenæs, 1993). DEMOSTRAT employs a dual-lithology, depth-dependent diffusion equation to model deposition and erosion of sand and mud. Thus, sediment transport and erosion is modelled as a function of sediment input, slope, and diffusion coefficients. The program takes initial basin shape, compaction, tectonic and isostatic subsidence as well as eustatic sea-level variations into account (Fig. 17). Basement compaction and overload can also be considered. A summary of the input parameters used for DEMOSTRAT is given in Table 1.

The time interval covering the Spekk Formation (17.16 Myr) was subdivided into 600 time steps, which results in a temporal resolution of 28.6 kyr per modelled layer. The spatial (horizontal) resolution of the model is 400 m per cell, section length of 120 km divided in 300 columns. Thus, the model includes 180 000 cells.

Figure 18 illustrates the evolution of the modelled section; thickness and geometries are shown at the time of deposition. During deposition of the lower Spekk Formation the basin is filled relatively rapidly. Sands are confined to the basin margin while finer grain sizes are transported to the basin centre. Deposits onlap the western basin margin ca. 15–20 km west of the well locations. This compares well with the seismic data, which indicate a pinch out of the lower Spekk Formation approximately 15 km from the wells.

During deposition of the Rogn Sand the depositional pattern changed. Sands did not accumulate in the shore zone, but were transferred by mass

Table 1. Summary of the input parameters used in DEMOSTRAT. Compaction of sand and shale after Baldwin & Butler (1985).

Time interval	154–136.8 Myr (17.16 Myr)
Temporal resolution	600 steps (à 26.8 kyr)
Section length	120 km
Spatial resolution	300 columns (à 400 m)
Sea level	Haq *et al.* (1987)
Sediment input	From mainland (southeast)
Transport coefficients	Time-, depth-varying
Subsidence	Time-, location-varying
Isostasy (flexural rigidity)	1E23 (EET = 25 km)
Compaction	Sand: Baldwin & Butler (1985)
	Shale: Baldwin & Butler (1985)
Basement	500 m thickness
	80% Sand on erosion
	5% porosity
Overburden	1400 m

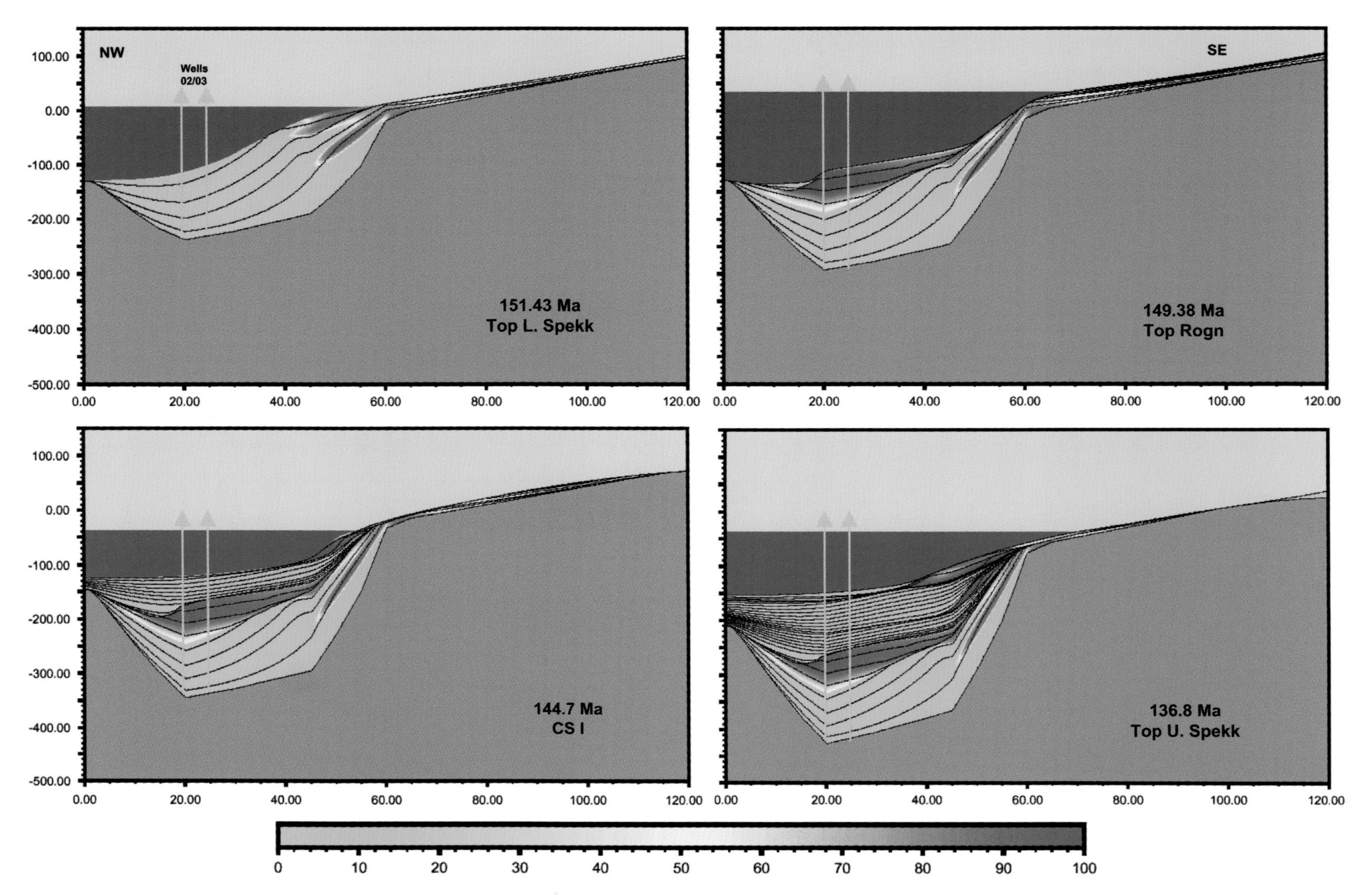

Fig. 18. Evolution of the modelled section. Colours indicate percentage sand fraction, distance is given in kilometres and elevation in metres relative to sea level at start of the model. The approximate positions of the IKU wells on the section are shown: CS, condensed section.

flow events towards the basin centre. Mass flows were probably triggered by fault movements. Thus, net erosion was modelled along the coast and sands accumulated as a wedge in deeper water.

During deposition of the upper Spekk Formation, the Froan High was overstepped by deposition. Sands are mainly restricted to the coastal zones. Only during the development of the two condensed sections during eustatic sea-level lowstands were sands distributed further into the basin. These model results suggest that the condensed sections developed in response to a lowstand bypassing of the basin. During the entire interval the modelled water depth at the well locations ranges between 160 m and 100 m which is compatible with the palaeowater depth estimates by Kjennerud and Gillmore (2001).

Modelled thickness, lithologies and sedimentation rates were extracted along the approximate well locations and compared to the drilling results (Fig. 19). The modelled thickness at well positions was realistically reproduced. However, modelled thickness of the upper Spekk Formation is somewhat lower than measured. Consequently, the modelled sedimentation rates are also slightly lower than the measured ones, but the trend of decreasing sedimentation rates are well reproduced (Fig. 19). Both, occurrence and age of the two condensed sections within the upper Spekk Formation match well with the well data.

Organic matter sedimentation

As a basis for the organic-sedimentation modelling, organic-geochemical, visual kerogen and palynofacies data from wells 6307/07-U-02 and 6307/07-U-03A were used to describe the depositional environment and to constrain the model settings. Based on these data, a simplified division of the Upper Jurassic sedimentary record into five units was made with respect to source and preservation of organic matter (Figs 20–23 and Table 2).

Core 6307/07-U-03 includes units I, II, and III, which cover the lower Spekk Formation (Figs 20 and 21). The variability in quantity and quality

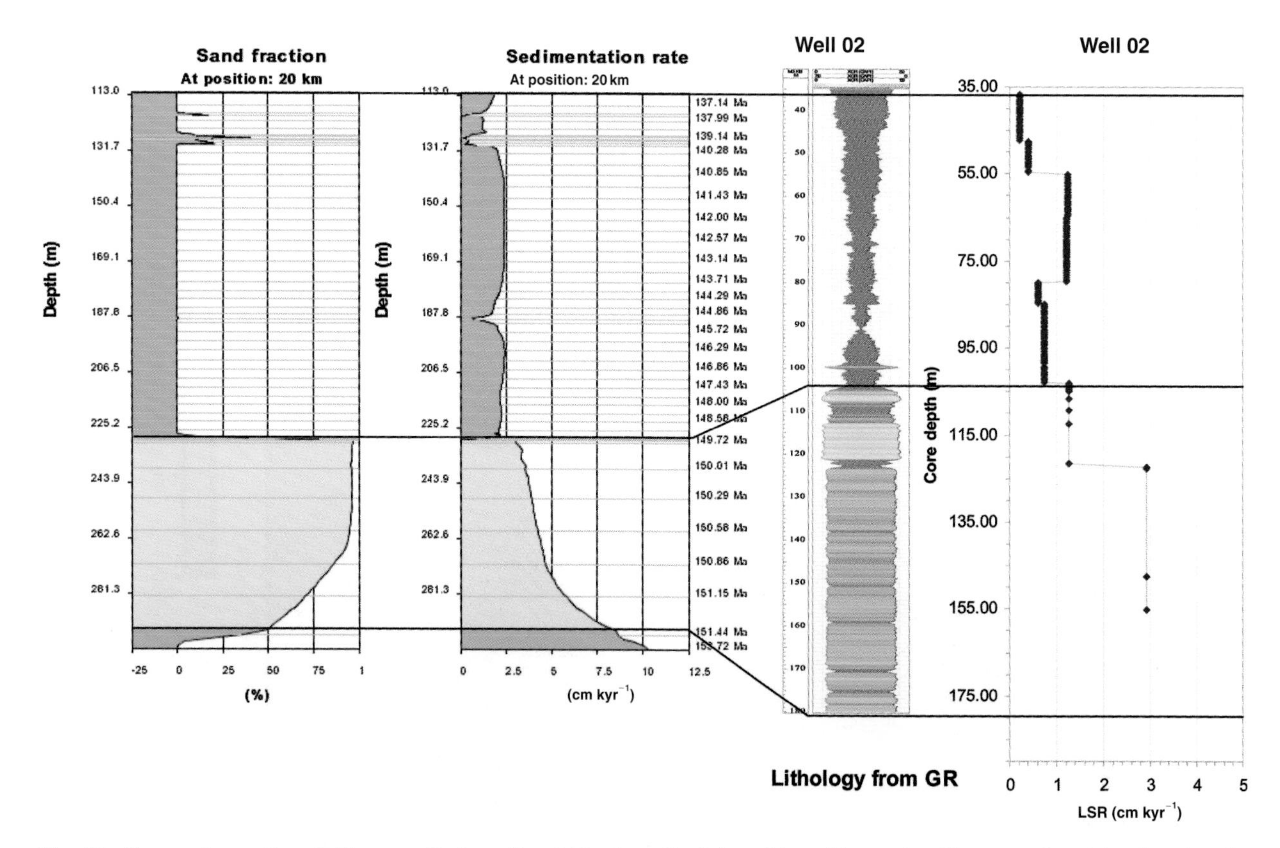

Fig. 19. Comparison of modelling results (position 20 km) on the left, with well 6307/07-U-02, on the right showing the Rogn Formation (lower part of the section, yellow colour coding) and the Spekk Formation (upper part of the section, green colour coding). Note that modelled thickness and sedimentation rates are uncompacted. GR = gamma ray.

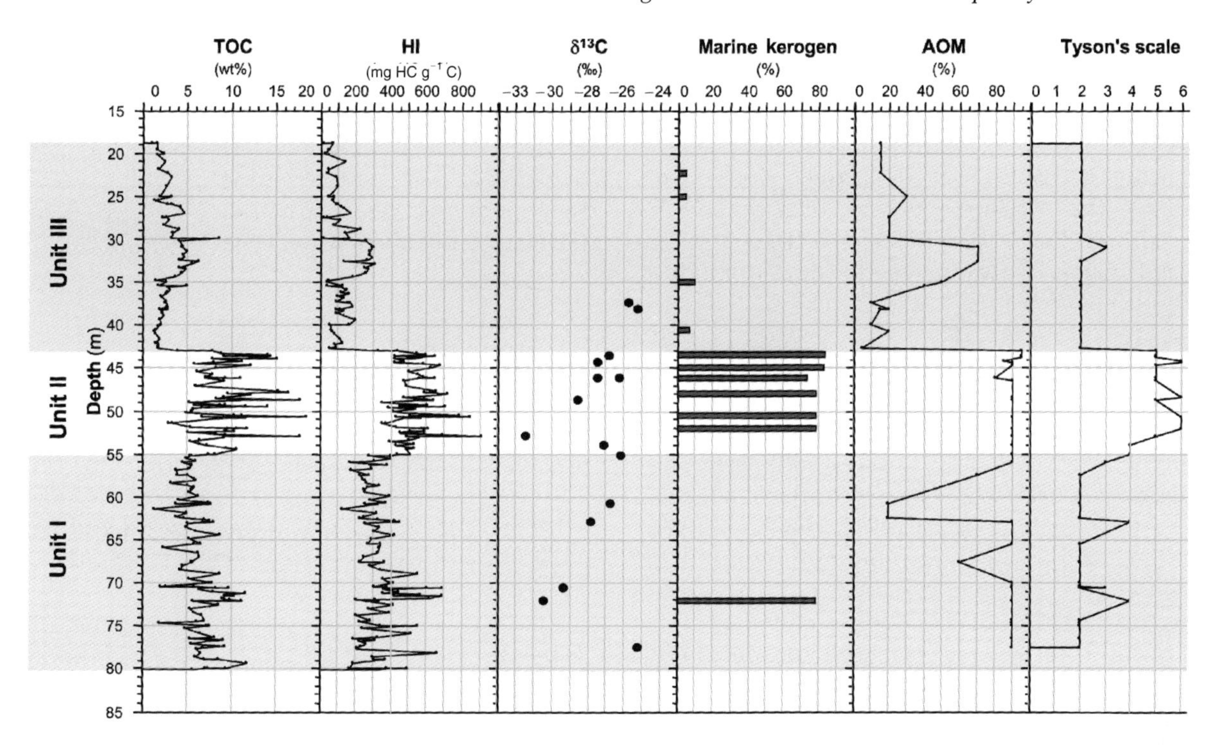

Fig. 20. Organic geochemical and microscopy data from well 6307/07-U-03, lower Spekk Formation: TOC, total organic carbon; HI, hydrogen index, δ^{13}C, organic carbon isotope ratio; marine kerogen, marine organic matter fraction (alginite + liptinite + fluorescing amorphous organic matter) estimated based on visual kerogen analyses; AOM, amorphous organic matter.

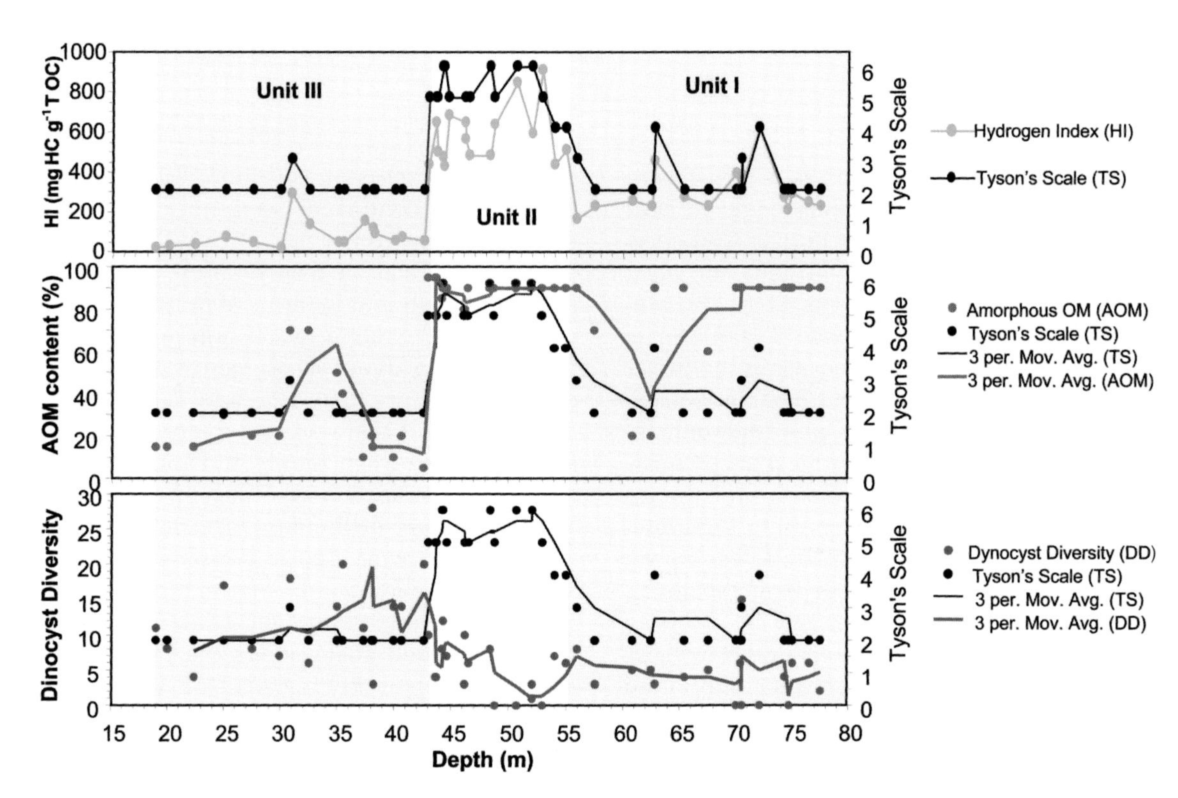

Fig. 21. Palynofacies and organic geochemical data from well 6307/07-U-03, lower Spekk Formation.

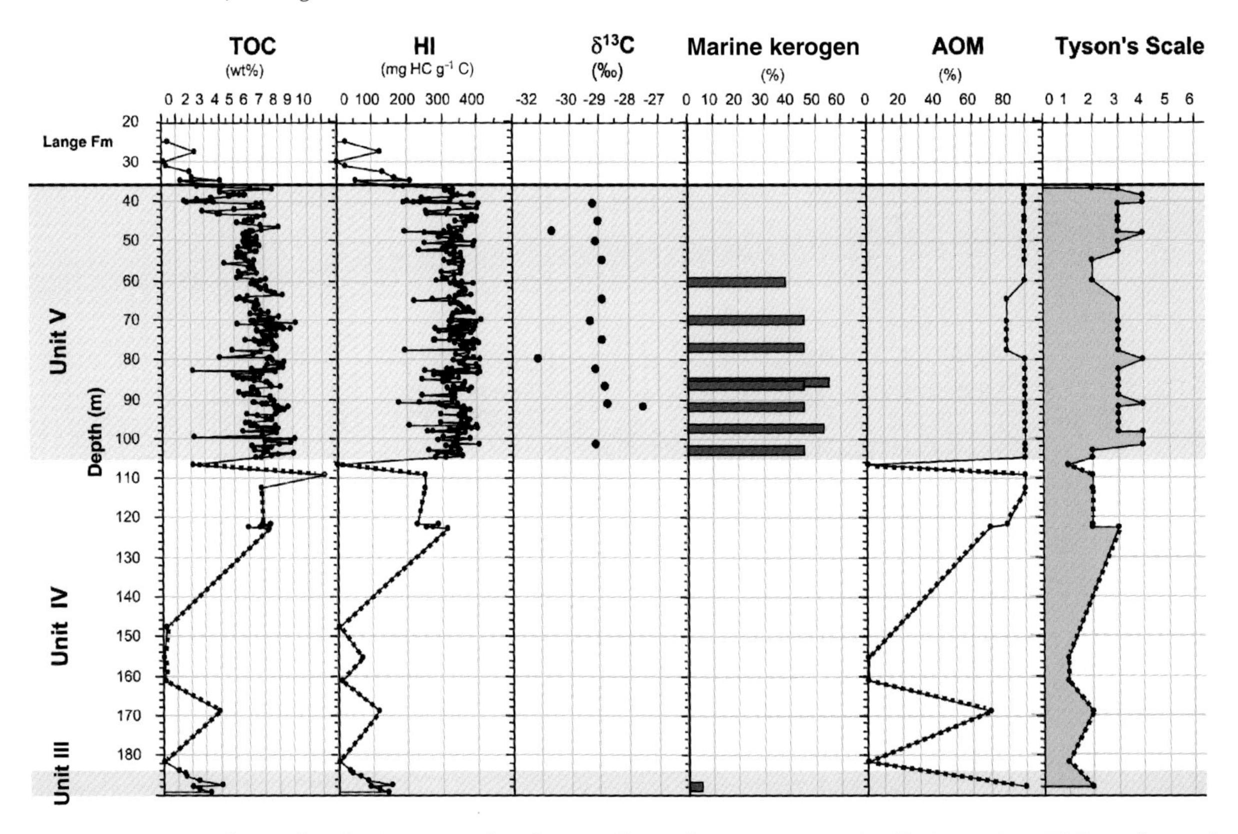

Fig. 22. Organic geochemical and microscopy data from well 6307/07-U-02, upper Spekk Formation: TOC, total organic carbon; HI, hydrogen index, $\delta^{13}C$, organic carbon isotope ratio; marine kerogen, marine organic matter fraction (alginite + liptinite + fluorescing amorphous organic matter) estimated based on visual kerogen analyses; AOM, amorphous organic matter.

of the organic matter in these units is significant. Highest OC contents and HI values occur in Unit II. In microscopy analyses this unit is characterized by distinct amounts of amorphous and algal organic matter (e.g. *Botryococcus* is relatively common) showing a bright fluorescence (Tyson's Scale 5–6, Fig. 21 and Table 2). Dynocysts in the lower Spekk Formation are generally characterized by low-diversity, high-dominance, and high-density assemblages which are often indicative of near-shore areas characterized by anoxia and fluctuating salinity. In terms of preservation the aforementioned trends indicate that the lower Spekk Formation and particularly Unit II of the lower Spekk Formation, is characterized by good to very good preservation conditions under oxygen-depleted conditions. This is further supported by the distinct odd/even predominance of the aliphatic hydrocarbon fraction (*n*-alkanes) and the total sulphur data (Mann & Zweigel, 2002b), both suggestive of anoxia. The total sulphur content is relatively high (>5% on average), and the mean C/S ratio which is significantly lower than 2.8 reflects an excess in sulphur. However, these preservation conditions seem not to have continued, and in the upper part of Unit III all geochemical data indicate a deterioration of the depositional environment in terms of preservation, and a more mixed/dynamic water column as indicated by increasing dynocyst diversity and increasing silt content (Figs 20 and 21, and Tables 2 and 3).

Units IV and V are present in core 6307/07-U02 and cover the Rogn Sands and the upper Spekk Formation, respectively (Figs 22 and 23). The lowermost part of the core also includes Unit III (uppermost lower Spekk Formation). The well sorted mass-flow sands of the Rogn Formation (Unit IV) have very poor OC contents, whereas in the intercalated shale layers the OC quality and quantity is distinctly higher. Unit V, covering the entire upper Spekk Formation, is generally more homogeneous than the lower Spekk Formation. It is characterized by organic-carbon values between 6 and 8% and HI values about 350 mg HC g^{-1} C. In the few intercalated shales of Unit IV and in Unit V preservation conditions are good, but not

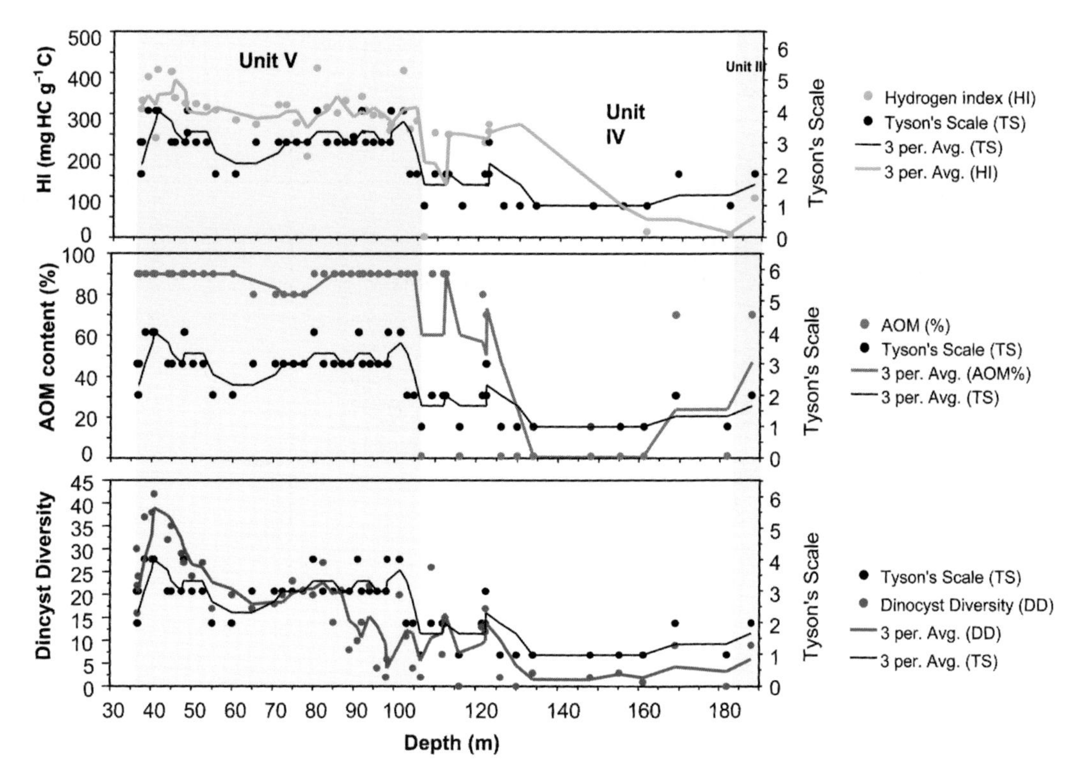

Fig. 23. Palynofacies and organic geochemical data from well 6307/07-U-02, upper Spekk Formation/Rogn Sands AOM, amorphous organic matter.

as exceptional as in parts of Unit I and Unit II, as indicated by fluorescence values and dynocyst diversity. Unit V seems to represent the more general pattern of the Spekk Formation elsewhere at the mid-Norwegian Shelf.

In the overlying Lange Formation on top of Unit V only geochemical bulk data, and no microscopy or biomarker data, exist indicating that organic-carbon quality and quantity decrease dramatically. This interval is not included in the modelling study. Also Lipinsky *et al.* (2003) and Langrock *et al.* (2003) infer good preservation conditions in an oxygen-depleted environment based on trace metal enrichment and visual kerogen data of the Spekk Formation, a drastic decrease in organic-C content, and a change in depositional conditions during deposition of the overlying Lange Formation from a restricted marine to a more open marine, well-ventilated environment.

Input parameter estimation

For estimation of the palaeoprimary productivity (PP) and the approximate terrestrial organic-matter input we had to decipher the marine and terrestrial OC fractions from the total OC content. We used HI values, visual kerogen, and biomarker data for estimating the proportions of the marine/aquatic (amorphous fluorescing + alginite + liptinitic material) and terrestrial (amorphous non-fluorescing, vitrinitic + inertinitic material) OC fractions (Mann & Zweigel, 2002b; Table 3).

For the PP calculation from MOC we had to estimate water depth, sedimentation rate and density of the sediments at the well positions (cf. equation 8). These data were taken from the stratigraphic model at positions corresponding to the location of well 6307/07-U02 and -U03 (Table 4). Primary productivity was subsequently calculated for the well positions for Unit I, II, III and V of the Spekk Formation, based on rearrangement of equation (8). The derived PP values are generally in good agreement with literature data of the Late Jurassic and Cretaceous Atlantic (e.g. Bralower & Thierstein, 1984; Stein, 1986) taking into account that the wells 6307/07-U02 and -U03 were located in a very near-shore, shallow and temporarily restricted environment. The calculated PP values, particularly for Unit II, are quite high but not unreasonably high, as we have indications for temporary enrichment of algae in these units (algal blooms, cf. Tables 2 and 3). In addition the geochemical and microscopic

Table 2. Visual kerogen analyses of selected samples from wells 6307/07-U-02 and 6307/07-U-03.

Well	Depth (m)	Am	Al	Lm	W	C	Fmin	Fmax	SUM	Particle size	Preservation palynomorph	Thermal maturation index	Remarks
6307/07-U-02													
Unit V	60.50	75	10	5	5	5	35	40	100	F-M-L	Fair-Good	3-4	Dominated by amorphous OM (including a fair fraction of fluorescent material. Algal OM represented by tasmanitids.)
	65.05	75	5	10	5	5	30	35	100	F-M-L	Fair	3-4	Comparable to sample E5107, but with a somewhat increased proportion of liptinitic and a smaller proportion of algal OM.
	70.15	75	10	5	5	5	45	50	100	F-M-L	Fair	3-4	Comparable to sample E5107.
	77.00	75	10	5	5	5	45	50	100	F-M-L	Fair	3-4	Comparable to sample E5107.
	85.84	70	10	10	5	5	60	70	100	F-M-L	Fair	3-4	Dominated by amorphous, mainly dull orange fluorescent, organic matter.
	86.34	70	5	10	5	5	50	60	95	F-M-L	Fair	3-4	Dominated by amorphous, mainly dull orange fluorescent, organic matter.
	91.85	70	5	10	5	5	50	60	95	F-M-L	Fair	3-4	Dominated by amorphous, mainly dull orange fluorescent, organic matter.
	97.37	70	5	10	5	5	60	70	95	F-M-L	Fair	3-4	Dominated by amorphous, mainly dull orange fluorescent, organic matter.
Unit IV	102.90	70	5	10	5	5	50	60	95	F-M-L	Fair	3-4	
	188.00	65	5	15	5	5		0	95	F-M-L	Fair	3-4	
6307/07-U-03A													
Unit III	22.35	50	5	10	20	15	0	0	100	F-M-L	Fair	3-4	The amorphous material shows no fluorescence
	25.00	50	5	20	15	10	0	0	100	F-M-L	Fair	3-4	The amorphous material shows no fluorescence
	35.00	60	10	10	10	10	0	0	100	F-M-L	Fair	3-4	Increased proportion of fresh-water algae (*Botryococcus*) compared to the overlying samples
	40.51	65	5	15	10	5	0	5	100	F-M-L	Fair	3-4	Dominated by non-fluorescent amorphous organic matter

Unit II	43.31	65	15	10	5	5	95	100	100	F-M-L	Fair	3-4	Dominated by amorphous material showing a dull yellow/orange fluorescence. High proportion of algal material. Fluorescent AOM.
	43.50	45	40	5	5	5	95	100	100	F-M-L	Fair	3-4	Characterized by high proportion of bright yellow fluorescent algal material, and amorphous material showing dull yellow/orange fluorescence.
	45.00	75	10	5	5	5	95	100	100	F-M-L	Fair	3-4	Dominated by amorphous OM showing dull yellow/orange fluorescence.
	46.20	60	15	15	5	5	95	100	100	F-M-L	Fair		Increased proportion of algal and liptinitic OM compared to sample E5326 above.
	48.00	60	20	10	5	5	95	100	100	F-M-L	Fair	3-4	Fairly high proportion of yellow fluorescent algal material together with dominantly amorphous organic matter.
	50.50	60	20	10	5	5	95	100	100				Fairly high proportion of yellow fluorescent algal material together with dominantly amorphous organic matter.
	52.00	60	20	10	5	5	95	100	100	F-M-L	Fair	3-4	Comparable to samples E5339 and E5349
Unit I	72.09	65	15	10	5	5	95	100	100	F-M-L	Fair	3-4	
	72.09	80	3	3	5	5	95	100	96	F-M-L	Fair	3-4	Dominated by yellow/orange fluorescent amorphous matter. Some yellow fluorescent algal material, few polymorphs.

Am Amorphous material (% of kerogen).
Al Algal material (% of kerogen).
Lm Liptinitic material (% of kerogen).
W Woody material (% of kerogen).
C Coaly fragments (% of kerogen).
F Fine.
M Medium.
L Large.
Fmin, Fmax Minimum/maximum % fluorescent material in Am.

Table 3. Averaged values for the five units of the Spekk Formation. The estimates of marine and terrestrial organic carbon fractions are based on the analytical and microscope data shown in Figs 20–23. Ctot = total organic carbon, MOC = marine organic carbon, tTOC = total terrestrial organic carbon, HI = Hydrogen Index.

Units	Ctot (wt%)	tTOC (wt%)	MOC (wt%)	HI (mg HC g^{-1}C)	Preservation	Remarks
V. Upper Spekk	6–8	3–4	3–4	340	Good	Some freshwater input (*Botryococcus*), open marine
IV. Rogn Sands	7 shale/ 0.2 sand	3–5 shale/ 0.2 sand	2 shale/ 0.0 sand	250 shale/ ≤100 sand	Good in shale/ poor in sand	More or less pure sand unit, only very few thin shale layers
III. Upper lower Spekk	2–(5)	1.9–(3.5)	0.1–(1.5)	100 (250)	Fair–good	Increasing silt content, increasing dynocyst diversity
II. Rich lower Spekk	10(5–18)	3	7	500 (maximum up to 900)	Very good	Algal blooms (*Botryococcus*) especially in Unit II, restricted environment, very low dynocyst diversity, significant freshwater input
I. lower lower Spekk	5–8(12)	3.5–5	1.5–3(7)	200 (maximum 600)	Good–very good	

Table 4. Calculated PP values for well 6307/07-U02 and 03. OC = organic carbon, PP = primary productivity.

	Unit I	Unit II	Unit III	Unit IV	Unit V
Water depth (m)*	140	125	105	115	40–80
Dry bulk density (g cm^{-3})*	0.67	0.67	0.67	1.35	0.67
Sedimentation rate (cm kyr^{-1})*	10	9	8	???	2
Sand fraction (%)*	0.1	0.1	0.1	97–100	0.1
Total organic carbon (Ctot wt%)*	8	10	~2.5	0.2/7	7
Marine organic carbon (wt%)†	3–4	7	0.2	0/2	3.5
Terrestrial organic carbon (wt%)†	3–5	3	~2.4	0.2/3–5	3.5
Preservation factor (%)‡	3–6	5–10	~*0.1*	–	1.5–3
Primary productivity (g C m^{-2} yr^{-1})	74–150	150–250	91	n.a.	90–100

*Data derived from basin-fill model.
†Data from geochemical and visual kerogen analyses from well 6307/07-U03 and 02.
‡Estimated values based on organic geochemical, microscopy and palynofacies data.

data indicate very good preservation conditions (Table 3). Therefore relatively high preservation factors were applied for this unit.

The total terrestrial organic carbon (tTOC) fraction was estimated based on visual kerogen data, palynofacies analyses and HI values. The tTOC content is about 3% in Units I, II, III and V (cf. Tables 3 and 4). In Unit IV, Rogn Formation, the proportion of the terrestrial fraction in the few shale layers was about 3–3.5% too, whereas in the sands only very little residual or no OC was deposited. Generally, the main part of the terrestrial OC fraction in all units consists of non-fluorescent amorphous organic matter with a minor fraction of macerals (microscopically identifiable particles such as woody and coal particles, liptinites, etc., Table 2). In contrast to these discrete organic particles, whose occurrence is often associated with a coarser sediment type (silt and sand), the amorphous organic fraction is enriched in the fine sediment/shale fraction.

End-member HI values were calculated for the different pure OC fractions (marine organic carbon, MOC; particulate terrestrial organic carbon, pTOC; residual/soil organic carbon, SOC) based on estimates of the different amounts of the OC fractions at the well position. These end-members were assumed to be valid for the entire modelling interval, except for parts of Unit I and the whole of Unit II. As indicated by the HI values, palynofacies and kerogen data, distinct enrichment of algae (e.g. *Botryococcus*) suggests a significantly enhanced HI value for the MOC during these periods. A summary of all input parameters derived from the geochemical data and used in OF-Mod is shown in Table 5.

Results and discussion

Figure 24 shows the modelled distribution of the total OC content and HI values. A variable pattern of OC and HI is clearly visible in the lower part of the basin fill (i.e., in Units I, II and III (lower Spekk Formation)). Highest OC values occur in Unit II as a response to the increased PP modelled, but also due to the range of preservation factors (PF) applied. Lower OC values were modelled in Unit III, where a reduced PF was used to account for environmental changes, and lowest OC values in the sands of the Rogn Formation (Unit IV) and within the coastal sands. Nearly the same pattern is observed for the HI distribution except that the highest values do not occur where highest OC values were modelled, but more basinward (cf. Fig. 24a and b). This is due to the defined preservation conditions. Because fully anoxic conditions do not apply in very shallow surface water (Fig. 25), preservation decreases towards the shoreline and therefore also organic matter quality (HI).

When comparing the modelled data corresponding to the well position to the data from wells 6307/07-U-02 and -03, a very good match is observed (Fig. 26). The PFs in the ranges proposed in Table 4 roughly mimic the higher frequency variations observed in the measured data, even though well data indicate an even higher variability. In this context the preservation factor seems to be one of the most sensitive parameters in this case study. Sedimentation rates, which are generally relatively low in the Late Jurassic (decompacted sedimentation rates vary between 2 and 10 cm kyr^{-1}), could not have caused such a high variability in the resulting OC and HI values. Besides the additional input of algal organic matter (*Botryococcus*-type matter) during deposition of Unit II, no further variability in PP was modelled and a time constant 'basic' PP was applied in all other units (PP coastal to open ocean: 100–80 g C m^{-2} yr^{-1}; cf. Table 5). Geochemical data point more to variations of preservation conditions and a variable depositional environment as clearly indicated by the Tyson's fluorescence scale and dynocyst diversity (Figs 21 and

Table 5. Summary of input parameters used in OF-Mod.

			Remarks
Marine primary productivity (PP)	**PP** coastal (g C m^{-2} yr^{-1})	100	
	Distance to open ocean (km)	80	
	PP open ocean (g C m^{-2} yr^{-1})	100	
	Time varying	No	
	δ**PP** (g C m^{-2} yr^{-1})	100	δPP occurs only between 152.3 and 153.0 Ma (Unit II)
	Distance from shore (km)	1	
	Extension of δPP (km)	60	
	Time varying	Yes	
	Preservation	**Anoxic**	In Units I–III and V PF varies randomly in ranges as shown in Table 4
	Time varying	Yes	
Terrestrial organic matter	pTOC (%)	0.2	
	SOC (%)	2.5	
	Time varying	No	
Organic matter properties	HI MOC (mg HC g^{-1} C)	500	HI MOC varies randomly between 500 and 800 in the time interval from 152.3 to 154 Ma (Unit I and II)
	HI pTOC (mg HC g^{-1} C)	200	
	HI SOC (mg HC g^{-1} C)	50	
	Time varying	Yes	

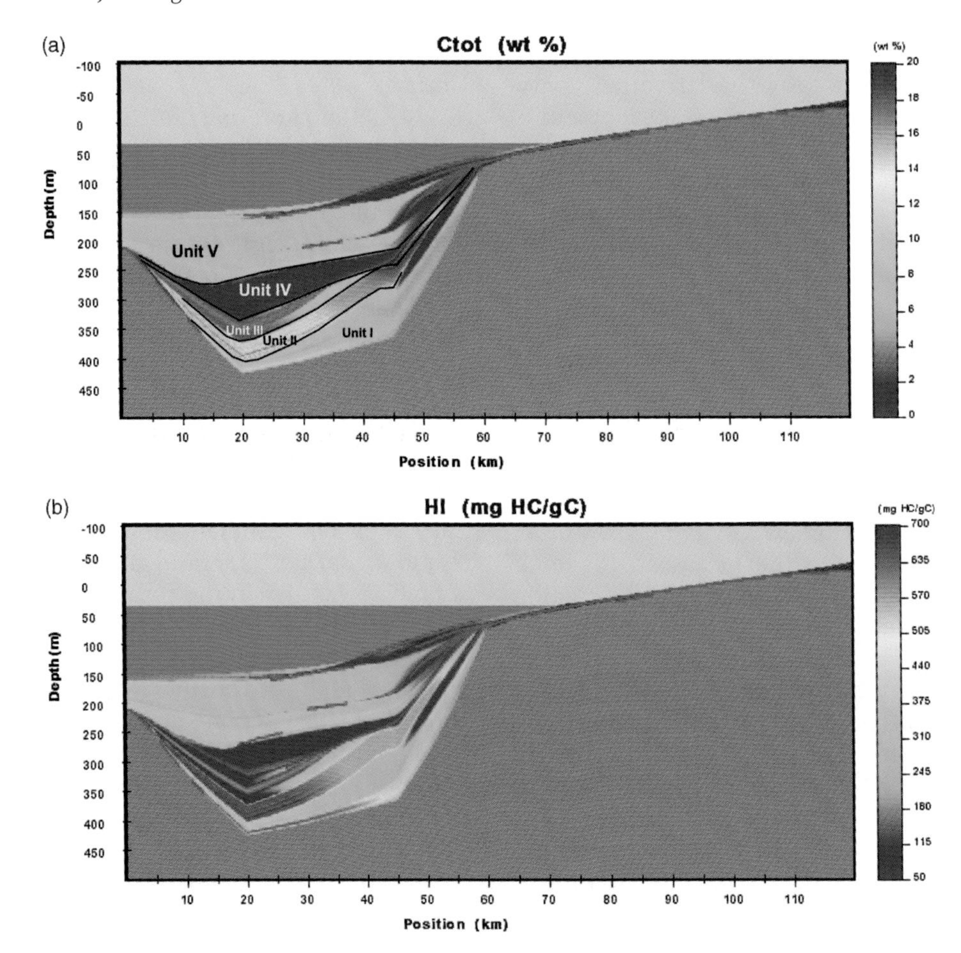

Fig. 24. (a) Distribution of total organic carbon (Ctot) and (b) Hydrogen Index (HI). Pink lines indicate unconformities.

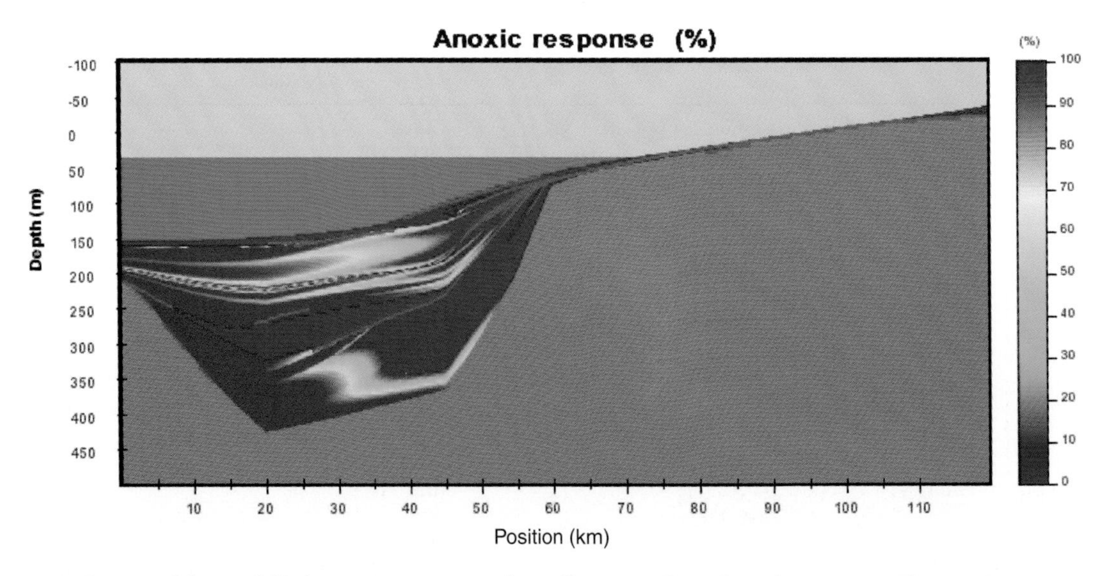

Fig. 25. Distribution of the modelled anoxic response at the sediment surface along the transect. The anoxic response shows where the defined preservation conditions affect organic-carbon deposition at the transect. Because preservation conditions do not occur in the photic zone (water depth <70 m), the shallower parts of the transect show no anoxic response (blue colours). Additionally, a thickness of 50 m for a transition zone with dysoxic conditions was defined. Consequently, full application of the defined preservation conditions occurs only below 120 m water depth (red colours) and therewith only in the deepest part of the Froan Basin.

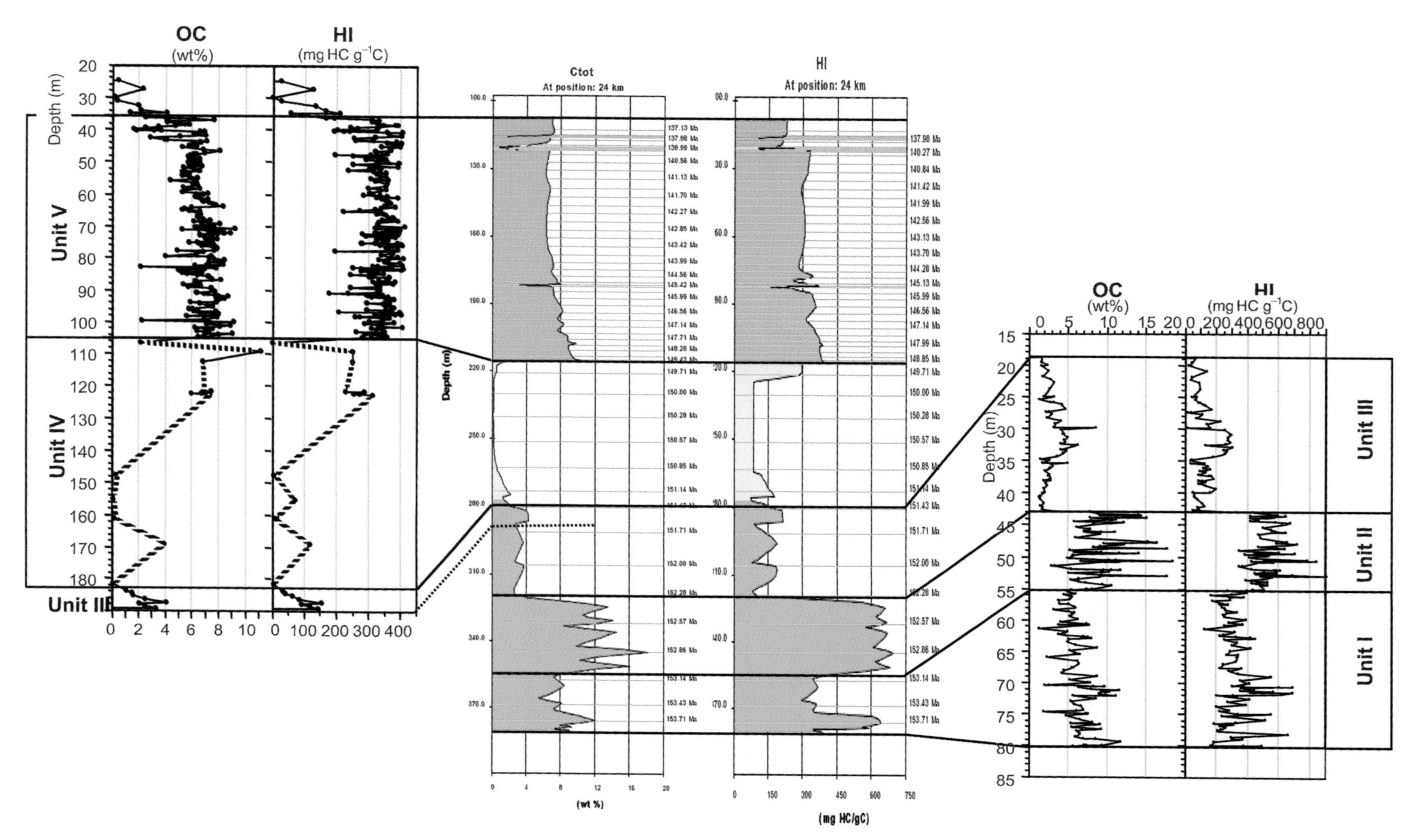

Fig. 26. Comparison of measured (in black) and modelled organic carbon (OC) and hydrogen index values (in colours) from wells 6307/07-U-02 and 6307/07-U-03.

23). In addition, the change in depositional environment and the decrease in preservation in Unit III are accompanied by increasing silt content and a complete change of sedimentation from shale to sands in Unit IV (Rogn Formation). Here, the lowest PFs were applied. The exceptional preservation conditions of parts of Unit I and Unit II could not be re-established later during deposition of Unit V, where the swell restricting the Froan Basin from the open Norwegian Sea was overstepped by sedimentation and more open-marine conditions developed. This is indicated by moderate dynocyst diversities and lower Tyson's scale and HI values. Based on this, the moderate PF values applied for this unit seem reasonable in comparison to the distinctly higher values applied in Unit II.

Another important aspect is that the extremely OC-rich sediments of the lower Spekk Formation (partly Unit I and Unit II) are most probably a very local phenomenon. Algal (*Botryococcus* type) material was transported by fresh-water supplied by rivers or continental runoff into the surface water of the semi-restricted Froan Basin. The lower density fresh-water flushing over the more saline sea water may have favoured stratification of the water column in addition to enhanced oxygen consumption due to a higher organic-matter content, and hence resulted in very good preservation conditions. In the upper Spekk Formation (Unit V) only few indications for such input of fresh-water algal material have been reported (cf. Table 2; Monteil, 1997) and may explain the worse preservation during deposition of this interval (lower carbon flux and less intense salinity stratification). Generally, Unit V seems to be more representative of the Spekk Formation as described elsewhere in the mid-Norwegian Shelf. In other wells further to the north along the mid-Norwegian shelf little evidence for such input of fresh-water algal matter has been documented, even though the wells are also located in a very nearshore deltaic environment (e.g. well 6407/09-08, Monteil, 1997; well 6814/04, Langrock *et al.*, 2003).

With OF-Mod it was possible to account for this local phenomenon by applying the δPP parameter. We could mimic the input of additional algal organic material via run-off by positioning a δPP cell directly at the coastline and letting it decrease into the direction of the Frøya High, or with increasing distance to the source during deposition of Unit II. Beyond that we used a constant PP value based on the back-calculated marine OC values from the well for the entire modelling of the Spekk Formation (cf. Table 5). In our opinion this PP value (PP coastal to open ocean: 100–80 $g\,C\,m^{-2}\,yr^{-1}$; cf. Table 5) represents a general PP estimate for the Late Jurassic organic-rich sediments of the Norwegian Shelf at least for near-shore environments. This is in contrast to the results from Langrock and Stein (2004) who estimated productivities from the same well, 6307/02, ranging between 20–30 $g\,C\,m^{-2}\,yr^{-1}$. Reasons for this difference remain unclear since their conclusions are not justified by more detailed data. Nevertheless, Langrock *et al.* (2003) also identified the variation in preservation conditions as a major controlling factor and found the supply of terrestrial organic matter to be of minor importance.

CONCLUDING REMARKS

The process-based modelling software OF-Mod that models organic/source-rock facies delivers results in terms of source-rock thickness, and the distribution of quantity and quality of organic content (TOC, HI, source-rock potential). Generally, the numerical modelling enables a multi-variable parameter cross check of all (previously defined) aspects influencing organic sedimentation. It can thereby handle much more complex scenarios than any conceptual model approach. Two very important and completely new aspects in this context are that OF-Mod considers (1) the OC source fractions, that is, marine and terrestrial, and (2) organic-sedimentation modelling is combined with basin-fill stratigraphy. The source fraction approach enables differentiation between the specific processes affecting each of the OC types on their way from their sources to the site of final accumulation. Furthermore, it enables application and refinement of end-member mixing models to achieve more information on organic-matter properties (i.e., HI, but also possibly carbon-isotope composition ($\delta^{13}C$) and other prospects in the future). On the other hand the combination of the organic sedimentation with the basin-fill stratigraphy permits evaluation of a realistic spatial distribution of a source-rock unit in a basin.

When defining and constructing depositional settings through application of the aforementioned processes and parameters, there always is some uncertainty regarding the definition of the input parameters, as many bounding conditions are not exactly known and data coverage is usually limited. This applies especially to source-rock models in frontier areas. However, the process-based approach in general, and the possibility to test the complexity and interactions of different parameters and processes in particular, enables a much more holistic and extensive evaluation of a source-rock unit, even in cases of very limited well control.

The case study from the Late Jurassic Spekk Formation of the mid-Norwegian Shelf presented here documents that OF-Mod is able to reproduce realistically the environmental conditions during deposition of the Spekk Formation and to adequately reproduce the subsurface well data. Also the local conditions during deposition of the lower 'rich' Spekk Formation can be accounted for with processes and modules included in OF-Mod. The modelling allowed further identification of the most important controlling factors: such as, in the lower Spekk Formation the combination of increased algal input and increased/variable preservation conditions, whereas in the more uniform upper Spekk Formation, the less pronounced preservation conditions most decisively influencing the OC content and quality.

ACKNOWLEDGEMENTS

The Norwegian Research Council, ENI and ConocoPhillips supported the development of the OF-Mod 2D software and this case study of the Late Jurassic mid-Norwegian Shelf. ENI, ConocoPhillips and Total provided further financial support for the development of the OF-Mod 3D version. IFP is gratefully acknowledged for providing the synthetic basin-fill model from DIONISOS, on which some of the shown example plots are based. We would like to thank Robert Drysdale for his careful review of our English. Johannes Wendebourg, Sadat and Dave Waltham are gratefully acknowledged for their constructive contributions to an earlier version of the manuscript.

NOTE

1. The kitchen area is the part of a basin where the source rock generates hydrocarbons (i.e., the parts of the basin where the source rock is in a respective depth and temperature regime). In the case where a source rock shows lateral facies variability, then particularly the source rock type/quality of the area where hydrocarbon generation takes place is of importance for the type of hydrocarbons generated (i.e., oil prone and gas prone).

REFERENCES

Baldwin, B. and **Butler, C.O.** (1985) Compaction curves. *Am. Assoc. Petrol. Geol. Bull.*, **69(4)**, 622–626.

Barron, E.J., Fawcett, P.J. and **Peterson, W.H.** (1995) A 'simulation' of mid-Cretaceous climate. *Paleoceanography*, **10(5)**, 953–962.

Bergamaschi, B.A., Tsamakis, E., Keil, R.G., Eglinton, T.I., Montlucon, D.B. and **Hedges, J.I.** (1997) The effect of grain size and surface area on organic matter, lignin and carbohydrate concentration, and molecular compositions in Peru Margin sediments. *Geochim. Cosmochim. Acta*, **61**, 1247–1260.

Berger, W.H., Fischer, K., Lai, C. and **Wu, G.** (1987) *Ocean Productivity and Organic Carbon Flux. I. Overview and Maps of Primary Production and Export Production.* San Diego, Univ. California: SIO Reference, 87–30.

Berger, W.H., Smetacek, V. and **Wefer, G.** (1989) *Productivity of the Ocean: Past and Present.* New York, Wiley & Sons. 471 pp.

Bessereau, G., Guillocheau, F. and **Huc, A.Y.** (1995) Source rock occurrence in a sequence stratigraphic framework: The example of the Lias of the Paris Basin. In: *Paleogeography, Paleoclimate, and Source Rocks* (Ed. A.Y. Huc), *Am. Assoc. Petrol. Geol.*, **40**, 273–301.

Betts, J.N. and **Holland, H.D.** (1991) The oxygen content of ocean bottom waters, the burial efficiency of organic carbon, and the regulation of atmospheric oxygen. *Palaeogeogr. Palaeoclimatol. Palaeoecol.*, **97**, 5–18.

Betzer, P.R., Showers, W.J., Laws, E.A., Winn, C.D., di Tullio, G.R. and **Kroopnick, P.M.** (1984) Primary productivity and particle fluxes on a transect to the equator at 153° W in the Pacific Ocean. *Deep-Sea Res.*, **31(1)**, 1–11.

Bralower, T.J. and **Thierstein, H.R.** (1984) Low productivity in slow deep-water circulation in Mid-Cretaceous oceans. *Geology*, **12**, 614–618.

Brumsack, H.J. (1980) Geochemistry of Cretaceous black shales from the Atlantic Ocean (DSDP Legs 11, 14, 36 and 41). *Chem. Geol.*, **31**, 1–25.

Carpentier, B., Huc, A.Y. and **Bessereau, G.** (1991) Wireline logging and source rocks – Estimation of organic carbon content by the Carbolog method. *The Log Analyst*, **32(3)**, 279–297.

Carpentier, B., Huc, A.Y., Guilhaumou, N. and **Ramsey, M.H.** (1993) Geological and geochemical modeling, an approach for understanding organic cyclic sedimentation in evaporitic sequences. Application to the Mulhouse Basin (France). *Org. Geochem.*, **20(8)**, 1153–1164.

Dean, W.E. and **Gardner, J.V.** (1998) Pleistocene to Holocene contrasts in organic matter production and preservation on the California continental margin. *Geol. Soc. Am. Bull.*, **110(7)**, 888–899.

Degens, E.T., Emeis, K.C., Mycke, B. and **Wiesner, M.G.** (1986) Turbidities: The principle mechanism yielding black shales in the early deep Atlantic Ocean. *Geol. Soc. Special Publ.*, **21**, 361–376.

Gradstein, F.M., Agterberg, F.P., Ogg, J.G., Hardenbol, J. and **Backstrom, S.** (1999) On the Cretaceous time scale. *Neues Jahrbuch Geol. Palaeontol. Abhandlungen*, **212 (1–3)**, 3–14.

Granjeon, D., Joseph, P., Assier-Rzadkiewicz, S. and **Bassant, P.** (1999) Application of 3D fluvial and turbiditic sediment transport laws in stratigraphic modelling of siliciclastic and carbonate formations. In: *Proceedings of IAMG '99 – The 5th Annual Conference of the International Association for Mathematical Geology* (Eds S.J. Lippard, A. Næss and R. Sinding-Larsen). IAMG, Trondheim, 515–520.

Griffith, C.M. and **Dyt, C.** (2001) Six years of Sedsim exploration applications. *Am. Assoc. Petrol. Geol. Annual Convention*, A 75.

Handoh, I.C., Bigg, G.R. and **Jones, E.J.W.** (2003) Evolution of upwelling in the Atlantic Ocean basin. *Palaeogeogr. Palaeoclimatol. Palaeoecol.*, **202**, 31–58.

Haq, B., Hardenbol, J. and **Veil, P.R.** (1987) Chronology of fluctuating sea levels since the Triassic. *Science*, **235**, 1156–1167.

Hartnett, H.E. and **Devol, A.H.** (2003) Role of a strong oxygen-deficient zone in the preservation and degradation of organic matter: A carbon budget for the continental margins of northwest Mexico and Washington State. *Geochim. Cosmochim. Acta*, **67(2)**, 247–264.

Hartnett, H.E., Keil, R.G., Hedges, J.I. and **Devol, A.H.** (1998) Influence of oxygen exposure time on organic carbon preservation in continental margin sediments. *Nature*, **391**, 572–574.

Hedges, J.I. and **Keil, R.G.** (1995) Sedimentary organic matter preservation: An assessment and speculative synthesis. *Marine Chem.*, **49**, 81–115.

Hedges, J.I. and **Oades, J.M.** (1997) Comparative organic geochemistry of soils and marine sediments. *Org. Geochem.*, **27(7/8)**, 319–361.

Henrichs, S.M. and **Reeburgh, W.S.** (1987) Anaerobic mineralization of marine sediment organic matter: Rates and the role of Anaerobic processes in the oceanic carbon economy. *Geomicrobiol. J.*, **5(3/4)**, 191–237.

Huessner, H., Roessler, J., Betzler, C., Petschick, R. and **Peinl, M.** (2001) Testing 3D computer simulation od carbonate platform groth with REPRO: The Miocene Llucmajor carbonate platform (Mallorca). *Palaeogeogr. Palaeoclimatol. Palaeoecol.*, **175**, 239–247.

Huc, A.Y., Lallier-Verges, E., Bertrand, P., Carpentier, B. and **Hollander, D.J.** (1992) Organic matter response to change of depositional environment in Kimmeridgian shales, Dorset, U.K. In: *Organic Matter: Productivity Accumulation, and Preservation in Recent and Ancient Sediments* (Eds J.K. Whealan and J.W. Farrington), Columbia University Press, New York, pp. 469–486.

Ittekkot, V., Haake, B., Bartsch, M., Nair, R.R. and **Ramaswamy, V.** (1992) Organic carbon removal in the sea: The continental connection. In: Upwelling systems: evolution since the early Miocene (Eds C.P. Summerhayes, W.L. Prell and K.C. Emeis), *Geological Society Special Publication*, London, 167–176.

Johnson-Ibach, L.E. (1982) Relationship between sedimentation rate and total organic carbon content in ancient marine sediments. *Am. Assoc. Petrol. Geol. Bull.*, **66(2)**, 170–188.

Katz, B.J. and **Pratt, L. M.** (1993) Source rocks in a sequence stratigraphic framework. *Am. Assoc. Petrol. Geol. Stud. Geol.*, **37**, 247 pp.

Keil, R.G., Tsamakis, E., Fuh, B.C., Giddings, J.C. and **Hedges, J.I.** (1994) Mineralogical and textural controls on the organic composition of coastal marine sediments: Hydrodynamic seperation using SPLITT-fractionation. *Geochim. Cosmochim. Acta*, **58(2)**, 879–893.

Keil, R.G., Tsamakis, E., Giddings, J.C. and **Hedges, J.I.** (1998) Biochemical distribution (amino acids, neutral sugars, and lignin phenols) among size-classes of modern marine sediments from the Washington coast. *Geochim. Cosmochim. Acta*, **62(8)**, 1347–1364.

Kjennerud, T. and **Gillmore, G.K.** (2001) Reconstruction of late Jurassic palaeobathymetry in the Froan Basin for use in organic facies modelling. Sintef Report 24.4476.00/03/01.

Knies, J. (1999) Late Quarternary paleoenvironment along the northern Barents and Kara seas continental margin: A multi parameter analysis. *Berichte zur Polarforschung*, **304**, 159.

Knies, J. and **Mann, U.** (2002) Depositional environment and source rock potential of Miocene strata from the central Fram Strait: Introduction of a new computing tool for simulation organic facies variations. *Marine Petrol. Geol.*, **19**, 811–828.

Knies, J. and **Stein, R.** (1998) New aspects of organic carbon deposition and its paleoceanographic implications along the northern Barents Sea margin during the lasr 30,000 years. *Paleoceanography*, **13(4)**, 384–394.

Knies, J., Hald, M., Ebbesen, H., Mann, U. and **Vogt, C.** (2003) A deglacial-middle Holocene record of biogenic sedimentation and paleoproductivity changes from the northern Norwegian continental shelf. *Paleoceanography*, **18(4)**, 1096.

Krokstad, W. and **Monteil, E.** (1995) Models of Late Jurassic organic sedimentation, with focus on the Mid-Norwegian Continental Shelf; Volume 1: Biostratigraphy. Sintef Report 23.2445.00/02/95.

Langrock, U. and **Stein, R.** (2004) Origin of marine petroleum source rocks from the Late Jurassic to Early Cretaceous Norwegian Greenland Seaway – Evidence for stagnation and upwelling. *Marine Petrol. Geol.*, **21**, 157–176.

Langrock, U., Stein, R., Lipinsky, M. and **Brumsack, H.-J.** (2003) Late Jurassic to Early Cretaceous black shale formation and paleoenvironment in high northern latitudes – examples from the Norwegian Greenland Seaway. *Paleoceanography*, **18(3)**, DOI: 10.1029/2002PA000867.

Lipinski, M., Warning, B. and **Brumsack, H.-J.** (2003) Trace metal signatures of Jurassic/Cretaceous black shales from the Norwegian and the Barents Sea. *Palaeogeogr. Palaeoclimatol. Palaeoecol.*, **190**, 459–475.

Littke, R., Baker, D.R., Leythaeser, D. and **Rullkötter, J.** (1991) Keys to the depositional history of the Posidonia Shale (Toarcian) in the Hils Syncline, northern Germany. In: *Modern and Ancient Continental Shelf Anoxia* (Eds R.V. Tyson and T.H. Pearson), *Geol. Soc. Special Publ.*, 311–333.

Littke, R., Baker, D.R. and **Rullkoetter, J.** (1997) Deposition of petroleum source rocks. In: *Petroleum and Basin Evolution* (Eds D.H. Welte, B. Horsfield and D.R. Baker), Springer Verlag, Berlin, pp. 273–333.

Mallick, R.K. and **Raju, S.V.** (1995) Application of wireline logs in characterization and evaluation of generation potential of Paleocene-Lower Eocene source rocks in parts of Upper Assam Basin, India. *The Log Analyst* (May–June 1995), 49–63.

Mann, U. and **Stein, R.** (1997) Organic facies variations, source rock potential, and sea level changes in Cretaceous black shales of the Quebrada Ocal, Upper Magdalena Valley, Colombia. *Am. Assoc. Petrol. Geol. Bull.*, **81(4)**, 556–576.

Mann, U. and **Zweigel, J.** (2002a) Organic facies modelling (OF-Mod): Documentation of the simulation program. Part I, revised version. *Sintef Report* 24.4476.00/01/02.

Mann, U. and **Zweigel, J.** (2002b) Organic facies modelling of the Late Jurassic Spekk formation. *Sintef Report* 24.4507.00/01/02.

Mayer, L.M. (1994) Surface area control of organic carbon accumulation in continental shelf sediments. *Geochim. Cosmochim. Acta*, **58**, 1271–1284.

Monteil, E. (1997) Models of Late Jurassic Organic Sedimentation, with Focus on the Mid-Norwegian Continental Shelf. Vol.6: Palynofacies analysis and palaeoenvironmental interpretation. Final report. Sintef Report 23.2445.00/04/97.

Müller, P.J. and **Suess, E.** (1979) Productivity, sedimentation rate, and sedimentary organic matter in the oceans.- I.Organic matter preservation. *Deep-Sea Res.*, **26A**, 1347–1362.

Nordlund, U. (1996) Formalizing geological knowledge – With an example of modeling stratigraphy using fuzzy logic. *J. Sediment. Res.*, **66**, 689–698.

Parrish, J.T. (1995) Paleogeography of Corg-rich rocks and the preservation versus production controversy. In: *Paleogeography, Paleoclimate, and Source Rocks* (Ed. A.Y. Huc), *Am. Assoc. Petrol. Geol. Stud. Geol.*, **40**, 1–20.

Pasley, M.A., **Riley, G.W.** and **Nummedal, D.** (1993) Sequence stratigraphic significance of organic matter variations: Example from the Upper Cretaceous Mancos Shale of the San Juan Basin, New Mexico. In: *Source Rocks in a Sequence Stratigraphic Framework* (Eds B.J. Katz and L.M. Pratt), *Am. Assoc. Petrol. Geol. Stud. Geol.*, Tulsa, 221–241.

Passey, Q.R., **Creaney, S.**, **Kulla, J.B.**, **Moretti, F.J.** and **Stroud, J.D.** (1990) A practicle model for organic richness from porosity and resistivity logs. *Am. Assoc. Petrol. Geol. Bull.*, **74(12)**, 1777–1794.

Pedersen, T.F. and **Calvert, S.E.** (1990) Anoxia vs. productivity: What controls the formation of organic-carbon-rich sediments and sedimentary rocks? *Am. Assoc. Petrol. Geol. Bull.*, **74(4)**, 454–466.

Prahl, F.G., **Ertel, J.R.**, **Goni, M.A.**, **Sparrow, M.A.** and **Eversmeyer, B.** (1994) Terrestrial organic carbon contributions to sediments on the Washington margin. *Geochim. Cosmochim. Acta*, **58(14)**, 3035–3048.

Rachold, V. and **Hubberten, H.W.** (1999) Carbon isotope composition of particulate organic material of east Siberian rivers. In: *Land-Ocean Systems in the Siberian Arctic: Dynamics and History* (Eds H. Kassens, H.A. Bauch, I. Dmitrenko, H. Eiken, H.W. Hubberten, M. Melles, J. Thiede and L. Timokhov), Springer Verlag, pp. 223–238.

Ransom, B., **Kim, D.**, **Kastner, M.** and **Wainwright, S.** (1998) Organic matter preservationon continental slopes: Importance of mineralogy and surface area. *Geochimica et Cosmochimica Acta*, **62(8)**, 1329–1345.

Ringås, J.E., **Grading, M.** and **Leith, D.A.** (1997) Models of Late Jurassic organic sedimentation, with focus on the Mod-Norwegian continental shelf, Volume 2: Sequence stratigraphic evolution, Kimmeridgian to Ryazanian. *IKU-Report* 23.2445.00/00/09/97.

Rivenæs, J.C. (1993) *A Computer Simulation Model for Siliciclastic Basin Stratigraphy.* Doktor Ingeniør Thesis, The Norwegian Institute of Technology, University of Trondheim.

Sarnthein, M., **Winn, K.**, **Duplessy, J.C.** and **Fontugne, M.R.** (1988) Global variations of surface ocean productivity in low and mid latitudes: Influence on CO2 reservoirs of the deep ocean and atmosphere during the last 21000 years. *Paleoceanography*, 3, 361–399.

Schluenz, B. and **Schneider, R.** (2000) Transport of terrestrial organic carbon to the oceans by rivers: Re-estimating flux and burial rates. *International Journal of Earth Science*, **88**, 599–606.

Schwarzkopf, T.A. (1993) Model for prediction of organic carbon content in possible source rocks. *Marine Petroleum Geol.*, **10**, 478–491.

Sellwood, B.J., **Valdes, J.P.** and **Price, G.D.** (2000) Geological evaluation of multiple general circulation model simulations of Late Jurassic palaeoclimate. *Palaeogeogr. Palaeoclimatol. Palaeoecol.*, **156(1–2)**, 147–160.

Showers, W.J. and **Angle, D.G.** (1986) Stable isotopic characterization of organic carbon accumulation on the Amazon continental shelf. *Continental Shelf Res.*, **6(1/2)**, 227–244.

Skarbø, O., **Bakke, S.**, **Jacobsen, T.**, **Krokstad, W.**, **Lundschien, B.**, **Myhr, M.B.**, **Rise, L.**, **Schou, L.**, **Smelror, M.**, **Verdenius, J.**, **Vigran, J.** and **Århus, N.** (1988) Shallow drilling off Møre-Trøndelag 1988 Main Report. *Sintef Report* 21.3434.00/03/88.

Steckler, M.S., **Swift, D.J.P.**, **Syvitski, J.P.**, **Goff, J.A.** and **Niedoroda, A.W.** (1996) Modeling sedimentology and stratigraphy of continental margins. *Oceanography*, **9(3)**, 183–188.

Stein, R. (1986) Organic carbon and sedimentation rate – Further evidence for anoxic deep-water conditions in the Cenomanian/Turonian Atlantic Ocean. *Marine Geol.*, **72**, 199–209.

Stein, R. (1991) Accumulation of organic carbon in marine sediments. *Lecture Notes in Earth Science*, Springer-Verlag, Berlin, 34, 217 pp.

Syvitski, J.P.M. and **Hutton, E.W.H.** (2001) 2D SEDFLUX 1.0C: An advanced process-response numerical model for the fill of marine sedimentary basins. *Comp. Geosci.*, **27**, 731–753.

Tetzlaff, D.M. (1989) SEDO: A simple clastic sedimentation program for use in training and education. In: Quantitative Dynamic Stratigraphy (Ed. T.A. Cross), Prentice-Hall, pp. 401–415.

Tyson, R.V. (1995) *Sedimentary Organic Matter: Organic Facies and Palynofacies.* Chapman and Hall, London, 615 pp.

Tyson, R.V. (1996) Sequence-stratigraphical interpretation of organic facies variations in marine siliclastic systems: General principles and application to the onshore Kimmeridge Clay Formation, UK. S. P. In: *Sequence Stratigraphy in British Geology* (Eds S.P. Hesselbo and D.N. Parkinson), *Geological Society Special Publication*, **103**, 75–96.

Tyson, R.V. (1997) Models of Late Jurassic organic sedimentation, with focus on the Mid-Norwegian continental shelf. Main Report. *Sintef Report* 23.2445.00/08/97.

Tyson, V. and **Follows, B.** (2000) Palynofacies prediction of distance from sediment source: A case study from the Upper Cretaceous of the Pyrenees. *Geology*, **28(6)**, 569–571.

Villanueva, J., **Grimalt, J.O.**, **Cortijo, E.**, **Vidal, L.** and **Labeyrie, L.** (1997) A biomarker approach to the organic matter deposited in the North Atlantic during the last climatic cycle. *Geochim. Cosmochim. Acta*, **61(21)**, 4633–4646.

Wagner, T. (2000) Control of organic carbon accumulation in the late Quaternary equatorial Atlantic (Ocean Drilling Program sites 664 and 663): Productivity versus terrigenous supply. *Paleoceanography*, **15(2)**, 181–199.

Warrlich, G., **Waltham, D.** and **Bosence, D.** (2001) Quantitative 3-D facies and sealevel history prediction from stratigraphic computer modeling. *Am. Assoc. Petrol. Geol. Annual Convention*, A 211.

Westerhausen, L., **Poynter, J.**, **Eglinton, G.**, **Erlenkeuser, H.** and **Sarnthein, M.** (1993) Marine and terrigenous origin of organic matter in modern sediments of the equatorial East Atlantic: The $\delta^{13}C$ and molecular record. *Deep-Sea Research*, **40(5)**, 1087–1121.

Wignall, P.B. (1991) Model for transgressive black shales? *Geology*, **19**, 167–170.

Spec. Publ. Int. Assoc. Sedimentol. (2008) **40**, 275–286

Spatial data templates: combining simple models of physical processes with stochastic noise to yield stable, archetypal landforms

PETER A. BURROUGH[1]

Department of Physical Geography, Faculty of Geosciences, Utrecht University, P.O. Box 80.115, 3508 TC, Utrecht, The Netherlands (E-mail: peter@unclogged.co.uk)

ABSTRACT

Francis Galton (1822–1911), one of the fathers of mathematical statistics, used a mechanical device known as the Quincunx as a template for simulating the mathematics of the normal distribution and its role in regression. This paper shows that with suitable modifications, a spatial version of the Quincunx template may be used for exploring commonalities in stochastic models of recognizable, generic landform types, such as alluvial fans and volcanoes. The procedures followed may increase understanding of the complex processes of landform generation.

Keywords Geocomputation, sedimentary landscape modelling, stochastic processes, Quincunx, cellular automata.

INTRODUCTION

Recently, there have been rapid increases in our ability to create analogue and digital models of landform development that elucidate the ways in which landscape change may occur. Analogue models often require the building of large physical structures, but numerical models that capture the essence of landscape change processes can be run relatively cheaply on personal computers. While detailed numerical models of large areas still need sizeable databases and powerful computing, useful insights into the dynamic processes operating in landscapes can be gleaned from modular, computational tool kits.

During the past 20 years many papers have reported various aspects of the mathematical modelling of hydrological and geomorphological processes. A recent review by Pike (2002) lists more than 6000 papers dealing with surface modelling published in the period 1900–2002. Much of this work has been stimulated by the need to improve our understanding of sedimentological or hydrological processes in terms of the interactions between the various agents responsible for landscape change, such as tectonic activity and relief (sources of potential energy) and the surface transport of sediment and fluids (Beven, 1996; Burrough, 1998; Harmon & Doe, 2001; Mitas & Mitasova, 2001; Clevis, 2003).

Improvements in computer technology have encouraged the development of more complex numerical runoff, erosion and sedimentation models. Increased availability of digital surface data has also spawned new mathematical models, and more applications, including studying the surfaces of extra-terrestrial bodies such as Mars and Titan.

One great advantage of numerical modelling compared with analogue modelling is that it allows complex phenomena such as landscape development to be studied at many different levels of spatial and temporal resolution (Casti, 1997; Stewart, 1997; Favis-Mortlock *et al.*, 2001). We are no longer constrained by our object of study being too large to comprehend, or too small to see, or changing too slowly. In theory, virtual landscapes

[1]Currently Visiting Professor of Geography, Oxford University Centre for the Environment, Oxford University, UK (OUCE).

can be developed, studied and displayed at any required level of resolution.

Today, the main constraints to the numerical modelling of erosion and sedimentation are neither computer power, nor the availability of data in the form of digital elevation models, but are to do with our lack of understanding of how essentially simple stochastic processes may combine with deterministic processes in time and space to produce complex, self-organized forms. It is interesting to note that of the 6000 papers listed by Pike (2002), fewer than 20 articles explicitly cite the concepts of stochastic modelling, and only some 80 consider the application of fractal concepts in the context of multiscale patterns and processes.

The main problems associated with modelling of dynamic processes such as erosion and sedimentation or the effects of tectonics on the shape and forms of landscapes include:

- incomplete knowledge – therefore using different data sets from the same area may lead to different results and conclusions;
- equifinality – different data sets and procedures may lead to similar results and therefore it is difficult to establish the main processes at work (Beven, 1996);
- landscapes are too large to study in a laboratory and therefore physical models (e.g. scale models) are incomplete replicas;
- it is difficult to obtain a complete picture of all the processes at work and the temporal and spatial scales at which they operate and interact – many models assume linear behaviour and ignore processes that are nonlinear and include a multiplicity of feedback loops, including considerable sensitivity to small differences in the values of the initial control parameters.

Even with the simplest models there may be many unknowns. The conceptual basis of the model will almost certainly be incomplete, different algorithms may yield different results (Jones, 2002), control parameters and boundary conditions may not be fully specified and will include errors and uncertainties, and finally, the data will include errors of definition, measurement and location. Of these, probably only the latter have been improved by recent technical developments such as GPS (global positioning systems). Some mathematical studies (e.g. Burrough & McDonnell, 1998; Heuvelink, 1998) have demonstrated the difficulties of applying classic error propagation through even quite simple linear, empirical, regression models. Karssenberg (2002) illustrates an alternative procedure using Monte Carlo methods, but this is still computationally demanding.

GENERIC TOOLS FOR MODELLING HYDROLOGICAL AND GEOMORPHOLOGICAL PROCESSES: THE SPATIAL DATA TEMPLATE

Parallel to model development, there have been considerable advances in the development and application of modular software packages, not only for geographical information systems (GIS) but also for statistics (Bivand & Gebthardt, 2000) and geostatistics (Pebesma & Wesseling, 1998), and for hydrological and geomorphological modelling (Gallant & Wilson, 1996; Burrough, 1998; Pullar, 2003). Implicit in the development and application of these tool kits is the underlying assumption that they will function in a similar way even when applied to data from many different scales, sources and even study areas. In short, there is a fundamental assumption of a common theoretical but usually unspecified base that links data, sampling, numerical procedures, algorithms and applications.

The assumption of a common theoretical base is well tested in statistics, but in geocomputation (the generic term for numerical landscape modelling) there have as yet been few investigations of the effects of choice of algorithm or data set or level of resolution on the results of numerical modelling of erosion or sedimentation. For example, Beven (1996) noted that virtually all hydrological analyses and modelling of surface forms start with some form of catchment topography – this is usually a digital elevation model. Until recently, however, few models have incorporated feedbacks between hydrological and geomorphological processes. This is attributed to the different nature of the disciplines and also to the differences in time scales over which observable processes operate. Favis-Mortlock *et al.* (2001) point out the cultural differences inherent in erosion modelling by different groups of workers in Europe, North America and the tropics. They also comment on the 'sobering conclusions of validations of erosion models using common data sets'.

For most study areas, it is rare for multiple data sets to have been acquired for the numerical modelling of geomorphological and hydrological processes; data may have been gathered by many different sampling techniques over a wide range of spatial and temporal scales, and may contain many outliers, all of which make it difficult to separate the different components of the resulting patterns (Burrough, 1983a, 1983b). Consequently, we may not be able unambiguously to disentangle patterns in data arising from different processes, nor to assess the contributions of non-linearity due to sensitivity to boundary conditions or feedback.

It would improve insight into the computing of spatial processes if we could develop a set of standard, but essentially unambiguous and uncomplicated data sets that have a known and well understood behaviour with respect to the processes being modelled. Given their spatial character, I call these initial data sets 'spatial data templates' (SDT).

A typical example SDT for a given section of landscape is the wetness index (Burrough & MacDonnell, 1998) that is computed from a raster digital elevation model (DEM). This SDT not only defines the area and the processes used to compute it, but forms a basic structure on which many other forms of spatial analysis are worked out. This SDT is a deterministic structure with well defined topological links. In reality, however, the SDT includes contributions from uncertainties such as elevation errors and effects of the choice of algorithm (Jones, 2002). When this is so, as it is for most natural landscapes, the simple SDT may underrepresent the sources of variation. It is also unlikely that the levels of uncertainty are uniform over the whole area.

In addition, change in natural landforms such as fans, deltas and volcanoes may also reflect positive and negative feedback loops. So, rather than attempting to model how any specific landscape may have been created deterministically following a single set of linear rules, the focus in this paper is on the way in which the combination of simple non-linear interactions, feedback loops and small levels of uncertainty may lead to characteristic, stable, recognizable and emergent landforms. The approach considers the uncertainties and errors in the initial surface, and the uncertainties in the processes as part of the feedback system that leads to the development of characteristic landforms, such as alluvial fans, braided rivers, deltas and volcanoes. In all cases the development of a stable landform is the result of several interacting components that may have different temporal and spatial scales (Burrough, 1983a, 1983b). The aim is to represent each factor by a simple SDT together with a very simple probabilistic model and to explore how they interact to create recognizable, self-generating forms.

Consider overland flow and the development of river networks, the underlying processes are well understood and there are several popular algorithms for deriving overland flow paths (e.g. Beven, 1996; Gallant & Wilson, 1996; Burrough, 1998; Burrough & MacDonnell, 1998; Jones, 1998). Most of these methods create a set of flow lines over a gridded DEM that follow the steepest downhill path from a given cell to one of its eight downhill neighbours on the grid. None of these algorithms immediately gives rise to dispersion patterns because algorithms that follow the steepest downhill path automatically cause drainage convergence. They do, however, create a version of a stable SDT. In raster GIS-based hydrological models, the grid-based algorithms usually confine the width of the flow channel to a single grid cell. Accumulating flow over the steepest downhill path leads to estimates of the size of the catchment upstream of a given cell (Fig. 1a) which can be expressed only for the cells located on the simple, deterministic single-cell flow path. Consider the increase in knowledge that results from adding a small amount of noise to the elevation surface of the SDT (Fig. 1b), repeating the computations for 100 or more realizations of the stochastic component of the process and displaying the joint results. With this approach the variations in estimated wetness index reflect a basic SDT component plus a stochastic contribution.

The stochastic approach to determining the variations in size and location of drainage channels automatically searches out the most likely flow paths, giving narrower or broader zones that reflect the probability of flow discharging through any given cell (Fig. 1b). Note that where the amplitude of surface noise is large with respect to the deterministic variations in elevation (the area of application is predominantly flat) the location of drainage channels may vary widely and stream avulsions are frequent. When global variations from the SDT dominate, the locations of drainage channels scarcely vary.

A SIMPLE, SPATIAL DATA TEMPLATE – THE QUINCUNX

The results shown in Fig. 1 reflect only the balance between a chosen spatial data template (SDT) of

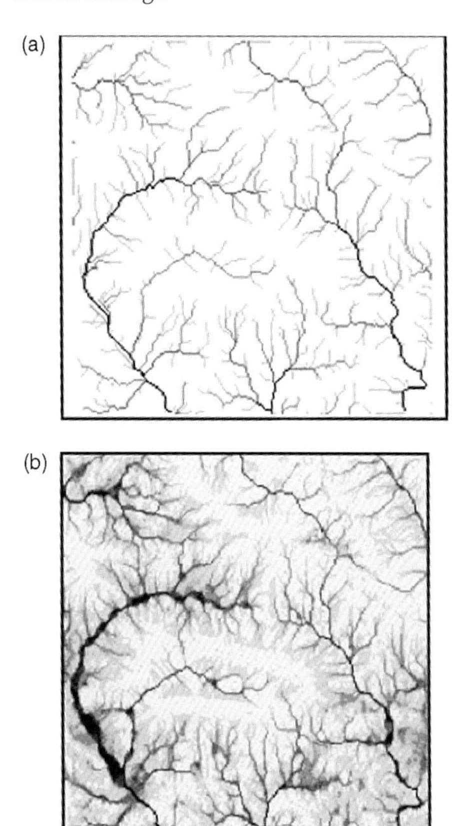

Fig. 1. (a) A spatial data template of wetness index derived from a gridded digital elevation model (DEM) for single grid cells, implying that the stream is only as wide as the grid cell: darkness of grey scale indicates magnitude of upstream catchment. (b) Stochastic modelling of wetness indices based on the spatial data template takes account of stochastic variations in elevation of DEM cells (pooled results of 100 realizations).

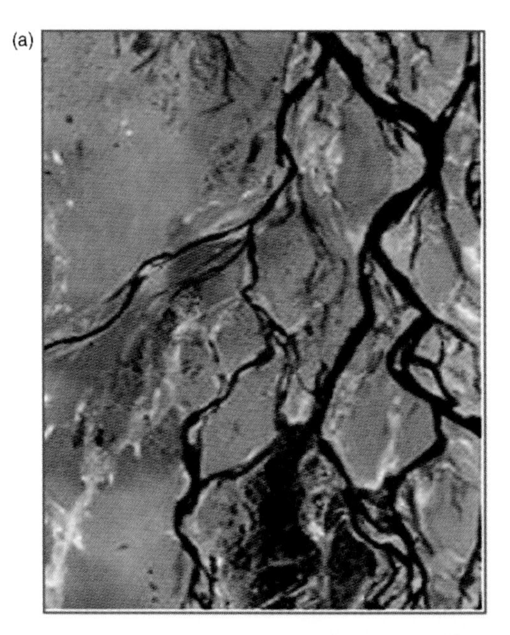

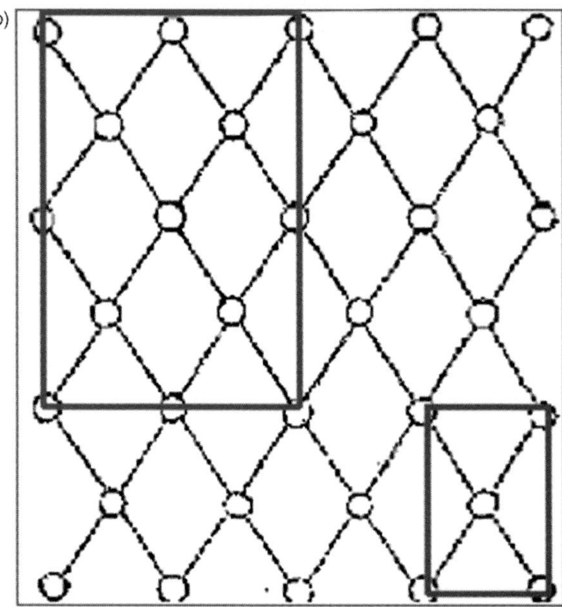

Fig. 2. (a) Braided-stream channels, showing bifurcations and avulsions over a range of scales. (b) The classic Quincunx used as a spatial data template for an alluvial fan, showing nested levels of bifurcation on a regular lattice.

the selected area plus a realistic, but rough estimate of noise. Now it would be useful to find a spatial landform model for which the SDT is generic, i.e., a landform model for which change in pattern is caused only by change in a set of defining parameters.

The first step in creating a generic model based on a 'representative' surface is to select a simple SDT that permits the derivation of a stochastically based drainage topology with the options to incorporate feedback based on random changes. One such model that can deal with convergent and divergent flow paths and multiple scales of Fig. 2a is given in Fig. 2b by the 'Quincunx'. Francis Galton (1822–1911) used the Quincunx (Fig. 3) as a mathematical tool (i.e., a computer) to generate a probabilistic, but physical model of the normal distribution and demonstrate its role in regression (Stewart, 1997). The Quincunx, or Galton's Board, is a device that allows a ball to drop through an array of n rows of nails stuck in an inclined surface. The nails are equally spaced both horizontally and vertically and when the ball hits a nail it is equally likely to fall to the left or the right. The ball then lands on a nail in the next row where the process is repeated, until after passing through all rows it is collected in a bucket at the bottom. For each nail on a Galton's Board (Quincunx), the probabilities of a solid particle following the left-hand or right-hand

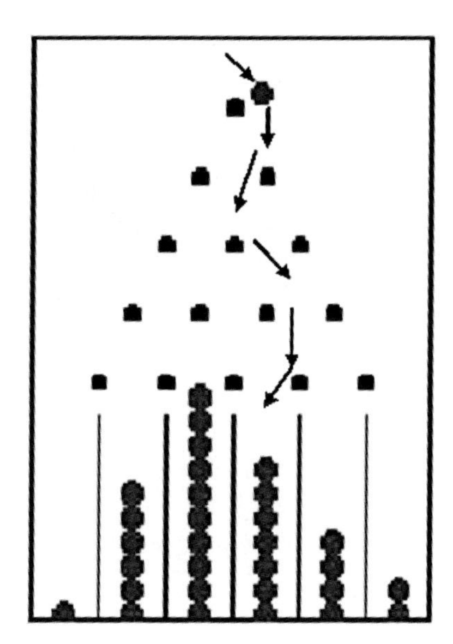

Fig. 3. Galton's Board or pin table as a template for simulating the Poisson distribution. At each pin, a falling ball has a 50–50 chance of falling to left or right. Buckets at the bottom of the board indicate the most likely routes. Given sufficient realizations, the distribution of balls per bucket yields a normal distribution.

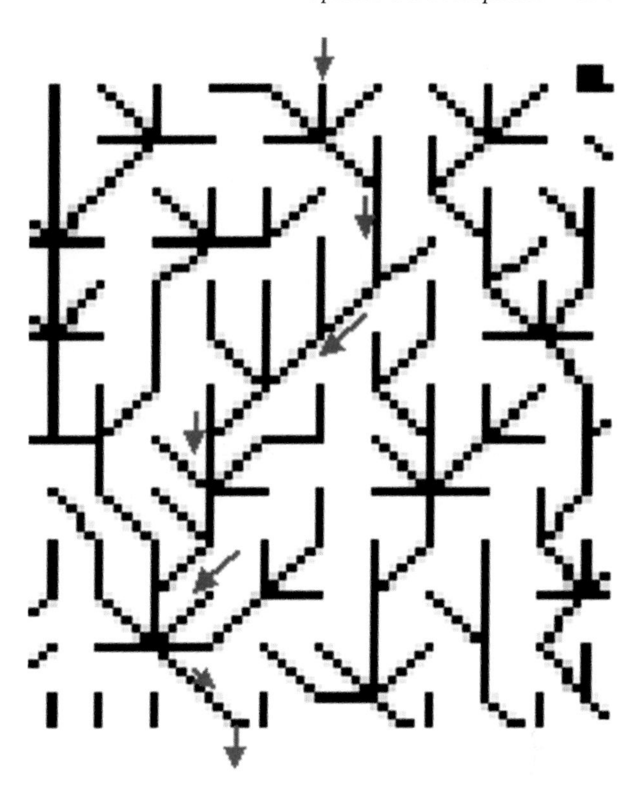

Fig. 4. A single realization of a computer simulation of the path of a cellular automaton spatial data template that yields connected flow paths over a tilted regular grid surface. Each iteration, the surface is coloured with low level random noise to mimic the binomial action of the Quincunx. The arrow points to the input cell at the top and the lines give the connections via the steepest downhill descent to the next row of 'pins'. Every realization of this process yields a separate, independent, and unique set of the flow paths.

path around any given nail are equal and sum to 1.

$$Pr_{\text{left}} = Pr_{\text{right}} = 0.5 \tag{1}$$

The probability of a ball following any particular path through the Quincunx, however, depends on the cumulative probability summed over all n rows. This cumulative probability can be more or less than 0.5, depending on the cumulative probabilities that have accrued over all rows. For example, the probability of a ball reaching any given pin depends on the route it takes through the rows (Fig. 3). The probability of a ball reaching either of the two pins on the second row is 0.5; the probability for a ball reaching each of the three pins on the third row is 0.25, 0.5 and 0.25; the probability of a ball reaching each of the four pins on the fourth row is 0.125, 0.375, 0.375, 0.125, and so on.

After sufficient balls have traversed the board, their cumulative distribution over the buckets provides a realization of the binomial distribution.

Because each row in the Galton's Board is a separate experiment, the piles of balls in the buckets represent the sum of n random variables. This illustrates the central-limit theorem, which states that the distribution of the sum of n random variables approaches the normal distribution when n is large. The more rows of nails on the board, the better the approximation to the normal distribution.

Although Fig. 4 shows a single possible path of simple convergent flow over a connected SDT surface, many paths are possible. Therefore a probabilistic flow regime can be generated by repeating the experiment many times and recording the results, as shown by Fig. 1b. Each repeat of the experiment computes and updates the probability that any given path will be traversed by a ball. The joint probability surface defines the Poisson distribution, which is also the surface reached by the falling stones.

A CELLULAR AUTOMATON MODEL OF THE QUINCUNX

If the Quincunx is laid out in a raster GIS (such as PCRaster – Wesseling *et al.*, 1996) it can easily be programmed as a cellular automaton (Wolfram, 2004). In this context, the cellular automaton implies that for every iteration of the model, the value of a given grid cell is computed from values of the surrounding cells, where the derivation may follow a simple mathematical relationship. This provides a feedback mechanism whereby new, local values are derived purely by local interactions. The whole set of these interactions may lead to self-organizing patterns that are capable of having a physical interpretation (Wolfram, 2004).

USING THE QUINCUNX TO MODEL EROSION AND SEDIMENTATION

The basic Quincunx has several attractive properties as a simple model of sedimentation. It reproduces the divergence seen in nature, it has a balanced binomial reaction to stochastic inputs, it has a credible statistical pedigree and incorporates fractal behaviour in that it may reflect many levels of spatial resolution. The following discussion investigates Quincunx as a prototype model of fan building and explores what additions need to be made to it in order to be able to successfully simulate the self-organizing dynamic processes of fan and delta development.

A purely stochastic Quincunx returns realizations that are independent and that contain no feedback: as there is no accumulation of material on the surface of the board, each realization starts out from the same premise. In reality, this will not be so, because in a landscape the changes brought about by one realization may shape the situation for the next (Tucker & Bras, 2000; Tucker *et al.*, 2001). On a real fan, material may be left behind, or knocked into motion, so that the flow path becomes modified with every realization. Adding a small, but different forcing level of random noise to the surface between one run and the next will modify the situation further because the accumulation of serial results causes differences through feedback.

Because of its simplicity, the Quincunx is really a very stupid model in the sense that it only reacts to unrelated events and has no intelligence. Intelligence is defined as the capability to retain an ordered set of information about the processes that have been interacting with it. Through feedback, all landscapes retain a memory of the past processes that have shaped them – if they did not there would be no need for sciences such as geology, hydrology and geomorphology. Clearly any realistic model of fan formation must include a memory of the way a channel might become blocked or eroded. The incorporation of 'memory' in the 'process' may strongly affect the probabilities of a given unit of sediment arriving at any given location on which material is deposited.

The resulting, limiting properties of a simple Quincunx can be summarized as follows.

1. All balls entering the top of the Quincunx are the same size and weight – this means that for every cycle of the model, each package of sediment or falling stone delivered to the fan is the same, which is clearly unrealistic.
2. Each left–right binary switch is equally likely: this implies that there are no biased irregularities in the surface that will consistently nudge the sediment package to the right or left of the flow. Again, in reality this may not be realistic due to anisotropy in the underlying surface caused by geological forces (see Holmes' Fig. 17.35; Duff, 2002).
3. There is no interaction between the balls and the pins, which means that flow cannot be blocked, neither can new channels be created.
4. Each impact between a ball and pin is independent of all other interactions between all other balls and pins – this means that the path of one ball cannot influence others.
5. The inertia of the balls is not taken into account.
6. All balls leave the system. There is no memory of the paths taken, except in the probability distribution contained in the final buckets.

While a stupid Quincunx must have the above properties to be unbiased, each of the above limiting properties provides a way in which the ideal Quincunx can be modified to match real events. For example, it is unlikely that all particles falling on an alluvial fan will be of the same size and weight or that the sediment load of a discharging river is always the same. Local factors, such as rock shoulders or easily eroded soil may influence the interactions between a travelling parcel of sediment or a rock and the substrate over which it is falling. In reality there is even the chance that a rock falling onto a surface may break into smaller stones or even come to rest on the fan where it may interfere with the trajectories of subsequent stones. All these factors can be seen as essentially stochastic and they enhance the role of chance in the development of the fan.

The simple Quincunx implies that the inertia of the balls plays no part, while in nature, the kinetic energy of stones or a flood peak may force continuation of the flow along a direction that is other than a simple switching of direction. The alluvial deposits left behind by material that has been transported downstream are the landscape's memory of what has passed along that route.

The following discussion addresses what results when extra, but simple, components such as the following are added to the original Quincunx SDT.

1. Record the cumulative paths taken by all particles (adding 'memories'). This includes allowing a particle to get blocked on the Quincunx surface (deposition) or to be knocked into motion (erosion).
2. Modify the transport direction probabilities as may be the case with diffusion (Freeman, 1991) so that at each nail three or more flow paths are possible.

Add serial correlation to the paths by treating the trajectory of the falling particle as following a Markov chain. This illustrates how the inertia of a particle describes its momentum in a given stochastically determined direction (diffusion), thereby increasing its chance of continuing in that direction. In addition, let the sizes and inertia of stones or loads of sediment vary for each realization.

SOME SIMPLE EXPERIMENTS OF FAN GENERATION AND THEIR RESULTS

Alluvial fans

Experiments embodying simple modifications of the Quincunx were implemented using the PCRaster dynamic modelling language, developed by the Department of Physical Geography, Utrecht University (Wesseling *et al.*, 1996). PCRaster is a raster-based GIS and dynamic modelling package that permits numerical, iterative models to incorporate both stochastic processes and feedback loops. Tables 1 and 2 are examples of simple simulation models written in PCRaster to demonstrate how the language is used to generate and report a time series of spatial events.

The PCRaster codes used for the generation of a fan are presented in Tables 1 and 2. Figure 5 presents the results of very simple fan-generation experiments which combine the aforementioned processes. Except for the display in Fig. 5a of the paths taken by five realizations of the binary model without serial correlation or memory, each model is iterated for 2000 times (realizations) and the results are added cumulatively to provide the memory of the processes. Each realization includes a new, but low level of surface noise, to randomize the initial conditions.

Table 1. Listing of PCRaster model for a simple Quincunx.

```
#QC1.mod
#Simple Quincunx model of a stochastic fan (Poisson)

binding
  #Inputs
  Pin = Pin1.msk;   #Top pin for input of balls
  Tc1 = Tc1.msk;    #Accumulator buckets for stones
  Qcm = Qc1.msk;    #Mask to seed random routes

#Outputs
  Qcflux = Qcflux1;   #Cumulative numbers of stones
                      in buckets

areamap
  Qc1.msk;

timer
  1 2000 1;
  rep1 = 1,1 + 1 .. endtime;
  #rep1 = endtime;

initial
  Qccumx1=0;

dynamic
  Randomld = ldd(if(uniform(Qcm) le 0.5, 1, 3));
  Ldpath = lddrepair(Randomld);
  Qcflux = accufractionflux(Ldpath,Pin,Tc1);
  report(rep1) Qccumx1 = Qccumx1+Qcflux;
  report(rep1) LQCflux = ln(Qccumx1);
```

In these experiments, the probabilities of the flow paths have been modified in three ways. First (Fig. 5a), the simple binary switch (simple Quincunx) has been used to generate five alternative scenarios. The cumulative results for the simple Quincunx (not shown) yield a broad, shallow fan that spreads out over the whole width of the modelling area. Avulsions (channel switching) may occur at every step.

Secondly, Fig. 5b shows the result of controlled diffusion plus weak random surface noise for five directions at each node (strong left, moderate left, straight on, moderate right, strong right). The weight given to 'straight ahead' dominates the resulting pattern, and avulsions are less frequent than in Fig. 5a. As illustrated (and expected) this concentrates the sediment deposition in the middle of the area.

The third experiment (Fig. 5c) computes the post-node transport direction as the sum of the

Table 2. PCRaster script for simulating 2000 realizations of alluvial fan development including five diffusion directions from each cell, plus serial correlation (Markov chain) to model forward momentum of sediment flow. Feedback loops in the dynamic section of the model script describe the interactions between a simple Quincunx SDT and a correlated, random noise surface (Rnds2 and Rnds3).

```
#QC3COR.mod
  #Simple ternary Quincunx model of stochastic fan (Poisson)
  # with serial correlation and 5 weighted directions

binding
  #Inputs
  Pin = Pin1.msk;   #Top pin for input of balls
  Tc1 = Tc1.msk;   #Accumulator buckets for stones
  Qcm = Qc1.msk;   #Mask to seed random routes
  B1 = 0.49;   #Boundaries for flow directions
  B2 = 0.51;
  B4 = 0.350;
  B6 = 0.650;
  Ac1 = 0.995;   #Flow direction correlation parameter
  #Outputs
  Qccumx3 = Qccum3z0;   #Cumulative numbers of
                          stones in buckets
  Qcflux3 = Qcflx3z0;   #Flow paths of balls

areamap
  Qc1.msk;

timer
  1 2000 1;
# rep1 = 1,10+10..endtime;
# rep1 = 1,1+1..endtime;
  rep1=endtime;
initial
  Qccumx3=0;
  #Rnds3=uniform(Qcm);  # Initial randomization
                          of directions
  Rnds3=.5;
dynamic
  Rld = Rnds3;
  Randomld=ldd(if(Rld le B4,4, if(Rld le B1, 1,
  if(Rld le B2, 2, if(Rld le B6,3,6)))));
  Ldpath=lddrepair(Randomld);
  Qcstate,Qcflux= accufractionstate,
    accufractionflux(Ldpath,Pin,Tc1);
  report(rep1) Qcflux3 = Qcflux;
  report(rep1) Qccumx3 = Qccumx3+Qcstate;

Rnds2=uniform(Qcm); #noise
Rnds3=(Ac1*Rnds3+(1-Ac1)*Rnds2);  #correlated
                                    randomness
```

strong nodal diffusion of the previous experiment and the serial autocorrelation (Markov chain) with the pre-node transport direction. The stronger the autocorrelation, the less likely that the flow path will be affected by channel-direction switching at the node. This is illustrated by Fig. 5d, which displays flow patterns in Siberian peat lands to relate the strong autocorrelation models to reality. All three simulations in Fig. 5 were modelled over a tilted, but planar surface.

Figure 6 provides a view of a simulated fan in which the deposition probabilities have been enhanced by a break of slope (less steep) near the lower edge of the simulation. The flow of 'stones' is also constrained in a channel to mimic conditions in the Scottish mountain scene. This illustrates another feature of the models in which the form and slope angle of the underlying template is also seen to modify the locations and forms of the fans generated.

Simulating a volcano

To illustrate how changing the parameters of the underlying SDT controls the form of the simulated fan, consider the form of a volcano. For example, the SDT of a volcano can be thought of as a large, circular fan. Material (ash, magma) is ejected through a pipe and follows gravitational forces which distribute material away from the vent. Once deposited, the magma/basalt flows harden and provide a feedback barrier for the next lava flow, which must take another route. If the process is iterated a sufficient number of times, a realistic model of a volcano develops spontaneously (Figs 7 and 8) along the lines of the drawings of volcanoes (Holmes' diagrams in Figs 13.3 and 13.5 in Duff, 2002).

Figure 8 shows the results of modelling a volcano in this way, using PCRaster code that is similar to that for fans, with the addition of a circular DEM Quincunx template instead of an inclined slope. The PCRaster code used for modelling the volcano is presented in Table 3. The results are remarkable and are at least as convincing as Holmes' diagrams, providing results that suggest the simple stochastic feedback model may go a long way to providing sufficient understanding of volcano building for basic studies in quantitative landform generation.

DISCUSSION AND CONCLUSIONS

Stochastic processes of dispersion (embodied as sets of most probable paths over a Quincunx or similar SDT) go a long way towards providing a stochastic, but simple theory of the generation of archetypal landforms, namely channel belts, alluvial fans and volcanoes. The ability to write the model in a dynamic generic computer

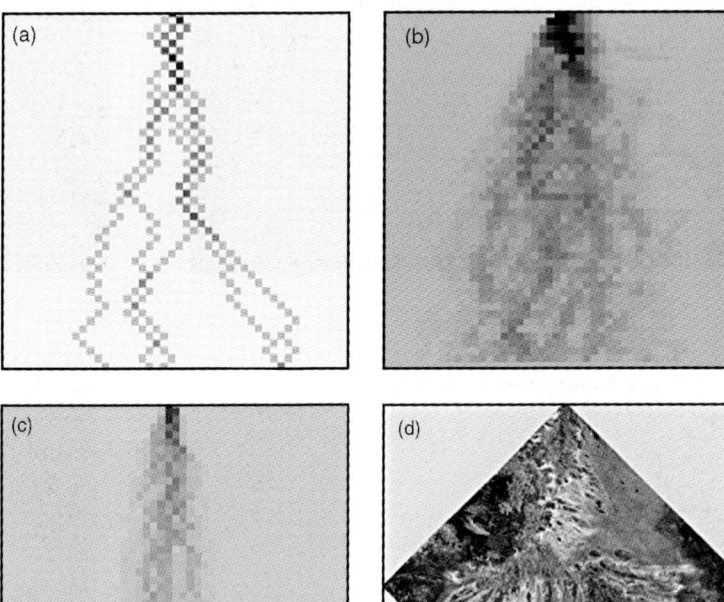

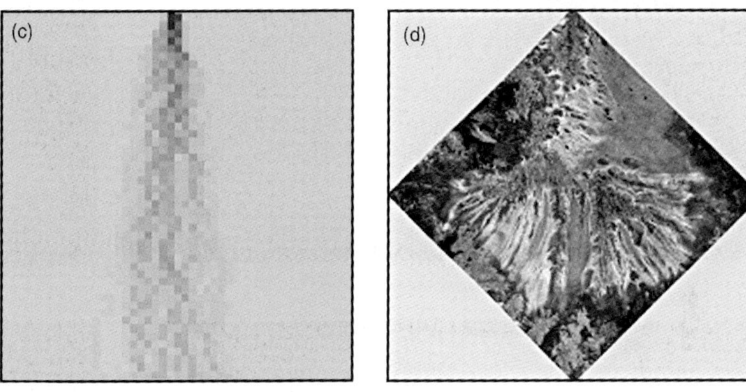

Fig. 5. Fan simulation with binary Quincunx and feedback memory (flow from top to bottom of figure). (a) Five independent realizations of simulated paths according to simple Quincunx with a record of paths traversed. (b) As in (a) but cumulative results with strong serial correlation. Sediment tends to converge and remain on chosen, random locations. (c) Serial Markov chain with diffuse discharge from point source: fan tends to converge. (d) Delta fan in Siberian peat wilderness exhibiting strongly autocorrelated flow and large dispersion of flow paths. (Courtesy W. Bleuten.)

Fig. 6. Stochastic model of rock fan created by stones falling randomly down a gulley. The abrupt change of slope near the bottom of the hillside slows the stones and initiates fan development, both on the mountain and in the Quincunx (top).

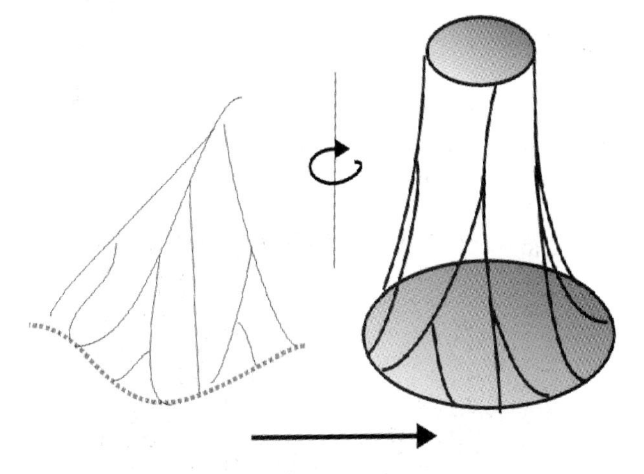

Fig. 7. Transforming the spatial data template (SDT) of a fan to a cone requires only that it to be rolled up around the z axis so that the probabilities of magma being ejected around the vent are spread equally.

language that uses the concepts of geomorphology, hydrology and physical geology makes the model-building exercise and the addition of extra modules within the reach of all, providing insight into how combinations of simple process can yield self-organizing archetypal forms.

The dynamic, probabilistic models also suggest that far from increasing instability in

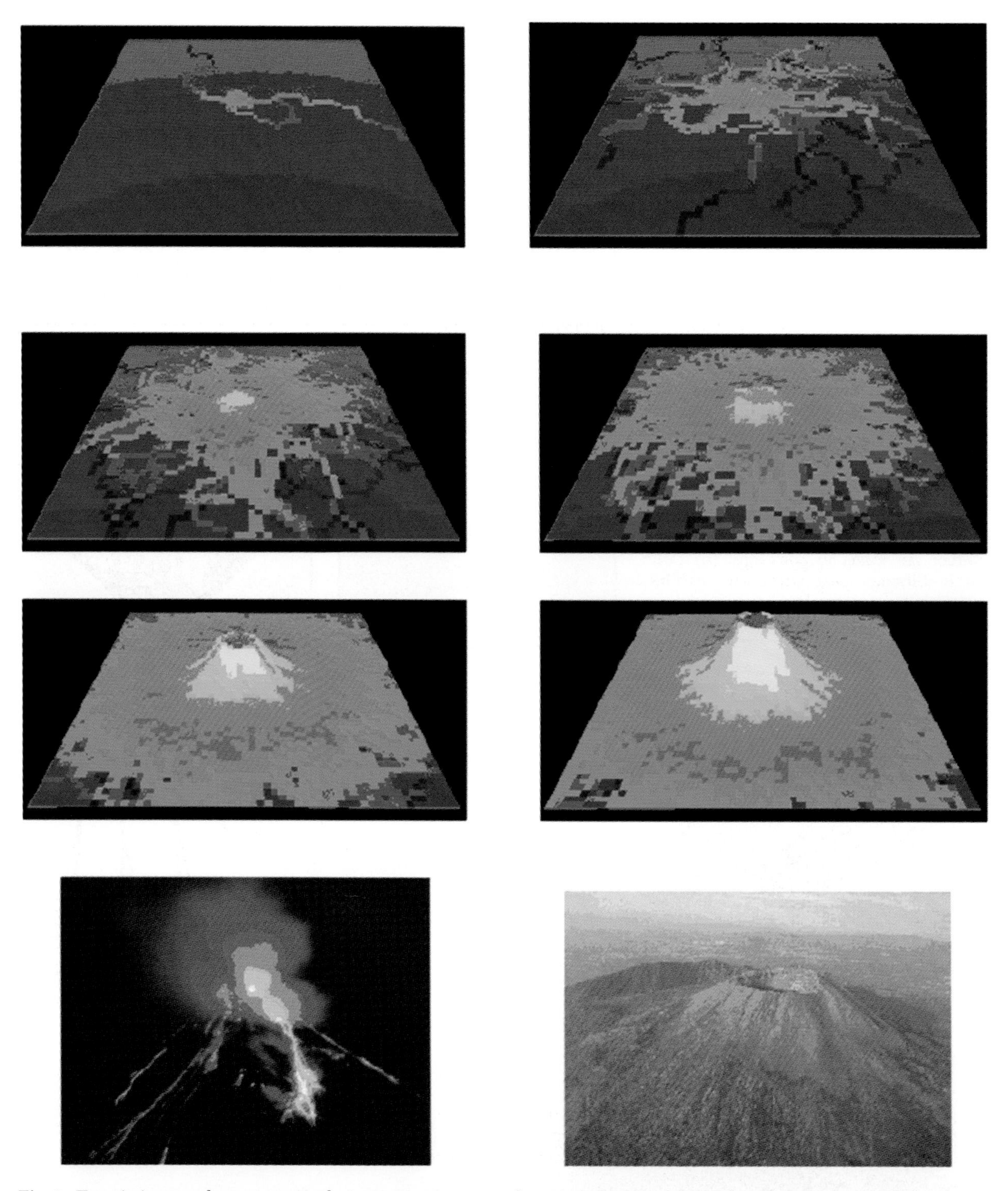

Fig. 8. Top six images show respectively 1, 10, 50, 100, 250 and 400 iterations in the simulation and growth of the volcanic cone based on random paths taken by the lava. Each outflow of lava may modify the path taken by the subsequent flow. This process shows a strong autocorrelation, as seen in bottom images. (Photograph bottom right taken by Raymond Sluiter, Utrecht.)

Table 3. PCRaster code for volcano simulation.

```
# VOLCANO1.MOD - model for simple volcano formation

binding
  # inputs
  dem=volcbas2.map;   # initial dem
  lavain=input.map;   # location of magma pipe
  lavaload=100;   # average lava load delivered per eruption
  base=-50.0;   # base level (metres a.s.l)
  depcnv=0.05;   #depth of lava deposition for DEM update

  # outputs
  dem1=volcano0;   # new dem after deposition
  depos2=lava0000;   # deposition output maps
  ld=lddmap;   # local drain direction output - surface flow paths

areamap
  volcbas2.map;

timer
  1 250 1;

initial
  depos=0;
  dem1 = dem;
  # lava source
  sedmsk=boolean(lavain);

dynamic
  # initialize the driving DEM - add a smoothed random error and report surface
  report dem1 = dem1 + windowaverage(normal(boolean(volcbas2.map))*0.2,120);

  # dem1 = dem1 + normal(boolean(volcbas2.map))*0.3;

  # build topology of volcano
  ld = lddcreate(dem1,1E35,1E35,1E35,1E35);

  #add random component to lavaflow load delivered by the lava flow
  #lavaflow load must be 0 or more
  lavaflow = cover((normal(sedmsk)*5+lavaload),0);
  lavaflow = if(lavaflow lt 0,0,lavaflow);

  #Compute transport capacity as function of slope (tc in range .9-1)
    synslp = sin(atan(slope(dem1)));
  # Maximum slope is 45 degrees (sin45 = 0.7071)
    synslp = if(synslp gt 0.7071, 0.7071, synslp);

  # Transport capacity of lava flow is a function of elevation difference
  # and surface resistance

  transcap = 0.9 + synslp*0.14142;

  # compute fluxes as volcano discharges lava over cone from cell to cell
    flux1 = accufractionflux(ld,lavaflow,transcap);
    # amount added to each cell from upstream cells

  # report cumulative deposition on all paths as natural log to aid display
    depos = (depos+flux1);
    report depos2=if(depos ge 0, ln(depos+1), 0);

  # convert fresh lava layer to depth
  # (model build up of cone in first term)
    dem1 = (if(lavain eq 1, (dem1+0.15*flux1*depcnv), (dem1+flux1*depcnv)));
```

the landscape-forming processes, incorporating uncertainty in the physical processes of erosion and sedimentation tends to lead to clear, stable results. The models demonstrate that this stability arises from the cumulative sets of most likely realizations. Put another way, the landforms resulting from a set of most likely outcomes incorporate a strong stability through negative feedback loops, which can also be seen as a kind of hysteresis in the landscape. Perturbation of a dynamic, stochastic process does not always lead to unstable conditions, but may return a disturbed system to a stable configuration. It is interesting to note that the addition of small amounts of random noise to the feedback loops of the models also provides a means of achieving stability.

Finally, it appears that the underlying similarity in the form of elementary fans and volcanoes is also a result of different scales of probability. The fact that symmetry is not always apparent in a landscape is a result of interactions between different processes and different levels of resolution, leading to variation and unevenness in the spatial and temporal distribution of the contributing processes.

REFERENCES

Beven, K. (1996) Equifinality and Uncertainty in Geomorphological Modelling. In: *The Scientific Nature of Geomorphology: Proceedings of the 27th Binghamton Symposium in Geomorphology*, 27–29 September 1996 (Eds B.L. Rhoads and C.E. Thorne), J. Wiley & Sons Ltd, Chapter 12, pp. 289–313.

Bivand, R. and **Gebhardt, A.** (2000) Using the R statistical data analysis language on GRASS 5.0 GIS data base files. *Comput. Geosci.*, **26**, 1043–1052.

Burrough, P.A. (1983a) Multi-scale sources of spatial variation in soil. I. The application of Fractal concepts to nested levels of soil variation. *J. Soil Sci.*, **34**, 577–597.

Burrough P.A. (1983b) Multi-scale sources of spatial variation in soil. II. A non-Brownian Fractal model and its application to soil survey. *J. Soil Sci.*, **34**, 599–620.

Burrough, P.A. (1998) *Dynamic Modelling and Geocomputation.* In: *Geocomputation: A Primer* (Eds P.A. Longley, S.M. Brooks, R. McDonnell and B. MacMillan), Wiley, Chichester, UK.

Burrough, P.A. and **McDonnell R.A.** (1998) *Principles of Geographical Information Systems*, Oxford University Press, Oxford, UK.

Casti, J.L. (1997) *Would-be-Worlds*, J. Wiley, New York.

Clevis, Q.J.W.A. (2003) Three-dimensional modelling of thrust-controlled foreland basin stratigraphy. *Geologica Ultraiectina No. 226*, Utrecht University, The Netherlands.

Duff, P. McL. D. (2002) *Holmes' Principles of Physical Geology*, Thornes, Cheltenham, UK.

Favis-Mortlock, D., **Boardman, J.** and **MacMillan, V.** (2001) *The Limits of Erosion Modelling: Why we Should Proceed with Care* (Eds R.S.Harmon and W.W. Doe), Kluwer Academic/Plenum Publishing, New York, pp. 477–516.

Freeman, G.T. (1991) Calculating catchment area with divergent flow based on a rectangular grid. *Comput. Geosci.*, **17**, 413–422.

Gallant, J.C. and **Wilson, J.P.** (1996) TAPES-G: A terrain analysis program for the environmental sciences. *Comput. Geosci.*, **22**, 713–722.

Harmon, R.S. and **Doe, W.W.** (Eds) (2001) *Landscape Erosion and Evolution Modelling*, Kluwer Academic, Dordrecht.

Heuvelink, G.B.M. (1998) *Error Propagation in Environmental Modelling.* Taylor & Francis Ltd, London.

Jones, K.H. (1998) A comparison of algorithms used to compute hill slope as a property of the DEM. *Comput. Geosci.*, **24**, 315–323.

Jones, R. (2002) Algorithms for using a DEM for mapping catchment areas of stream sediment samples. *Comput. Geosci.*, **28**, 1051–1060.

Karssenberg, D.J. (2002) Thesis Chapter 4: Adding functionality for modelling error propagation in a dynamic, 3D spatial environmental modelling language. Nederlands Geographical Studies 305, Utrecht University, The Netherlands.

Mitas, L. and **Mitasova, H.** (2001) Multiscale soil erosion simulations for land management. In: *Landscape Erosion and Evolution Modelling* (Eds R.S. Harmon and W.W. Doe), Kluwer Academic, pp. 321–347.

Pebesma, E.J. and **Wesseling, C.G.** (1998) GSTAT: A program for geostatistical modelling, prediction and simulation. *Comput. Geosci.*, **24**, 17–31.

Pike, R.J. (2002) A bibliography of Terrain Modelling (Geomorphometry), the Quantitative Representation of Topography – Supplement 4.0 Open File Report 02-46, USGS, Washington.

Pullar, D. (2003) Simulation modelling applied to runoff modelling using MapScript. *Trans. GIS*, **7**, 267–283.

Stewart, I. (1997) *Does God Play Dice?*, Penguin, London, 401 pp.

Tucker, G.E. and **Bras, R.L.** (2000) A stochastic approach to modelling the role of rainfall variability in drainage basin evolution. *Water Resour. Res.*, **36**, 1953–1964.

Tucker, G.E., **Lancaster S.**, **Gasparini, N.** and **Bras, R.** (2001) The channel-hillslope integrated landscape development model (CHILD), In: *Landscape Erosion and Evolution Modelling* (Eds R.S. Harmon and W.W. Doe), Kluwer Academic.

Wesseling, C.G., **Karssenberg, D.-J.**, **Burrough, P.A.** and **van Deursen, W.P.A.** (1996) Integrating dynamic environmental models in GIS: The development of a dynamic modelling language. *Trans. Geogr. Info. Systems*, **1**, 40–48.

Wolfram (2004) http://www.wolframscience.com/nksonline.

Spec. Publ. Int. Assoc. Sedimentol. (2008) **40**, 287–306

Models that talk back

JOHN C. TIPPER

Geologisches Institut, Albert-Ludwigs-Universität, Albertstrasse 23B, D-79104 Freiburg, Germany (E-mail: john.tipper@geologie.uni-freiburg.de)

ABSTRACT

Modelling is a critically important part of all scientific work. Its goals are (1) to gain understanding of the systems being studied, and (2) to predict how those systems are likely to behave under given input conditions. Two substantially distinct styles of modelling can be identified, based on the degree to which the model and its parent system are behaviourally equivalent; a model and its parent system are defined as being behaviourally equivalent to each other if they can readily be trusted to behave in a satisfactorily similar way, for the very great majority of possible input states. The first modelling style involves the use of models that cannot be trusted to be behaviourally equivalent to their parent systems; these models typically are ones that rely on numerous auxiliary hypotheses and unconstrained parameters. Though they often give superficially impressive results, these models are unlikely ever to be capable of saying much that is significant about the systems concerned. The second modelling style is feasible only for models that can reasonably be believed to be close to being behaviourally equivalent to their parent systems. These models may often seem oversimplified, yet they are capable of giving considerable insight into the nature of their parent systems; they justifiably can be referred to as 'models that talk back'. Published examples of the two styles of modelling are analysed in this paper. The first style is illustrated by an exercise in landscape evolution modelling. The model used gave apparently realistic predictions of present-day landscapes and drainage patterns, ones that appeared to lend support to a previously published hypothesis about the evolution of a highland area. However, the model gave these predictions under demonstrably unrealistic conditions. The example of the second style of modelling concerns the patterns of cyclicity found in stratigraphic successions. The model used predicted that cyclic successions generally should be considerably more complete than they usually are reported to be. The inherent simplicity of this model meant that this 'too-complete' dilemma had to be confronted; it could not simply be explained away. The confrontation of the dilemma then led directly to the identification of several new hypotheses worthy of investigation.

Keywords Model, sedimentation, landscape evolution, stratigraphy, cycle.

INTRODUCTION

Every model that a scientist makes of a system being studied is a practical expression of a particular set of ideas about that system. These ideas – they can also be dignified by the terms 'hypothesis' or 'theory', depending on the amount of evidence that exists to support them – are expressed in the model in a form in which they can readily be worked with; this form may be physical, or mathematical, or graphical, or even verbal. There are two reasons for studying a system with the help of models. Either it may be hoped that experimentation with those models will help in understanding the structure and function of the system, or it may be believed that the models will be capable of predicting how the system would behave under given input conditions. Understanding and prediction are rightly described as the twin goals of scientific modelling.

Understanding clearly must precede prediction, for no sensible scientist would ever rely on a model to predict the behaviour of a system that is not yet understood. But how can any system – natural or artificial – ever be known to be properly understood, or at least be known to be understood well enough to allow a model of it to be used for prediction? To this question there is a range of answers, from the principled to the pragmatic. The principled answer is that a system should be taken to be understood only when it can be fully represented in terms of fundamental physical laws; the pragmatic answer is that a system can be taken to be understood as soon as any model of it has been made that can be shown to be capable of predicting its behaviour for a sufficiently wide range of input conditions. The principled position renders unnecessary any use of modelling for the purpose of gaining understanding; it foresees modelling only for prediction. The pragmatic position foresees modelling both for understanding and for prediction; it is the position earth scientists have to adopt.

The systems with which this paper is concerned are natural sedimentation systems, i.e., systems involving natural processes of erosion, transport and deposition. One example of such a system would involve the processes operating in some newly uplifted mountain area; this system might be modelled using what are now usually referred to as landscape evolution models (e.g. Flemings & Jordan, 1989; Tipper, 1991, 1992; Jordan & Flemings, 1991; Kooi & Beaumont, 1994; Slingerland *et al.*, 1994; Tucker & Slingerland, 1994; Braun & Sambridge, 1997; Syvitski *et al.*, 1998; Weltje *et al.*, 1998). A second example would involve the processes operating in a subsiding sedimentary basin; this system might be modelled using one of the many sediment-distribution modelling packages developed in the last two decades (e.g. Helland-Hansen *et al.*, 1988; Strobel *et al.*, 1990; review in Paola, 2000). Models of natural sedimentation systems are used both for understanding and for prediction, depending on how well understood the system being studied is judged to be.

This paper has three objectives: (1) to look at some aspects of the theory and practice of modelling that seem not to be fully appreciated by many earth scientists who make and use models of natural sedimentation systems; (2) to identify two contrasting styles of modelling; (3) to illustrate these modelling styles, using as examples two modelling exercises drawn from the recent literature. Finally, some general observations about the use of models and modelling are presented.

SOME ASPECTS OF MODELLING IN GENERAL, WITH PARTICULAR REFERENCE TO NATURAL SEDIMENTATION SYSTEMS

The model and the system

The definition of a scientific model adopted in this paper – 'a practical expression of a particular set of ideas about a system' – is by no means the only one that could have been chosen. Alternatives to it can readily be found in any general text on scientific modelling (e.g. Vemuri, 1978) and in papers in numerous specialist volumes (e.g. Cross, 1990; Harbaugh *et al.*, 1999; Merriam & Davis, 2001). The advantage of this definition is that it is absolutely straightforward. There is no place in it for waffle about subtle but supposedly significant differences between different forms of model (numerical, physical, etc.); nor is there place for delicately teased out distinctions between 'models' and 'simulations' and 'scenarios' (cf. Greenwood, 1989); nor is there place for details that ultimately are only of secondary importance, for instance details about different types of parameterization (distributed-parameter, lumped-parameter, etc.). The definition says simply that I – the scientist – am studying a system, that I have developed certain ideas about it (for example, about its structure and function), and that I have chosen to express those ideas in some particular practical form. If my ideas and my expression of them are both correct, then the system and the model will behave identically to each other, for all of the possible input states (Fig. 1a).

When looked at closely, every natural system is unique. However, few natural systems are so singular that they must be modelled in an entirely unique way. Usually it is possible to group together systems that are broadly similar, then to apply the same general model to each of the systems within that group. The application of the same general model to a group of broadly similar systems is made possible by parameterization. As an example – a deliberately artificial one – consider a group of river systems, each composed of a single channel meandering over a floodplain; to these systems might be being applied a numerical transport model based partly

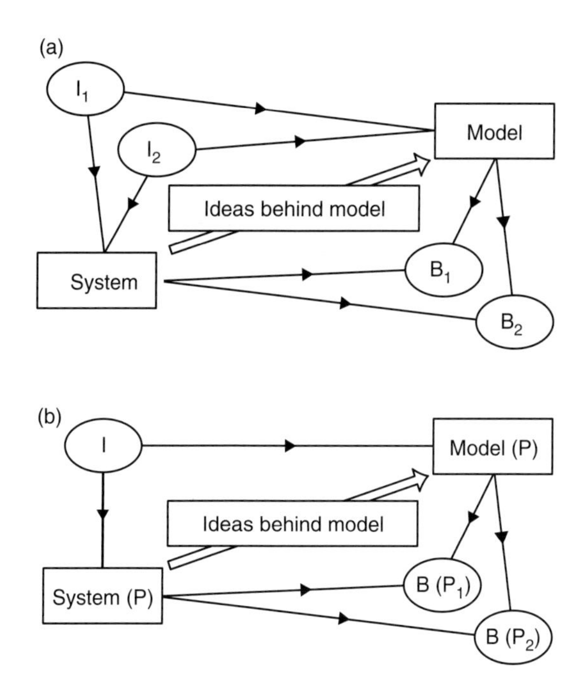

Fig. 1. Relationships between input value (I) and behaviour (B) for the simplest possible modelling framework, for a model that is correct. (a) No parameters in either system or model. System and model behave identically, for all allowable input values. (b) System and model are both parameterized, with identical parameter values (P). System and model behave identically for the same input values, for all allowable parameter values.

on Chezy's equation (Hsü, 1989). One difference that would certainly exist between the individual systems within the group would be the nature of the floodplain materials into which the channels were cut. This difference would be recognized within the model by means of a parameter (Chezy's coefficient) that would take into account the effects on river flow that different bed and bank materials are known to have. To apply the model to one particular system, the value set for Chezy's coefficient would be the one most appropriate to that system's bed and bank material. The model would then behave identically to the system, for all the allowable input states, provided of course that it was otherwise correct (Fig. 1b).

The testing of models

There can be no such thing as a perfect model of a natural system – that would be one that could be relied on always to behave in an identical way to its parent system. The best to be hoped for is that a model is and its parent system behave in satisfactorily similar ways. This raises automatically the question of how models of natural systems can be tested. How is it possible to judge how satisfactory the behaviour of such a model is and hence how satisfactory are the ideas behind it are?

The obvious strategy to use in testing a model would seem to be to allow it and its parent system to run under exactly the same conditions. If the behaviour of the model appears to differ significantly from that of the system, the confidence to be had in the model is decreased ('test failed'); if the behaviour of the model appears to match that of the system, that confidence is increased ('test passed'). This strategy is attractively straightforward, but problems inevitably arise whenever an attempt is made to implement it in practice. These problems involve: (1) the access that is necessary to the parent system, (2) the significance of apparent differences, and (3) the phenomenon of non-uniqueness (equifinality). These problems are particularly severe for the types of models with which earth scientists commonly work, i.e., field-scale models of natural systems operating over long periods of time.

The problem of access

Just how severe the problem of access is for natural systems can readily be appreciated once the two irreducible requirements of the testing strategy have been spelled out. First, the input state of the parent system must be known exactly, in order that the correct input and parameter values can be set for the model. Second, the behaviour of the parent system must be capable of being observed and measured in as much detail as is the behaviour of the model. The degree of access necessary in order for these requirements to be fulfilled is quite unattainable for any natural system. Such access can only ever be achieved for what can be thought of as pseudo-natural experiments – and then only approximately. An example of a pseudo-natural experiment would be the flooding of a lake behind a newly built dam. This might allow the testing of some model for sedimentation associated with sea-level rise. Invaluable though they are, pseudo-natural experiments clearly have their drawbacks. First, they are essentially unrepeatable (for economic and environmental reasons). Second, they seldom can be run for long enough under sufficiently controlled conditions. Third, the input state cannot usually be set at will.

An alternative to the use of pseudo-natural experiments involves laboratory-based testing. An artificial version of the natural system in question is constructed in the laboratory; then the behaviour of the model being tested is subsequently compared with the behaviour of this artificial system. Tests carried out in this way are repeatable, unlike those based on pseudo-natural experiments. They can also be carried out for any of the input states that are allowable for the system, provided of course that the state is realizable in the laboratory. There are numerous descriptions of models of sedimentation systems being tested in this way. The best known are numerical models for fluvial transport and deposition, which commonly are tested against artificial river systems set up in laboratory flumes (see references in Bridge, 2003). Work of this type has given significant insight into the processes active in natural river systems, even though the range of flow conditions capable of being produced in even the largest flume is but a fraction of that found in natural rivers.

Laboratory-based testing has two principal drawbacks. The first is that the artificial systems constructed in a laboratory are themselves models – they are expressions of what the scientist who has made them thinks the parent systems are like. Thus laboratory-based testing is nothing more than a comparison of two different types of model of the same parent system. The second drawback to laboratory-based testing concerns the scaling that has to be introduced into every artificial system in order that the system's behaviour can properly be compared with behaviour observed in nature (Yalin, 1971). Such scaling is often impossible to achieve. Partially scaled laboratory-based systems certainly can produce intriguing results, results that often stimulate important ideas about how natural systems work (e.g. references in van Heijst *et al.*, 2001). But no artificial system that is at all incompletely scaled should ever be used to test a model of a natural system.

The problem of behavioural differences

In the testing strategy considered here, the criterion used to declare 'test passed' is that the behaviour of a model be found to match the behaviour of the parent system. The criterion used to declare 'test failed' is that these behaviours differ significantly. But what does 'match' mean in this context? And how small can behavioural differences be that still must be deemed significant? Two contrasting interpretations of the word 'match' can be put forward – one strict and one liberal. The strict interpretation is that 'match' really does mean 'match' – the behaviour of the model must therefore be completely indistinguishable from that of the system. In that case, no test will ever be passed (cf. Brown, 1992)! The liberal interpretation is that behavioural differences should be deemed insignificant whenever they lie within the range of expected natural and experimental variation. This lets differences be discounted that are attributable to measurement error and sampling variation. Sadly, this liberal interpretation is no less unsatisfactory than the strict one, for it takes into account only those aspects of behaviour that actually were measured during the test. What about aspects that were not measured, either deliberately (because they were thought not to be relevant) or accidentally (because they had not been identified before the test began)? Surely these should be taken into account too! The following conclusion is inescapable: the behaviour of a model can never be shown to match that of a parent system, no matter how strictly or how liberally the word 'match' is interpreted. Therefore, testing can never increase the confidence to be had in a model; it can only decrease it.

The problem of non-uniqueness

Non-uniqueness (equifinality) is the bugbear of all modelling. It exists (a) whenever several intrinsically different models behave in essentially the same way for one particular combination of input and parameter values ('intermodel non-uniqueness'), and (b) whenever one model behaves in essentially the same way for several combinations of input and parameter values ('intramodel non-uniqueness'). Non-uniqueness is an intractable problem, for there is never any objective basis for choosing between modelling alternatives that appear equally satisfactory. A decision in favour of one particular model (or in favour of one particular combination of input and parameter values) can only ever be made subjectively, using what have been referred to as 'extra-evidential considerations' (Oreskes *et al.*, 1994, p. 642).

This impossibility of deciding objectively between equally satisfactory alternatives is a tempting invitation to stop testing once the first model of a system has been developed that matches that system's behaviour acceptably (or once the first acceptable combination of input and parameter values has been identified). Many scientists accept this invitation with evident relief, asserting as they do so that there is a negligible chance of two different models of a natural system behaving in essentially the same way (or of two different combinations of input and parameter values giving essentially the same result). This assertion is entirely unwarranted: well documented examples of non-uniqueness are readily found in the literature (Burton *et al.*, 1987; Kendall & Lerche, 1988). In fact, there is only one reputable course of action to take when a model is found in a test to behave in a similar way to its parent system: this is to continue with the testing, using as many other models of the system as can reasonably be formulated, and as many combinations of input and parameter values as can reasonably be thought of (cf. Chamberlin, 1897). This, sadly, is a recipe for unending work – the 'worse dilemma' alluded to in Oreskes *et al.* (1994, p. 642).

The real purpose of testing

It is important to be clear on what testing can and cannot show. Most importantly, it cannot show that the model being tested is true. As so convincingly argued by Oreskes *et al.* (1994), models of natural systems are inherently incapable of being verified, or of being validated, or of being confirmed. Nor – perhaps surprisingly to many scientists – can testing show that a model is false. Certainly it can decrease the confidence we have in a model; that is what a 'test-failed' result does. But there is always the possibility that the data with which the model's predictions are being compared are biased against the model in some way, perhaps because of the way they were collected or interpreted; this is the effect referred to by Eldredge & Gould (1972) as 'the cloven hoofprint of theory' (see also the quotation from Darwin, 1861, reproduced in Medawar, 1969, p. 11). In fact the real purpose of testing a model should never be to try to show that it is true or false; it should always be simply to probe for weaknesses – in effect to search for aspects of the parent system that are not yet properly understood.

Auxiliary hypotheses

The description so far given of modelling has presented it as if it could be carried out *in vacuo*, with perfect knowledge. But modelling cannot be carried out in that way, at least not for natural systems. Natural systems are open systems, and they are always incompletely known; therefore, they must always be modelled against a background of partial knowledge. This partial knowledge is augmented by the modeller with what are termed 'auxiliary hypotheses'. These are 'the additional assumptions, inferences, and input parameters required to make a model work' (Oreskes *et al.*, 1994, p. 642). It must be stressed that ideas incorporated in auxiliary hypotheses have an entirely distinct status to those expressed in the model itself. Those latter ideas – they are conveniently referred to as the 'principal hypothesis' – should be thought of as the model's basic rules.

The most widespread use of auxiliary hypotheses in modelling is in the specification of initial and boundary conditions. Some such hypotheses are required in almost every modelling exercise. The credibility of the initial and boundary condition hypotheses used in a modelling exercise is largely a function of how well the system being modelled can be observed and measured. For instance, a model of how a particular modern reservoir will fill with sediment in the coming decade will certainly have initial and boundary condition hypotheses that are accurate and detailed; these hypotheses will be highly credible ones. In contrast, a model of the sedimentation systems in a foreland basin of Tertiary age will surely rely on initial and boundary condition hypotheses that are highly speculative at best. These hypotheses will depend on how the known geology of the basin is interpreted, and they will also reflect the modeller's personal views on how the sedimentation systems in the basin are likely to have been driven. Were they driven largely by tectonics, for instance, or did climate and sea-level change also have roles to play? Obviously there is a danger here that the principal and auxiliary hypotheses will become inextricably confounded.

Auxiliary hypotheses are also commonly used for specifying values of model parameters. This is done whenever the true values cannot be measured directly, for whatever reason. Care is always needed in using auxiliary hypotheses to specify parameters, to avoid three common pitfalls. The first of these occurs when particular values are chosen solely on grounds of mathematical neatness or computational expediency, or sometimes because of a desire for simplicity (see discussion in Oreskes *et al.*, 1994, note 25, p. 645). An example would be when all the members of a set of related parameters are given the same value purely because there is no convincing rationale by which they can be given different ones. The second common pitfall occurs when parameter values are chosen so as to maximize the fit of a model to its parent system (see references in Oreskes *et al.*, 1994, note 19, p. 644). Parameters that can be treated in this way are so-called 'free parameters'. They are ones that play no intrinsic or satisfactorily explained role in the ideas of which the model is the expression. Modelling exercises in which free parameters are used to force a fit between a model and its parent system contribute nothing to the understanding of that system; they demonstrate only that the model obeys the wishes of its maker. The third common pitfall involves the uncritical reliance on higher-level hypotheses, in particular on ones that somehow have managed to attain fact-like status. An example would be the use of the Haq–Vail global sea-level chart to determine parameter values for a regional sea-level modelling exercise. This chart (Haq *et al.*, 1988) is one of the most prominent examples of a hypothesis that commonly is taken as fact despite still being essentially untested (Miall, 1992; Tipper, 1993, 1994; Miall & Miall, 2001). Uncritically accepted higher-level hypotheses are dangerous in that they tend to confer fact-like status on every lower-level hypothesis that is based on them. This status is entirely unjustified, even if that lower-level hypothesis actually is correct.

The presence of auxiliary hypotheses alters radically the way in which modelling results must be viewed. This is seen most easily by looking at how the results of simple pass/fail tests have to be interpreted (Fig. 2). Consider first the 'test failed' result – the model in question is found to behave in a way that is significantly different to its parent system. This could mean that there are errors in the principal hypothesis; however, it could also mean that there are errors in one or more of the auxiliary hypotheses. Similarly, the 'test passed' result could mean that the principal and auxiliary hypotheses are all correct; or it could mean that somewhere there are cross-cancelling errors (Oreskes *et al.*, 1994, p. 642). These could be in the principal hypothesis, or they could be in the auxiliary hypotheses, or they could be in both. It is never possible to determine that such cross-cancelling errors exist, nor is it possible to distinguish between effects due to errors in the principal hypothesis and effects due to errors in auxiliary hypotheses. That is true no matter what the test result is.

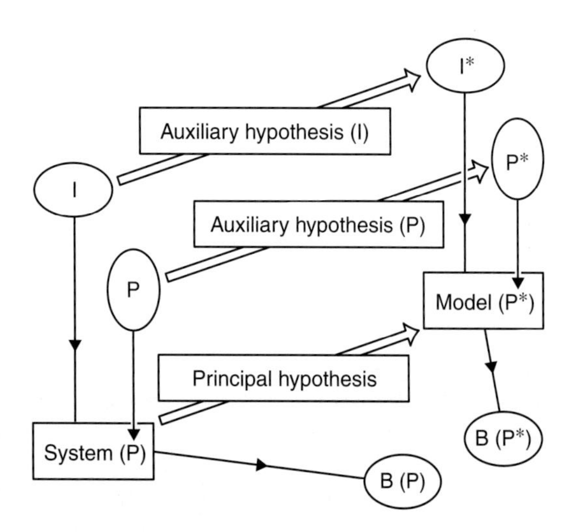

Fig. 2. Modelling framework with auxiliary hypotheses about input value and parameter value. System has input value I and parameter value P; model uses hypothesized input value I* and hypothesized parameter value P*; I ≠ I*, P ≠ P*. System and model will generally not behave identically, even if model is correct.

Inverse modelling ('inversion')

The type of modelling so far described here is commonly termed 'forward modelling'. It is also possible to carry out 'inverse modelling'; this is also termed 'quasi-backward modelling' or 'inversion' (Lessenger & Lerche, 1999). Inverse modelling involves identifying the input and/or parameter values that result in a given model reproducing some specified target behaviour. This target behaviour is usually a simplified version of a pattern of behaviour produced by the model's parent system.

The aim of inverse modelling is nothing other than the solution of a conventional optimization problem (Bornholdt *et al.*, 1999). It is carried out using three computational components (Fig. 3): (1) a model of the required system, (2) a comparator, and (3) an adjuster. The model is any normal scientific model; its behaviour is observed as it is run forward in time with given input and parameter values. The comparator takes this observed behaviour and compares it against the target behaviour; as output it returns the value of some chosen measure of discrepancy. The adjuster takes this particular discrepancy value and considers it together with other values obtained from previous runs of the model, runs that were made with different input and/or parameter values. The adjuster calculates how best to adjust the model's input and parameter values so that the discrepancy between the observed and the target behaviour is reduced. This modelling–comparison–adjustment process continues automatically until some acceptably low value of the discrepancy measure has been obtained, preferably one that seems to be a global minimum. The

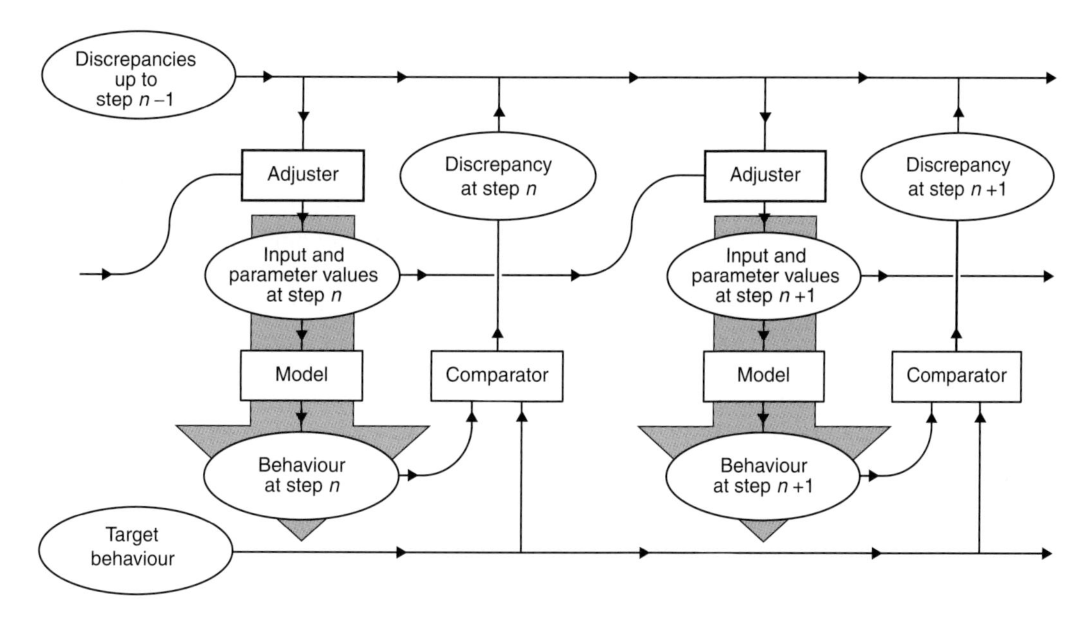

Fig. 3. Computational process involved in inverse modelling, for two steps. See text for further explanation. Note that the model is always run forward (shaded arrow).

input and parameter values finally found are interpreted as being those that were responsible for the pattern of parent system behaviour from which the target behaviour was derived.

Inverse modelling is sometimes presented as the Holy Grail of modelling. After all, what could be more important than to obtain estimates of the input and parameter values that evidently were responsible for some given pattern of natural behaviour – and best of all to obtain these estimates in a fully automatic way? This thoroughly rosy perspective has problems, unsurprisingly. These can be appreciated immediately by remembering just what type of model is used in every inverse modelling exercise – it is a normal scientific model, with all of the weaknesses that normal scientific models have. It is not fundamentally 'a different type of model', as claimed by Cross & Lessenger (1999, p. 69). The inescapable conclusion is that the results of inverse modelling should be viewed in the same way as all other modelling results – with a healthy degree of scepticism.

TWO CONTRASTING STYLES OF MODELLING

It was stressed earlier that there is no such thing as a perfect model of a natural system. Yet certainly there are some models that will always be preferred over others: these are the ones that clearly are able to be used for prediction immediately, with only the very minimum of testing. The identifying characteristic of models such as these is that they can readily be trusted to be behaviourally equivalent to their parent systems. (By this is meant that they can readily be trusted to behave in a satisfactorily similar way to their parent systems, for the very great majority of possible input states.) In contrast to these are the models that cannot readily be trusted to be behaviourally equivalent to their parent systems. These latter models cannot be used for prediction, at least not until they have been subjected to a considerable amount of testing. A conceptual measure of the amount of testing a model needs is given by the proportion of the allowable input states for which the model and the system cannot be trusted to show the same behaviour. What is here termed the 'necessary-testing curve' is a graph of the amount of testing that a model needs plotted against the degree to which the model and its parent system are behaviourally equivalent; a form for this curve is suggested in Fig. 4.

The necessary-testing curve separates two contrasting styles of modelling. Below the curve is modelling that is inherently speculative; above the curve is modelling designed always to err on the side of safety. The below-the-curve style of modelling tests models less frequently than necessary, it uses them for prediction in situations where they cannot be trusted to behave in the same way as their parent systems, and it includes exploratory investigations that are

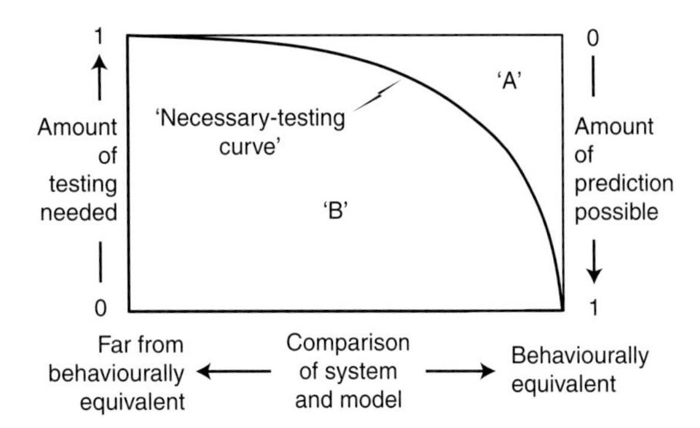

Fig. 4. The necessary-testing curve. The amount of testing a model needs is measured by the proportion of the allowable combinations of input and parameter values for which it cannot be trusted to show the same behaviour as the system; the amount of prediction possible from a model is measured by the proportion of the allowable combinations for which it can be so trusted. 'A' and 'B' indicate the two distinct styles of scientific modelling – above-the-curve modelling and below-the-curve modelling respectively.

essentially inductive (Baconian experimentation). The above-the-curve style tests models more frequently than really is necessary, it uses them for prediction less frequently than really would be possible, and it includes exploratory investigations that are strictly deductive (Kantian experimentation). The below-the-curve style of modelling – it corresponds to what Medawar termed 'academic play' (1969, p. 38) – is prevalent whenever models are used that cannot be trusted to be behaviourally equivalent to their parent systems; these models typically are ones that rely on numerous auxiliary hypotheses and contain numerous free parameters. The above-the-curve style of modelling is feasible only for models that can reasonably be believed to be close to being behaviourally equivalent to their parent systems. Such models often seem oversimplified, yet they are capable of providing considerable insight into the nature of the parent systems. These models can justifiably be referred to as 'models that talk back' – an expression that I borrow gratefully from Toffoli & Margolus (1987).

BELOW-THE-CURVE MODELLING – LANDSCAPE EVOLUTION IN SOUTHEASTERN AUSTRALIA

The modelling exercise

An example of below-the-curve modelling is furnished by a recent exercise in landscape evolution modelling (van der Beek *et al.*, 1999; abbreviated to 'VBL'), the principal object of which was to offer support to an existing hypothesis about the evolution of the highland area of southeastern Australia. This hypothesis, due originally to Lambeck & Stephenson (1986), suggests that the present-day topography of this area – here referred to simply as the 'Highlands' – is due solely to the denudation and isostatic rebound of an old (late Palaeozoic) mountain belt; there is therefore no need to postulate that episodes of uplift occurred in the Highlands during the Mesozoic or Cenozoic, as is the case in some competing hypotheses. For the purpose of this present paper, it is only the modelling exercise itself that is important; the underlying geomorphological questions are essentially irrelevant.

The principal hypothesis used in the modelling exercise is the set of general ideas about landscape evolution that are expressed in 'CASCADE' (Braun & Sambridge, 1997; VBL, fig. 3); this is a modelling package that aims to show how landscapes evolve in time as a result of the action of commonly occurring geomorphic processes. CASCADE takes account of the following processes: (1) fluvial transport, bedrock incision and alluvial deposition, (2) hillslope transport by slow and continuous processes such as soil creep, (3) hillslope transport by landsliding, (4) isostatic response to denudational unloading. These processes are represented using seven parameters: (1) a fluvial transport parameter, (2) a precipitation parameter (the mean annual precipitation), (3) a length scale for bedrock incision, (4) a length scale for alluvial deposition, (5) a hillslope diffusion coefficient, (6) a threshold slope for landsliding, (7) a parameter describing the flexural rigidity of the lithosphere in the area being modelled. The geometry of an area to which CASCADE is being applied is represented as a mesh of triangles, with the elevations of the vertices serving to define the land-surface topography. The values of the parameters can be set separately at each vertex, and they can also be altered in time; therefore there are no less than $7 \times M \times N$ potentially independent parameter values in a CASCADE model, where M is the number of time-steps used and N is the number of vertices.

The Highlands area is approximately 600×600 km in size (Fig. 5). It was represented as a mesh of 14 641 triangles, with an average vertex spacing of about 5 km. The vertex elevations were initially set to reproduce the required hypothesis about the topography of the area at the end of the Palaeozoic.

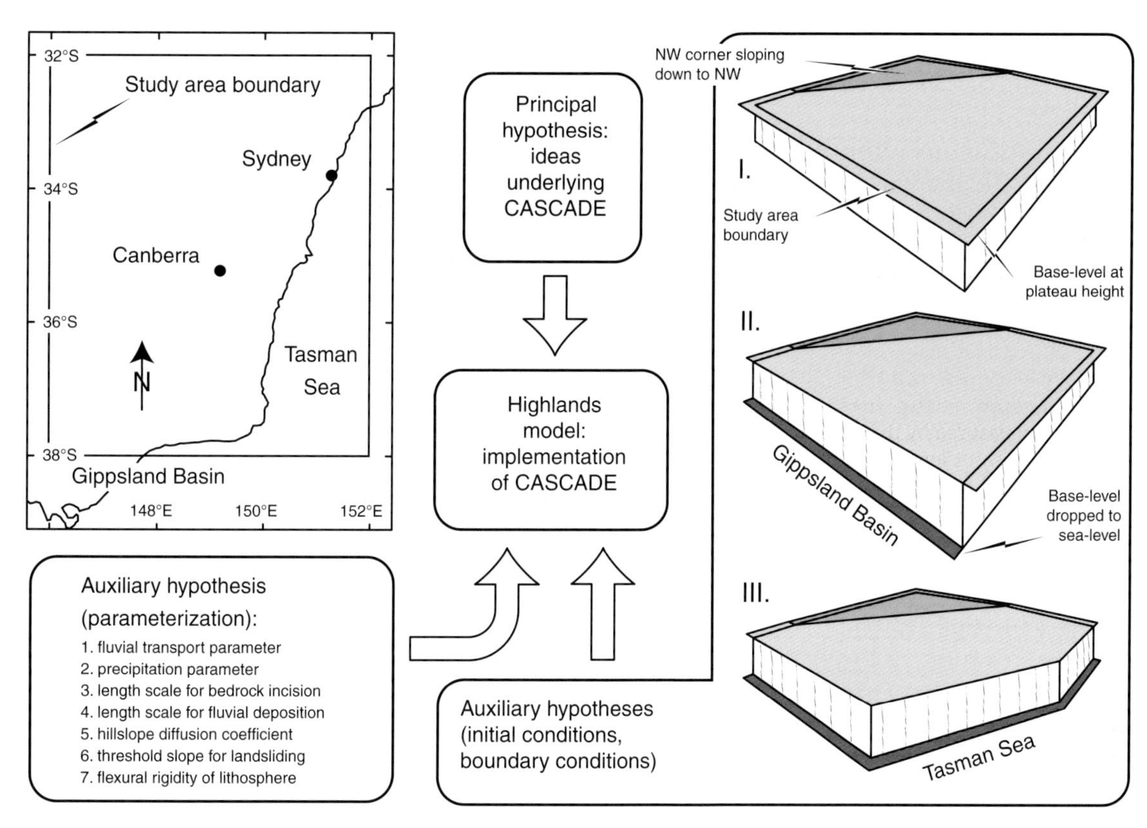

Fig. 5. CASCADE model of southeastern Australian highlands; details in van der Beek *et al.* (1999). Study area (location map at top-left) is approximately 600 × 600 km in size. Modelling framework is made up of principal hypothesis and three auxiliary hypotheses; see text for details. Right-hand panel shows oblique views of the study area taken from the southeast: I, view initially (end of Palaeozoic); II, view after first boundary condition change (formation of Gippsland Basin); III, view after second boundary condition change (rifting of Tasman Sea).

This hypothesis – it constitutes the first of three auxiliary hypotheses – is that the Highlands at the end of the Palaeozoic were a more-or-less uniformly high plateau, the northwestern corner of which sloped gently down to the northwest. The second auxiliary hypothesis concerns the boundary conditions used by the model; these were set so that initially there would be little or no land-surface gradient normal to any of the study area margins, except in the sloping northwestern corner. Two changes were made subsequently, as the model ran (Fig. 5). The first change permitted the development of high north–south land-surface gradients at the southern margin of the study area; this simulated the drop in base level at that margin that would have resulted from the formation of the Gippsland Basin. The second change simulated rifting in the Tasman Sea, together with the drop in base level that that rifting would have produced. The rifting itself was simulated by removing a triangular section of the mesh in the southeastern corner of the study area; the drop in base level was simulated by permitting the development of high margin-normal gradients at the study area's eastern and southeastern margins. The model was run in a series of time-steps simulating 250 million years of landscape evolution. The first boundary condition change was implemented 'between 150 and 140 Ma' and the second 'between 100 and 90 Ma' (VBL, p. 161).

The third of the auxiliary hypotheses concerns the model's parameters. Each of these was taken to be fixed in its value, both spatially and temporally. The fixed values used were ones that previously had been obtained by comparing the model's predictions with present-day morphology and denudation rates (van der Beek & Braun, 1998). The apparent circularity involved here was defended using the following argument (VBL, p. 162): '… the evolution of landforms in CASCADE is independent of process rates and the absolute parameter values mainly serve to translate "model time" to "real time" '.

The results of the exercise: two contrasting assessments

The results of the modelling exercise were presented largely in the form of snapshots showing predicted topographies and drainage patterns (VBL, fig. 5). It was stressed that the landscapes depicted were of course synthetic ones '... which do not have an obvious relationship to real landforms' (VBL, p. 162). Their smallest features were necessarily no smaller than the intervertex spacing used in the model's mesh (c. 5 km). Yet even considering these limitations, the results seemed impressive. For instance: '... the predicted present-day drainage ... shows some remarkable analogies to the drainage patterns observed in southeastern Australia' (VBL, p. 163), and 'the model predictions thus appear to be in reasonable agreement with the available geomorphological information ...' (VBL, p. 165). The conclusion drawn was that a high plateau existing in the Highlands area at the end of the Palaeozoic could certainly have evolved into the present-day landscape there solely as the result of denudation and isostatic rebound.

A radically different assessment of the modelling results is possible, namely that they show very little. Certainly they do not – *indeed they cannot* – show what was claimed for them. This latter assessment is based primarily on the fact that the third of the model's auxiliary hypotheses is demonstrably unrealistic – this is the one by which the values of all of the parameters were taken to be fixed, both spatially and temporally. The rejection of this hypothesis then renders the second auxiliary hypothesis invalid – this is the one concerning the changing boundary conditions.

The rationale for adopting fixed parameter values

The decision to adopt spatially and temporally fixed values for all of the parameters might seem simply to be an example of the use of Ockham's Razor – the scientist's metaphysical commitment to simplicity (Oreskes *et al.*, 1994). Yet it was not that at all, for the developers of the Highlands model were more than ready to admit there was 'no reason why these parameters should be fixed in space and time' (VBL, p. 162). Indeed, they stressed that some of the processes to which the parameters relate 'can obviously vary both spatially and temporally' (VBL, p. 162). In fact, the rationale for adopting fixed parameter values was entirely pragmatic: '... *since* we present first-order results here, and *since* the variability of these parameters is completely unconstrained, we adopt fixed values for this study' (VBL, p. 162, italics added here for emphasis). Something had to be done, and why not that!

In assessing this rationale, it is necessary first to clarify the sense in which the word 'first-order' is being used here. (This is an adjective with a number of special-context meanings that also is used in a general context – but then often with an unfortunate lack of care.) When used in a general context, 'first-order' properly has the sense of 'just adequate', for instance in the following explanation of the term 'first-order measurement' (Weinberg, 1997, italics added here for emphasis): 'The term "first-order measurement" is used ... to describe a measurement *just adequate to the task* of getting things built. First-order measurements ... support the "rule of thumb" and the rough sketch, as well as the "quick and dirty" or "seat of the pants" estimate.' If 'just adequate' was indeed the sense intended in the Highlands modelling exercise, then the results of that exercise should be seen only as demonstrating that synthetic landscapes similar to those in the present-day Highlands can be produced by running a CASCADE-based model. The adoption of fixed values for any or all of the parameters would then be amply justified, for all that ultimately would be important would be that the required landscapes could somehow be produced.

It is hard to believe that the developers of the Highlands model had only this very limited objective in mind, for two compelling reasons. First, they wrote '... we show what the consequences of certain assumed scenarios can be' (VBL, p. 158; see also van der Beek & Braun, 1999, p. 4946); this surely indicates an intention to use the model they had developed as a tool for studying the evolution of the Highlands area, not just to show that CASCADE-based models could work. Second, they had already spent considerable time determining which particular set of parameter values would be most appropriate for modelling the Highlands (van der Beek & Braun, 1998); this work would have been quite unnecessary if just any parameter values would have sufficed in the end. Yet if the objective in carrying out the modelling exercise was really to show how the landscape of the Highlands could have evolved in time *from a particular initial state using a particular set of parameter values*, what justification could there possibly have been for adopting single fixed values for the parameters for the entire study area and for the entire Mesozoic and Cenozoic? The sole justification offered – that the variability of the parameters is completely unconstrained – is plainly inadequate.

The effects of adopting fixed parameter values, particularly for the precipitation parameter

The fixing of the value for any one of the seven parameters can be expected to have had some effect on the landscape-evolution history predicted for the Highlands. However, the effect would certainly not have been equally great for each parameter. The effect of fixing the flexural rigidity parameter would have been the least, simply because that parameter is a solid-earth parameter. Solid-earth parameters can be expected to be more-or-less spatially uniform in their value at any one time (at least over an area the size of the Highlands); they can also be expected to vary slowly and steadily in time. Of the other parameters, five are earth-surface parameters: the fluvial transport parameter, the length scales for bedrock incision and alluvial deposition, the hillslope diffusion coefficient, and the threshold slope for landsliding. Earth-surface parameters can be expected to vary more than solid-earth parameters, and their fixing should therefore have correspondingly greater effects. The final parameter – the mean annual precipitation – is a climate parameter; it can be expected to vary considerably, both spatially and temporally. It is the fixing of this precipitation parameter that can be expected to have had the greatest effect on the predicted landscape-evolution history.

An idea of how substantial this effect is likely to have been can be obtained by listing some of the surface processes that are strongly controlled by either the rate, the seasonal distribution, or the type of precipitation. These include: (1) the processes of local surface-water flow and infiltration (these control the amount of surface run-off, the soil-water saturation, and the height of the groundwater table); (2) the processes affecting the development and spread of vegetation (this tends both to stabilize and to destabilize land surfaces, and also is a source of organic acids that play a crucial role in weathering); (3) the processes involved in soil formation and in physical and chemical weathering (the control that these processes have on the development of the regolith is discussed in Taylor & Eggleton, 2001); (4) the processes affecting slope stability and hence the activity of slope movement processes; (5) the processes associated with river systems (rivers are by far the most efficient means of facilitating long-range drainage and sediment transport, and they also create directly their own erosional and depositional landforms). Without an appropriate supply of water, many of these processes do not work; other processes do work, but at entirely different rates and with entirely different results. It is not sensible to think of modelling landscape evolution without first having accepted the possibility of spatial and temporal variation in precipitation. And it is extremely unwise to start to model the evolution of the landscape in a particular area without first having ascertained if significant spatial and temporal variations in precipitation are likely to have occurred there.

Did the precipitation in the modelled area vary significantly?

There is strong evidence that the mean annual precipitation in the Highlands area was probably far from constant in time during the Mesozoic and Cenozoic. There is also strong evidence that it was unlikely always to have been spatially uniform. Two lines of evidence bear on the probable non-constancy of precipitation in time (Fig. 6): one concerns the changing global climate during the Mesozoic and Cenozoic, the other concerns the changing palaeolatitude of southeastern Australia. The global climate during the Mesozoic and Cenozoic was characterized by the alternation of warm and cool climate modes (Frakes *et al.*, 1992): a

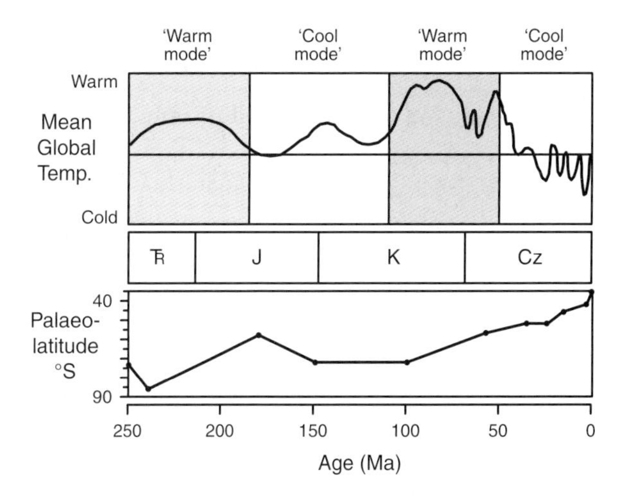

Fig. 6. Variation in time of mean global temperature (after Frakes *et al.*, 1992) and variation in time of palaeolatitude of modelled area (data from Schmidt & Clark, 2000, fig. 10).

warm mode lasting from the Triassic to the middle Jurassic was followed by a cool mode lasting until the early Cretaceous, which was followed by a warm mode lasting until the early Eocene, which gave way to the cool mode culminating in the modern ice age. Certainly there will be few scientists who would claim that the mean global temperature directly controls the mean annual precipitation at every point on the Earth's surface. Equally, however, there will be few who would suggest that the mean annual precipitation in any particular area is likely to be wholly independent of the current global climate state. A similar line of argument applies to the very substantial changes in the palaeolatitude of the Highlands area during the Mesozoic and Cenozoic (Schmidt & Clark, 2000; fig. 10). The area is presently centred at a latitude of about 35°, but it was situated close to the South Pole during the early Triassic; only during the first half of the Cretaceous was its palaeolatitude constant for any length of time, and even then its orientation with respect to the prevailing wind systems was continually changing. Again, there will be few scientists who would claim a direct correlation between latitude and mean annual precipitation (especially in times of changing global climate). Equally, however, there will be few who would suggest that mean annual precipitation is entirely independent of latitude, either in amount or in type. Most scientists will surely want to assume that the mean annual precipitation in the Highlands area did vary significantly in time during the Mesozoic and Cenozoic, rather than that it did not. The adoption of a temporally fixed value for the precipitation parameter therefore has to be rejected.

The evidence bearing on the spatial distribution of precipitation in the Highlands area is limited but incontrovertible – it concerns the precipitation pattern at the present-day (Fig. 7). This pattern is very far from being spatially uniform: there is a fivefold difference between the highest mean annual precipitation (over 2000 mm, in the Victorian Alps) and the lowest (400 mm, in central New South Wales). There are also substantial differences between the seasonal distributions of precipitation in different parts of the Highlands (Commonwealth Bureau of Meteorology, 2004). The spatial differences in precipitation are reflected in the spatial distribution of different vegetation types (Australian Native Vegetation Assessment, 2001), as would be expected. Clearly there is no good reason to assume that the present-day precipitation pattern existed in the past (or, at

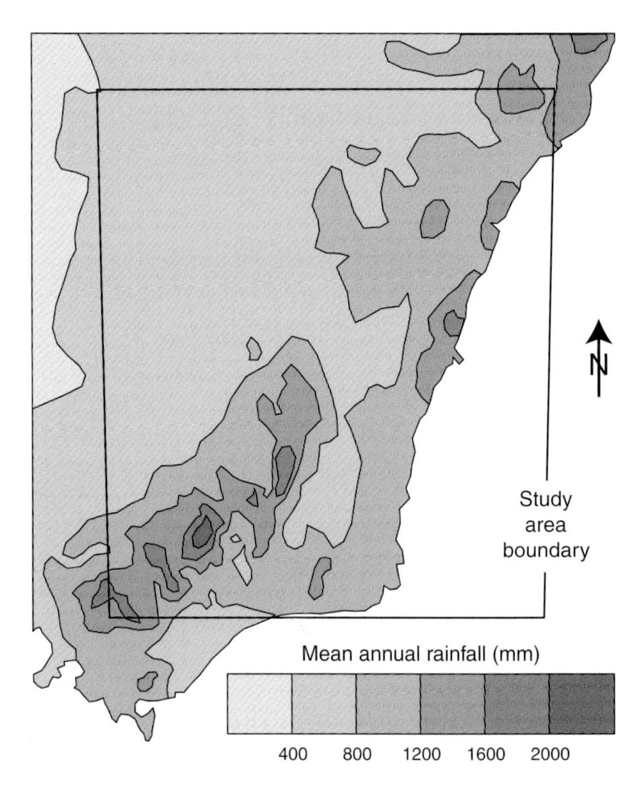

Fig. 7. Present-day mean annual rainfall in southeastern Australia. Data from Commonwealth Bureau of Meteorology (2004); averages are for 30-year period 1961 to 1990.

least, prior to the Quaternary), but equally there is no good reason to assume that precipitation in the past was ever spatially uniform (particularly bearing in mind the changes in the area's palaeolatitude). The adoption of a spatially fixed value for the precipitation parameter therefore has to be rejected.

The auxiliary hypothesis concerning boundary conditions

The rejection of the hypothesis of fixed parameter values has an immediate knock-on effect: it leads automatically to the rejection of the hypothesis concerning boundary conditions. The reason for this is that there are two changes in boundary conditions foreseen in this latter hypothesis, each of which has to be made at exactly the right moment in time. These changes correspond to the formation of the Gippsland Basin and to the rifting of the Tasman Sea. These events took place 'between 150 and 140 Ma' and 'between 100 and 90 Ma' (VBL, p. 161, i.e., after 40% and 60% of the total time-span of the Mesozoic and Cenozoic had elapsed). The conversion of real time to model

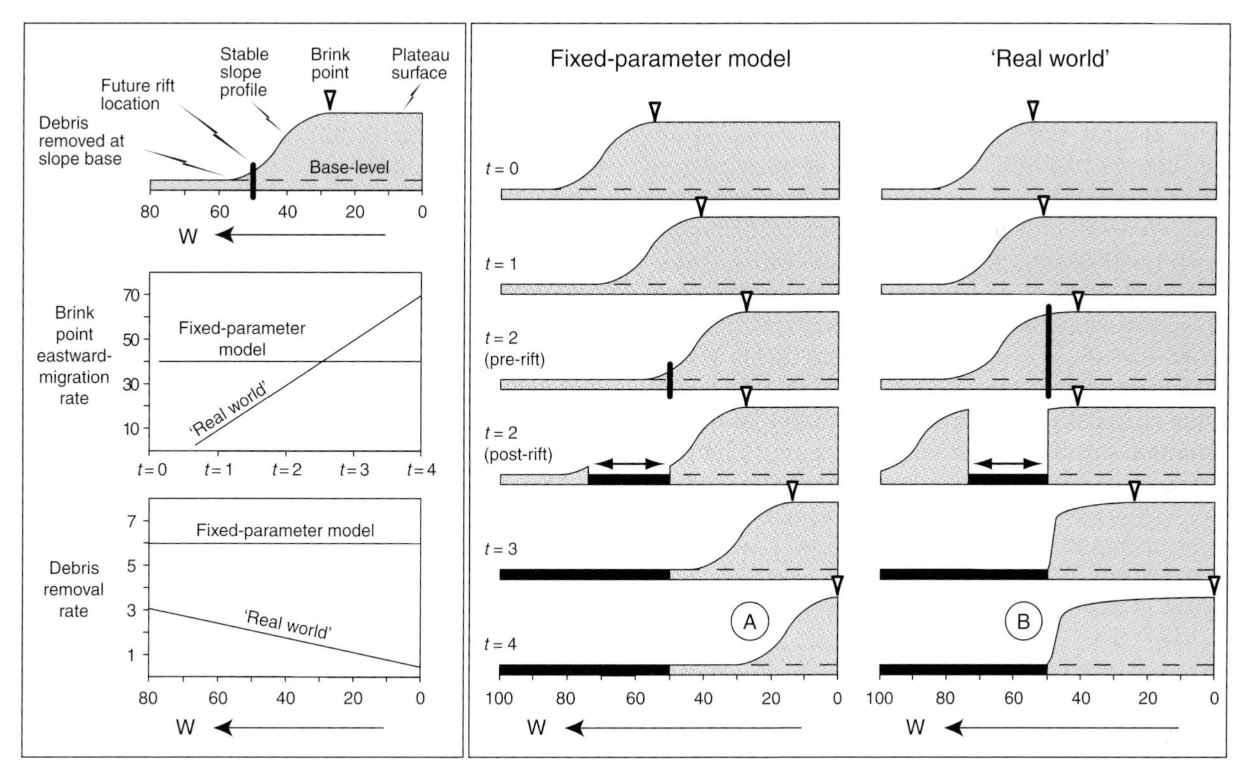

Fig. 8. Effects of inappropriate use of temporally and spatially fixed parameters, demonstrated for two-dimensional land-surface evolution system. Sketch at top of left-hand panel shows details of system definition. Locations are measured in arbitrary units west of zero. Plateau is cut back in time from west to east, with debris being removed at slope base (base-level). System has two parameters: (1) the eastward-migration rate of the brink point at the plateau edge; (2) the debris removal rate. A rifting event is incorporated in the system; it is known to take place at 50% of real-time and to be sited at 50 units west. Right-hand panel shows evolution in time of land surface (1) for a fixed-parameter model of the system, and (2) for a variable-parameter model ('Real World'). Graphs in left-hand panel show (1) temporal variation of brink point migration rate, and (2) spatial variation of debris removal rate. Assumption of linear conversion of real-time to model-time leads to rift being introduced at 50% of model-time (i.e., after time-step 2), both for the fixed-parameter model and for the 'Real World'. The result is that the final land-surface form produced by the fixed-parameter model differs greatly from that produced in the 'Real World' (compare 'A' with 'B'). To use the model to produce 'B', the rift must be introduced after time-step 1, i.e., after only 25% of model-time.

time would pose no difficulty if the spatial and temporal fixing of the parameter values really were justified. The evolution of landforms in CASCADE is independent of process rates, therefore the correct conversion would be achieved automatically by making the boundary condition changes after exactly 40% and 60% of the total model run-time had elapsed. This is what the developers of the Highlands model did. But the rejection of the hypothesis of fixed parameter values changes this situation entirely, as shown in Fig. 8. Real time now no longer converts linearly to model time. Making the boundary condition changes after 40% and 60% of the total model run-time therefore makes the model rifting events happen at the wrong points in real time. It also makes the model rift basins appear in landscapes that are topographically quite unlike the ones in which the corresponding real basins would have appeared.

The value of below-the-curve modelling

The Highlands modelling exercise produced realistic landscapes, despite being based on two demonstrably invalid auxiliary hypotheses. So what did it really show? And what do its results really mean?

One possible answer is that what is seen in this exercise is simply an example of non-uniqueness. This entails believing (a) that the general ideas about landscape evolution expressed in CASCADE are complete and correct, *and* (b) that the hypothesis about the plateau-like topography of the Highlands at the end of the Palaeozoic is correct, *and*

(c) that the Highlands model would have behaved in exactly the same way for the right combination of boundary conditions and parameter values as it did for the particular wrong combination that happened to be specified in the invalid auxiliary hypotheses. A second possible answer is that what is seen here is an example of errors fortuitously cancelling each other out. Errors introduced by using the invalid auxiliary hypotheses have been cancelled out *either* (a) by errors in CASCADE, *or* (b) by errors in the hypothesis about the initial topography of the Highlands, *or* (c) by other errors in the invalid auxiliary hypotheses, *or* (d) by some unknown combination of all three. A third possible answer is that the modelling exercise – a classic example of the below-the-curve style of modelling – was intrinsically incapable of showing anything other than that a CASCADE-based model for the Highlands could produce realistic landscapes. The results of the exercise then say nothing about the rightness or wrongness of any of the auxiliary hypotheses – or indeed about the rightness or wrongness of CASCADE itself.

I would suggest that it is this third answer that should be preferred here, even though it is in many ways the least pleasant to accept. The reason for urging this preference is simply to stress that in modelling exercises it always pays to be critical – all the more so when the exercise has clearly been designed to offer support to some already-held belief (cf. Oreskes *et al.*, 1994, p. 644). This 'safety-first' attitude is the attitude characteristic of the scientist – 'a critical man, a skeptic, hard to satisfy' (Medawar, 1969, p. 2). It can never be safe to depend on the results of modelling exercises carried out using numerous auxiliary hypotheses and effectively free parameters. Such exercises can readily produce realistic results under completely unrealistic conditions – as is shown so clearly in the case of the Highlands model.

ABOVE-THE-CURVE MODELLING – CYCLICITY IN STRATIGRAPHIC SUCCESSIONS

The model, and its theoretical background

An example of above-the-curve modelling is found in a recent paper on stratigraphic cyclicity (Tipper, 2000). The aim of the modelling exercise described there was to study the range of patterns of stratigraphic cyclicity that could be produced at a single point site as the result of variations in sediment budget and sedimentation capacity.

Sediment budget and sedimentation capacity are generalizations of the ideas of sediment supply and accommodation. The sediment budget at a site is defined as the difference between the amount of sediment being imported into the site and the amount being exported. The sedimentation capacity of a site is defined as *either* the amount of space at the site in which sediment can be deposited *or* the amount of space at the site from which sediment can be released into transport by erosion. Sedimentation capacity is measured by the vertical distance between base level and the lithic surface; base level is taken to be the surface corresponding to the locally stable equilibrium of the sedimentation system concerned. The definition of sedimentation capacity in terms of deviation from a proper equilibrium makes possible the formulation of what effectively are general rules for sedimentation (Fig. 9). From these can be predicted the sedimentation history of any site for which the variations in time of sediment budget and sedimentation capacity are known; from that history can then be predicted the resulting stratigraphic succession.

The model used in the exercise is a strict implementation of the budget-capacity theory of sedimentation; therefore, the principal hypothesis is simply the budget-capacity theory itself. An auxiliary hypothesis – there is just one – describes how the sediment budget and sedimentation capacity at the site are assumed to vary in time. The

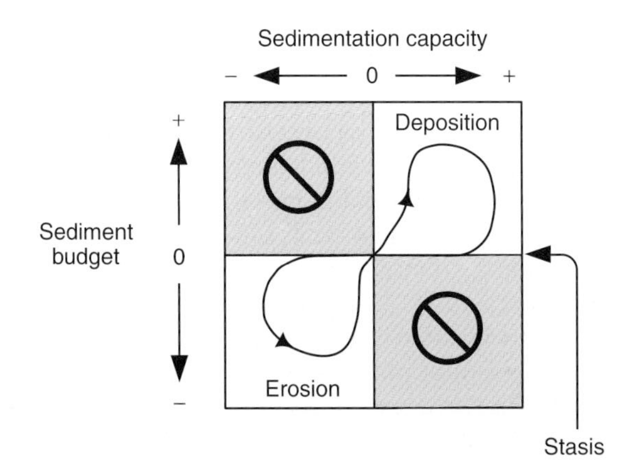

Fig. 9. Budget-capacity theory of sedimentation. Deposition takes place when both budget and capacity are positive, erosion takes place when both budget and capacity are negative, and stasis takes place whenever budget is zero. Time-arrowed trajectory shows hypothetical pattern of cyclic sedimentation (deposition–stasis–erosion).

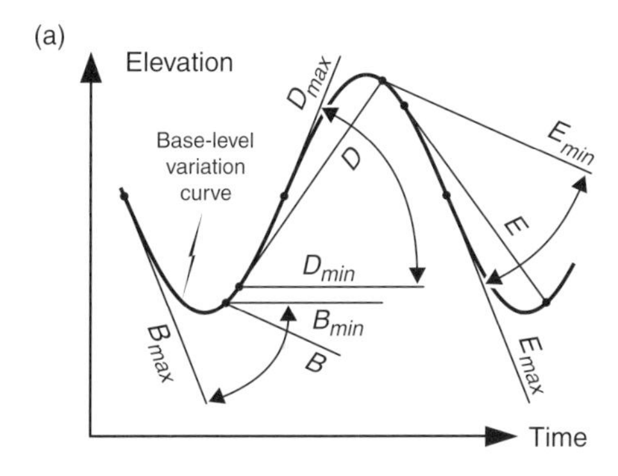

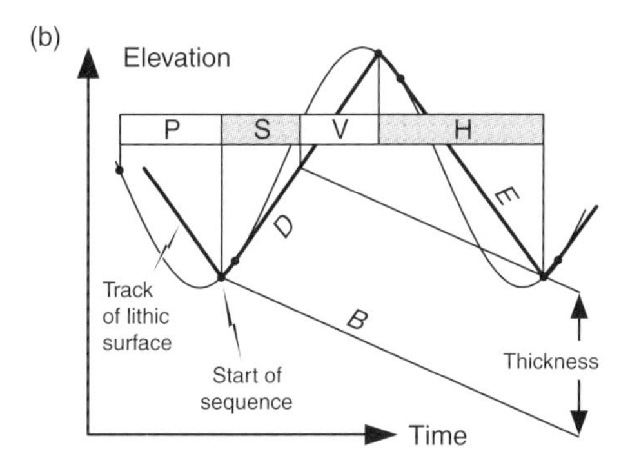

Fig. 10. Definition diagrams for one-dimensional cyclic sedimentation model; details in Tipper (2000). (a) Elevation–time plot showing the three variables and their maximum and minimum values: B is the dimensionless subsidence rate; D is the dimensionless limiting deposition rate; E is the dimensionless limiting erosion rate. (b) Elevation–time plot showing the four form descriptors: P is the phase lag, S is the sequence proportion, V is the vacuity proportion, H is the hiatus proportion. For the cycle shown, $P = 114°$, $S = 0.25$, $V = 0.25$, $H = 0.50$.

sediment budget at the site is taken to be always just enough to keep the site at base level, except when either of two prescribed budget limits is reached (Fig. 10a). One of these limits (the maximum permitted budget surplus) controls the maximum attainable rate of deposition. The other limit (the maximum permitted budget deficit) controls the maximum attainable rate of erosion. The variation in time of the sedimentation capacity at the site is driven by change in base level and by tectonic subsidence. The base-level is assumed to vary sinusoidally in time, with a fixed amplitude and wavelength; the rate of tectonic subsidence is assumed to be constant.

It should be noted that the modelling being carried out here is deliberately restricted in its intention. The model itself is a strictly one-dimensional implementation of the budget-capacity theory of sedimentation, and the auxiliary hypothesis presupposes a timespan over which the assumptions of constant subsidence rate and sinusoidal base level variation can be considered reasonable. The model does not pretend to be a model of stratigraphic cyclicity on a basin-wide scale, nor does it pretend to offer insight into long-term patterns of cyclicity.

There are three primary variables in this model: the tectonic subsidence rate, and the two prescribed budget limits. These variables can be expressed in terms of the amplitude and wavelength of the base-level sinewave, resulting in a set of three secondary variables, all of them dimensionless. These secondary variables are the dimensionless subsidence rate (B), the dimensionless limiting deposition rate (D), and the dimensionless limiting erosion rate (E). Variable B is expressed as a percentage of the maximum subsidence rate for which any type of cyclic sedimentation is possible.

Cyclic sedimentation patterns are produced in this model for all combinations of B, D, and E for which $0\% < B < 100\%$, $B\% \leq D \leq (100 + B)\%$, and $0\% < E \leq (100- B)\%$. The cycles corresponding to the different combinations differ from each other only in detail, and they can all be described using the same set of four form descriptors (Fig. 10b). These are: (1) the phase lag, P; (2) the sequence proportion, S; (3) the vacuity proportion, V; and (4) the hiatus proportion, H. Descriptor P measures the time in a cycle at which deposition starts; S measures the proportion of a cycle's wavelength corresponding to preserved deposition; V measures the proportion of the wavelength corresponding to deposition that is later removed; H measures the proportion corresponding to erosion. Of these descriptors, S is of particular interest; it is equivalent to the conventional measure of stratigraphic completeness (Tipper, 1998), at the resolution of the cycle wavelength.

The modelling exercise and its overall results

The aim of the modelling exercise – to study the range of possible patterns of stratigraphic cyclicity – required first the generation of the sedimentation cycles corresponding to all the possible B–D–E combinations. Then the form descriptors (P, S, V, H) had to be calculated for each of these cycles. The analysis of these form descriptors led to a number of interesting results. For instance, it proved possible (1) to recognize four distinct families of cycle types, which have different degrees of stability, (2) to show that sequence boundaries in cyclic successions need not be closely tied to inflection points in base-level variation curves, and (3) to develop a strategy for reconstructing sedimentation conditions from preserved patterns of cyclicity. These results and their interpretation are described in detail in Tipper (2000), where also are discussed the effects of relaxing some of the assumptions incorporated in the model's auxiliary hypothesis. These effects turn out to be relatively minor in their magnitude.

Cyclic stratigraphic successions: the 'too complete' dilemma

One entirely unexpected result came out of the modelling. It concerned the range of values of S that cycles can have – in effect, the expected completeness of cyclic stratigraphic successions. Put simply, cyclic successions seemed usually to have to be far too complete: the great majority of B–D–E combinations gave rise to cycles that had values of S that were very much higher than the completeness values usually claimed for real stratigraphic successions (e.g. Sadler, 1981). This result is best appreciated on S–V–H plots (Fig. 11); these are triangular plots that show how the time in a cycle is partitioned between sequence, vacuity and hiatus. These plots show clearly that only when B is just slightly greater than zero is it possible to obtain cycles in which S is realistically low, i.e., in some cases no greater than about 0.01. This leads to the following dilemma: if cycles are in fact produced in the way that the model would suggest, how can it be possible for cyclic successions to be realistically incomplete?

There are several escape routes from this dilemma. One possibility is that real stratigraphic successions are considerably less incomplete than they are currently believed to be, perhaps because spatio-temporal masking has not been allowed for when data for completeness estimation are being collected (Tipper, 1998). A second possibility is that real stratigraphic successions are indeed as incomplete as they are believed to be, *but only on average*, and that cyclic successions are inherently more complete than non-cyclic ones. A third possibility is that cyclic successions can be highly incomplete only when there is compound base-level fluctuation; they therefore cannot be produced by a model in which only sinusoidal base-level fluctuation is permitted. A fourth possibility is that cyclic successions are only ever produced at sites where the tectonic subsidence rate is low and the amplitude of base-level variation high, exactly as predicted by the model. The values of B at these sites will automatically be low (and therefore the incompleteness will automatically be high), simply because B is a version of the tectonic subsidence rate that has been scaled in terms of the amplitude of the base-level variation.

The first two of these escape routes accept the completeness result from the modelling exercise

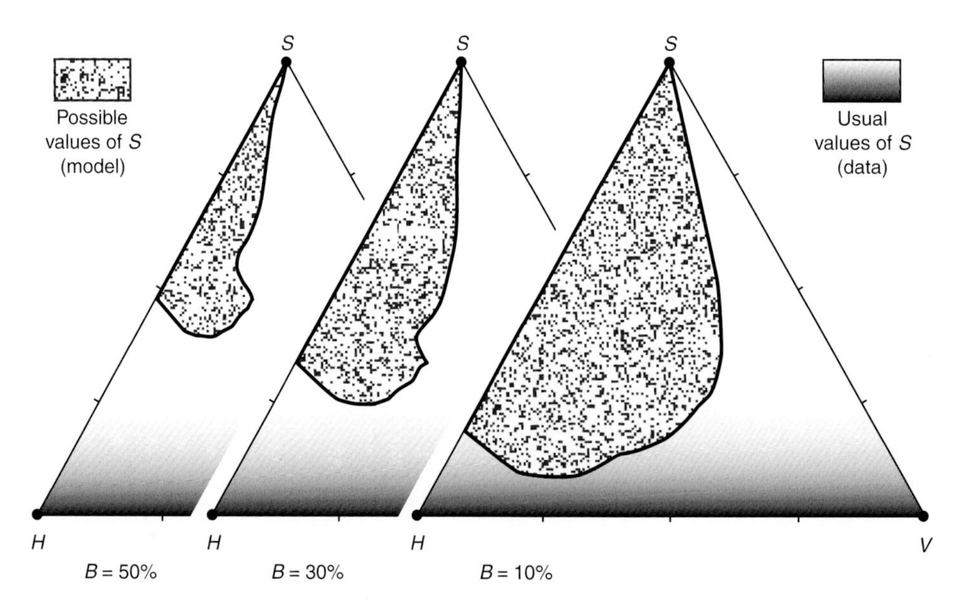

Fig. 11. S–V–H plots. Stippled areas show values of S that can be obtained from the cyclic sedimentation model for selected values of B (10%, 30%, 50%). Shaded areas close to H–V axis show values of S typically found for real stratigraphic successions.

and instead challenge aspects of conventional stratigraphic wisdom. Both of them lead to hypotheses worth investigating: (1) that real stratigraphic successions are considerably less incomplete than they are currently believed to be, and (2) that cyclic successions are inherently more complete than non-cyclic ones. The third escape route was already partly investigated as part of the original modelling exercise (Tipper, 2000). Apparently the addition of high-frequency components of base-level fluctuation does not generally lead to significant decreases in completeness. An exception seems to be when (1) the high-frequency components have higher amplitudes than the lower-frequency ones on which they are superimposed, *and* (2) the sediment budget is sufficient to keep the lithic surface within reach of fast, short-term base-level falls. This exception demands a 'shorter-wavelengths-mean-higher-amplitudes' assumption about compound base-level fluctuation, a hypothesis that is diametrically opposite to the 'shorter-wavelengths-mean-lower-amplitudes' assumption commonly accepted by stratigraphers, for instance in the Haq–Vail sea-level chart (Haq *et al.*, 1988). This hypothesis may also be worth investigating.

The fourth of the escape routes accepts the completeness result from the modelling exercise and also the conventional wisdom about stratigraphic incompleteness. It would therefore seem to be the logical route to take. Yet even this 'route of least resistance' leads to a hypothesis worth looking at, in this case the one that was once referred to as the 'intuitive but undemonstrated notion that completeness is well correlated with thickness' (McShea & Raup, 1986, p. 573). This hypothesis is certainly not valid over long time-spans (Tipper, 1987), but it seems that it might nonetheless be useful in interpreting cyclic successions. This can be appreciated by remembering that the thickness of any individual cycle in a succession is controlled ultimately by B (Fig. 10b). If cyclic successions that are highly incomplete are formed only when B is extremely low, those highly incomplete successions should be expected to be made up of individual cycles that are very thin indeed. Successions made up of cycles that are slightly thicker should be expected to be slightly less incomplete, and so on.

The value of above-the-curve modelling

The cyclicity modelling exercise was carried out within a remarkably austere framework. There was the principal hypothesis expressed in the model (i.e., the budget-capacity theory of sedimentation), and there was the auxiliary hypothesis concerning how the model's three primary variables would be allowed to vary in time. There were no other auxiliary hypotheses or free parameters. The three variables were independent, meaning that all possible sedimentation conditions could readily be represented in a three-dimensional space. The variables also were bounded, meaning that there was a region within this space outside of which cyclic sedimentation could not take place. All of the patterns of stratigraphic cyclicity that could possibly be produced by the model could therefore be studied by looking at the forms of cyclic sedimentation that occur in this region.

The strength of this style of modelling is clearly shown by considering the response that had to be made to the discovery that cyclic successions seemed usually to be far too complete. Either there had to be errors in the principal hypothesis, or there had to be errors in the auxiliary hypothesis, or there had to be problems with some aspects of conventional stratigraphic wisdom. The first possibility could immediately be ruled out, simply because the budget-capacity theory of sedimentation is a straightforward expression of mass balance in a system with a single locally stable equilibrium. The second possibility could not be ruled out – there could indeed be errors in the auxiliary hypothesis. But errors there seemed unlikely, especially as it could be shown that the effects of relaxing some of the stronger assumptions in the auxiliary hypothesis appeared not to affect the completeness result to any significant extent (Tipper, 2000). That left only the possibility that there were problems with conventional stratigraphic wisdom. Some aspects of this were duly challenged, leading in every case to the identification of new hypotheses worth investigating. There were no other auxiliary hypotheses in this exercise that could be hidden behind to explain away the 'too-complete' dilemma; nor were there free parameters that could be manipulated to make it disappear. The model had talked back, and the scientist had had to listen.

MODELS AND MODELLING: SOME FINAL OBSERVATIONS

What really is 'heuristic'?

In concluding their deservedly influential paper, Oreskes and her co-authors posed the question

'What good are models?' (1994, p. 644). Their answer – that the primary value of models is heuristic – has since become widely quoted (e.g. van der Beek *et al.*, 1999). The general drift of this answer is surely acceptable, for most scientists certainly regard models primarily as being tools to be used in a process of discovery. More controversial, however, are some of the specific suggestions given to show what 'heuristic' is evidently to mean in the context of modelling. One suggestion was that models can corroborate hypotheses 'by offering evidence to strengthen what may be already partly established through other means'. A second was that models can be used for sensitivity analysis. A third was that models can be used 'for exploring "what if" questions' (Oreskes *et al.*, 1994, p. 644). All these suggestions are worthy ones, yet it would seem inadvisable to accept them at their face value unless the style of modelling being used has been taken carefully into account.

Consider first the suggestion that models can offer evidence to corroborate hypotheses. The danger inherent in this can readily be seen by looking at the Highlands modelling exercise. That exercise – below-the-curve modelling *par excellence* – apparently offered evidence to corroborate an existing hypothesis, evidence that the developers of the model in question gratefully accepted. Yet all of the evidence had been obtained as the model was run under what fortunately could be shown to be entirely unrealistic conditions. If the evidence had been obtained as the model was run under other conditions – *conditions that were equally unrealistic but could not be shown to be unrealistic* – then that evidence would not have been capable of being discredited. And it might then well have been used – quite wrongly – to corroborate the hypothesis. It must be stressed that the results of below-the-curve modelling should never be used as evidence to corroborate any hypothesis, under any circumstances.

Next, consider the suggestion that models can be used for sensitivity analysis, i.e., for investigating the degree to which the behaviour of a model is influenced by variation in the values of individual input variables and parameters. This suggestion is certainly a most appropriate one, especially for models that involve a not-overwhelming number of parameters. The only problem with it is that sensitivity analysis is something that should be carried out properly or not at all; it is valueless if carried out piecemeal, or if it is restricted to only a small part of the possible parameter or input space. Proper sensitivity analysis can be extraordinarily time-consuming, especially if the model being used is a particularly compute-intensive one. Certainly there are computational techniques that are relatively efficient under some circumstances (e.g. Bagirov & Lerche, 1999, and references therein), but even these techniques are likely to be of limited value for realistic models of natural sedimentation systems. Only in the simplest and most straightforward circumstances is it ever feasible to investigate a model's behaviour exhaustively, for all the possible combinations of input and parameter values. Such exhaustive investigation is of considerable value whenever it can be carried out, for it shows above all what the model in question (and hence the system in question) cannot do. This can lead to the challenge of conventional wisdom, as was shown in the cyclicity modelling exercise.

Finally, what about the suggestion that an appropriate use of models is for exploring 'what if' questions? This is what is sometimes plausibly termed "'what if" testing' (Watney *et al.*, 1999). Great play is often made of the value of this type of work, particularly – one hesitates to say this – in modelling exercises that are more than usually below-the-curve in style. Yes, 'what if' testing certainly can be valuable, but only if there is known to be a one-to-one mapping from the 'if' to the 'what'. The results of 'what if' testing have little or no meaning when no such mapping exists – as will surely be the case for most models of natural sedimentation systems.

Modelling and scientific enquiry

Scientific enquiry has been described in the following way, I believe aptly: '[It] begins as a story about a Possible World – a story which we invent and criticize and modify as we go along, so that it ends by being, as nearly as we can make it, a story about real life' (Medawar, 1969, p. 59). Modelling lies at the very heart of scientific enquiry, for it is ultimately through modelling that the story about the Possible World is invented, is criticized, and is modified. To best carry out this modelling, we need models that are as close as possible to being behaviourally equivalent to their parent systems. It is these models that have the power to challenge what we currently believe, and therefore to get us most effectively to that final story about real life. They are the models that talk back.

ACKNOWLEDGEMENTS

I thank the organizers of the Utrecht symposium for their invitation to present a keynote lecture there. They were not to know what would result! I thank also the reviewers for their comments on the manuscript: Andrew Miall, Peter van der Beek, and Lynn Watney.

REFERENCES

Australian Native Vegetation Assessment (2001) audit.ea.gov.au/ANRA/vegetation/docs/Native_vegetation/nat_veg_fig8_popup.cfm

Bagirov, E. and **Lerche, I.** (1999) Probability and sensitivity analysis of two-dimensional basin modeling results. In: *Numerical Experiments in Stratigraphy: Recent Advances in Stratigraphic and Sedimentologic Computer Simulations* (Eds J.W. Harbaugh, W.L. Watney, E.C. Rankey, R. Slingerland, R.H. Goldstein and E.K. Franseen), *SEPM Spec. Publ.*, **62**, 35–68.

Bornholdt, S., Nordlund, U. and **Westphal, H.** (1999) Construction and application of a stratigraphic inverse model. In: *Numerical Experiments in Stratigraphy: Recent Advances in Stratigraphic and Sedimentologic Computer Simulations* (Eds J.W. Harbaugh, W.L. Watney, E.C. Rankey, R. Slingerland, R.H. Goldstein and E.K. Franseen), *SEPM Spec. Publ.*, **62**, 85–90.

Braun, J. and **Sambridge, M.** (1997) Modelling landscape evolution on geological time scales: A new method based on irregular spatial discretization. *Basin Res.*, **9**. 27–52.

Bridge, J.S. (2003) *Rivers and Floodplains: Forms, Processes and Sedimentary Record.* Blackwell, Oxford.

Brown, D.M. (1992) The fidelity fallacy. *Groundwater*, **30**, 482–483.

Burton, R., Kendall, C.G.ST.C. and **Lerche, I.** (1987) Out of our depth: On the impossibility of fathoming eustatic sea level from the stratigraphic record. *Earth-Science Reviews*, **24**, 237–277.

Chamberlin, T.C. (1897) The method of multiple working hypotheses. *J. Geol.*, **5**, 837–848.

Commonwealth Bureau of Meteorology (2004) www.bom.gov.au/climate/environ/other/seas_all.shtml

Cross, T.A. (Ed.) (1990) *Quantitative Dynamic Stratigraphy.* Prentice Hall, Englewood Cliffs.

Cross, T.A. and **Lessenger, M.A.** (1999) Construction and application of a stratigraphic inverse model. In: *Numerical Experiments in Stratigraphy: Recent Advances in Stratigraphic and Sedimentologic Computer Simulations* (Eds J.W. Harbaugh, W.L. Watney, E.C. Rankey, R. Slingerland, R.H. Goldstein and E.K. Franseen), *SEPM Spec. Publ.*, **62**, 69–83.

Eldredge, N. and **Gould, S.J.** (1972) Punctuated equilibria: An alternative to phyletic gradualism. In: *Models in Paleobiology* (Ed. T.J.M. Schopf). Freeman Cooper, San Francisco, pp. 82–115.

Flemings, P.B. and **Jordan, T.E.** (1989) A synthetic stratigraphic model of foreland basin development. *J. Geophys. Res.*, **94**, 3851–3866.

Frakes, L.A., Francis, J.E. and **Syktus, J.I.** (1992) *Climate Modes of the Phanerozoic.* Cambridge University Press, Cambridge.

Greenwood, H.J. (1989) On models and modeling. *The Canadian Mineral.*, **27**, 1–14.

Haq, B.U., Hardenbol, J. and **Vail, P.R.** (1988) Mesozoic and Cenozoic chronostratigraphy and cycles of sea-level change. In: *Sea-Level Changes: An Integrated Approach* (Eds C.K. Wilgus, B.S. Hastings, H. Posamentier, J. Van Wagoner, C.A. Ross and C.G.St.C. Kendall), *SEPM Spec. Publ.*, **42**, 71–108.

Harbaugh, J.W., Watney, W.L., Rankey, E.C., Slingerland, R., Goldstein, R.H. and Franseen, E.K. (Eds) (1999) Numerical experiments in stratigraphy: Recent advances in stratigraphic and sedimentologic computer simulations. *SEPM Spec. Publ.*, **62**, 362 pp.

Helland-hansen, W., Kendall, C.G.ST.C., Lerche, I. and **Nakayama, K.** (1988) A simulation of continental basin margin sedimentation in response to crustal movements, eustatic sea level change, and sediment accumulation rates. *Math. Geol.*, **20**, 777–802.

Hsü, K.J. (1989) *Physical Principles of Sedimentology.* Springer-Verlag, Berlin.

Jordan, T.E. and **Flemings, P.B.** (1991) Large-scale stratigraphic architecture, eustatic variation, and unsteady tectonism: A theoretical evaluation. *J. Geophys. Res.*, **96**, 6681–6699.

Kendall, C.G.ST.C. and **Lerche, I.** (1988) The rise and fall of eustasy. In: *Sea-Level Changes: An Integrated Approach* (Eds C.K. Wilgus, B.S. Hastings, H. Posamentier, J. Van Wagoner, C.A. Ross and C.G.St.C. Kendall), *SEPM Spec. Publ.*, **42**, 3–17.

Kooi, H. and **Beaumont, C.** (1994) Escarpment evolution on high-elevation rifted margins: Insights derived from a surface processes model that combines diffusion, advection, and reaction. *J. Geophys. Res.*, **99**, 12191–12209.

Lambeck, K. and **Stephenson, R.** (1986) The post-Palaeozoic uplift history of southeastern Australia. *Austr. J. Earth Sci.*, **33**, 253–270.

Lessenger, M. and **Lerche, I.** (1999) Inverse modeling. In: *Numerical Experiments in Stratigraphy: Recent Advances in Stratigraphic and Sedimentologic Computer Simulations* (Eds J.W. Harbaugh, W.L. Watney, E.C. Rankey, R. Slingerland, R.H. Goldstein and E.K. Franseen), *SEPM Spec. Publ.*, **62**, 29–31.

Mcshea, D.W. and **Raup, D.M.** (1986) Completeness of the geological record. *J. Geol.*, **94**, 569–574.

Medawar, P.B. 3(1969) *Induction and Intuition in Scientific Thought.* American Philosophical Society, Philadelphia.

Merriam, D.F. and **Davis, J.C.** (Eds) (2001) *Geologic Modeling and Simulation: Sedimentary Systems.* Kluwer, New York.

Miall, A.D. (1992) Exxon global cycle chart: An event for every occasion? *Geology*, **20**, 787–790.

Miall, A.D. and **Miall, C.E.** (2001) Sequence stratigraphy as a scientific enterprise: The evolution and persistence of conflicting paradigms. *Earth-Science Rev.*, **54**, 321–348.

Oreskes, N., Shrader-frechette, K. and **Belitz, K.** (1994) Verification, validation, and confirmation of numerical models in the earth sciences. *Science*, **263**, 641–646.

Paola, C. (2000) Quantitative models of sedimentary basin filling. *Sedimentology*, **47** (Suppl. 1), 121–178.

Sadler, P.M. (1981) Sediment accumulation rates and the completeness of stratigraphic sections. *J. Geol.*, **89**, 569–584.

Schmidt, P.W. and **Clark, D.A.** (2000) Paleomagnetism, apparent polar-wander path, & paleolatitude. In: *Billion-year Earth History of Australia and Neighbours in Gondwanaland* (Ed. J.J. Veevers), pp. 12–17. GEMOC Press, Sydney.

Slingerland, R.L., Harbaugh, J.W. and **Furlong, K.P.** (1994) *Simulating Clastic Sedimentary Basins.* Prentice-Hall, Englewood Cliffs.

Strobel, J., Soewito, F., Kendall, C.G.ST.C., Biswas, G., Bezdek, J. and **Cannon, R.** (1990) Interactive simulation (SED-pak) of clastic and carbonate sedimentation in shelf to basin settings. In: *Quantitative Dynamic Stratigraphy* (Ed. T.A. Cross), pp. 433–444. Prentice Hall, Englewood Cliffs.

Syvitski, J.P., Morehead, M.D. and **Nicholson, M.** (1998) HYDROTREND: A climate-driven hydrologic-transport model for predicting discharge and sediment load to lakes or oceans. *Comp. Geosci.*, **24**, 51–68.

Taylor, G. and **Eggleton, R.A.** (2001) *Regolith Geology and Geomorphology.* Wiley, Chichester.

Tipper, J.C. (1987) Estimating stratigraphic completeness. *J. Geol.*, **95**, 710–715.

Tipper, J.C. (1991) Modelling the fill of sedimentary basins. *Expl. Geophys.*, **22**, 397–400.

Tipper, J.C. (1992) Landforms developing and basins filling: Three-dimensional simulation of erosion, sediment transport, and deposition. In: *Three-Dimensional Computer Graphics in Modeling Geologic Structures and Simulating Geologic Processes* (Eds R. Pflug and J.W. Harbaugh), pp. 155–170. Springer, Berlin.

Tipper, J.C. (1993) Testing the Exxon global cycle chart. *Die Geowissenschaften*, **11**, 380–384.

Tipper, J.C. (1994) Some tests of the Exxon global cycle chart. *Math. Geol.*, **26**, 843–855.

Tipper, J.C. (1998) The influence of field sampling area on estimates of stratigraphic completeness. *J. Geol.*, **106**, 727–739.

Tipper, J.C. (2000) Patterns of stratigraphic cyclicity. *J. Sedi. Res.*, **70**, 1262–1279.

Toffoli, T. and **Margolus, N.** (1987) *Cellular Automata Machines.* MIT Press, Cambridge, MA.

Tucker, G.E. and **Slingerland, R.L.** (1994) Erosional dynamics, flexural isostasy, and long-lived escarpments: A numerical modeling study. *J. Geophys. Res.*, **99**, 12229–12243.

Van der beek, P.A. and **Braun, J.** (1998) Numerical modelling of landscape evolution on geological time-scales: A parameter analysis and comparison with the south-eastern highlands of Australia. *Basin Res.*, **10**, 49–68.

Van der beek, P.A. and **Braun, J.** (1999) Controls on post-mid-Cretaceous landscape evolution in the southeastern highlands of Australia: Insights from numerical surface process models. *J. Geophys. Res.*, **104**, 4945–4966.

Van der beek, P.A., Braun, J. and **Lambeck, K.** (1999) Post-Palaeozoic uplift history of southeastern Australia revisited: Results from a process-based model of landscape evolution. *Austr. J. Earth Sci.*, **46**, 157–172.

Van heijst, M.W.I.M., Postma, G., Meijer, X.D., Snow, J.N. and **Anderson, J.B.** (2001) Quantitative analogue flume-model study of river-shelf systems: Principles and verification exemplified by the Late Quaternary Colorado river-delta evolution. *Basin Res.*, **13**, 243–268.

Vemuri, V. (1978) *Modeling of Complex Systems.* Academic Press, New York.

Watney, W.L., Rankey, E.C. and Harbaugh, **J.W.** (1999) Perspectives on stratigraphic simulation models: Current approaches and future opportunities. In: *Numerical Experiments in Stratigraphy: Recent Advances in Stratigraphic and Sedimentologic Computer Simulations* (Eds J.W. Harbaugh, W.L. Watney, E.C. Rankey, R. Slingerland, R.H. Goldstein and E.K. Franseen). *SEPM Spec. Publ.*, **62**, 3–21.

Weinberg, G.M. (1997) *Quality Software Management: First-order Measurement.* Dorset House, New York.

Weltje, G.J., Meijer, X.D. and **De Boer, P.L.** (1998) Stratigraphic inversion of siliciclastic basin fills: A note on the distinction between supply signals resulting from tectonic and climatic forcing. *Basin Res.*, **10**, 129–153.

Yalin, M.S. (1971) *Theory of Hydraulic Models.* Macmillan, London.

Index

Note: Page numbers in *italics* refer to figures, those in **bold** refer to tables